D0203351

Elements of Differential Topology

Elements of Differential Topology

Anant R. Shastri

CRC Press is an imprint of the
Taylor & Francis Group an **informa** business
A CHAPMAN & HALL BOOK

CRC Press
Taylor & Francis Group
6000 Broken Sound Parkway NW, Suite 300
Boca Raton, FL 33487-2742

International Standard Book Number: 978-1-4398-3160-1 (Hardback)

Visit the Taylor & Francis Web site at
http://www.taylorandfrancis.com

and the CRC Press Web site at
http://www.crcpress.com

FOREWORD

The mathematical community was startled in 1956 by John Milnor's discovery of a smooth manifold homeomorphic but not diffeomorphic to the 7-dimensional sphere [M3]. A few years later Michel Kervaire found a closed topological 10-dimensional manifold that does not support a smooth structure [K]. Thus was differential topology born as a separate field of manifold study. Of course there were earlier important topological theorems that used differentiable techniques, e.g., Morse Theory, Pontryagin-Thom transversality, and Whitney's strong embedding theorem. However, the Kervaire-Milnor examples showed that these techniques could not be directly applied to the study of topological manifolds in general.

At about the same time that the Kervaire-Milnor examples were discovered, important new differentiable techniques were developed; in particular, the Smale-Hirsch immersion theorem, Smale's h-cobordism theorem [Sm1], and Haefliger's embedding theorem ([H]). These made it possible for Kervaire and Milnor in 1963 to classify, up to diffeomorphism, all smooth manifolds homeomorphic to the n-dimensional sphere $\mathbb{S}^n$ when $n > 4$ ([K-M]). Thus was established the new field of surgery. This field rapidly matured over the next 7 years through the work of Browder, Novikov, Sullivan, and Wall to become an effective method for classifying, up to diffeomorphism, all closed simply connected manifolds of a given homotopy type X and of dimension greater than 4 (so-called high dimensional manifolds) ([Bd], [W]). Surgery theory is also effective for classifying high-dimensional non-simply connected manifolds provided certain groups $Wh(\pi_1(X))$ and $L_n(\pi_1(X))$ can be calculated. Much progress has been made calculating these groups over the last 40 years and this is currently a focus of research. Building on this earlier work Kirby and Siebenmann, in their 1976 monograph ([K-S]), accomplished the daunting task of indirectly extending the above differential topology techniques so as to apply to topological manifolds. In particular, they established effective versions of the h-cobordism theorem and of surgery theory that are valid for high-dimensional topological manifolds. They also established a smoothing theory to answer the basic question: Does a given topological manifold support a smooth structure and if so (up to diffeomorphism) how many?

We further remark that the Kervaire-Milnor exotic spheres, i.e., smooth n-dimensional manifolds homeomorphic but not diffeomorphic to $\mathbb{S}^n$, have also been of much interest to differential geometers through the study of

Riemannian manifolds of positive sectional curvatures. The classical sphere theorem, due independently to Berger and Klingenberg 1960 ([B], [Kl]), showed that if the sectional curvatures of a closed simply connected n-dimensional manifold are strictly 1/4-pinched, then the manifold is homeomorphic to $\mathbb{S}^n$. And recently Brendle and Schoen have shown, under this same 1/4-pinching assumption, that the manifold is actually diffeomorphic to $\mathbb{S}^n$, by using Hamilton's Ricci flow theory ([B-S]). A bit earlier use of this theory by Perelman positively solved perhaps the oldest problem in topology, Poincare's Conjecture; i.e., any simply connected closed 3-dimensional manifold is homeomorphic to $\mathbb{S}^3$.

Professor Shastri's book gives an excellent point of entry to this fascinating area of mathematics by providing the basic motivation and background needed for the study of differential geometry, algebraic topology, and Lie groups.

His book is accessible to a serious first year graduate student reading it either independently or as a text in a graduate course. If such a student also familiarizes himself/herself with the basics of algebraic topology, through cup products, then he would be prepared to understand the important results outlined above.

A major strength of Professor Shastri's book is that detailed arguments are given in places where other books leave too much for the reader to supply on his/her own. This, together with the large quantity of accessible exercises makes this book particularly reader friendly as a stable text for an introductory course in differential topology.

F. Thomas Farrell
Binghamton, New York
Autumn, 2010

PREFACE

This book is intended for a preparatory course for the vast and elegant theories in topology developed by Morse, Thom, Smale, Whitney, Milnor, etc. It grew out of several years of teaching at my department, a third-semester course in Differential Topology. The selection of topics broadly follows the classical book of Milnor, *Topology from the Differentiable Viewpoint,* [M1] from which I myself learned differential topology. One may see here quite a bit of similarity with the book *Differential Topology* [G-P] by Guillemin and Pollack. That was the book from which I used to teach my course initially and which is also modeled on Milnor's book. Two other books I have been influenced by are Kosinski's book *'Differentiable Manifolds'* [K] and John Lee's book *Manifolds* [L].

This book assumes that the reader has gone through a semester course each in real analysis, multivariable calculus, and point-set-topology. The entire book or parts of it can be adopted as text for M.Sc./B.Tech./M.Tech./Ph.D. students. The exercises in each chapter with solutions/hints at the end make this book self-readable by any interested student. I have included a 'section-wise dependence tree' which may help a teacher to make his/her course plan.

The first two chapters offer a quick review of differential and integral calculus of several variables. They also serve as a ready reference to fundamental results to be used throughout the rest of the book. They include standard material such as inverse and implicit function theorems, change of variable formula for integration, Sard's theorem, etc. As a precursor to the study of manifolds, we discuss the Lagrange multiplier method with complete proof and include interesting examples.

Chapter 3 deals with smooth manifolds as submanifolds in a Euclidean space. Basic notions of tangent space, immersions, embeddings, transversality, etc. and the stability properties of some of these notions are discussed. In Chapter 4, we introduce the notion of orientability and develop the algebraic machinery of differential forms that is necessary for the study of integration on manifolds, and then present the general form of the Stokes' theorem and a little bit of De Rham cohomology.

In Chapter 5 we introduce the notion of abstract smooth manifolds. A fundamental gluing lemma is introduced, which is used again and again in the construction of new manifolds out of the old ones. As an immediate application of this lemma, we give an elementary proof of the classification of 1-dimensional manifolds. To my knowledge, there is some novelty in this proof. This chapter concludes with Whitney's (easy) embedding theorems.

In Chapter 6, we begin with the normal bundle, tubular neighborhoods and orientation on normal bundles laying down the foundation for homotopical aspects of manifolds. We then go on to study vector fields and isotopies, thereby clearing the ground for bringing in constructional tools of differential topology.

In Chapter 7, we discuss intersection theory, which is central to the theme of the book. Technically useful results such as relative transverse homotopy theorems are proved with complete detail. We directly discuss oriented intersection numbers, relate this to the degree of a map, the concept of winding number, index of a vector field, etc. We discuss various equivalent definitions of the Euler characteristic, except the standard definition itself, viz., alternative sum of the face numbers of a triangulation. This chapter includes big results such as the Jordan-Brouwer separation theorem, the Borsuk-Ulam theorem, the Hopf degree theorem, the Gauss-Bonnet theorem etc.

In Chapter 8, we introduce the Morse functions. For a submanifold of a Euclidean space, existence of 'linear' Morse functions is established. We then introduce the notion of attaching handles and connected sum, prove handle presentation theorem for compact manifolds due to Smale and present a proof of classification of compact smooth surfaces. Again, to my knowledge, there is some novelty here. Here the reader will meet two more definitions of Euler characteristic, one through handle presentation and another directly through the Morse function and see a proof of Poincaré-Hopf index theorem.

Chapter 9 deal with the basics of Lie groups. It is primarily included as a rich source of examples of manifolds and most of it can be read independently of the rest of the book. Part of it needs a little more background such as working knowledge of covering space theory and topological groups.

Acknowledgement

Though hand-written primitive form of these notes were there since 1990, these were latexed down for the first time in preparation for the first Annual Foundation School-2004 of the Advanced Training in Mathematics sponsored by the National Board for Higher Mathematics, Department of Atomic Energy,, Govt. of India. This school was organized by me at the Department of Mathematics, Indian Institute of Technology Bombay and I was also the coordinator for the topology component. In this four week training program aimed at fresh research scholars in the country, lectures in Topology were shared among four people: A. J. Parameswaran, R. R. Simha, R. S. Kulkarni and myself. Of course, these notes have gone through several revisions to arrive at the present form. I have received encouragement and feedback from C. S. Aravinda, Parameswaran Sankaran and other friends.

The revision efforts were supported *twice* by the Curriculum Development Programme of Indian Institute of Technology Bombay.

As a student, I learned differential topology from M. S. Raghunathan, R. R. Simha and Gopal Prasad through 'coffee table discussions.' Interaction with several students especially B. Subhash and colleague Debraj Chakravarti has helped in the improvement of the presentation of the material. Mahuya Datta of I.S.I. Kolkatta has gone through several versions of these notes and had pointed out various inaccuracies with suggestions for better presentation. Nevertheless, shortcomings that still persist are all my own. Prof. Farell has written a nice FOREWORD which is friendly to the author and valuable to the reader. I am deeply indebted to all these people.

Thanks are due to CRC Press for publishing these notes and for doing an excellent job of converting it into a book.

Anant R. Shastri
Department of Mathematics
Indian Institute of Technology, Bombay
Powai, Mumbai
Autumn 2010

Contents

List of Symbols

Symbols	Section no.
$\mathbb{N}$	1.1
$\mathbb{Z}$	1.1
$\mathbb{Q}$	1.1
$\mathbb{R}$	1.1
$\mathbb{C}$	1.1
$\mathcal{C}^r$	1.1
$\mathcal{C}^\infty$	1.1
MVT	1.1
WMVT	1.1
$B_\delta(\mathbf{x})$	1.2
$\bar{B}_\delta(\mathbf{x})$	1.2
$\mathbb{D}^n_r$	1.2
$\mathbb{D}^n$	1.2
$\mathbb{S}^{n-1}_r$	1.2
$\mathbb{S}^{n-1}$	1.2
$D_\mathbf{v}(f)$	1.2
$D_i(f)$	1.2
$D_x(f)$	1.2
∇f	1.2
Df	1.3
RHS	1.4
$D_{1,2}f$	1.4
$M(m,n;\mathbb{R})$	1.5
$M(n;\mathbb{R})$	1.5
$GL(n;\mathbb{R})$	1.5
$M(m,n;\mathbb{C})$	1.5
$GL(n;\mathbb{C})$	1.5
$Df_\mathbf{x}$	1.5
IFT	1.6
LHS	1.6
ImFT	1.6
LMM	1.7
$\operatorname{supp} f$	1.9
$V_{k,n}$	3.4
$\mathcal{R}_k(m,n;\mathbb{R})$	3.4
$\mathfrak{sl}(n,\mathbb{R})$	3.4
$O_{p,q}(\mathbb{R})$	3.4
$SO_{p,q}$	3.4
$f \pitchfork Z$	3.5
$\Omega(X)$	4.2
$\Omega(f)$	4.2
$H^*_{DR}(X)$	4.4
$\mathbb{P}^n$	5.1
$\mathbb{C}P^n$	5.2
$D(X)$	5.6
$\mathcal{T}(X)$	5.6
$G_{k,n}$	5.8
$N(X)$	6.1
$\mathcal{N}_\mathbb{N}(X)$	6.1
IVP(σ)	6.3
$I(f,Z)$	7.2
$I(X,Z)$	7.2
$I(X;Y)$	7.2
$\deg\ f$	7.3
$W(f,z)$	7.5
$I(X)$	7.8
$L(f)$	7.8
$\chi(K)$	7.9
$M_1\#M_2$	8.3
$\mathbb{T}_{g,h}$	8.5
$\mathbb{P}_{k,h}$	8.5
$\mathbb{H}$	9.1
$Sp(n,\mathbb{C})$	9.1

Sectionwise Dependence Tree

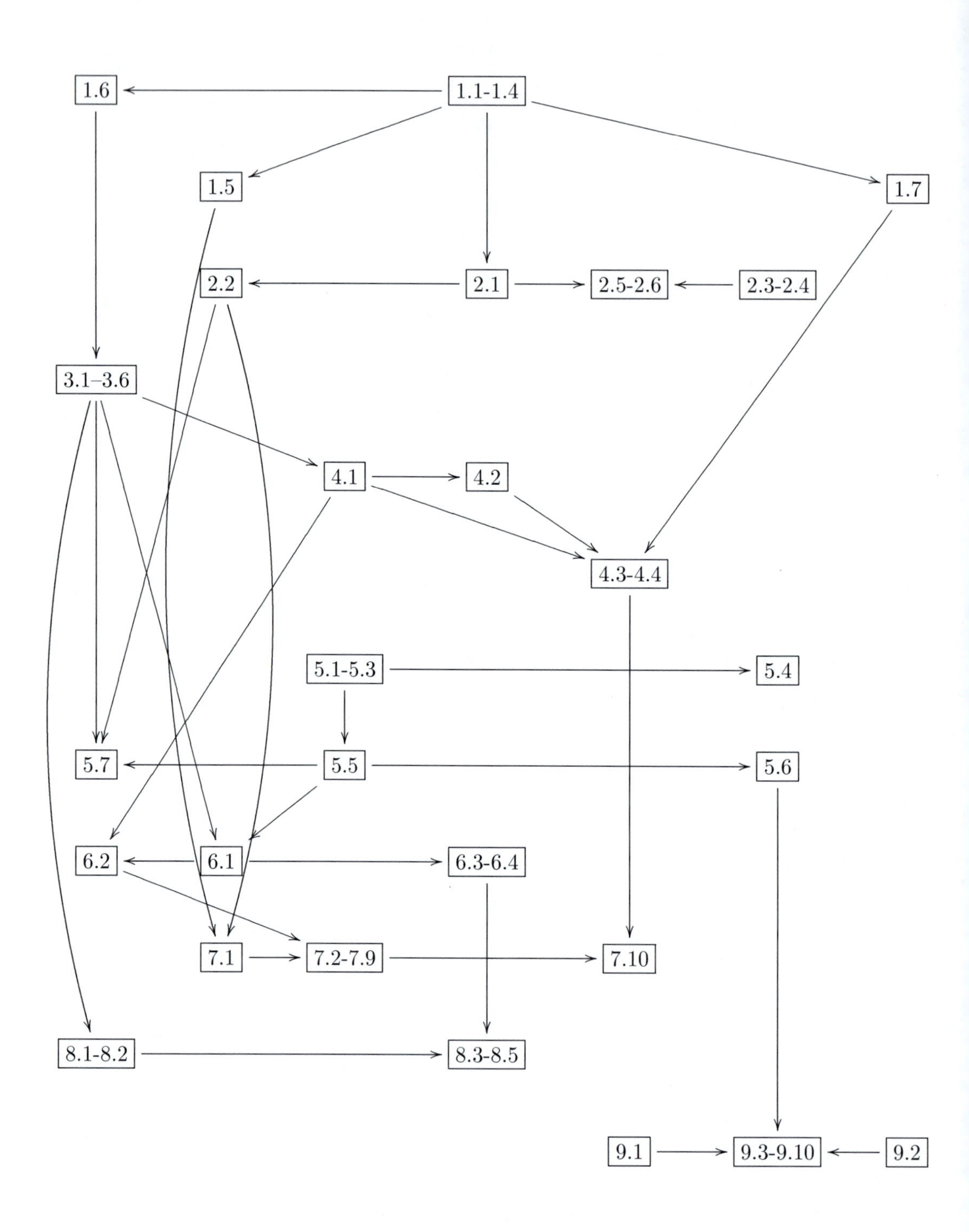

Chapter 1

Review of Differential Calculus

This chapter is a self-contained review of some basics of the differential calculus of several variables. We assume that you are fairly familiar with these results and just need to browse through this chapter to get familiar with some standard notation/results to be used in the rest of the book. However, if you are learning some of them for the first time, you may need to go through it slowly. In any case, whenever you have some difficulty in later chapters, often it just helps to come back to this chapter for the relevant clarification.

1.1 Vector Valued Functions

We shall use the notation $\mathbb{N}, \mathbb{Z}, \mathbb{Q}, \mathbb{R}, \mathbb{C}$ to denote the set of positive integers, the ring of integers, and the fields of rational numbers, the real numbers, and the complex numbers, respectively.

Definition 1.1.1 Let $r \in \mathbb{N}$ be a nonnegative integer. Let U be an open interval in $\mathbb{R}$. We say $f : U \to \mathbb{R}$ is a $\mathcal{C}^r$-function if f possesses all derivatives of order $\leq r$ and these derivatives are all continuous throughout U. (For $r = 0$ this is taken to mean that f is continuous.) We say f is $\mathcal{C}^\infty$ on U if it is $\mathcal{C}^r$ for all $r \geq 0$. The set of all real valued $\mathcal{C}^r$-functions on U is denoted by $\mathcal{C}^r(U)$. Similarly $\mathcal{C}^\infty(U)$ denotes the set of all $\mathcal{C}^\infty$ functions. We say f is *smooth* if it is of class $\mathcal{C}^r$, for some $r \geq 1$. The purpose of introducing this somewhat vague terminology is twofold:
(i) often the actual value of $r \geq 1$ is not so important.
(ii) most of the properties that we need to discuss merely require that $r = 1$; only once in a while we need to have higher values of r.

Remark 1.1.1 In the literature, the terminology "*f is a continuously differentiable function*" means that f is a $\mathcal{C}^1$-function. We shall also use this without hesitation.

Recall that a complex valued function is continuous if and only if its real and imaginary parts are continuous. Motivated by this, we can immediately adopt the following definition.

Definition 1.1.2 Let U be an open interval. A vector valued function $f : U \to \mathbb{R}^n$ where, $f(t) = (f_1(t), \ldots, f_n(t))$ is *differentiable* at $t = t_0 \in U$ if and only if each one of its component functions f_i is differentiable at t_0. In this case, we shall use the notation

$$D(f)(t_0) := (f_1'(t_0), \ldots, f_n'(t_0))$$

for the derivative of f at t_0. If f is differentiable at all points $t \in U$, we then say that f is differentiable on U. We shall use the notation $\mathcal{C}^r(U;\mathbb{R}^m)$ to denote the set of all functions $f : U \to \mathbb{R}^m$ with each component $f_i \in \mathcal{C}^r(U)$.

Remark 1.1.2 As can be easily verified, this definition is quite satisfactory in the sense that all the standard properties of differentiability of scalar valued function of a real variable have parallels here also. For instance, we have the chain rule for differentiation of composite of two functions viz., for any real valued differentiable functions $g : V \to U$ and $f : U \to \mathbb{R}^n$, we have $f \circ g$ is differentiable on V and

$$D(f \circ g)(t) = D(f)(g(t))g'(t) = g'(t)(f_1'(g(t)), \ldots, f_n'(g(t))).$$

Thus, the theory of vector valued functions is a straightforward extension of the theory of scalar valued functions. Having said this, we should however take some precaution in dealing with vector valued functions. Indeed, the Mean Value Theorem (MVT) is one of the most fundamental results about smooth real valued functions. Alas! It is no longer true in the case of vector valued functions:

Example 1.1.1 Consider $f(x) = (\cos x, \sin x)$. Then $f(2\pi) - f(0) = 0$. On the other hand, $f'(x) = (-\sin x, \cos x)$, and hence $\|f'(x)\| = 1$, for all x. Therefore, there can be no x_0 such that $f(2\pi) - f(0) = 2\pi f'(x_0)$.

Remark 1.1.3 Nevertheless, the situation is not so bad, since we can always salvage something useful out of the corresponding results for real valued functions. As an illustration, here we give a result that is an easy consequence of the Mean Value Theorem in 1-variable calculus. We shall refer to it as the Weak Mean Value Theorem (WMVT).

Theorem 1.1.1 *Let $f : [a,b] \to \mathbb{R}^n$ be a continuous map, which is differentiable in the open interval, $a < b \in \mathbb{R}$. Then there exists $t_0 \in (a,b)$ such that*

$$\|f(b) - f(a)\| \leq (b-a)\|D(f)(t_0)\|. \tag{1.1}$$

Proof: Let $\mathbf{v} = f(b) - f(a)$ and consider the function $g(t) = \mathbf{v} \cdot f(t)$. Apply MVT to g to obtain $t_0 \in (a,b)$ such that

$$\mathbf{v} \cdot \mathbf{v} = \mathbf{v} \cdot (f(b) - f(a)) = g(b) - g(a) = (b-a)g'(t_0) = (b-a)(\mathbf{v} \cdot D(f)(t_0)).$$

Therefore, by Cauchy-Schwartz's inequality, it follows that

$$\|f(b) - f(a)\| = \|\mathbf{v}\| \leq \|(b-a)D(f)(t_0)\| = (b-a)\|D(f)(t_0)\|,$$

which is nothing but (1.1). ♠

Exercise 1.1 Verify the following statements:
(a) A vector valued function $f : \mathbb{R} \to \mathbb{R}^n, f = (f_1, f_2, \ldots, f_n)$ is continuous if and only if each $f_i : \mathbb{R} \to \mathbb{R}$ is continuous.
(b) The function f as above is differentiable at $t = t_0$ if and only if

$$\lim_{t \to t_0} \frac{f(t) - f(t_0)}{t - t_0}$$

exists.

1.2 Directional Derivatives and Total Derivative

We are interested in differential properties of vector valued functions of several real variables. In the case of one variable, the domains of definition of our functions were, at least to begin with, open intervals. The analogue of this in several variables will be open subsets. Let us first recall some standard notations and concepts.

Given a point $\mathbf{x} \in \mathbb{R}^n$ and $\delta > 0$, the open ball and the closed ball of radius δ around $\mathbf{x}$ are defined respectively as follows:

$$B_\delta(\mathbf{x}) = \{\mathbf{y} \in \mathbb{R}^n \;:\; \|\mathbf{y} - \mathbf{x}\| < \delta\}, \;\; \bar{B}_\delta(\mathbf{x}) = \{\mathbf{y} \in \mathbb{R}^n \;:\; \|\mathbf{y} - \mathbf{x}\| \leq \delta\}.$$

We shall also use the notation:

$$\mathbb{S}_r^{n-1} = \{\mathbf{x} \in \mathbb{R}^n \;:\; \|\mathbf{x}\| = r\}; \; \mathbb{D}_r^n = \{\mathbf{x} \in \mathbb{R}^n \;:\; \|\mathbf{x}\| \leq r\}; \; \mathbb{S}^{n-1} = \mathbb{S}_1^{n-1}; \; \mathbb{D}^n = \mathbb{D}_1^n.$$

A subset $U \subset \mathbb{R}^n$ is said to be open in $\mathbb{R}^n$ if for each $\mathbf{x} \in U$ there exists $\delta = \delta_{\mathbf{x}} > 0$ such that $B_\delta(\mathbf{x}) \subset U$. A subset is closed if its complement is open.

So let us begin with a function $f : U \to \mathbb{R}^m$, where U is an open subset of $\mathbb{R}^n$ for some $m, n \geq 1$. Considerations similar to the one in the previous section offer us an easy approach to convert the study of f into the study of its components f_i's, where $f = (f_1, f_2, \ldots, f_m)$. So, let us concentrate on the case $m = 1$.

Can we do a similar simplification in the domain of the function as well? That is, can we restrict the function to line segments parallel to one of the coordinate axes and passing through a given point in U and talk about the differentiability of these restricted functions? No doubt this can be a useful idea but we need to be cautious. Let us examine an example before we proceed.

Example 1.2.1 Define a function $f : \mathbb{R}^2 \to \mathbb{R}$ by

$$f(x, y) = \begin{cases} \dfrac{xy}{x^2 + y^2}, & \text{if} \quad (x, y) \neq (0, 0), \\ 0, & \text{if} \quad (x, y) = (0, 0). \end{cases} \tag{1.2}$$

This map has the property that for each fixed y, it is continuous for all x and for each fixed x it is continuous for all y. However, it is not continuous at $(0, 0)$ even if we are ready to redefine its value at $(0, 0)$. This is checked by taking limits of f as $x \to 0$ along the lines $y = mx$ for different values of m.

Remark 1.2.1
(i) Note that in the above example, we have taken a rational function in two variables, which is homogeneous of degree 0. Such functions provide a variety of counterexamples to illustrate the failure of certain phenomena for functions of several variables. As you will soon see most of our counterexamples in this chapter are based on rational functions which are homogeneous, possibly with different weights assigned to different variables.
(ii) This however should not discourage us completely. As a first shot, we shall give some more thought to this idea of restriction along line segments in the domain which leads us to the notion of directional derivatives and partial derivatives.

Definition 1.2.1 Let $\mathbf{v}$ be a unit vector in $\mathbb{R}^n$, U be an open subset and let $\mathbf{x} \in U$ be any point. Then the line $\{\mathbf{x} + t\mathbf{v} \;:\; t \in \mathbb{R}\}$ which passes through $\mathbf{x}$ contains an open segment around $\mathbf{x}$ completely contained in U. Therefore, given a function $f : U \to \mathbb{R}$, we can talk about the differentiability at $\mathbf{x}$ of the function obtained by restricting f to this line segment.

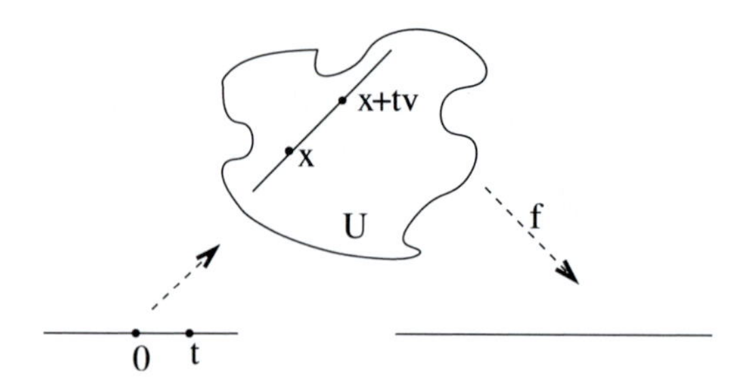

Figure 1 The directional derivative.

More precisely, consider the function $g(t) = f(\mathbf{x} + t\mathbf{v})$, which is defined in an open interval containing 0. We say that the *directional derivative* of f in the direction of $\mathbf{v}$ at $\mathbf{x}$ exists and is equal to $g'(0)$ if $g'(0)$ exists; in this case, we denote $g'(0)$ by $D_{\mathbf{v}}(f)(\mathbf{x})$. Thus,

$$D_{\mathbf{v}}(f)(\mathbf{x}) = \lim_{t \to 0} \frac{f(\mathbf{x} + t\mathbf{v}) - f(\mathbf{x})}{t}. \tag{1.3}$$

If the derivative exists for all $\mathbf{x} \in U$, we obtain a function $\mathbf{x} \mapsto D_{\mathbf{v}}(f)(\mathbf{x})$ defined on U. We shall denote this function by $D_{\mathbf{v}}f$. In the special case when $\mathbf{v} = \mathbf{e}_i$, the ith standard basis element for the vector space $\mathbb{R}^n$, the directional derivative is called the *partial derivative* of f with respect to the ith variable. We also use the simpler notation $D_i f$ in this case. Moreover, when the variables are denoted by x, y, z, etc., as in the case $n = 2$, or 3, we use the notations $\dfrac{\partial f}{\partial x}, \dfrac{\partial f}{\partial y}, \dfrac{\partial f}{\partial z}$, etc. The notations $D_x f, D_y f, D_z f$, etc. are also in use. Whenever all the partial derivatives exist, we shall denote the vector $(D_1 f, \ldots, D_n f)$ by the symbol ∇f (read "del f" or "grad f"). This is called the *gradient* of f and is also denoted by grad f.

Remark 1.2.2 Directional derivatives share most of the natural properties of the derivatives of a function of one variable. One is tempted to define the notion of differentiability in terms of directional derivatives, or even more economically, in terms of partial derivatives. Namely, tentatively, let us say that f is 'T-differentiable' at p if it possesses all the directional derivatives at p. Caution is needed here in the sense that we should check whether such a "T-differentiable" function, as expected, is automatically continuous or not. Alas, this is not the case, as illustrated by the following example. Therefore, we discard this tentative definition.

Example 1.2.2 Consider the function defined by

$$f(x, y) = \begin{cases} \dfrac{x^2 y}{x^4 + y^2}, & \text{if} \quad (x, y) \neq (0, 0) \\ 0, & \text{if} \quad (x, y) = (0, 0). \end{cases} \tag{1.4}$$

Check that at $(0, 0)$ all the directional derivatives exist. Yet the function is not continuous at $(0, 0)$ since along parabolas $y = mx^2$, if we take the limit as $x \to 0$, we get different limits. So, such a function cannot be declared as differentiable. Check that the partial derivatives are not continuous at $(0, 0)$, nor even bounded near $(0, 0)$.

Remark 1.2.3 In case of three variables, it is customary to denote grad f by the sum $\frac{\partial f}{\partial x}\mathbf{i} + \frac{\partial f}{\partial y}\mathbf{j} + \frac{\partial f}{\partial z}\mathbf{k}$. Implicit in this is the notation $\mathbf{i} = (1, 0, 0), \mathbf{j} = (0, 1, 0), \mathbf{k} = (0, 0, 1)$.

We can generalize this practice for any number of variables by using the standard basis elements:

$$\mathbf{e}_1 = (1, 0, \ldots, 0), \mathbf{e}_2 = (0, 1, 0, \ldots, 0), \ldots, \mathbf{e}_n = (0, 0, \ldots, 1).$$

For a function f of n-variables we can then write

$$(\nabla f)(x) := \sum_i D_i f(x) \mathbf{e}_i.$$

Remark 1.2.4 Recall that when we have a differentiable function f of one variable, by the Increment Theorem, we have,

$$f(x+h) = f(x) + f'(x)h + \varepsilon(h)h \tag{1.5}$$

where the error function $\varepsilon(h) \to 0$ as $h \to 0$. Indeed, (1.5) is equivalent to the differentiability of the function f at x : simply by rearranging the terms in (1.5) we get,

$$\left| \frac{f(x+h) - f(x) - hf'(x)}{h} \right| = |\varepsilon(h)| \to 0 \tag{1.6}$$

as $h \to 0$. It turns out that (1.6) can be adopted to give a very satisfactory definition of differentiability of a real valued function of several real variables.

Definition 1.2.2 Let U be an open subset of $\mathbb{R}^n$, $\mathbf{x} \in U$ and let $f : U \to \mathbb{R}$ be a function. We say f is *differentiable* at $\mathbf{x}$ if there exists $\alpha = (\alpha_1, \alpha_2, \ldots, \alpha_n) \in \mathbb{R}^n$ such that for all $\mathbf{h} = (h_1, h_2, \ldots, h_n) \in \mathbb{R}^n \setminus \{\mathbf{0}\}$ with $\mathbf{x} + \mathbf{h} \in U$, we have,

$$\lim_{\mathbf{h} \to \mathbf{0}} \varepsilon(\mathbf{h}) = \lim_{\mathbf{h} \to \mathbf{0}} \frac{f(\mathbf{x}+\mathbf{h}) - f(\mathbf{x}) - \sum_{i=1}^n \alpha_i h_i}{||\mathbf{h}||} = 0. \tag{1.7}$$

We say α is the total derivative of f at $\mathbf{x}$ and denote it by $D(f)(\mathbf{x})$ or sometimes by $D(f)_{\mathbf{x}}$. If this happens at all points of U, then we say f is *differentiable* in U. Also, the assignment $\mathbf{x} \mapsto \alpha$ defines a vector valued function

$$Df : U \to \mathbb{R}^n$$

called the *total derivative* or simply the derivative of f.

Remark 1.2.5
(1) Note that if $n = 1$, then we get back the old definition.

$$Df = f'.$$

(2) Let $\mathbf{u}$ be unit vector. By putting $\mathbf{h} = t\mathbf{u}$ in (1.7), and letting $t \to 0$, we immediately deduce that

$$D_{\mathbf{u}}(f)(\mathbf{x}) = \sum \alpha_i u_i = D(f)(\mathbf{x}) \cdot \mathbf{u}. \tag{1.8}$$

In particular, we conclude that if the total derivative exists then all the partial derivatives exist and the components of the total derivatives are nothing but the partial derivatives. Thus,

$$D(f)(\mathbf{x}) = \nabla f(\mathbf{x}). \tag{1.9}$$

(3) All the standard laws of differentiation such as additivity, the Leibniz rule, etc., are valid and can be verified in a straightforward manner. We have considered only some of these properties below.

Theorem 1.2.1 Increment Theorem: *Let $f : U \to \mathbb{R}$ be a function where U is an open subset of $\mathbb{R}^n$ and let $\mathbf{x} \in U$. Set $z = f(\mathbf{x})$ and $\Delta z = f(\mathbf{x}+\mathbf{h}) - f(\mathbf{x})$. Then f is differentiable at $\mathbf{x}$ if and only if there exists a vector $\alpha \in \mathbb{R}^n$ and an error function ε defined in a neighborhood of $\mathbf{0} \in \mathbb{R}^n$ such that*

$$\Delta z = \alpha \cdot \mathbf{h} + ||\mathbf{h}||\varepsilon(\mathbf{h}) \tag{1.10}$$

and $\varepsilon(\mathbf{h}) \to 0$ as $\mathbf{h} \to \mathbf{0}$.

Proof: In fact, this is just a rewording of Definition 1.2.2. ♠

Taking the limit as $\mathbf{h} \to \mathbf{0}$ in (1.10) yields:

Theorem 1.2.2 Continuity Theorem: *If $f : U \to \mathbb{R}$ is differentiable at $\mathbf{x} \in U$, then f is continuous at $\mathbf{x}$.*

Theorem 1.2.3 Chain Rule (Simpler Case): *Let V be an open subset of $\mathbb{R}^m$. Let $f : (a, b) \longrightarrow V$ and $g : V \to \mathbb{R}$ be such that $f = (f_1, f_2, \ldots, f_m)$ is differentiable at $t_0 \in (a, b)$ and g is differentiable at $f(t_0) \in V$. Then the composite function $g \circ f$ is differentiable at t_0 and its derivative at t_0 is given by*

$$\frac{d(g \circ f)}{dt}(t_0) = \sum_{j=1}^{m} \frac{\partial g}{\partial x_j}(f(t_0))\frac{df_j}{dt}(t_0). \tag{1.11}$$

Proof: Put $\mathbf{y} = f(t_0)$. By the increment theorem we have,

$$f(t_0 + k) - f(t_0) = kf'(t_0) + k\epsilon_1(k); \tag{1.12}$$

$$g(\mathbf{y} + \mathbf{h}) - g(\mathbf{y}) = Dg(\mathbf{y}) \cdot \mathbf{h} + \|\mathbf{h}\|\epsilon_2(\mathbf{h}) \tag{1.13}$$

with $\epsilon_1(k) \to 0$ as $k \to 0$ and $\epsilon_2(\mathbf{h}) \to 0$ as $\mathbf{h} \to 0$. (Note that ϵ_1 is a vector valued function.)

We must find $\epsilon(t)$ such that $\epsilon(t) \to 0$ at $t \to 0$ and

$$g \circ f(t_0 + k) - g \circ f(t_0) = Dg(f(t_0)) \cdot f'(t_0) + k\epsilon(k). \tag{1.14}$$

First of all, observe that f is continuous at t_0 and hence we can put $\mathbf{h} = \mathbf{h}(k) = f(t_0+k) - f(t_0)$ for sufficiently small k. Then we have $\mathbf{h}(k) \to 0$ as $k \to 0$. Next, we observe that

$$\lim_{k \to 0^{\pm}} \frac{||\mathbf{h}(k)||}{k} = \pm\|f'(t_0)\|$$

In particular, $\frac{||\mathbf{h}(k)||}{k}$ is a bounded function. We now take

$$\epsilon(k) = D(g(\mathbf{y})) \cdot \epsilon_1(k) + \frac{||\mathbf{h}(k)||}{k}\epsilon_2(\mathbf{h}(k))$$

which clearly tends to 0 as $k \to 0$. Now, substitute $\mathbf{h} = \mathbf{h}(k)$ in (1.13) to obtain (1.14). ♠

Remark 1.2.6 Remark 1.2.5(2) and Theorem 1.2.3 give enough indication that total derivative and directional derivatives are closely related. This is what we want to investigate further.

Theorem 1.2.4 *Let $f : U \to \mathbb{R}$ be a function such that its partial derivatives all exist and are all bounded in U. Then f is continuous in U.*

Proof: We shall write down the proof for $n = 2$ and leave the general case as an exercise to the reader.

Let $M > 0$ be such that $|f_x(x,y)| < M$, $|f_y(x,y)| < M$ for all $(x,y) \in U$. Let (x_0, y_0) be a point in U. Let (x,y) be sufficiently near (x_0,y_0) so that the rectangle with sides parallel to the axes and with vertices $(x,y), (x_0,y_0)$ is contained in U. Then we have,

$$|f(x,y) - f(x_0,y_0)| \leq |f(x,y) - f(x,y_0)| + |f(x,y_0) - f(x_0,y_0)|. \tag{1.15}$$

Using the mean value theorem for f_x and f_y we can find a point c_1 between y_0 and y and a point c_2 between x_0 and x such that $f(x,y) - f(x,y_0) = f_y(x,c_1)|y - y_0|$ and $f(x,y_0) - f(x_0,y_0) = f_x(c_2,y_0)|x - x_0|$. Plugging these on the right-hand-side (RHS) of (1.15) above yields:

$$|f(x,y) - f(x_0,y_0)| \leq M(|y - y_0| + |x - x_0|)$$

which tends to 0 as $(x,y) \to (x_0,y_0)$. ♠

Example 1.2.3 We may improve upon Example 1.2.2 as follows. Take $g(x,y) = \sqrt{x^2+y^2}f(x,y)$, where f is given as in Example 1.2.2. Then the function g is continuous also at $(0,0)$ and has all the directional derivatives vanish at $(0,0)$. That means that the graph of this function has the xy-plane as a plane of tangent lines at the point $(0,0,0)$. Once again (see Remark 1.2.2), we may be tempted to award such "nice" geometric behavior of the function and admit it to be "differentiable" at $(0,0)$. However, it is not differentiable at $(0,0)$, according to the definition that we have adopted. For

$$\frac{g(x,y) - g(0,0)}{\|(x,y)\|} = f(x,y)$$

has no limit at $(0,0)$. We hope that this example illustrates the subtlety of the situation in the following theorem.

Theorem 1.2.5 *If $f : U \to \mathbb{R}$ is such that all its partial derivatives exist in U and are continuous at $p \in U$, then f is differentiable at p. Moreover, the derivative Df itself is a continuous function on U.*

Proof: Here again we shall write down the proof for $n = 2$ and leave the general case as an exercise. Taking $p = (x_0, y_0)$ concentrating attention inside an open rectangle around p contained in U, we have,

$$\left.\begin{array}{rl} & |f(x,y) - f(x_0,y_0) - (\Delta x f_x(x_0,y_0) + \Delta y f_y(x_0,y_0))| \\ \leq & |f(x,y) - f(x,y_0) - \Delta y f_y(x_0,y_0)| + |f(x,y_0) - f(x_0,y_0) - \Delta x f_x(x_0,y_0)| \\ = & |\Delta y|\ |f_y(x,c_1) - f_y(x_0,y_0)| + |\Delta x|\ |f_x(c_2,y_0) - f_x(x_0,y_0)| \text{ (by MVT)} \\ \leq & ||(\Delta x, \Delta y)||\ (|f_y(x,c_1) - f_y(x_0,y_0)| + |f_x(c_2,y_0) - f_x(x_0,y_0)|). \end{array}\right\} \tag{1.16}$$

Note that c_1 lies between y_0 and y; also c_2 lies between x_0 and x. Thus, $c_1 \to y_0$ and $c_2 \to x_0$ as $(x,y) \to (x_0,y_0)$. Hence, by continuity of f_x and f_y, it follows that the quantity in the bracket on the RHS above tends to 0 as $||(\Delta x, \Delta y)|| \to 0$. Thus, dividing out by $||(\Delta x, \Delta y)||$ and then taking limit as $||(\Delta x, \Delta y)|| \to 0$, we obtain

$$\frac{|f(x,y) - f(x_0,y_0) - (\Delta x f_x(x_0,y_0) + \Delta y f_y(x_0,y_0))|}{||(\Delta x, \Delta y)||} \to 0$$

as required. The last assertion of the theorem follows immediately if you recall that a vector valued function is continuous if all its components are so, as the components of $Df = \nabla f$ are nothing but the partial derivatives. ♠

The following corollary has the flavor of Riemann's removable singularity for functions of one complex variable.

Theorem 1.2.6 Removable Singularity: *Let U be an open subset of $\mathbb{R}^n$, $p \in U$ and $f : U \to \mathbb{R}$ a continuous function. Suppose Df exists in $U \setminus \{p\}$ and $\lim_{q\to p} Df(q) = L$ exists. Then Df_p exists and is equal to L.*

Proof: For $n = 1$ this is a direct consequence of MVT. Since this is a nonstandard result, we shall write down the proof. Choose $\delta > 0$ such that $[p-\delta, p+\delta] \subset U$. By the mean value theorem applied to $f : [p-\delta, p] \to \mathbb{R}$ for each $q \in [p-\delta, p)$ there exists $q < x < p$ such that $f(q) - f(p) = f'(x)(q-p)$. Now take the limit as $q \to p$ inside $[p-\delta, p)$ we obtain that the left-hand derivative of f exists and is equal to $\lim_{x\to p} f'(x) = L$. Similarly the right-hand derivative exists and is equal to L and we are done.

We shall now write down the proof for the case $n = 2$, the general case being similar. Taking $p = (x_0, y_0)$ concentrating attention inside an open rectangle around p contained in U, we merely reproduce the inequality (1.16), except that we have to replace $f_x(x_0, y_0), f_y(x_0, y_0)$ respectively by $L(\mathbf{e}_1), L(\mathbf{e}_2)$. Therefore,

$$\begin{array}{cl} & |f(x,y) - f(x_0,y_0) - L(\nabla tax, \Delta y)| \\ = & |f(x,y) - f(x_0,y_0) - (\Delta x L(\mathbf{e}_1) + \Delta y L(\mathbf{e}_2)| \\ \leq & ||(\Delta x, \Delta y)|| \, [|f_y(x, c_1) - L(\mathbf{e}_2)| + |f_x(c_2, y_0) - L(\mathbf{e}_1)|]. \end{array} \tag{1.17}$$

Here c_1 lies between y_0 and y and c_2 lies between x_0 and x. Since $\lim_{q\to p} Df(q) = L$, we have

$$\lim_{q\to p} f_x(q) = L(\mathbf{e}_1); \lim_{q\to p} f_y(q) = L(\mathbf{e}_2).$$

Therefore, the expression in the square bracket tends to 0 as $q = (x, y) \to p = (x_0, y_0)$. This implies Df_p exists and is equal to L. ♠

Remark 1.2.7 One way we would like to use this result is the following. Often in topology maps on a disc are defined using polar coordinates. Since the polar coordinates have a built-in singularity at the origin, the smoothness of the maps so defined need to be checked carefully.

Recall that the disc $\mathbb{D}^n$ can be thought of as the quotient space of $\mathbb{S}^{n-1} \times [0,1]$, where we identify all $(\mathbf{v}, 0), \mathbf{v} \in \mathbb{S}^{n-1}$ to a single point. This indeed gives polar coordinate representation of the disc.

Now suppose we have a $\mathcal{C}^1$-function $f : \mathbb{S}^{n-1} \times [0,1] \to \mathbb{R}$ such that $f(\mathbf{v}, 0) = 0$ for all $\mathbf{v} \in \mathbb{S}^{n-1}$. Let $\hat{f} : \mathbb{D}^n \to \mathbb{R}$ be the function defined by $\hat{f}[\mathbf{v}, r] = f(\mathbf{v}, r)$. Since the quotient map $(\mathbf{v}, r) \mapsto [\mathbf{v}, r]$ is a local diffeomorphism at all points $(\mathbf{v}, t) \in \mathbb{S}^{n-1} \times (0, 1]$, it follows that $\hat{f}$ is a $\mathcal{C}^1$-function on $\mathbb{D}^n \setminus \{0\}$. Therefore, $\hat{f}$ will be differentiable at 0 if $\lim_{r\to 0} D\hat{f}(\mathbf{v}, r)$ exists, i.e, independent of $\mathbf{v}$ as a single linear map $\mathbb{R}^n \to \mathbb{R}$.

Remark 1.2.8 Theorem 1.2.5 fully justifies the need to study the partial derivatives. Indeed, often in literature, the conditions of this theorem are taken as the definition of differentiability. Though, such a definition, automatically excludes functions that may be differentiable but having a derivative which is not continuous, it has the tremendous advantage of being extremely simple to work with and being sufficient in many practical purposes. For instance, this definition very easily extends to higher order differentiation. All that we have to do is to consider partial derivatives of partial derivatives and so on, since each partial derivative is a new scalar function. For instance, for a function f of 2-variables, there are four possible partial derivatives of second order, viz. $D_1(D_1), D_1(D_2 f), D_2(D_1 f)$ and $D_2(D_2 f)$. These are also written as $\frac{\partial^2 f}{\partial x^2}, \frac{\partial^2 f}{\partial x \partial y}, \frac{\partial^2 f}{\partial y \partial x}, \frac{\partial^2 f}{\partial y^2}$, respectively. There is a further simplification of notations, such as $D_{1,2} f$ to denote $D_1(D_2 f)$, etc. It is not true, in general that $D_{1,2}(f) = D_{2,1}(f)$. Finally, when fewer than three variables are involved, such as for a function $f(x,y,z)$, there are the classical notation f_x, f_{xy}, etc. to denote respectively, $\frac{\partial f}{\partial x}, \frac{\partial f}{\partial y \partial x}$, etc. Pay attention to the order in which the variables with respect to which the differentiation is being taken. The theorem below gives a sufficient condition under which we can permute the order of taking mixed partial derivatives.

Definition 1.2.3 Let r be a nonnegative integer. Let U be an open set in $\mathbb{R}^n$. As in Definition 1.1.1, we say $f : U \to \mathbb{R}$ is a $\mathcal{C}^r$-function if f possesses all partial derivatives of order $\leq r$ and these partial derivatives are all continuous throughout U. We say f is $\mathcal{C}^\infty$ on U if it is $\mathcal{C}^r$ for all $r \geq 0$. The set of all real valued $\mathcal{C}^r$-functions on U is denoted by $\mathcal{C}^r(U)$. Similarly $\mathcal{C}^\infty(U)$ denotes the set of all $\mathcal{C}^\infty$ functions.

Remark 1.2.9 We say a function is *smooth* if it is in $\mathcal{C}^r$ for $r \geq 1$. It is a convenient practice to adjust the value of r according to the context.

Theorem 1.2.7 Clairaut's Theorem:[1] *Let U be an open subset of $\mathbb{R}^2$. Let $f : U \to \mathbb{R}$ be such that both f_{xy} and f_{yx} exists and are continuous in U. Then $f_{xy} = f_{yx}$ on U.*

Remark 1.2.10 In order to prove this, we need to exploit the mean value theorem thoroughly. On the way we shall be able to get a "second order mean value theorem" for functions of two (or more) variables. Moreover, we shall be able to give a proof of a statement that is slightly more powerful than Clairaut's theorem.

Theorem 1.2.8 Second Order Mean Value Theorem : *Let f be defined in an open subset U of $\mathbb{R}^2$, and possess $D_1 f$, $D_{2,1} f$ in U. Let the rectangle R with vertices $(a,b), (a+h,b), (a+h,b+k)$, and $(a,b+k)$ be contained in U. Then there exists a point $(x,y) \in R$ such that*

$$hkD_{2,1}f(x,y) = \Delta,$$

where Δ is the double difference given by

$$\Delta := \Delta f := f(a+h,b+k) - f(a+h,b) - f(a,b+k) + f(a,b). \tag{1.18}$$

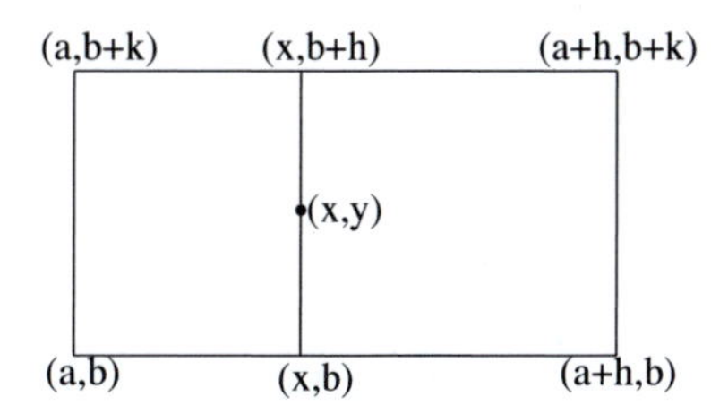

Figure 2 Second order mean value theorem.

Proof: Put $g(t) = f(t,b+k) - f(t,b)$. Then by the mean value theorem (MVT) for one variable functions, there exists $0 \leq s_1 \leq 1$ such that

$$\Delta = g(a+h) - g(a) = hg'(a+s_1 h). \tag{1.19}$$

Take $x = a + s_1 h$. Observe $g'(t) = D_1 f(t,b+k) - D_1 f(t,b)$. Now define $G(r) = D_1 f(x,r)$ and observe that $G'(r) = D_{2,1} f(x,r)$. Apply the MVT again to get $0 \leq s_2 \leq 1$ such that

$$kG'(b+s_2 k) = G(b+k) - G(b) = D_1 f(x,b+k) - D_1 f(x,b) = g'(x). \tag{1.20}$$

Now, multiply both sides of (1.20) by h and take $y = b + s_2 k$. ♠

[1] Alexis Claude Clairaut (1713–1765), contemporary of D'Alembert, a prodigious Parisian, is well known for his contributions in astronomy.

Theorem 1.2.9 Commutativity of Mixed Partial Derivatives: *Let f be defined in an open subset U of $\mathbb{R}^2$, and possess $D_1 f$, $D_{2,1}f$, and $D_2 f$ in U. Suppose $D_{2,1}f$ is continuous at some point $(a,b) \in U$. Then $D_{1,2}f$ exists at (a,b) and we have,*

$$D_{1,2}f(a,b) = D_{2,1}f(a,b). \tag{1.21}$$

Proof: By the continuity of $D_{2,1}f$, given $\epsilon > 0$, there exists a rectangle R contained in U such that

$$|D_{2,1}f(x,y) - D_{2,1}f(a,b)| < \epsilon,$$

for all $(x,y) \in R$. Then,

$$\left| \frac{\Delta f}{hk} - D_{2,1}f(a,b) \right| < \epsilon$$

where Δ is defined as in (1.18). Fix h and let $k \to 0$. Since $D_2 f$ exists, we obtain,

$$\lim_{k\to 0} \frac{\Delta f}{hk} = \frac{D_2 f(a+h,b) - D_2 f(a,b)}{h}.$$

Plug this in the above inequality to get,

$$\left| \frac{D_2 f(a+h,b) - D_2 f(a,b)}{h} - D_{2,1}f(a,b) \right| \leq \epsilon.$$

This proves that $D_{1,2}f(a,b)$ exists and is equal to $D_{2,1}f(a,b)$ as required. ♠

Example 1.2.4 You are familiar with examples of one variable functions having a discontinuous first derivative, e.g.,

$$f(x) = x^2 \sin\left(\frac{1}{x}\right). \tag{1.22}$$

Such functions can be used to construct examples of functions of several variables which are differentiable but have discontinuous partial derivatives. For instance, take $g(x,y) = f(x)$ where f is as above.

Example 1.2.5 Consider the function

$$g(x,y) = \begin{cases} \dfrac{x^2 - y^2}{x^2 + y^2}, & (x,y) \neq (0,0), \\ 0, & (x,y) = (0,0). \end{cases} \tag{1.23}$$

Put $f(x,y) = xyg(x,y)$. Then g_x and g_y exist everywhere and

$$f_x = yg + xyg_x; \quad f_y = xg + xyg_y.$$

Therefore,

$$f_{xy}(0,0) = \lim_{y\to 0} \frac{f_x(0,y) - f_x(0,0)}{y} = \lim_{y\to 0} \frac{yg(0,y)}{y} = -1.$$

Interchanging x and y, we see $f_{yx}(0,0) = 1$. If one of the functions f_{xy} and f_{yx} were continuous at $(0,0)$, then both would have the same value at $(0,0)$. Therefore, Theorem 1.2.9 tells you that neither f_{xy} nor f_{yx} is continuous. One can directly verify that being a homogeneous rational function of degree zero, f_{xy} takes a constant value on the line $y = mx$ and these constants are different for different m. That is enough to conclude that f_{xy} (and similarly f_{yx}) is not continuous at $(0,0)$.

Theorem 1.2.10 Taylor's Theorem: *Let $f : U \to \mathbb{R}$ be a C^r-function, where U is a convex open subset of $\mathbb{R}^n$. Let $\mathbf{0} \in U$. Then for each point $\mathbf{x} \in U$, we have*

$$f(\mathbf{x}) = f(\mathbf{0}) + \sum_{k=1}^{r-1} \frac{1}{k!} \sum D_{i_1 i_2 \ldots i_k} f(\mathbf{0}) x_{i_1} x_{i_2} \cdots x_{i_k} + R(\mathbf{x}) \tag{1.24}$$

where the remainder term $R(\mathbf{x})$ satisfies

$$\lim_{\mathbf{x}\to 0} \frac{R(\mathbf{x})}{\|\mathbf{x}\|^{r-1}} = 0. \tag{1.25}$$

Proof: Fix $\mathbf{x} \in U$ and define $g(t) = f(t\mathbf{x})$. Then by Taylor's theorem for one variable, there exists $t_{\mathbf{x}} \in (0,1)$ such that

$$g(1) = g(0) + \sum_{k=0}^{r-1} \frac{g^{(k)}(0)}{k!} + \frac{g^{(r)}(t_{\mathbf{x}})}{r!}. \tag{1.26}$$

By repeated application of the chain rule to find successive derivatives of $g(t) = f(t\mathbf{x})$, we find that

$$\left.\begin{array}{l} g^{(1)}(1) := g'(1) = \sum_i D_i f(\mathbf{x}) x_i; \quad g^{(2)}(1) = \sum D_{i,j} f(\mathbf{x}) x_i x_j; \cdots, \\ g^{(k)}(1) = \sum D_{i_1 i_2 \ldots i_k} f(\mathbf{x}) x_{i_1} x_{i_2} \cdots x_{i_k}; \quad \cdots \end{array}\right\} \tag{1.27}$$

for $k \leq r-1$ and

$$g^{(r)}(t_{\mathbf{x}}) = \sum D_{i_1 i_2 \ldots i_r} f(t_{\mathbf{x}}\mathbf{x}) x_{i_1} x_{i_2} \cdots x_{i_r}. \tag{1.28}$$

Substituting these, in (1.26) and taking

$$R(\mathbf{x}) = \frac{g^{(r)}(t_{\mathbf{x}})}{r!} \tag{1.29}$$

we get the first part of the statement. Since,

$$\lim_{\mathbf{x}\to 0} D_{i_1 i_2 \ldots i_r} f(t_{\mathbf{x}}\mathbf{x}) = D_{i_1 i_2 \ldots i_r} f(\mathbf{0}); \ \& \lim_{\mathbf{x}\to 0} \frac{x_{i_1} x_{i_2} \cdots x_{i_r}}{\|\mathbf{x}\|^{r-1}} = 0$$

for all $i_1, i_2, \ldots, i_r$, (1.28) yields (1.25). ♠

We end this section with an analogous result, which is obtained by a simple application of the chain rule. We shall have some opportunities to use this later.

Theorem 1.2.11 *Let f be a C^r function in a convex neighborhood U of $0 \in \mathbb{R}^n$. Then there exist C^{r-1} functions g_i on U such that $\frac{\partial f}{\partial x_i}(0) = g_i(0)$ and*

$$f(\mathbf{x}) = f(0) + \sum_{i=1}^{n} x_i g_i(\mathbf{x}), \tag{1.30}$$

for $\mathbf{x} = (x_1, \ldots, x_n) \in U$.

Proof: By the fundamental theorem of integral calculus of one variable, we have

$$f(\mathbf{x}) = \int_0^1 \frac{df(t\mathbf{x})}{dt} = \int_0^1 \sum_{i=1}^{n} \frac{\partial f(t\mathbf{x})}{\partial x_i} x_i dt.$$

Therefore, we define $g_i(\mathbf{x}) = \displaystyle\int_0^1 \frac{\partial f(t\mathbf{x})}{\partial x_i} dt$ and check that all the required properties are satisfied. ♠

Exercise 1.2

1. In each of the following cases draw a picture of the level surfaces for f and that for ∇f at some select points of the surface. Check that ∇f is perpendicular to the surface at all these points.
(a) $f(x,y,z) = x^2 + 2y^2 + 3z^2$; (b) $f(x,y,z) = x^2 + y^2 - z^2$;
(c) $f(x,y,z) = y - x^2$.

2. Find all the directional derivatives for f at P in each of the following cases:
(a) $f = x^2 - 9y^2$; $P = (1,2)$; (b) $f = (x^2 + y^2 + z^2)^{-1/2}$; $P = (1,1,1)$;
(c) $f = e^x \cos y,$; $P = (1,\pi)$.

3. Simplify the expression for $g \circ f$ in each of the following cases.
(a) $f(x) = (\cos x, \sin^2 x); g(u,v) = u^2 + v^2$;
(b) $f(x) = (x, x^2, x^3), g(u,v,w) = e^u + \cos v + w^2$.

4. Use the chain rule and the Leibniz rule, etc., to find the derivative of these functions:
(a) $e^{\cos^2 x^2}$ (b) $e^{-5\tan x} \sin^3 2\pi x$; (c) $f(\cos x, \sin x)$.

5. Let f, g be smooth real valued functions. Derive the following formulae:
(a) $\nabla(fg) = g(\nabla f) + f(\nabla g)$; (b) $\nabla(f^n) = nf^{n-1}\nabla f$.
(c) $\nabla(f/g) = (1/g^2)(g\nabla f - f\nabla g)$;
(d) $\nabla^2(fg) = g\nabla^2 f + 2\nabla f \nabla g + f\nabla^2 g$.

6. The divergence $\operatorname{div} f$ of a smooth function $f : \mathbb{R}^n \to \mathbb{R}^n$ is defined by
$$\operatorname{div} f = \sum_i \frac{\partial f_i}{\partial x_i}.$$
In the vector notation, writing $f = \sum_i f_i \mathbf{e}_i$ and $\nabla = \sum_i \frac{\partial}{\partial x_i}$ we can write
$$\operatorname{div} f = \nabla \cdot f.$$
Find the divergence of the following functions:
(a) $\dfrac{x\mathbf{i} + y\mathbf{j} + z\mathbf{k}}{(x^2 + y^2 + z^2)^{3/2}}$; (b) $(\cos x \cosh y)\mathbf{i} + (\sin x \sinh y)\mathbf{j}$.

7. Show that
(a) $\operatorname{div}(kf) = k \operatorname{div} f$, k a constant;
(b) $\operatorname{div}(\alpha f) = \alpha \operatorname{div} f + \alpha \cdot \nabla f$, α, a scalar function;
(c) $\operatorname{div}(f\nabla g) = f\nabla^2 g + (\nabla f)\cdot(\nabla g)$;
(d) $\operatorname{div}(f\nabla g) - \operatorname{div}(g\nabla f) = f\nabla^2 g - g\nabla^2 f$.

8. For a smooth function $f : \mathbb{R}^3 \to \mathbb{R}^3$, $f = (f_1, f_2, f_3)$, we define the curl of f by the formula
$$\operatorname{curl} f = \nabla \times f = \left(\frac{\partial f_3}{\partial y} - \frac{\partial f_2}{\partial z}\right)\mathbf{i} + \left(\frac{\partial f_1}{\partial z} - \frac{\partial f_3}{\partial x}\right)\mathbf{j} + \left(\frac{\partial f_2}{\partial x} - \frac{\partial f_1}{\partial y}\right)\mathbf{k}$$
Compute the curl of the following functions:
(a) $x\mathbf{i} + xy\mathbf{j} + (x^2 + y^2 + z^2)\mathbf{k}$; (b) $\dfrac{x\mathbf{i} + y\mathbf{j} + z\mathbf{k}}{(x^2 + y^2 + z^2)^{3/2}}$.

9. Establish the following identities where f, g, etc. denote vector valued functions and α, β denote scalar functions:
(a) $\operatorname{curl}(f + g) = \operatorname{curl} f + \operatorname{curl} g$; (b) $\operatorname{curl} \operatorname{grad} \alpha = 0$;
(c) $\operatorname{div}(f \times g) = g \cdot \operatorname{curl} f - f \cdot \operatorname{curl} g$; (d) $\operatorname{div}(\operatorname{curl} f) = 0$;
(e) $\operatorname{curl}(\alpha f) = (\operatorname{grad} \alpha) \times f + \alpha \operatorname{curl} f$; (f) $\operatorname{div}(\beta \nabla\alpha \times \alpha\nabla\beta) = 0$.

10. Let $f : U \to \mathbb{R}$ be as in the above theorem, $r \geq 2$. Suppose $f(0) = 0$ and $Df(0) = 0$. Then show that there are $\mathcal{C}^{r-2}$ functions τ_{ij} such that $\tau_{ij}(0) = \frac{\partial^2 f}{\partial x_i \partial x_j}(0)$ and $f(\mathbf{x}) = \sum_{ij} \tau_{ij}(\mathbf{x}) x_i x_j, \quad \mathbf{x} \in U$.

11. A smooth function $f : \mathbb{R}^n \to \mathbb{R}$ is called *homogeneous of degree k* if $f(t\mathbf{x}) = t^k f(\mathbf{x})$ for all $\mathbf{x} \in \mathbb{R}^n$ and $t \in \mathbb{R}$. Typical examples of such functions are homogeneous polynomials of degree k. Prove the following **Euler's Identity:**

$$kf(x) = \sum_{i=1}^{n} x_i \frac{\partial f}{\partial x_i}. \tag{1.31}$$

12. Deduce (1.30) directly from (1.24).

1.3 Linearity of the Derivative

We shall denote the space of all $m \times n$ matrices with real entries by $M(m, n; \mathbb{R})$ or by the more suggestive notation $\mathbb{R}^{m \times n}$ to indicate that it can be identified with the Euclidean space of dimension mn. When $m = n$, we shall use the simpler symbol $M(n; \mathbb{R})$ also. We shall use the notation $GL(n, \mathbb{R})$ for the (open) subspace of all invertible $n \times n$ matrices with real entries. The corresponding notations $M(m, n; \mathbb{C}), M(n; \mathbb{C}), GL(n, \mathbb{C})$, etc., with $\mathbb{R}$ replaced by $\mathbb{C}$ has the obvious meaning.

Consider a smooth function $f : U \to \mathbb{R}^m$ at a point $\mathbf{x} \in U$ where, $U \subset \mathbb{R}^n$ is an open set. It is convenient to have the convention of writing elements in $\mathbb{R}^n$ as column vectors. Since this will take up a lot of typed space, we shall write them as $(x_1, \ldots, x_n)^t \in \mathbb{R}^n$. With this convention, if the components of f are $f_1, f_2, \ldots, f_m$ then $f(\mathbf{x}) = (f_1(\mathbf{x}), \ldots, f_m(\mathbf{x}))^t$.

On the other hand, if $L : \mathbb{R}^n \to \mathbb{R}$ is a linear map, it is represented by the row vector $(L(\mathbf{e}_1), \ldots, L(\mathbf{e}_n))$. In this way, the dual space $(\mathbb{R}^n)^*$ of linear maps $\mathbb{R}^n \to \mathbb{R}$ gets identified with $\mathbb{R}^n$ but treated as row vectors.

Following the theme in Definition 1.1.2, we now make:

Definition 1.3.1 Let U be a nonempty open subset of $\mathbb{R}^n$. A vector valued function $f : U \to \mathbb{R}^m$ is said to be differentiable at $\mathbf{x} \in U$ if and only if each component f_i is differentiable at $\mathbf{x}$. In that case, the derivative of f at $\mathbf{x}$ is defined to be the $m \times n$ matrix obtained by writing the row vectors $Df_i, i = 1, 2, \ldots, m$, one below the other.

We would like to interpret the derivative $Df(a)$ of f at $a \in U$ as a linear map $\mathbb{R}^n \to \mathbb{R}^m$. In the simplest case, viz., $m = n = 1$, i.e., when f is a real valued function defined on an interval, the derivative $f'(a)$ at a is a real number. Recall that each real number r corresponds to a unique linear map $h_r : \mathbb{R} \to \mathbb{R}$ viz., $x \mapsto rx$. Hence, we have no difficulty in thinking of $f'(a)$ as a linear map from $\mathbb{R}$ to $\mathbb{R}$. The graph of the affine linear map

$$x \mapsto f(a) + f'(a)(x - a) \tag{1.32}$$

is nothing but the tangent line to the graph of f at the point $(a, f(a))$. Also observe that the RHS of (1.32) is nothing but the linear approximation to f near a.

Likewise, when $n = 1, \ m \geq 2$, we can identify the total derivative of f at $x = a$ with a $m \times 1$ matrix or a *column vector* $Df(a) = (f_1'(a), \ldots, f_m'(a))^t$. This in turn is identified with the linear map $\mathbb{R}^1 \to \mathbb{R}^m$, which sends $\mathbf{e}_1$ to $(f_1'(a), \ldots, f_m'(a))$. Observe that the graph of this linear map parallely shifted so as to pass through $(a, f(a))$ is the tangent hyperplane to the graph of the function at the point $(a, f(a))$.

Next, we consider the case when $n \geq 2$ and $m = 1$. As an example consider a linear function $L : \mathbb{R}^3 \to \mathbb{R}$, given by

$$L(x, y, z)^t = ax + by + cz. \tag{1.33}$$

For each point $p \in \mathbb{R}^3$, the total derivative $DL(p)$ of L at p is the linear map $\mathbb{R}^3 \to \mathbb{R}$ given by the 1×3 matrix (a, b, c). On the other hand, the gradient $\nabla L(a)$ of L is the vector $a\mathbf{i} + b\mathbf{j} + c\mathbf{k} \in \mathbb{R}^3$ which can be identified with the row vector (or the 1×3 matrix) (a, b, c). Thus, DL is the constant function on $\mathbb{R}^3$ with a single value in the space of linear maps $\mathbb{R}^3 \to \mathbb{R}$, which can be identified with the space of 1×3 matrices $M(1, 3; \mathbb{R})$ with real coefficients. Likewise, if $Q(x, y, z) = x^2 + y^2 + z^2$, then the total derivative DQ is a function from $\mathbb{R}^3$ to $M(1, 3; \mathbb{R})$ given by

$$(x, y, z)^t \mapsto (2x, 2y, 2z). \tag{1.34}$$

Finally, consider the case $m, n \geq 2$. Then $Df(\mathbf{x})$ is the matrix obtained by writing the row vectors $Df_1, \ldots, Df_m$ one below the other in that order. Incidentally this happens to be the same matrix obtained by writing the column vectors $D_1 f, \ldots, D_n f$ next to each other in that order. The corresponding linear map is seen to send $\mathbf{e}_j$ to the vector $\sum_{i=1}^{n} D_j f_i \mathbf{e}_i$. We can now formulate a direct definition of differentiability of a map $f : U \to \mathbb{R}^n$ on an open set $U \subset \mathbb{R}^m$. This, we shall state as a theorem, in view of Definition 1.1.2.

Theorem 1.3.1 *Let $f : U \to \mathbb{R}^n$ be a function and $a \in U$ any point where U is an open subset of $\mathbb{R}^m$. Then f is differentiable at a if and only if there exists a linear map $L = L_a : \mathbb{R}^m \to \mathbb{R}^n$ such that the error function has the property*

$$\lim_{\|h\| \to 0} \epsilon(h) = \lim_{\|h\| \to 0} \frac{\|f(a + h) - f(a) - L(h)\|}{\|h\|} = 0. \tag{1.35}$$

Proof: Assuming f is differentiable at a, we set $L : \mathbb{R}^n \to \mathbb{R}^m$ to be the linear map defined by the matrix obtained by writing the row vectors $Df_1(a), \ldots, Df_m(a)$ one below the other. If ϵ_i are the corresponding error functions for f_i as in (1.7) we take $\epsilon = (\epsilon_1, \ldots, \epsilon_m)^t$ and verify that (1.35) is satisfied. The above argument is completely reversible and hence the converse follows. ♠

Definition 1.3.2 Given $f : U \to \mathbb{R}^m$ where U is an open subset of $\mathbb{R}^n$, if f possesses total derivative at every point of U we obtain a function $Df : U \to M(m, n; \mathbb{R}) = \mathbb{R}^{m \times n}$ defined by $x \mapsto Df(x), x \in U$. This is called the *total derivative of f*. If Df is differentiable at $a \in U$ then we say f is *twice differentiable at a*. Once again if this is true for all $a \in U$ then we say f is twice differentiable in U and the second derivative $D^2 f$ is then a map $U \to M(m \times n, n; \mathbb{R}) = \mathbb{R}^{mn \times n}$. More generally, if f is $(k-1)$-times differentiable and if $D^{k-1} f : U \to \mathbb{R}^{mn^{k-1} \times n}$ is differentiable, then we say f is k-times differentiable.

Remark 1.3.1 Note that $D^k f$ exists implies that $D^{k-1} f$ is continuous and therefore $f \in C^{k-1}(U)$. Further, it also follows that $D^k f$ is continuous if and only if $f \in \mathcal{C}^k(U; \mathbb{R}^m)$.

The following theorem gives the chain rule for the derivative of composite of two such functions. Note that if we have to show that two given matrices are equal then it amounts to showing that their corresponding entries are equal. Then Theorem 1.2.3 discussed above completes the proof. We leave the details of the proof to the reader.

Theorem 1.3.2 Chain Rule: *Let U be an open subset of $\mathbb{R}^n$, V be an open subset of $\mathbb{R}^m$. Let $f : U \to V$ and $g : V \to \mathbb{R}^l$ be such that f is differentiable at $\mathbf{x} \in U$ and g be differentiable at $f(\mathbf{x}) \in V$. Then $g \circ f$ is differentiable at $\mathbf{x}$ and the derivative is given by*

$$D(g \circ f)(\mathbf{x}) = (Dg)(f(\mathbf{x})) \circ (Df)(\mathbf{x}). \tag{1.36}$$

Remark 1.3.2 Going back to the case, $n = 1$ and $l = 1$, note that Dg is a matrix of size $1 \times m$ and Df is a matrix of size $m \times 1$ and their product $D(g \circ f)$ is a matrix of size 1×1, which can be thought of a real number. This yields the statement of Theorem (1.2.3) discussed in the previous section. We could have expressed this in vector notation by the dot product of two column vectors:

$$D(g \circ f)(\mathbf{x}) \;=\; [(Dg)(f(\mathbf{x}))]^t \bullet (Df)(\mathbf{x}). \tag{1.37}$$

Finally, we also note that, in the above theorem, if we interpret the quantities in (1.36) as the corresponding matrices then the composition corresponds to the matrix multiplication.

Remark 1.3.3 There is an analogue of the mean value theorem for real valued functions of several variables, proved by simply converting the problem into one variable. From this, we can then derive the analogue of the weaker version of the mean value theorem for vector valued functions as seen in Theorem 1.1.1.

Theorem 1.3.3 *Let U be a convex domain in $\mathbb{R}^n$ and $f : U \to \mathbb{R}^m$ be a continuously differentiable function. Let $\mathbf{a}, \mathbf{b} \in U$ be any two points.*
(i) *If $m = 1$, then there exists $\mathbf{x_0} \in [\mathbf{a}, \mathbf{b}]$ such that*

$$f(\mathbf{b}) - f(\mathbf{a}) = D(f)(\mathbf{x_0})(\mathbf{b} - \mathbf{a}). \tag{1.38}$$

(ii) *In general, there exists $\mathbf{x_0} \in [\mathbf{a}, \mathbf{b}]$ such that*

$$\|f(\mathbf{b}) - f(\mathbf{a})\| \leq \|D(f)(\mathbf{x_0})\|\|\mathbf{b} - \mathbf{a}\|. \tag{1.39}$$

Proof: Consider the function $\gamma : [0, 1] \to \mathbb{R}^n$, defined by

$$\gamma(t) = t\mathbf{b} + (1 - t)\mathbf{a} \tag{1.40}$$

and put $g = f \circ \gamma$.
(i) If $m = 1$, apply the MVT to g to obtain $t_0 \in [0, 1]$, such that

$$f(\mathbf{b}) - f(\mathbf{a}) = g(1) - g(0) = g'(t_0) \tag{1.41}$$

Put $\mathbf{x}_0 = \gamma(t_0)$. Then by chain rule,

$$g'(t_0) = D(f)(\mathbf{x_0})(\mathbf{b} - \mathbf{a}). \tag{1.42}$$

Now (1.41) and (1.42) together give (1.38).
(ii) Apply Theorem 1.1.1 to the function $g : [0, 1] \to \mathbb{R}^m$ as above. ♠

Example 1.3.1 Let us compute the derivative of some specific maps:
(a) Consider the function $\phi(\mathbf{x}) = \mathbf{x} \bullet \mathbf{x} = \sum_i x_i^2$ on $\mathbb{R}^n$. Clearly, its partial derivatives all exist and are continuous and hence $D\phi$ exists and is determined by the partial derivatives. Therefore, $D\phi_{\mathbf{x}} = (2x_1, 2x_2, \ldots, 2x_n) = 2\mathbf{x}^t$. (Remember that $\mathbf{x}$ is a column vector.) Observe that $D\phi_{\mathbf{x}} : \mathbb{R}^n \to \mathbb{R}$ as a linear map is actually given by

$$D\phi_{\mathbf{x}}(\mathbf{y}) = 2\mathbf{x} \bullet \mathbf{y} = 2\sum_i x_i y_i.$$

(b) On $M(n;\mathbb{R})$ consider the function $\sigma(A) = AA^t$. The space $M(n, \mathbb{R})$ is identified as the Euclidean space $\mathbb{R}^{n \times n}$; the only difference being the variables are indexed by ordered pairs $(i, j), 1 \leq i, j \leq n$. Therefore, at any point $B \in \mathbb{R}^{n \times n}$, the derivative $D\sigma_B$ is a linear map from $\mathbb{R}^{n\times n} \to \mathbb{R}^{n\times n}$. The $(i,j)^{\text{th}}$-coordinate functions of f are nothing but $\sum_k a_{ik}a_{jk}$. Each of these is therefore a quadratic in the variables and hence is continuously differentiable any number of times. Thus, the function f is also differentiable any number of times. Here we shall find its first derivative and this amounts to finding the partial derivatives of $\sigma_{ij} = \sum_k a_{ik}a_{jk}$. With respect to the variable a_{pq} this is zero unless $p = i, j$. If $p = i \neq j$ then $\frac{\partial \sigma_{ij}}{\partial a_{pq}} = a_{jq}$ Similarly, if $p = j \neq i$, it is a_{iq}. If $p = i = j$ then it is $2a_{pq}$. We can now replace a_{ij} by b_{ij} to get the value of the respective derivatives at $A = B$. The matrix representing $D\sigma_B$ is of size $n^2 \times n^2$ and its entries can now be written down. Here is the answer for $n = 2$:

$$D\sigma_B = \begin{bmatrix} 2b_{11} & 2b_{12} & 0 & 0 \\ b_{21} & b_{22} & b_{11} & b_{12} \\ b_{21} & b_{22} & b_{11} & b_{12} \\ 0 & 0 & 2b_{21} & 2b_{22} \end{bmatrix}$$

Here is an alternative approach. We know that $D\sigma_B : M(n,\mathbb{R}) \to M(n,\mathbb{R})$ is a linear map. A linear map is completely determined if we know its values on all unit vectors. The value of $D\sigma_B$ on any unit vector A is the directional derivative of $D\sigma$ at B in the direction of A. Therefore, by the definition of the directional derivative we have

$$\begin{aligned} D\sigma_B(A) &= \lim_{s\to 0} \frac{f(B+sA) - f(B)}{s} = \lim_{s\to 0} \frac{(B+sA)(B+sA)^t - BB^t}{s} \\ &= \lim_{s\to 0} \frac{s(AB^t + BA^t) + s^2 AA^t}{s} \end{aligned}$$

Therefore,

$$D\sigma_B(A) = AB^t + BA^t. \tag{1.43}$$

Thus we have determined $D\sigma$ at all points of $M(n;\mathbb{R})$ as a linear map from $M(n;\mathbb{R})$ to $M(n;\mathbb{R})$. Its matrix representation can be recovered from (1.43) by substituting

$$A = E_{(ij),(kl)},$$

which are $n^2 \times n^2$ matrices with entries equal to 1 only at $((ij),(kl))^{\text{th}}$ place and zero elsewhere.

(c) Let us now consider $\psi : M(n;\mathbb{R}) \to \mathbb{R}$ given by $\psi(X) = \det X$. This is a polynomial function in the entries x_{ij} of the matrix X of degree n. Once again it is enough to determine the partial derivatives. In the first method as in the above case, we need to use the full expansion formula for the determinant. There is a simpler way, viz., we can use the Laplace expansion formulae

$$\psi(X) = \det X = \sum_j x_{ij}X_{ij}$$

where X_{ij} is the $(ij)^{\text{th}}$ cofactor of X. Observe that for each fixed i and j the cofactor X_{ij} is a polynomial in x_{kl}'s but does not involve the variable x_{ij}. Therefore, we can compute the partial derivative of $\det X$ with respect to x_{ij} treating X_{ij} as constants and hence $\frac{\partial \psi}{\partial x_{ij}} = X_{ij}$. One can write $\operatorname{grad}\psi$ as a n^2-row vector with its $(ij)^{\text{th}}$ entry as X_{ij}. With a little more caution, we can also write it as $n \times n$-matrix: $D\psi_X = adj(X)^t$. However, caution is required in interpreting $adj(X)^t$ as a linear map on $M(n;\mathbb{R})$. It is neither by left multiplication nor by right multiplication:

$$D(\psi)_X(Y) = \sum_{ij} X_{ij}y_{ij} = Tr(adj(X)Y). \tag{1.44}$$

In particular, for $X = Id_n$, we have

$$D(\det)_{I_n}(Y) = Tr(Y).$$

We can also employ the second method here to directly compute

$$\lim_{s \to 0} \frac{\det(X + sY) - \det(X)}{s}$$

and arrive at (1.44). However, we need to use the multilinearity of det meticulously here. We leave this to you as a joyful exercise.

Exercise 1.3

1. Compute the total derivative of the functions in each of the following cases.
   (a) $f : M(n, \mathbb{R}) \times M(n, \mathbb{R}) \to M(n, \mathbb{R})$ given by $f(A, B) = A + B$.
   (b) $g : M(n, \mathbb{R}) \times M(n, \mathbb{R}) \to M(n, \mathbb{R})$ given by $g(A, B) = AB$.
   (c) $h : M(n, \mathbb{R}) \to M(n, \mathbb{R})$ given by $h(A) = A^2$.

2. Let $\tau : M(m, n; \mathbb{R}) \to M(n, m; \mathbb{R})$ denote the transpose: $A \mapsto A^t$. Prove that $D(\tau)_A = \tau$, for all $A \in M(m, n; \mathbb{R})$.

3. **Leibniz's rule:** Let U be an open subset of $\mathbb{R}^m$, and $\alpha, \beta : U \to M(n; \mathbb{R})$ be any two smooth maps. Put $\gamma(x) = \alpha(x) \cdot \beta(x)$. Show that

$$D(\gamma)_x = D(\alpha)_x \cdot \beta(x) + \alpha(x) \cdot D(\beta)_x. \tag{1.45}$$

   (Note that $D(\alpha), D(\beta)$ take values in $M(n; \mathbb{R})^m$ which admits action by $M(n; \mathbb{R})$ both on the left and on the right, viz.,

$$A \cdot (B_1, \ldots, B_m) = (A \cdot B_1, \ldots, A \cdot B_m); \quad (A_1, \ldots, A_m) \cdot B = (A_1 \cdot B, \ldots, A_m \cdot B)$$

   where $A_i \cdot B_j$ etc. denote the matrix multiplication. You have to interpret the right hand side of (1.45) in this sense.)

4. Let $f : \mathbb{R}^n \to \mathbb{R}^n, g : \mathbb{R}^n \to \mathbb{R}$ be smooth maps. Derive the formula

$$D^2(g \circ f)_{\mathbf{x}} = D(f)^t_{\mathbf{x}} \cdot D^2(g)_{f(\mathbf{x})} \cdot D(f)_{\mathbf{x}} + D(g)_{f(\mathbf{x})} \cdot D^2(f)_{\mathbf{x}}. \tag{1.46}$$

5. Show that $\operatorname{div} f$ is nothing but the trace of the linear map Df. Conclude that div remains unchanged under an orthogonal change of coordinates.

6. Consider the vector field $f(x, y, z) = x\mathbf{j}$. Compute its curl with respect to the standard right-handed Cartesian coordinate system $\{\mathbf{i}, \mathbf{j}, \mathbf{k}\}$. Compute the same with respect to the system $\mathbf{i}^* = -\mathbf{i}, \mathbf{j}^* = \mathbf{j}, \mathbf{k}^* = \mathbf{k}$ also. What do you conclude?

7. Show that the curl remains unchanged under an orientation preserving orthogonal change of coordinates, i.e., the curl is an $SO(3)$-invariant.

1.4 Inverse and Implicit Function Theorems

The edifice of differential topology is built on the base theme: *a differentiable map closely imitates the behavior of its tangent map, in a sufficiently small neighborhood.* The very first instance of this is the *Inverse Function Theorem.* If $f : (a,b) \to \mathbb{R}$ is a continuously differentiable function such that $f'(c) \neq 0,\ c \in (a,b)$, then we know that in a sufficiently small neighborhood U of c, f is strictly monotonic; equivalently, f is one-to-one on U. Moreover, the inverse of f, defined over $f(U)$ is also differentiable. Such a behavior holds in higher dimensions as well. This is going to be the topic of this section, the importance of which cannot be overemphasized.

The basic idea in the proof is to use the contraction mapping principle. The first part of this theorem is equivalent to saying that a certain system of equations can be solved uniquely in a sufficiently small neighborhood of 0, viz.,

$$y_i = f_i(x_1, \ldots, x_n); \quad f = (f_1, \ldots, f_n)$$

under suitable conditions on f. Experience tells us that if you want to show the existence of certain solutions the contraction mapping principle comes quite handy e.g., remember the proof of Picard's iteration method for the solution of the first order linear differential equations.

Definition 1.4.1 Let (X,d) be a metric space. A function $\phi : X \to X$ is called a *contraction* if there exists a constant $0 < c < 1$ such that for all $x, y \in X$ we have

$$d(\phi(x), \phi(y)) \leq c\, d(x,y).$$

Theorem 1.4.1 Contraction Mapping Principle:
Let (X,d) be a complete metric space[2]*. Then every contraction $\phi : X \to X$ has a unique fixed point in X, i.e., there exists a unique point $y \in X$ such that $\phi(y) = y$.*

Proof: Start with any point x_0. Define

$$x_1 = \phi(x_0), x_2 = \phi(x_1), \ldots, x_n = \phi(x_{n-1}), \ldots.$$

Observe that $d(x_{n+1}, x_n) \leq c^n d(x_1, x_0)$ for some $0 < c < 1$. Since $\sum_n c^n < \infty$ it follows that given $\epsilon > 0$ we can find n_0 such that for $m > n > n_0$:

$$d(x_m, x_n) \leq \sum_{k=n}^{m-1} c^k d(x_1, x_0) < \epsilon d(x_1, x_0).$$

Therefore $\{x_n\}$ is a Cauchy sequence. Since X is complete, this sequence has a limit point $y \in X$. Also observe that any contraction is a continuous function. Therefore,

$$\phi(y) = \phi(\lim_n x_n) = \lim_n \phi(x_n) = \lim_n x_{n+1} = y.$$

Finally if $y_i \in X, i = 1,2$ are such that $\phi(y_i) = y_i$, then

$$d(y_1, y_2) = d(\phi(y_1), \phi(y_2) \leq c\, d(y_1, y_2)$$

with $c < 1$. This is meaningful only if $y_1 = y_2$. ♠

Remark 1.4.1 Since $\mathbb{R}^n$ is a complete metric space, every closed ball in it is a complete metric space. This is what we need in the following theorem.

[2] Recall that a metric space X is complete if every Cauchy sequence in X is convergent in X.

Theorem 1.4.2 Inverse Function Theorem (IFT): *Let $E \subset \mathbb{R}^n$ be an open set, $0 \in E$ and let $f \in \mathcal{C}^1(E, \mathbb{R}^n)$ be such that $Df(0)$ is invertible. Then there exist open neighborhoods U of 0 and V of $f(0)$ such that*
(i) $f : U \to V$ is a bijection;
(ii) $g = f^{-1} : V \to U$ is differentiable.
(iii) $D(f^{-1})$ is continuous on V.

Proof: (i) Put $A = Df(0)$ and consider $\hat{f} = A^{-1} \circ f$. Then $\hat{f} \in (E; \mathbb{R}^n)$ and $D(\hat{f})(0) = Id$. Now if we prove the conclusion of the theorem with f replaced by $\hat{f}$ then it will also hold for f. Thus without loss of generality we may assume that $D(f)(0) = Id$.

Choose $0 < \epsilon \leq \frac{1}{2}$. Let $0 < \delta < 1$ be such that the open ball $U = B_\delta(0)$ is contained in E and

$$\|Df(x) - Id\| < \epsilon \tag{1.47}$$

for all $x \in U$. This is possible by the continuity of Df at 0. Put $V = f(U)$. First we claim that f is injective on U.

For any $y \in V$, let us define $\phi_y : U \to \mathbb{R}^n$ by

$$\phi_y(x) = x + (y - f(x)). \tag{1.48}$$

Then differentiating with respect to x, we have,

$$D(\phi_y)(x) = Id - Df(x) \tag{1.49}$$

for all $x \in U$. Hence, for all $x \in U$, we have,

$$\|D(\phi_y)(x)\| \leq \|Id - Df(x)\| \leq \epsilon < 1. \tag{1.50}$$

Now for $x_1, x_2 \in U$, by Theorem 1.3.3, we have

$$\|\phi_y(x_1) - \phi_y(x_2)\| \leq \|D(\phi_y)(x')\| \|x_1 - x_2\|, \tag{1.51}$$

for some $x' \in [x_1, x_2] \subset U$. Combining with (1.50), this gives

$$\|\phi_y(x_1) - \phi_y(x_2)\| \leq \epsilon \|x_1 - x_2\|. \tag{1.52}$$

(That means ϕ_y is a contraction mapping on U.)

Now suppose $f(x_1) = f(x_2) = y$ with $x_1, x_2 \in U$. Then

$$\phi_y(x_i) = x_i, \quad i = 1, 2.$$

From (1.52), we conclude that $x_1 = x_2$. Therefore, $f|_U$ is injective.

Next, we must show that $f(U) = V$ is open. Fix $y_0 \in V, y_0 = f(x_0), x_0 \in U$. Let $r > 0$ be such that the closed balls $B_r(x_0) \subset U$. It suffices to show that $B_{r\epsilon}(y_0) \subset V$. If $y \in B_{r\epsilon}(y_0)$, then

$$\|\phi_y(x_0) - x_0\| \leq \|y - y_0\| < r\epsilon \leq \frac{r}{2}. \tag{1.53}$$

Therefore, if $x \in B_r(x_0)$, then

$$\|\phi_y(x) - x_0\| \leq \|\phi_y(x) - \phi_y(x_0)\| + \|\phi_y(x_0) - x_0\| \leq \frac{\|x - x_0\| + r}{2} \leq r. \tag{1.54}$$

Therefore, ϕ_y maps the closed ball $B_r(x_0)$ into itself.

By contraction mapping principle (Theorem 1.4.1), there is $a \in B_r(x_0)$ such that $\phi_y(a) = a$. This means $f(a) = y$. So, $y \in f(U) = V$. Therefore, $B_{r\epsilon}(y_0) \subset V$. This proves that V is open.

(ii) (The proof of this part is exactly similar to the proof of the statement for one variable. Nevertheless, we write down the details here.)

Recall that the space of all invertible $n \times n$ matrices is an open set in the space of all $n \times n$ matrices. Since the map $x \mapsto D(f)(x)$ is continuous at x_0 and since $D(f)(x_0)$ is invertible, it follows that we can choose $\delta > 0$ such that for all $x \in U = B_\delta(x_0)$, we have $D(f)(x)$ is invertible. Now to show that g is differentiable, and also $Dg(y) = (Df(x))^{-1}$, consider $y \in V$ and $y + k \in V$. Put $g(y) = x, g(y+k) = x+h, T = (Df(x))^{-1}$. Then, $f(x+h) = y+k, f(x) = y$ and we have,

$$\begin{aligned} & \frac{\|g(y+k) - g(y) - T(k)\|}{\|k\|} \\ \leq \ & \|T\| \frac{\|Df(x)[g(y+k) - g(y)] - k\|}{\|k\|} \\ \leq \ & \|T\| \frac{\|Df(x)(h) - (f(x+h) - f(x))\|}{\|h\|} \frac{\|h\|}{\|k\|}. \end{aligned}$$

We now want to take the limit as $\|k\| \to 0$ on the left-hand-side (LHS). On the RHS, we have

$$\lim_{\|h\| \to 0} \frac{\|Df(x)(h) - (f(x+h) - f(x))\|}{\|h\|} = 0.$$

Therefore, it suffices to show that as $\|k\| \to 0$, we have, $\|h\| \to 0$ and $\dfrac{\|h\|}{\|k\|}$ remains bounded.

We have, $\phi_y(x+h) - \phi_y(x) = h + [f(x) - f(x+h)] = h - k$. Therefore, $\|h - k\| = \|\phi_y(x+h) - \phi_y(x)\| \leq \|h\|/2$ (by (1.52)) and hence, $\|h\| \leq 2\|k\|$. This completes the proof of all the claims. Thus, g is differentiable and $D(g) = T = D(f)^{-1}$.

(iii) Finally, the continuity of $D(g)$ is obvious from the continuity of $D(f)$, and the continuity of the inverse operation on invertible square matrices. ♠

Remark 1.4.2 Let us re-examine the above proof a little bit.

(a) In (i) we started with $\epsilon \leq 1/2$, which was used in (1.52) to conclude that ϕ_y is a contraction mapping. We could have achieved this with any $0 < \epsilon < 1$. However, later in (1.54), we would only get $\phi_y(B_r(x_0)) \subset B_{2\epsilon r}(x_0)$, which is not enough unless $\epsilon \leq 1/2$.

Now consider an example: $f(x) = x + x^2$. Then $Df(x) = f'(x) = 1 + 2x$ and hence (1.47) is satisfied with $\delta = \frac{1}{4}$. The theorem then concludes that f is a 1-1 mapping in the interval $(-1/4, 1/4)$.

On the other hand, since $f'(x) = 1 + 2x > 0$ in the interval $(-1/2, \infty)$, it follows that f is strictly monotonically increasing in this interval and hence is a 1-1 mapping. Thus, we see that the open set U given by the theorem is not optimum even if we restrict ourselves to open balls around the point under consideration.

(b) The second point that we would like to make is that the proof of the theorem is existential, i.e., it does not tell you how to get the inverse map (unlike for instance, Picard's iteration method for solution of ODE's). However, there are special situations in which we may be able to write down the inverse map. In the above example, we can write down the inverse map by inspection. viz., $g(y) = \sqrt{w + \frac{1}{4}} - \frac{1}{2}$. More generally, if f is given by a convergent power series,

$$f(x) = \sum_n a_n x^n, \qquad a_0 = 0, \ a_1 \neq 0$$

then we can formally write down the inverse of this power series. It turns out that this formal inverse power series $g(x) = \sum_n b_n x^n$ is also of positive radius of convergence and the function defined by it is the inverse of f. (See [Sh2] for more details.) Such a method is available for analytic functions of several variables also.
(c) Alternatively, for holomorphic functions, Cauchy's residue theory allows one to write down an integral formula for the inverse. This is then available for real valued functions as well provided they are given by convergent power series by taking the real part of the integral of the complexified function. (See [Sh2] for more details.)
(d) However, the space of $\mathcal{C}^1$ functions is much much larger than the space of analytic functions and certainly the method of power series would not work there.
(e) The inverse function theorem is valid even in a Banach spaces. Far reaching generalizations have been obtained using the good old "Newton's method" by Nash [N], [Ham], [Ho], et al.

Definition 1.4.2 We shall set up the notation: Let $\pi_1 : \mathbb{R}^n \times \mathbb{R}^m \to \mathbb{R}^n$ and $\pi_2 : \mathbb{R}^n \times \mathbb{R}^m \to \mathbb{R}^m$ be the respective projections and let $\eta_1 : \mathbb{R}^n \to \mathbb{R}^{n+m}$ and $\eta_2 : \mathbb{R}^m \to \mathbb{R}^{n+m}$ be the inclusion maps given by $\mathbf{u} \mapsto (\mathbf{u}, 0)$ and $\mathbf{v} \mapsto (0, \mathbf{v})$ respectively. Given a linear map $A : \mathbb{R}^n \times \mathbb{R}^m \to \mathbb{R}^r$, we put $A_i = A \circ \eta_i$. Clearly then $A = A_1 \circ \pi_1 + A_2 \circ \pi_2$. We have,

Lemma 1.4.1 Suppose that $r = n$ and A_1 is invertible. Then for every $v \in \mathbb{R}^m$, there exists a unique $\mathbf{u} \in \mathbb{R}^n$ such that $A(\mathbf{u}, \mathbf{v}) = 0$, given by the formula

$$\mathbf{u} = -A_1^{-1} A_2(\mathbf{v}).$$

Proof: Obvious.

Before taking up the nonlinear analogue of this result, which goes under the name "implicit function theorem", we would like to introduce the following important terminology.

Definition 1.4.3 Let $U \subset \mathbb{R}^n, V \subset \mathbb{R}^m$ be open subsets. A map $f : U \to V$ is called a *diffeomorphism* if it is a bijection and both f and f^{-1} are continuously differentiable. Clearly, then f^{-1} will also be a diffeomorphism. Given a diffeomorphism f, we also refer to it as a *change of coordinates* on U by treating the coordinate functions $f_1, \ldots, f_n$ of f as the new coordinates on U. We say, U and $f(U)$ are *diffeomorphic.*

Remark 1.4.3
(i) It follows immediately that $m = n$ once there is a diffeomorphism $f : U \to V$ where U, V are nonempty open sets. For, then $Df : \mathbb{R}^n \to \mathbb{R}^m$ will be a linear isomorphism.
(ii) Note that a diffeomorphism is necessarily a homeomorphism whereas, there are plenty of homeomorphisms that are not differentiable at all. Even if f is differentiable, the inverse of f need not be differentiable. The simplest example of this is the map $x \mapsto x^3$ on $\mathbb{R}$.
(iii) Clearly, composite of two diffeomorphisms is a diffeomorphism.
(iv) Typical examples of diffeomorphisms are $\exp : \mathbb{R} \to \mathbb{R}^+$, linear isomorphisms, translations, and so on. A diffeomorphism restricted to an open subset of the domain defines a diffeomorphism onto its image.
(v) Later on, we shall define diffeomorphisms of arbitrary subsets of Euclidean spaces. Obviously, "being diffeomorphic" is an equivalence relation among these sets. The central theme of differential topology is to classify subsets of Euclidean spaces up to diffeomorphism.

Theorem 1.4.3 Implicit Function Theorem (ImFT): *Let $E \subset \mathbb{R}^n \times \mathbb{R}^m$ be an open subset, $(\mathbf{a}, \mathbf{b}) \in E$ and let $f \in \mathcal{C}^1(E, \mathbb{R}^n)$ and $A := D(f)(\mathbf{a}, \mathbf{b})$. Suppose that $f(\mathbf{a}, \mathbf{b}) = 0$ and $A|_{\mathbb{R}^n \times 0}$ is invertible. Then there is an open neighborhood U of $(\mathbf{a}, \mathbf{b})$ in $\mathbb{R}^n \times \mathbb{R}^m$, a neighborhood W of $\mathbf{b}$ in $\mathbb{R}^m$ and a unique map $g \in \mathcal{C}^1(W, \mathbb{R}^n)$ such that for every point $\mathbf{y} \in W$, $f(g(\mathbf{y}), \mathbf{y}) = 0$. Moreover, $D(g)(\mathbf{b}) = -A_1^{-1} A_2$ where $A_1 = A|_{\mathbb{R}^n \times 0}$ and $A_2 = A|_{0 \times \mathbb{R}^m}$.*

Proof: Put $F(\mathbf{x},\mathbf{y}) = (f(\mathbf{x},\mathbf{y}),\mathbf{y}),\ (\mathbf{x},\mathbf{y}) \in E$. Then $F \in \mathcal{C}^1(E, \mathbb{R}^n \times \mathbb{R}^m)$ and we have,

$$DF(\mathbf{a},\mathbf{b}) = \begin{bmatrix} A_1 & A_2 \\ 0 & Id \end{bmatrix}.$$

Therefore, $DF(\mathbf{a},\mathbf{b})$ is invertible. By Theorem 1.4.2(IFT), it follows that, we have neighborhoods U and V of $(\mathbf{a},\mathbf{b})$ and $(\mathbf{0},\mathbf{b})$, respectively such that F is a diffeomorphism of U onto V. Put $W = \{\mathbf{y} \ : \ (\mathbf{0},\mathbf{y}) \in \mathbf{V}\}$. Then W is open in $\mathbb{R}^m$. Let G be the inverse of F. Write, $G(\mathbf{0},\mathbf{y}) = (g(\mathbf{y}), h(\mathbf{y})), \mathbf{y} \in W$. Then

$$(\mathbf{0},\mathbf{y}) = F \circ G(\mathbf{0},\mathbf{y}) = F(g(\mathbf{y}), h(\mathbf{y})) = (f(g(\mathbf{y}), h(\mathbf{y})), h(\mathbf{y})).$$

Therefore, $h(\mathbf{y}) = \mathbf{y}$ and $f(g(\mathbf{y}),\mathbf{y}) = \mathbf{0}$. Uniqueness of g follows from the injectivity of F. Finally, differentiating the equation $f(g(\mathbf{y}),\mathbf{y}) = 0$, with respect to $\mathbf{y}$, it follows that $A_1 D(g)(\mathbf{b}) + A_2 = 0$. This means, $D(g) = -A_1^{-1}A_2$. ♠

Remark 1.4.4 Think of the equation $f(\mathbf{x},\mathbf{y}) = 0$ as a system of n equations in $n+m$ variables. With $(\mathbf{a},\mathbf{b})$ as a given solution, the theorem tells you under what condition, we can solve for the variables $\mathbf{x}$ near $\mathbf{a}$ in terms of the variables $\mathbf{y}$, to obtain a unique continuously differentiable solution. The new function $g(\mathbf{y})$ so obtained is said to be given by the equation $f(\mathbf{x},\mathbf{y}) = 0$ implicitly. That is why the theorem is so named.

As a simple example consider the equation $x^2 + y^2 - 1 = 0$ defining the unit circle in $\mathbb{R}^2$. Observe that we can solve for x uniquely, in a small neighborhood of any point (a,b) such that $a \neq 0$. That is so, because, then $f_x(a,b) = 2a \neq 0$.

Theorem 1.4.4 Surjective Form of Implicit Function Theorem: *Let $U \subset \mathbb{R}^{n+m}$ be open, $f \in \mathcal{C}^1(U, \mathbb{R}^n)$, and let $\mathbf{a} \in U$ be such that $f(\mathbf{a}) = \mathbf{0}$ and $Df(\mathbf{a})$ is surjective. Then there exists a diffeomorphism $\phi : V \to \phi(V)$ of a neighborhood V of $\mathbf{0}$ in $\mathbb{R}^{n+m}$ onto a neighborhood of $\mathbf{a}$, such that $f\phi(x_1, \ldots, x_{n+m}) = (x_1, x_2, \ldots, x_n.)$*

Proof: In other words, the conclusion of the theorem is that after a change of coordinates in the domain, the function f coincides with the projection map in a small neighborhood of the origin.

By performing a translation, we may assume that $\mathbf{a} = \mathbf{0}$. Consider the $n \times (n+m)$ matrix A corresponding to $Df(\mathbf{0})$ which is of rank n. By performing column operations we can bring this to the form in which $a_{ij} = \delta_{ij}, 1 \leq i \leq n, 1 \leq j \leq n+m$. To sum it up, it follows that, we can first perform an affine linear change of coordinates $\mathbb{R}^{n+m}$ so that, with respect to the new coordinates, the point $\mathbf{a}$ is the origin and the given map $f = (f_1, \ldots, f_n)$ has the property that

$$\frac{\partial f_i}{\partial x_j}(\mathbf{0}) = \delta_{ij},\ i = 1, \ldots, n, j = 1, \ldots, n+m.$$

Consider the map $h : U \to \mathbb{R}^{n+m}$ defined by $h := (h_1, \ldots, h_{n+m})$ where,

$$h_i(x_1, \ldots, x_{n+m}) = \begin{cases} f_i(x_1, \ldots, x_{n+m}), & i \leq n, \\ x_i, & i \geq n+1. \end{cases}$$

Then $D(h)(\mathbf{0})$ is invertible and hence by the IFT 1.4.2, h has an inverse ϕ, which is continuously differentiable in a small neighborhood of $\mathbf{0}$. Writing

$$(x_1, \ldots, x_n) = \mathbf{x}, \quad (x_{n+1}, \ldots, x_{n+m}) = \mathbf{y},$$

we have,

$$(x_1, \ldots, x_{n+m}) = (\mathbf{x},\mathbf{y}) = h \circ \phi(\mathbf{x},\mathbf{y}) = (f \circ \phi(\mathbf{x},\mathbf{y}),\mathbf{y}).$$

Therefore, we have, $f \circ \phi(x_1, \ldots, x_{n+m}) = (x_1, \ldots, x_n)$ near $\mathbf{0}$. ♠

Theorem 1.4.5 Injective Form of Implicit Function Theorem: *Let $E \subset \mathbb{R}^n$ be an open subset, $\mathbf{0} \in E$ and let $f \in \mathcal{C}^1(E, \mathbb{R}^{n+m})$ be such that $f(\mathbf{0}) = \mathbf{0}$ and $Df(\mathbf{0})$ is injective. Then there exists a neighborhood U of $\mathbf{0} \in \mathbb{R}^{n+m}$ and a diffeomorphism ψ of U onto a neighborhood of $\mathbf{0} \in \mathbb{R}^{n+m}$ such that $\psi(\mathbf{0}) = \mathbf{0}$ and*

$$\psi f(x_1, \ldots, x_n) = (x_1, \ldots x_n, 0, \ldots, 0).$$

Proof: The proof here is somewhat dual to the proof of the surjective form considered above. First of all, by a linear change of coordinates in $\mathbb{R}^{n+m}$ we may assume that

$$\frac{\partial f_i}{\partial x_j}(\mathbf{0}) = \delta_{ij}, \ 1 \leq i \leq n+m, \ 1 \leq j \leq n.$$

(Perform row operations!) Define $F : E \times \mathbb{R}^m \to \mathbb{R}^{n+m}$ by $F(\mathbf{x}, \mathbf{y}) = f(\mathbf{x}) + (0, \mathbf{y})$. Then $DF(\mathbf{0})$ is invertible. Let ψ be the inverse of F on a suitable neighborhood of $\mathbf{0}$, etc. Then, since $F(\mathbf{x}, 0) = f(\mathbf{x})$, it follows that $\psi \circ f(\mathbf{x}) = \psi \circ F(\mathbf{x}, \mathbf{0}) = (\mathbf{x}, \mathbf{0})$, as required. ♠

Example 1.4.1 Take $f(x, y) = x^2 + y^2 - 1$. Let us work out explicitly the surjective form of ImFT for this function at the point $(1, 0)$. The conditions for ImFT are satisfied since $f(1, 0) = 0$ and $Df_a = (2, 0)$. The curve $x^2 + y^2 - 1 = 0$ (or $= \epsilon$) is a circle centered at the origin. This suggests that the map

$$(r, \theta) \mapsto (\sqrt{r+1}\cos\theta, \sqrt{r+1}\sin\theta)$$

to play the role of ϕ. We check that $\phi(0, 0) = (1, 0)$. And $f \circ \phi(r, \theta) = r$ as required. The function makes sense provided we restrict the domain $|r| < 1$ and will be a diffeomorphism in $-1 < r < 1, -\pi < \theta < \pi$. The same discussion goes through at the point $(0, 1)$ for the function ϕ changed to

$$(r, \theta) \mapsto (\sqrt{r+1}\sin\theta, \sqrt{r+1}\cos\theta).$$

Example 1.4.2 Consider the function $f(x, y) = y^2 - x^3$. The derivative vanishes identically only at $(0, 0)$. Therefore, the implicit function theorem is available at all points $\mathbf{a} \neq (0, 0)$. For instance, let $\mathbf{a} = (1, 1)$. Then $Df_{(1,1)} = (-3, 2)$. Now put $g(x, y) = 2x + 3y - 5$ and $\psi = (f, g)$. Then $\psi(1, 1) = (0, 0)$ and $D\psi_{(1,1)}$ is invertible. Therefore, by the inverse function theorem, in some neighborhood W of $(1, 1)$, we have $\psi : U \to V = \psi(W)$ a diffeomorphism. Put $\phi = \psi^{-1} : V \to W$. Then

$$(x, y) = \psi \circ \phi(x, y) = (f, g) \circ \phi(x, y) = (f(\phi(x, y)), g(\phi(x, y)))$$

implies that $f \circ \phi(x, y) = x$. Notice that ϕ is not unique.

Remark 1.4.5 The method that we have followed in the above two examples can be made into an algorithm to find a local diffeomorphism ψ whose inverse $\phi = \psi^{-1}$ fits the bill in Theorem 1.4.4 as follows: Complete the matrix of $Df(\mathbf{a})$ into an invertible matrix

$$\begin{pmatrix} Df(\mathbf{a}) \\ A \end{pmatrix}$$

where A is a $m \times (m + n)$ matrix. Let $g : \mathbb{R}^{n+m} \to \mathbb{R}^m$ be the affine linear map such that $g(\mathbf{a}) = 0$ and $Dg = A$. Put $\psi = (f, g)$. Then $D\psi(\mathbf{a})$ is invertible and hence in a neighborhood of $\mathbf{a}$ it is a diffeomorphism onto a neighborhood of $0 \in \mathbb{R}^{n+m}$, and $\psi(\mathbf{a}) = 0$. If $\phi = \psi^{-1}$, then we have

$$(\mathbf{x}, \mathbf{y}) = \psi \circ \phi(\mathbf{x}, \mathbf{y}) = (f, g) \circ \phi(\mathbf{x}, \mathbf{y})$$

which gives $f \circ \phi(\mathbf{x}, \mathbf{y}) = \mathbf{x}$, as required.

Combining the arguments employed in the above two theorems, we obtain the following more general result.

Theorem 1.4.6 Rank Theorem: *Let U be an open subset of $\mathbb{R}^n$, $F \in \mathcal{C}^1(U, \mathbb{R}^m)$, $\mathbf{a} \in U, \mathbf{b} = F(\mathbf{a})$ and let $DF(\mathbf{a})$ be of rank r. For any $N \geq r, 1 \leq i \leq r$, let $p_i : \mathbb{R}^N \to \mathbb{R}$ denote the projection map $(x_1, \ldots, x_N) \mapsto x_i$. Then there exist diffeomorphisms $\phi : V \to \phi(V)$, $\psi : W \to \psi(W)$ of suitable neighborhoods of the origins in $\mathbb{R}^n$ and $\mathbb{R}^m$ respectively, such that $\phi(\mathbf{0}) = \mathbf{a}, \psi(\mathbf{0}) = \mathbf{b}$, and*

$$p_i \circ \psi^{-1} \circ F \circ \phi(\mathbf{x}) = p_i(\mathbf{x}), \ 1 \leq i \leq r. \tag{1.55}$$

Furthermore, if $DF(\mathbf{x})$ is of rank r for all $\mathbf{x}$ in a neighborhood of $\mathbf{a}$, then ϕ, ψ, etc., can be chosen so that

$$\psi^{-1} \circ F \circ \phi(x_1, x_2, \ldots, x_n) = (x_1, \ldots, x_r, 0, \ldots, 0),$$

for points $(x_1, \ldots, x_n) \in V$.

Proof: By shifting the origins on either side, we will assume that $\mathbf{a} = \mathbf{0}, \ \mathbf{b} = \mathbf{0}$. Now consider $A = DF(\mathbf{0})$. Since the rank of this matrix is r, there exists linear isomorphisms B and C say, of $\mathbb{R}^m$, and $\mathbb{R}^n$ respectively, such that BAC is of the form

$$\begin{bmatrix} I_r & \mathbf{0} \\ \mathbf{0} & \mathbf{0} \end{bmatrix}.$$

Therefore, replacing $B \circ F \circ C$, by F, we will assume that the matrix $A = DF(\mathbf{0})$ itself is of the above form.

Consider the map $H : U \to \mathbb{R}^n$ defined by $H = (h_1, \ldots, h_n)$ where,

$$h_i(x_1, \ldots, x_n) = \begin{cases} f_i(x_1, \ldots, x_n), & 1 \leq i \leq r, \\ x_i, & r+1 \leq i \leq n. \end{cases}$$

Then it follows that the rank of $DH(\mathbf{0})$ is n. Hence, by IFT, in a small neighborhood of $\mathbf{0}$, H will be a diffeomorphism. Then we have, for $1 \leq i \leq r, \ p_i \circ F \circ H^{-1} = p_i \circ H \circ H^{-1} = p_i$ as required, in the first part.

Now assume that V, W, ϕ, ψ etc. have been chosen so as to satisfy (1.55). Note that we can always insist that the neighborhood V so chosen is convex. Put $\psi^{-1} \circ F \circ \phi =: G = (g_1, \ldots, g_m)$. It follows that the matrix of $DG(\mathbf{x})$ is of the form

$$\begin{bmatrix} I_r & 0 \\ A(\mathbf{x}) & B(\mathbf{x}) \end{bmatrix}.$$

Therefore, under the additional hypothesis that $DGF(\mathbf{x})$ is of rank r for all $\mathbf{x} \in V$, it follows that $B = 0$ for all $\mathbf{x} \in V$. This is the same as saying $\dfrac{\partial g_i}{\partial x_j} = 0$ for $i, j \geq r+1$. Since V is convex, this implies that $g_i, i \geq r+1$ depend only on $(x_1, \ldots, x_r)$. Therefore, $G(\mathbf{x}) = (p(\mathbf{x}), \alpha(p(\mathbf{x})))$, for some function $\alpha : W \to \mathbb{R}^{m-r}$, where W is a neighborhood of $\mathbf{0}$ in $\mathbb{R}^r$. Here p denotes the projection onto the first r factors. Let $L : W \times \mathbb{R}^{m-r} \to \mathbb{R}^m$ be the map defined by

$$(\mathbf{y}, \mathbf{z}) \mapsto (\mathbf{y}, \alpha(\mathbf{y}) + \mathbf{z}).$$

Then, L is a diffeomorphism in a neighborhood of $\mathbf{0}$. Moreover, $G(\mathbf{x}) = (p(\mathbf{x}), \alpha(p(\mathbf{x}))) = L(p(\mathbf{x}), \mathbf{0})$. Therefore, $L^{-1} \circ G(\mathbf{x}) = (p(\mathbf{x}), \mathbf{0})$. All that we need to do now is to replace ψ by $\psi \circ L$. ♠

Remark 1.4.6 Before winding up with the above theme we will now consider one more result of the above type. Recall that any invertible matrix can be written as a product of elementary matrices and permutation matrices. (This is an easy consequence of the *Gauss elimination method.*) We may ask the question whether there is a similar result with diffeomorphisms. First of all, we can hope to have such a result, locally only. With this modest modification in the question, the answer is in the affirmative: The prototype of an elementary matrix is a diffeomorphism of the type

$$\mathbf{x} = (x_1, \ldots, x_n) \mapsto (x_1, \ldots, x_{k-1}, x_k + \alpha(\mathbf{x}), x_{k+1}, \ldots, x_n), \tag{1.56}$$

where, α is a continuously differentiable map with certain properties. We will be satisfied with a result of the type, which ensures that every diffeomorphism is a composite of functions of the above form (1.56) and of course, of some permutations. We begin with:

Definition 1.4.4 By a *primitive mapping* $f : U \to \mathbb{R}^n$ of an open subset of $\mathbb{R}^n$, we mean a diffeomorphism f of U onto an open subset $f(U)$ and f is of the form (1.56) above.

Clearly, a diffeomorphism f is a primitive mapping if and only if $p_i \circ f = p_i$, for all i except perhaps for one say $i = k$. It follows that the matrix Df coincides with Id except in the kth row. Indeed, given any mapping f of the form (1.56), f is a local diffeomorphism if and only if $\dfrac{\partial \alpha}{\partial x_k} \neq -1$. Note that the inverse of a primitive mapping is again a primitive mapping.

The following theorem is what we were after:

Theorem 1.4.7 *Let U be an open set in $\mathbb{R}^n$ $\mathbf{0} \in U$, and $F \in \mathcal{C}^1(U, \mathbb{R}^n)$ be such that $F(\mathbf{0}) = \mathbf{0}$ and $DF(0)$ be invertible. Then there exists a neighborhood V of 0, in which we can express F as a composite*

$$F = P \circ G_n \circ \cdots \circ G_1, \tag{1.57}$$

where P is a permutation and $G_1, \ldots, G_n$ are primitive diffeomorphisms of the form

$$G_i(\mathbf{x}) = (x_1, \ldots, x_{i-1}, x_i + \alpha_i(\mathbf{x}), x_{i+1}, \ldots, x_n). \tag{1.58}$$

Proof: Let us put $F = F_1$. Inductively, for each $1 \leq k \leq n-1$, we shall find a permutation P_k, a map α_k, such that if G_k are defined as in (1.58), and if we put $F_{k+1} = P_k \circ F_k \circ G_k^{-1}$ then

$$p_i \circ F_{k+1} = p_i, \quad 1 \leq i \leq k. \tag{1.59}$$

Clearly, then F_n itself is primitive and we set $G_n = F_n$. Then it follows that

$$F = P_1 \circ P_2 \circ \cdots \circ P_{n-1} \circ G_n \circ \cdots \circ G_1.$$

By taking $P = P_1 \circ \cdots \circ P_{n-1}$, the conclusion (1.58) follows.

Inductively, consider the matrix $DF_k(\mathbf{0})$. Since this is invertible, from (1.59), it follows that there exists a $j \geq k$ such that $\dfrac{\partial(p_k \circ F_k)}{\partial x_j}(\mathbf{0}) \neq 0$. Take P_k to be the transposition interchanging j and k if $j \neq k$ and equal to Id if $j = k$. Let $H = P_k \circ F_k$. Then it follows that $(k,k)^{\text{th}}$ entry of $DH(\mathbf{0})$ does not vanish. Put $\alpha_k(\mathbf{x}) = p_k \circ H(\mathbf{x}) - x_k$. If G_k is defined as in (1.58), a simple computation show that $DG_k(\mathbf{0})$ is invertible and hence is a primitive diffeomorphism in a small neighborhood of $\mathbf{0}$. Since, $p_i \circ F_k = p_i$, and $p_i \circ P_k = p_i$ for $1 \leq i \leq k-1$, it follows that

$$p_i \circ F_{k+1} = p_i \circ P_k \circ F_k \circ G_k^{-1} = p_i \circ F_k \circ G_k^{-1} = p_i \circ G_k^{-1} = p_i, \quad i \leq k-1.$$

Moreover, for $i = k$, we have

$$p_k \circ F_{k+1} = p_k \circ H \circ G_k^{-1} = p_k \circ G_k \circ G_k^{-1} = p_k.$$

This completes the proof. ♠

Exercise 1.4

1. Verify that the following functions define a diffeomorphism of $\mathbb{R}^2 \to \mathbb{R}^2$ by expressing them as composites of primitive diffeomorphisms or otherwise.

   (a) $(x, y) \mapsto (x + y^2, y + x^3 + 3x^2y + 3xy^2 + y^3)$.

   (b) $(x, y) \mapsto (x + y^2 + 2x^3y + x^6, y + x^3)$.

2. Let U be an open subset of $\mathbb{R}^n$ and $f, g : U \to \mathbb{R}^m$ be two submersions such that $Ker\, Df = Ker\, Dg$ on U. Then for each $p \in U$, there is a neighborhood W of $f(p)$ in $\mathbb{R}^m$ and a diffeomorphism $h : W \to g(f^{-1}(W)) = W'$ such that $g = h \circ f$ on $V = f^{-1}(W)$.

1.5 Lagrange Multiplier Method

Near a maximum the decrements on both sides are in the beginning only imperceptible.

–J. Kepler

When a quantity is greatest or least, at that moment its flow neither increases nor decreases.

–I. Newton

If one is looking for a maximum or a minimum of some function of many variables subject to the condition that these variables are related by a constraint given by one or more equations, then one should add to the function whose extremum is sought the functions that yield the constraint equations each multiplied by undetermined multipliers and seek the maximum or minimum of the resulting sum as if the variables were independent. The resulting equations combined with constrained equations will serve to determine all unknowns.

–J. Lagrange

These epigrams have been reproduced from the beautiful book of Tikhomirov [Ti].

In this section we would like to study the above statement of Lagrange. Nowadays, we call it *Lagrange*[3] *Multiplier Method* (LMM). Recall that given a real valued function f on an open subset U of $\mathbb{R}^n$ and a point $p \in U$, we say p is a *local maximum* if there exists a neighborhood V of p in U such that $f(q) \leq f(p)$ for all $q \in V$. Similarly, p is called a *local minimum* if there is a neighborhood V of p such that $f(q) \geq f(p)$ for all $q \in V$. The point p is called a *local extremum point* if $f(p)$ is a local maximum or a local minimum. For a smooth function $f : U \to \mathbb{R}$, a necessary condition for a point $p \in U$ to be a local extremum of f is that all the 1st order partial derivatives of f vanish at p. This is deduced by merely restricting the function to various coordinate axes through the point p and then applying the corresponding result from one variable calculus. For future reference let us state and label this result here:

Lemma 1.5.1 Let $f : U \to \mathbb{R}$ be a smooth function defined on an open subset of $\mathbb{R}^n$. A necessary condition for $f(P)$ to be a local extremum of f at $P \in U$ is that $Df(P) = 0$.

[3]Joseph Louis Lagrange was born on Janaury 25, 1736, in Turin, Italy (and was named Giuseppe Lodovico Lagrangia) and lived in Paris most of his life. He is well known for his many mathematical discoveries, the calculus of variations being one of them.

The Lagrange Multiplier Method deals with such a necessary condition for local extremum of the function restricted to a subspace called the *constraint space.* Let us first make a little more concise statement out of the above quotation from Lagrange.

We are given a smooth function $f(x_1, \ldots, x_n)$ of n variables. The problem is to find the extremal values of this function on the constraint space G given by a number of equations $g_i(x_1, \ldots, x_n) = 0,\ i = 1, \ldots, k$. The LMM says that the points at which extrema of f may occur are contained in the space of solutions of

$$\frac{\partial \mathcal{L}}{\partial x_i} = 0,\ i = 1, \ldots, n; \quad \frac{\partial \mathcal{L}}{\partial \lambda_j} = 0,\ j = 1, \ldots, k; \tag{1.60}$$

where $\mathcal{L}$ is the Lagrange multiplier function defined by

$$\mathcal{L} := \mathcal{L}(\mathbf{x}, \Lambda) = f(\mathbf{x}) + \sum_{i=1}^{k} \lambda_i g_i(\mathbf{x}). \tag{1.61}$$

When we equate the partial derivatives $\frac{\partial \mathcal{L}}{\partial \lambda_i}$ to zero, we get back the constraint space itself. Thus, if $(x_1, \ldots, x_n, \lambda_1, \ldots, \lambda_k)$ is a solution of (1.60), it follows that $(x_1, \ldots, x_n)$ is a point on the constraint space. Conversely, if $(x_1, \ldots, x_n)$ belongs to the constraint space, and for $(\lambda_1, \ldots, \lambda_k)$ the point $(x_1, \ldots, x_n, \lambda_1, \ldots, \lambda_k)$ satisfies the first n equations of (1.60) then the other k equations are also satisfied automatically. Thus, we see that LMM is a clever way of converting a problem of extrema with constraints, to another problem of extrema without constraints, by increasing the number of variables.

Example 1.5.1 Before we proceed any further, let us examine a simple example where we have $f(x, y) = x$ and the constraint space is the cuspidal curve $y^2 = x^3$. This means $x^3 \geq 0$ and hence $x \geq 0$. Indeed the point $(0, 0)$ is on the curve and $f(0, 0) = 0$. Thus, the minimum is attained precisely at $(0, 0)$. Also we see that for any positive value of x, we can always find two solutions of $y^2 = x^3$ and hence it follows that f is not bounded above on the curve $y^2 = x^3$. So, it has no maximum. Now according to LMM, we must look for extrema of f among solutions of (1.60) where,

$$\mathcal{L} := \mathcal{L}(x, y, \lambda) = x + \lambda(y^2 - x^3). \tag{1.62}$$

This yields

$$1 - 3\lambda x^2 = 2\lambda y = y^2 - x^3 = 0, \tag{1.63}$$

which, alas, has no solutions at all.

Thus we see that the statement of LMM needs some technical correction. One simple way to do this is to put an additional condition on the constraint space:

Theorem 1.5.1 *Let G be the subspace of $\mathbb{R}^n$ given by some smooth equations $g_i(\mathbf{x}) = 0$, for $i = 1, 2, \ldots, k$, and f be a real valued smooth function on G. Suppose that for every $P \in G$, the set $\{\nabla g_1(P), \ldots, \nabla g_k(P)\}$ is independent. Then the local extrema, if any, of a given smooth function f defined on G has to be found among the solutions of equation (1.60), where $\mathcal{L}$ is as in (1.61).*

Proof: Put $g = (g_1, \ldots, g_k) : \mathbb{R}^n \to \mathbb{R}^k$. Then $G = \{\mathbf{x} \in \mathbb{R}^n : g(\mathbf{x}) = 0\}$. To say that $\{\nabla g_1(P), \ldots, \nabla g_k(P)\}$ is independent is the same as saying that $(Dg)_P : \mathbb{R}^n \to \mathbb{R}^k$ is surjective. Let now for some $P \in G$, $f(P)$ be a local extremum for f. In view of the discussion above, we have only to prove that there exists constants $\lambda_i, 1 \leq i \leq k$ such that

$$(\nabla f)_P + \lambda_1(\nabla g_1)_P + \cdots + \lambda_k(\nabla g_k)_P = 0. \tag{1.64}$$

Apply the surjective form of the ImFT (1.4.4) to the map $\mathbf{x} \mapsto g(\mathbf{x})$ to obtain a neighborhood V of 0 in $\mathbb{R}^n$ and a diffeomorphism $\phi : V \to \phi(V)$ to a neighborhood $\phi(V)$ of P such that $g \circ \phi(x_1, \ldots, x_n) = (x_1, \ldots, x_k)$.

Let W be the linear span of $\{\mathbf{v}_j := (D\phi)_{\mathbf{0}}(\mathbf{e}_j), \ : \ k+1 \leq j \leq n\}$. By the chain rule, it follows that $Dg_P \circ D\phi_0(\mathbf{e}_j) = 0, \ k+1 \leq j \leq n$. Therefore, $(\nabla g_i)_P \cdot \mathbf{v}_j = 0$ for $1 \leq i \leq k$ and $k+1 \leq j \leq n$. This means that $(\nabla g_i)_P \in W^\perp$, the orthogonal complement of W. Since these vectors are given to be independent, and since dim $W^\perp = k$, they actually form a basis for $W^\perp$.

On the other hand, since $\phi(0) = P$, it follows that 0 is a local extremum for $f \circ \phi$ restricted to $0 \times \mathbb{R}^{n-k}$. Therefore, by Lemma 1.5.1, $D(f \circ \phi)_0(\mathbf{e}_j) = 0, k+1 \leq j \leq n$. By the chain rule, this means that $(\nabla f)_P$ is perpendicular to $\mathbf{v}_j$ for all $k+1 \leq j \leq n$, i.e., $(\nabla f)_P \in W^\perp$. Therefore, (1.64) follows as required. ♠

Remark 1.5.1

(i) A global extremum in the interior of a constraint space is necessarily a local extremum. Thus to solve a global extremum problem, one uses LMM for the interior of the constraint space and gets all possible local solutions. To this set one adds all local extrema on the boundary of the constraint space. Often LMM itself can be used for the local extrema on the boundary of the constraint space as a problem with an increased number of constraints. This way, we keep cutting down the "dimension" of the space on which we need to look for the possible extrema. Hopefully, in a finite number of steps, this process will convert the constraint space into a finite set. We can then simply enumerate the value of the function at these finitely many points and find out the actual maximum or minimum.

(ii) In general, an extremum problem may not have any solutions and LMM does not address itself to the existence aspect at all. A result of Weierstrass, which says that a continuous function attains its extrema on a closed and bounded subset of a euclidean space, comes in handy in many situations, to ensure the existence. So, it is good if we have the constraint space as a closed and a bounded subset.

(iii) In some situations even if the constraint space is unbounded we can guarantee the existence of the extrema. For instance, suppose we know that the function tends to $+\infty$ as $||x|| \to +\infty$. Then this function will attain its minimum. (You may recall that this fact has been used in an elementary "canonical" proof of the Fundamental Theorem of Algebra. See for example [Sh2].)

(iv) The condition that ∇g_i are independent ensures that the constraint space is a smooth submanifold of the Euclidean space with its tangent space orthogonal to the gradient lines of the constraint function. (See Chapter 3 for more details.) In general, a constraint space may not be smooth all over and may have corners, e.g., it may be a cube or a polyhedron. Even in such cases the method can be employed for local extrema away from such corner points. At the corners, however, you will have to use further analysis of the situation.

(v) The condition that ∇g_i are independent is not always necessary. However, in the absence of this condition, weird things can happen. From a certain advanced point of view, this happens to be the more interesting case as it allows the study of singularities and degeneracies. In order to have the full force of LMM, it is necessary that we formulate the problem in the projective space of multipliers. Take $f = g_0$ and put $\mathcal{L} = \sum_{i=0}^{k} \lambda_i g_i$ in (1.61). We are now looking at the projective class of multipliers $[\lambda_0, \ldots, \lambda_k] \in I\!P^{k-1}$. [Two multipliers $(\lambda_1, \ldots, \lambda_k)$ and $(\lambda'_1, \ldots, \lambda'_k)$ are in the same class if and only if there exists a nonzero real number s such that $(\lambda_1, \ldots, \lambda_k) = s(\lambda'_1, \ldots, \lambda'_k)$. The extra condition that we put on multipliers is that they are not identically zero i.e., at least one $\lambda_i \neq 0$. For more details about the projective space, see Sections 5.1 and 5.2.] Thus, Theorem 1.5.1 takes the form:

Theorem 1.5.2 *The extrema of the function g_0 on the constraint space G are contained in the set of solutions of (1.61), where $\mathcal{L} = \sum_{i=0}^{k} \lambda_i g_i$.*

Equation (1.64) says that the vector $(\nabla f)_P = (\nabla g_0)_P$ belongs to the linear span of $\{(\nabla g_1)_P, \ldots, (\nabla g_k)_P\}$. This condition is now replaced by saying that the set of vectors, $\{(\nabla g_0)_P, \ldots, (\nabla g_k)_P\}$ is linearly dependent. Then the case that we have discussed in Theorem 1.5.1 corresponds to the restricted classes in which $\lambda_0 \neq 0$. The advantage in the new formulation is that we need not put the additional hypothesis that $\{(\nabla g_1)_P, \ldots, (\nabla g_k)_P\}$ are linearly independent. Geometrically, this amounts to allowing the constraint space to have singularities. We shall not go into more details here and be content with just discussing how this helps us to resolve the difficulty that we faced in Example 1.5.1.

Setting $\mathcal{L} = \lambda_0 x + \lambda_1(y^2 - x^3)$ with the only restriction that $(\lambda_0, \lambda_1) \neq (0,0)$, we then obtain $\lambda_0 - 3\lambda_1 x^2 = 2\lambda_1 y = y^2 - x^3 = 0$. This then allows the solution $\lambda_0 = 0$ together with $(x, y) = (0,0)$ as a probable solution, which we know happens to be the actual solution!

We can modify the above problem by taking $f(x,y,z) = x + z$, and the constraint space to be $g(x,y,z) = y^2 - x^3 = 0$. Then we get

$$\mathcal{L} = \lambda_0(x+z) + \lambda_1(y^2 - x^3), \ (\lambda_0, \lambda_1) \neq (0,0).$$

This yields that the probable solutions are inside $0 \times 0 \times \mathbb{R}$. Since obviously, $f(0,0,z) = z$ does not have any extremum values for $z \in \mathbb{R}$, it follows that $f(x,y,z) = x + z$ has no extremum values on the constraint space.

Example 1.5.2 Physical interpretation of grad:

Consider any linear function $\phi : \mathbb{R}^n \to \mathbb{R}, \ \phi(\mathbf{x}) = \sum_i \alpha_i x_i$, and the problem of finding its maxima on the unit sphere $\mathbb{S}^{n-1}$. The Lagrange multiplier function in this case is

$$\sum_{i=1}^{n} \alpha_i x_i - \lambda \left(\sum_{i=1}^{n} x_i^2 - 1 \right)$$

which leads to the solution that

$$x_i = \pm \frac{\alpha_i}{\sqrt{\sum_i \alpha_i^2}}, \quad i = 1, 2, \ldots, n. \tag{1.65}$$

Let us now apply this to a specific case where $\phi = Df_{\mathbf{0}}$, and $f : U \to \mathbb{R}$ is a smooth function in a neighborhood U of $\mathbf{0} \in \mathbb{R}^n$. To each $\mathbf{v} \in \mathbb{S}^{n-1}$ we can consider the path $(-\epsilon, \epsilon) \to U$ given by $t \mapsto t\mathbf{v}$ and look at the function $t \mapsto f(t\mathbf{v})$. The derivative of this map at 0 is nothing but $Df_0(\mathbf{v})$. Now $Df_0(\mathbf{x}) = \sum_i \frac{\partial f}{\partial x_i} x_i$. Therefore, from (1.65), it follows that the extrema of the function

$$\mathbf{v} \mapsto Df_0(\mathbf{v})$$

occur at $\pm \frac{\nabla f}{\|\nabla f\|}$. Thus, ∇f is the direction in which the increment in f is the maximum.

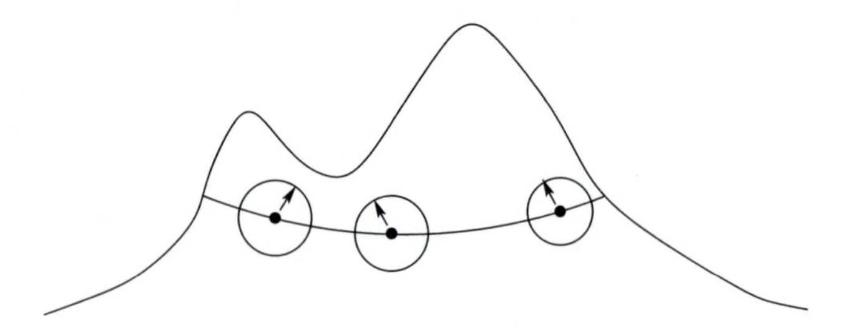

Figure 3 The hill is steepest in the direction of ∇f.

The figure shows the graph of f, for $n = 2$. At any point on the graph, the increment in f is maximum in the direction of the vector ∇f.

Example 1.5.3 Let x, y, z be nonnegative real numbers subject to the condition $x + y + z = 1$. Find the maximum value of $f(x, y, z) = xy + yz + zx - 2xyz$.

Observe that part of the constraint (viz., nonnegativity) is given by a number of inequalities. We shall ignore this part in the beginning and consider them only at a later stage. The other part of constraint is given by a linear equation. So, it seems easier to eliminate one of the variables and treat the problem with two variables only and without any constraint. I leave it to you to work out this problem in this method and see which one is actually easier. We shall carry out the LMM below.

Thus, here $\mathcal{L}(x, y, z, \lambda) = xy + yz + zx - 2xyz - \lambda(x + y + z - 1)$. Putting $\nabla\mathcal{L} = 0$, we obtain

$$x + y + z - 1 = 0$$

and

$$y + z - 2yz = x + z - 2xz = x + y - 2xy = \lambda.$$

This means $(y - x)(1 - 2z) = 0$, etc., and yields the following:

$$\left.\begin{array}{rclcrcl} y & = & x & \text{or} & z & = & 1/2 \\ z & = & y & \text{or} & x & = & 1/2 \\ x & = & z & \text{or} & y & = & 1/2. \end{array}\right\}$$

Observe that if none of x, y, z is equal to $1/2$ then $x = y = z = 1/3$ and $f(x, y, z) = 7/27$. So, consider the case when $z = 1/2$. Then $x = y$ and $f(x, y, 1/2) = (x + y)/2 = 1/4 < 7/27$. By symmetry, this is the case with $x = 1/2$ as well as with $y = 1/2$.

One should remember that the above set of solutions gives all possible local extrema, in the interior of the region under consideration. Since a global extrema which is in the interior of the constraint domain has to be among these, but not those which are on the boundary of the constraint domain, we should keep in mind to look at the boundary also separately.

So, let us look at one of the boundary pieces say given by $z = 0$. This is the same as adding another constraint to the above situation. Therefore, we may simply consider

$$\mathcal{L}(x, y, z, \Lambda) = f(x, y, z) - \lambda_1(x + y + z - 1) - \lambda_2 z.$$

But, in this case, we see that plugging simply $z = 0$ in the given problem makes it easier to handle. We get

$$\mathcal{L}(x, y, \Lambda) = xy - \lambda(x + y - 1).$$

Putting $\nabla\mathcal{L} = 0$ we get $y = x = \lambda = 1/2$. Hence, $f(1/2, 1/2, 0) = 1/4 < 7/27$. Other constraints such as $x = 0$, or $y = 0$ yield the same result because of the symmetry. The problem is not over yet. We have to consider the boundary constraints two at a time, say $y = z = 0$. But then $f(x, 0, 0) = 0 < 7/27$.

Therefore, the maximum value is $7/27$. Incidentally we have also found out that minimum value is 0.

Thus, we have proved an inequality:

$$0 \leq xy + yz + zx - 2xyz \leq \frac{7}{27}$$

whenever $x, y, z \geq 0$ and $x + y + z = 1$.

Example 1.5.4 The Inequality of Arithmetic and Geometric Means:
Given nonnegative real numbers $a_1, \ldots, a_n$ show that

$$a_1 a_2 \cdots a_n \leq \left(\frac{a_1 + a_2 + \cdots + a_n}{n} \right)^n .$$

We set $S = \sum_i a_i$ and $x_j = a_j / S$. Then $\sum_i x_i = 1$, $x_i \geq 0$ and we must prove that $x_1 x_2 \cdots x_n \leq (1/n)^n$. Thus, we have converted the given problem into an extremal problem:

$$\mathcal{L} = x_1 x_2 \cdots x_n - \lambda(x_1 + x_2 + \cdots + x_n - 1).$$

Putting $\nabla \mathcal{L} = 0$ gives

$$x_1 \lambda = x_2 \lambda = \cdots = x_n \lambda = x_1 x_2 \cdots x_n .$$

Since we are interested in the maximum of $x_1 \cdots x_n$ clearly we can assume that none of x_i is 0. Then $\lambda \neq 0$ and $x_1 = x_2 = \cdots = x_n = 1/n$. Therefore the maximum value $x_1 \cdots x_n$ is at this point which is equal to $(1/n)^n$ as required.

Example 1.5.5 Inequality of Geometric and Quadratic Means:
Given nonnegative real numbers $a_1, \ldots, a_n$ show that

$$(a_1 a_2 \cdots a_n)^{1/n} \leq \left(\frac{a_1^2 + \cdots + a_n^2}{n} \right)^{1/2} .$$

We set $S = (a_1^2 + \cdots + a_n^2)^{1/2}$ and $x_j = a_j / S$. Then $\sum_i x_j^2 = 1$. And we have to prove that

$$(x_1 x_2 \cdots x_n)^{1/n} \leq \sqrt{1/n}.$$

So, we take $f(x_1, \ldots, x_n) = x_1 x_2 \cdots x_n$ and maximize it subject to the constraint $\sum_i x_i^2 = 1$.

As before the maximum is seen to have attained at $x_1 = x_2 = \cdots = x_n = \sqrt{1/n}$ and is equal to $\sqrt{1/n}$.

This has a nice geometric interpretation: *Find the box of maximum size inscribed in a sphere.* The above solution tells us that this box is actually a cube and its volume is $(2r)^n \left(\frac{1}{n} \right)^{n/2}$, where r is the radius of the sphere.

Example 1.5.6 Cauchy's Inequality: *For arbitrary real numbers $a_1, \ldots, a_n, b_1, \ldots, b_n$ show that*

$$\sum_i a_i b_i \leq \left(\sum_i a_i^2 \right)^{1/2} \left(\sum_i b_i^2 \right)^{1/2} .$$

Put $A = \sqrt{\sum_i a_i^2}$. If all the $a_i = 0$ then the given inequality is obvious. So, we assume

$A \neq 0$. Put $x_i = a_i/A$ so that $\sum_i x_i^2 = 1$. We now fix $b_1, \ldots, b_n$ and put $B = \sqrt{\sum_i b_i^2}$. Observe that we may assume that $B \neq 0$. We have to prove that $\sum_i b_i x_i \leq B$ subject to the constraint $\sum_i x_i^2 = 1$. Thus,

$$\mathcal{L} = \sum_i b_i x_i - \lambda \left(\sum_i x_i^2 - 1 \right).$$

Equating $\nabla \mathcal{L} = 0$ we get

$$2\lambda x_i = b_i, i = 1, 2, \ldots, n.$$

Since some $b_i \neq 0$, we have $\lambda \neq 0$. Therefore, $1 = \sum_i x_i^2 = B^2/4\lambda^2$. This means that $\lambda = \pm B/2$. Correspondingly, we have $x_i = \pm b_i/B$. The negative sign corresponds to the value $-B$ whereas the positive sign gives B. Therefore, we get $-B \leq \sum_i b_i x_i \leq B$.

Example 1.5.7 Hölder's Inequality: *Fix real numbers $p, q > 1$ such that $\frac{1}{p} + \frac{1}{q} = 1$. Let a_i, b_i be any nonnegative real numbers. Then*

$$\sum_i a_i b_i \leq \left(\sum_i a_i^p \right)^{1/p} \left(\sum_i b_i^q \right)^{1/q}.$$

The proof is similar to the above after we set $x_j = a_j/A$, where $A = (\sum_i a_i^p)^{1/p}$, etc.

We shall end this section with an application of Lagrange Multiplier Method to Linear Algebra:

Theorem 1.5.3 *Given any symmetric matrix $A \in M_n(\mathbb{R})$ there exists an orthogonal matrix $P \in O(n)$ such that $P^{-1}AP$ is a diagonal matrix.*

Proof: Check that finding a matrix P as above is the same as finding an orthonormal basis $\{\mathbf{v}_1, \ldots, \mathbf{v}_n\}$ for $\mathbb{R}^n$ consisting of eigenvectors of A.

We start with the function $\phi : \mathbb{R}^n \setminus \{\mathbf{0}\} \to \mathbb{R}$ given by

$$\phi(\mathbf{x}) = \frac{\mathbf{x} \cdot A\mathbf{x}}{\mathbf{x} \cdot \mathbf{x}}.$$

Observe that ϕ is continuous and $\phi(\alpha \mathbf{x}) = \phi(\mathbf{x})$ for all real scalars. Thus, $\phi(\mathbb{R}^n \setminus \{0\}) = \phi(\mathbb{S}^{n-1})$. By Weierstrass' theorem, $\phi|_{\mathbb{S}^{n-1}}$ attains its maximum and minimum values. And by the above observation, ϕ attains its extrema on the whole of $\mathbb{R}^n \setminus \{\mathbf{0}\}$, and this will happen inside the unit sphere. Therefore, we can study the extrema of the function $f : \mathbb{S}^{n-1} \to \mathbb{R}$, given by

$$f(\mathbf{x}) = \mathbf{x} \cdot A(\mathbf{x})$$

in place of ϕ. As we shall see this simplification of the problem is a very important step since we have got rid of the denominator in the expression for ϕ. Now the idea is clear— an extremum value of f should give an eigenvalue of A. Indeed we shall soon see that it gives even the corresponding unit eigenvector.

So we set up

$$\mathcal{L}(\mathbf{x}, \Lambda) = f(\mathbf{x}) - \lambda \mathbf{x} \cdot \mathbf{x}.$$

The extrema of f with the constraint $\mathbf{x} \cdot \mathbf{x} = 1$ is to be found among the solutions of $\nabla \mathcal{L} = 0$.

Since $Df(\mathbf{x}) = 2A\mathbf{x}$, it follows that the local extrema are given by the solution of the equation

$$A\mathbf{x} = \lambda \mathbf{x}. \tag{1.66}$$

Remember that the existence of such a solution $\mathbf{x}$ on the unit sphere is guaranteed by the general consideration that a continuous map has to attain its maximum on a closed and bounded subset together with the principle that such a maximum is also a local maximum for all curves passing through the point.

Let such a solution of (1.66) be denoted by $\mathbf{v}_1$ with the corresponding eigenvalue μ_1. (Incidentally it may be noted that we have actually found $\mathbf{v}_1$ corresponding to the maximum eigenvalue or the minimum eigenvalue.)

Inductively, having found mutually orthogonal vectors $\mathbf{v}_1, \ldots, \mathbf{v}_k$ of unit length such that $A\mathbf{v}_i = \mu_i \mathbf{v}_i$, we now consider the LMM with the data:

$$f(\mathbf{x}) = \mathbf{x} \cdot A\mathbf{x};\ g(\mathbf{x}) = \mathbf{x} \cdot \mathbf{x} = 1;\ g_i(\mathbf{x}) = \mathbf{v}_i \cdot \mathbf{x} = 0,\ 1 \le i \le k.$$

For this we have to set

$$\mathcal{L}(\mathbf{x}, \Lambda) = f(\mathbf{x}) + \sum_{i=1}^{k} \lambda_i g_i(\mathbf{x}) + \lambda(g(\mathbf{x}) - 1)$$

and look for points $(\mathbf{x}, \lambda_1, \ldots, \lambda_k, \lambda)$ which satisfy $\nabla\mathcal{L} = 0$. This is the same as saying

$$2A\mathbf{x} = \sum_{1}^{k} \lambda_i \mathbf{v}_i + 2\lambda\mathbf{x}. \tag{1.67}$$

Since we are also having the constraints $g_i(\mathbf{x}) = 0 = \mathbf{x} \cdot \mathbf{v}_i$, by the symmetry property of A, it follows that $A\mathbf{x} \cdot \mathbf{v}_i = \mathbf{x} \cdot A\mathbf{v}_i = \mathbf{x} \cdot \mu_i \mathbf{v}_i = 0$, for all $1 \le i \le k$. Plugging in this information in (1.67) yields that $\lambda_i = 0$ and $A\mathbf{x} = \lambda\mathbf{x}$. Thus we have found another (not necessarily different from the old ones) eigenvalue of A which we shall denote by μ_{k+1}. The corresponding solution of $A\mathbf{x} = \mu_{k+1}\mathbf{x}$ can now chosen to be a unit vector and denoted by $\mathbf{v}_{k+1}$. The proof is completed by induction. ♠

Exercise 1.5

1. Examine the map $f(x, y, z) = xyz$ for extremum values on the unit sphere
$$x^2 + y^2 + z^2 = 1.$$
2. Examine the function $x^2 + y^2 + z^2$ for extremum values on the surface $z = xy + 1$.
3. Find the maximum value of $8x^2 + 4yz - 16z + 10$ on the ellipsoid
$$4x^2 + y^2 + 4z^2 = 16.$$
4. Consider the space $SL(2, \mathbb{R})$ of 2×2 matrices $\begin{bmatrix} a & b \\ c & d \end{bmatrix}$ over the real numbers and with determinant $ad - bc = 1$. Show that the Euclidean distance of $SL(2, \mathbb{R})$ from the origin, viz., minimum of $(a^2 + b^2 + c^2 + d^2)^{1/2}$ is $\sqrt{2}$. What about the same problem if $\mathbb{R}$ is replaced by $\mathbb{C}$?

1.6 Differentiability on Subsets of Euclidean Spaces

Given a subset X of some Euclidean space, a function $f : X \to \mathbb{R}$ and a point $x \in X$, in order to define the differentiability of f at x, we have insisted that the set X should be a neighborhood of x or at least contain a neighborhood of x in $\mathbb{R}^n$. Our aim here is to remove this restriction as far as possible and extend the notion of differentiability of functions. For instance, X could be a closed interval $[a, b]$ in $\mathbb{R}$ and the point may be one of the end points. Recall then that f is said to be differentiable at a if the "right-hand derivative" at a exists. Similarly, it is said to be differentiable at b if the "left-hand derivative" exists. When we consider subsets of $\mathbb{R}^n$ for $n \geq 2$, such simple-minded modifications would not be possible. We must remember that any modification/extension of the definition should conform with the old definition of differentiability when X is a neighborhood of x. Here is then a reasonably satisfactory answer to this problem.

Definition 1.6.1 Let $X \subset \mathbb{R}^N$ and $\mathbf{x}_0 \in X$ be any point. A function $f : X \to \mathbb{R}$ is said to be *differentiable (of class C^r)* at $\mathbf{x}_0$ if there exists an open subset U of $\mathbb{R}^N$ such that $\mathbf{x}_0 \in U$ and a function $\hat{f} : U \to \mathbb{R}$, which is differentiable (C^r) at $\mathbf{x}_0$ such that $\hat{f}|_{U \cap X} = f$. If f is differentiable at every point of X then we say f is differentiable on X.

Remark 1.6.1
(i) Clearly, if X itself is open in $\mathbb{R}^N$, then this new definition of differentiability coincides with the old one for functions defined on open subsets.
(ii) Now consider the case when $f : [a, b] \to \mathbb{R}$. Suppose that we have defined f to be differentiable at a if the right-hand derivative of f at a exists, viz.,

$$\lim_{h \to 0+} \frac{f(a+h) - f(a)}{h} = \alpha \tag{1.68}$$

exists. Clearly, this is implied by the existence of the derivative at a according to Definition 1.6.1. On the other hand, if f satisfies (1.68), consider

$$\hat{f}(x) = \begin{cases} f(a) - (a - x)\alpha, & x \leq a \\ f(x), & x \in [a, b]. \end{cases} \tag{1.69}$$

Then $\hat{f}$ is defined in $(-\infty, b]$ and is differentiable at a and $\hat{f}|_{[a,b]} = f$. Thus, the two definitions coincide here.
(iii) Also, we now have two different definitions of differentiability of a function defined on $\mathbb{R}^n$ itself, the new definition applied to the case when $\mathbb{R}^n$ is thought of as a subspace of $\mathbb{R}^{n+1}$ by a coordinate inclusion say,

$$(x_1, \ldots, x_n) \mapsto (x_1, \ldots, x_n, 0). \tag{1.70}$$

Verify that the two notions coincide.
(iv) It follows easily that sums, products, and composites of differentiable functions yield differentiable functions again—we have only to argue in the same old fashion with extended functions instead of the given functions.
(v) Observe that we have not defined the derivative of a function that is differentiable in the new sense on arbitrary subsets. No doubt if the domain of the given function is open then we can take the derivative to be the same as in the old sense. Can we take the derivative of f to be that of $\hat{f}$ at $\mathbf{x}_0$ in general? The trouble with this idea is the choice involved in $\hat{f}$ and that there is no guarantee that this derivative can be unique. One of the important aspects of the notion of manifolds is that it allows us to define the derivative of a differentiable function on a manifold, in an unambiguous way, as we shall see later. (The exercise 1.6 below will give you some idea about the conditions on the domain for the derivative of any functions to be defined uniquely.)

Definition 1.6.2 By a *diffeomorphism* $f : X \to Y$ of two subsets of Euclidean spaces we mean a bijection f such that both f and f^{-1} are smooth. If such a function exists, then we say X and Y are *diffeomorphic*. Compare Definition 1.4.3.

Remark 1.6.2 It is easy to see that "being diffeomorphic" is an equivalence relation. We remind you that one of the central theme of Differential Topology is to classify objects up to this equivalence. Observe that any diffeomorphism is a homeomorphism and hence nonhomeomorphic spaces cannot be diffeomorphic. The converse is far from being true but examples are not that easy to come by. We shall however establish easily the invariance of certain topological properties of spaces under diffeomorphisms, the corresponding result for homeomorphism being more difficult. The very first instance of this is the following result.

Theorem 1.6.1 Invariance of Domain: *Let $U \subset \mathbb{R}^n, V \subset \mathbb{R}^m$ be nonempty subsets which are diffeomorphic to each other.*
(a) *If U and V are both open then $m = n$.*
(b) *If $m = n$ and U is open, then V is also open.*

Proof: (a) Choose $x \in U$ and consider $D(f)_x$. Since f is a diffeomorphism, by the chain rule, it follows that $D(f)_x : \mathbb{R}^n \to \mathbb{R}^m$ is a linear isomorphism. Hence, $m = n$.
(b) Since U is open the two definitions of differentiability coincide here. By IFT 1.4.2, f is an open mapping. Therefore, $f(U) = V$ is open. ♠

Example 1.6.1

1. It is easy to see that any two open balls (closed balls) in $\mathbb{R}^n$ are diffeomorphic (use a translation followed by a scaling). Moreover, all open balls are diffeomorphic to the whole of $\mathbb{R}^n$. To see this consider the map $\phi : B^n \to \mathbb{R}^n$ given by

$$\phi(\mathbf{x}) = \frac{\mathbf{x}}{\sqrt{1 - \|\mathbf{x}\|^2}} \tag{1.71}$$

with its inverse map

$$\psi(\mathbf{y}) = \frac{\mathbf{y}}{\sqrt{1 + \|\mathbf{y}\|^2}}. \tag{1.72}$$

2. Consider the function $x \mapsto x^2$ on the unit interval $[0, 1]$. This is clearly a smooth map in the new sense, since it is the restriction of a smooth map on the whole of $\mathbb{R}$. Clearly it is a bijection and hence a homeomorphism. However, observe that the inverse function $x \mapsto \sqrt{x}$ is not differentiable at $x = 0$. Thus, this is not a diffeomorphism. Indeed, it can be seen that if $f : [a, b] \to X$ is a diffeomorphism where, X is any subset of $\mathbb{R}^n$, then $D(f)$ does not vanish at any point of $[a, b]$. (Exercise: Supply details.)

3. The phenomenon that we witnessed in (2) has nothing to do with the boundary. For instance, consider the map
$$x \mapsto x^3,$$
which is a homeomorphism of $\mathbb{R}$ with itself and is a $\mathcal{C}^\infty$ map. However, its inverse is not differentiable at 0.

Example 1.6.2 Let $\mathbf{H}^n$ denote the upper half-space in $\mathbb{R}^n$,

$$\mathbf{H}^n = \{(x_1, \ldots, x_n) \in \mathbb{R}^n \;:\; x_n \geq 0\}.$$

Let f be a smooth function defined in a neighborhood of $0 \in \mathbf{H}^n$. Let f_1, f_2 be any two extensions of f to a (common) neighborhood U of $\mathbf{0}$ in $\mathbb{R}^n$, which are differentiable at $\mathbf{0}$. Then

$$\lim_{\|\mathbf{h}\| \to 0} \frac{f_i(\mathbf{h}) - f_i(\mathbf{0}) - D(f_i)(\mathbf{0})(\mathbf{h})}{\|\mathbf{h}\|} = 0, \; i = 1, 2.$$

Therefore, for each $i = 1, 2$ we can take the limit, by restricting $\mathbf{h}$ to remain inside $\mathbf{H}^n$. But, for $\mathbf{h} \in \mathbf{H}^n$, $f_1(\mathbf{h}) = f(\mathbf{h}) = f_2(\mathbf{h})$. Hence, it follows that $Df_1(\mathbf{0}) = Df_2(\mathbf{0})$. Thus, the derivative of f at 0 can be defined unambiguously to be the linear map $D\hat{f}(\mathbf{0})$ where $\hat{f}$ is any differentiable extension of f in a neighborhood of $\mathbf{0}$.

Example 1.6.3 Let $n \geq 2$. Consider the unit sphere $\mathbb{S}^{n-1}$ in $\mathbb{R}^n$ given by

$$\mathbb{S}^{n-1} = \{(x_1, \ldots, x_n) \ : \ \sum_i x_i^2 = 1\}.$$

Let $N = (0, \ldots, 0, 1)$ denote the "north pole" and $U = \mathbb{S}^{n-1} \setminus \{N\}$.

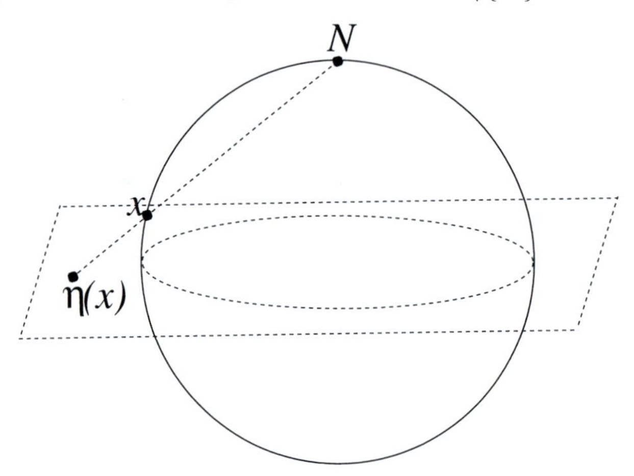

Figure 4 Stereographic projection.

Given any point $\mathbf{x} \in U$, there is a unique line $L_{\mathbf{x}}$ passing through $\mathbf{x}$ and N and this line is not parallel to the hyperplane $\mathbb{R}^{n-1} \times \{0\}$. Therefore, $L_{\mathbf{x}}$ intersects $\mathbb{R}^{n-1} \times \{0\}$ in a unique point. We shall denote it by $\eta(\mathbf{x})$. The function $\eta : U \to \mathbb{R}^{n-1} = \mathbb{R}^{n-1} \times 0$ is called a *stereographic projection.* Let us compute this explicitly. The line $L_{\mathbf{x}}$ can be parameterized as $t\mathbf{x} + (1-t)N, t \in \mathbb{R}$. We want to find the value of t for which this point belongs to $\mathbb{R}^{n-1} \times 0$. Therefore, we put the last coordinate equal to zero to obtain the equation $tx_n + 1 - t = 0$, i.e., $t = \frac{1}{1-x_n}$. (Note that $\mathbf{x} \in U$ implies $x_n \neq 1$ and hence this makes sense.) Thus

$$\eta(x) = \left(\frac{x_1}{1 - x_n}, \ldots, \frac{x_{n-1}}{1 - x_n} \right). \tag{1.73}$$

In order to compute the inverse map, we can reverse this geometric argument. Given any point $\mathbf{y} \in \mathbb{R}^{n-1} \times 0$, the line joining $\mathbf{y}$ and N has to meet the sphere in exactly two points, one of the points being N itself. The other point is clearly $\eta^{-1}(\mathbf{y})$. Following the same procedure, we first get the parameterization $t\mathbf{y} + (1-t)N, t \in \mathbb{R}$ of the line and then require that a point of the line to be on the sphere, which yields $t^2 \sum_i y_i^2 + (1-t)^2 = 1$. This is the same as $t[t(\sum_i y_i^2 + 1) - 2] = 0$. The solution $t = 0$ gives the point N. The solution $t = \frac{2}{1+\|\mathbf{y}\|^2}$ gives the point $\eta^{-1}(\mathbf{y})$. Therefore,

$$\eta^{-1}(\mathbf{y}) = \left(\frac{2y_1}{1 + \|\mathbf{y}\|^2}, \ldots, \frac{2y_{n-1}}{1 + \|\mathbf{y}\|^2}, \frac{\|\mathbf{y}\|^2 - 1}{1 + \|\mathbf{y}\|^2} \right).$$

It follows that both η and η^{-1} are smooth and hence η is a diffeomorphism. More generally, we can take any point $P \in \mathbb{S}^{n-1}$ and consider the "stereographic projection" from that point onto the hyperplane perpendicular to the vector p.

Remark 1.6.3

1. Consider the subset $X = [0,1] \times \{0\} \cup \{0\} \times [0,1]$. It is easy to see that this is homeomorphic to the closed interval $[-1,1]$. But there is no diffeomorphism between them! For suppose that $f : [-1,1] \to X$ is a homeomorphism with $f(p) = (0,0)$ say. Then the two half-intervals around p have to be mapped respectively onto horizontal and vertical segments of X. Suppose f is also smooth. It will then follow that the left-hand and right-hand derivatives of f at 0 are multiples of $\mathbf{e}_1, \mathbf{e}_2$, respectively; by the smoothness condition, they should be equal and hence both are zero. This means $Df(p) = 0$. Therefore, f cannot have a smooth inverse.

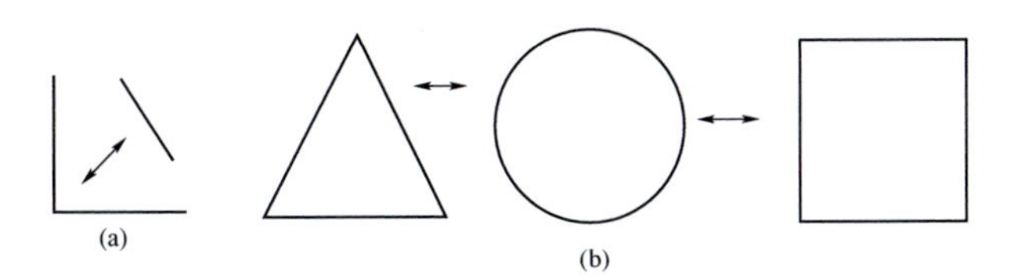

Figure 5 Nondiffeomorphic but homeomorphic subspaces.

2. The above argument can be used to prove that the boundary of a triangle in $\mathbb{R}^2$ cannot be diffeomorphic to a circle or the boundary of a square. Of course, it also can be used to show that a (full) triangle is not diffeomorphic to a disc or a (full) square, whereas they are all homeomorphic to each other. Thus, in differential topology, we have to be careful to distinguish between $I \times I$ and the unit disc D^2. And this is **not** just a minor irritation that one can ignore.

Exercise 1.6

1. Determine the maximal intervals on which these functions are diffeomorphisms:
(a) sin (b) cos (c) tan (d) $\sqrt{1-t^2}$.

2. Show that for $r > 0$ and $s = \sin r$, the map $(\mathbf{v}, t) \mapsto (\mathbf{v}, \sin t)$ is a diffeomorphism $\mathbb{D}^n_r \to \mathbb{D}^n_s$. [Hint: Use Theorem 1.2.6.]

3. Consider the quarter space
$$Q = \{(x,y) \in \mathbb{R}^2, x \geq 0,\ y \geq 0\}$$
and a real valued differentiable function f on it. Show that the derivative of f at $\mathbf{0}$ can be defined unambiguously, as a linear map $\mathbb{R}^2 \to \mathbb{R}$.

4. Let $X \subset \mathbb{R}^n$, and $\mathbf{0} \in X$. Suppose $\mathbf{v}_1, \mathbf{v}_2, \ldots, \mathbf{v}_n$ are independent vectors such that the line segments $[\mathbf{0}, \mathbf{v}_i] \subset X$, $\forall\ i$. Let $f : X \to \mathbb{R}$ be differentiable at 0. Show that $Df_{\mathbf{0}}$ is well defined.

5. Let X be a subset of $\mathbb{R}^2$, $\mathbf{0} \in X$. Let us denote the set of natural numbers by $\mathbb{N} = \{1, 2, \ldots\}$. Assume that $(\frac{1}{n}, \frac{k}{n}), (\frac{1}{n}, \frac{l}{n}) \in X$ for all $n \in \mathbb{N}$ where $k \neq \pm l$ are some non zero real numbers. Suppose $f : X \to \mathbb{R}$ is a continuously differentiable function and $f(\frac{1}{n}, \frac{k}{n}) = 0 = f(\frac{1}{n}, \frac{1}{n})$, $\forall\ n \in \mathbb{N}$. Show that $D(f)_{\mathbf{0}}$ is well defined and compute it.

6. Recall that for a subset $A \subset \mathbb{R}^n$, a point $a \in \mathbb{R}^n$ is called a *cluster point* of A (or a limit point of A) if there is a sequence $\{a_n\}$ of distinct points in $A \setminus \{a\}$, which converges to a. Assume that there are smooth curves $C_1, \ldots, C_n$ passing through a such that

the tangent vectors $\mathbf{v}_1, \dots, \mathbf{v}_n$ to these curves at a are independent. Further, suppose that a is a limit point of each $C_i \cap A$. Show that for any $f : A \to \mathbb{R}$, a differentiable map, Df_a is well defined.

7. Show that any two (nondegenerate) triangles in $\mathbb{R}^2$ are diffeomorphic.

8. A finite subset $\{p_0, \dots, p_n\} \subset \mathbb{R}^m$ is said to be affinely independent if the vectors $\{p_1 - p_0, \dots, p_n - p_0\}$ are independent. (Any singleton set is affinely independent.) By an n-simplex in $\mathbb{R}^m$, we mean the convex hull of any affinely independent subset with $n+1$ points. Show that any two n-simplexes in $\mathbb{R}^m$ are diffeomorphic.

9. Show that any two regular n-gons in $\mathbb{R}^2$ are diffeomorphic to each other.

10. Classify the following list of subsets of $\mathbb{R}^2$ into diffeomorphism types.

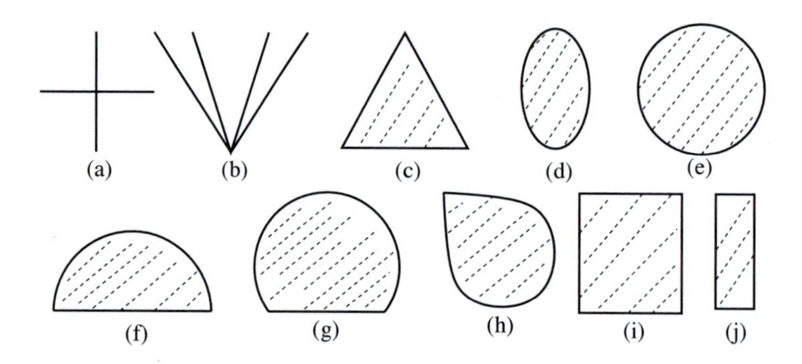

Figure 6 Classify into diffeomorphism types.

1.7 Richness of Smooth Maps

In this section, our aim is to establish the fact that smooth functions are "plenty" in a very loose sense.[4] (For a detailed treatment of this topic see [Hi].) Recall one of the most important $\mathcal{C}^\infty$ functions on $\mathbb{R}$:

$$x \mapsto e^{-1/x^2} \tag{1.74}$$

which you may have studied as an example of a nonzero smooth function with all its derivatives at 0 vanishing. (Verify this fact right now, if you have not seen this before.) The Taylor's expansion of this function at 0 is identically 0 and yet the function is nonvanishing except at 0. This and similar other functions become very useful for us and in fact contribute to the richness of the set of smooth functions.

[4]The space $\mathcal{C}([0,1];\mathbb{R})$ of all continuous real valued functions on the closed interval $[0,1]$ can be given a measure called Wiener measure with respect to which the subset of all continuous functions that are differentiable even at a single given point is of measure zero. Nevertheless, Theorem 1.7.2 below implies that under the "sup-norm" topology, the subset of smooth functions is dense in the space of all continuous functions.

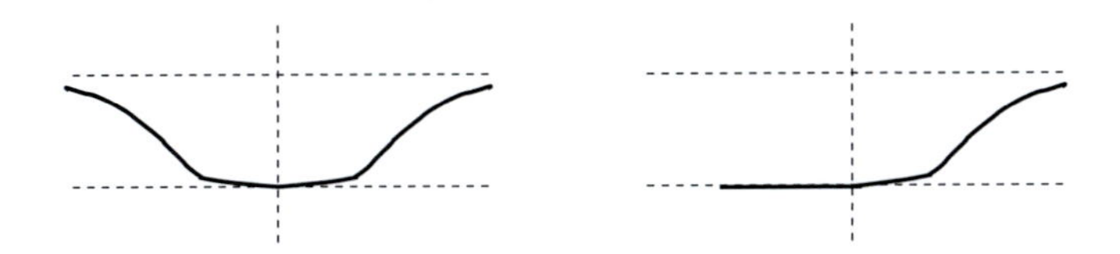

Figure 7 Patching up smooth maps.

Let us begin with any smooth function $f : \mathbb{R} \to \mathbb{R}$ taking only positive values except at 0 and having all its derivatives 0 at 0, for instance, as in (1.74). We consider the function g defined by

$$g(x) = \begin{cases} f(x), & x \geq 0, \\ 0, & x \leq 0. \end{cases} \tag{1.75}$$

Then $g^{(n)}(0) = 0$ for all $n \geq 0$ and $g(x) \geq 0$ for all x. Moreover, $g(x) = 0$ if and only if $x \leq 0$.

What we have done just now is a typical case of "patching up" of two smooth functions. Recall that if you have two continuous functions f_1 and f_2 that agree at a point $x = x_0$, then we could define another function g, which is f_1 for $x \leq x_0$ and f_2 for $x \geq x_0$. Of course, without any further assumptions, the function g is also continuous. However, if f_i were smooth we cannot say immediately that g is also smooth at x_0. This would require that all the derivatives (or at least as many as we are interested in) of f_1 and f_2 agree at x_0. The point is that "patching up" two smooth functions is possible and this is the secret of the richness of smooth maps. It may be recalled that holomorphic functions are rather too rigid: if two of them agree on a connected set with a limit point then they agree everywhere. So, one cannot "patch up" two distinct holomorphic functions. Recall that for any real or complex vector-valued function f defined on a topological space X, the *support* of f is defined by

$$\operatorname{supp} f = \operatorname{cl}(\{x \in X \ : \ f(x) \neq 0\}) \tag{1.76}$$

where cl denotes the closure of a set. The zero set $Z(f)$ of f is defined to be

$$Z(f) = \{x \in X \ : \ f(x) = 0\}.$$

Clearly $\operatorname{supp} f = \operatorname{cl}(X \setminus Z(f))$.

In this terminology, we have $\operatorname{supp} g = [0, \infty)$.

Next, we modify g to have compact support. Let now $0 < a < b$ be real numbers and define

$$h(x) = g(x - a)g(b - x). \tag{1.77}$$

Then check that the support of h is precisely $[a, b]$. Sketch the graph of h and observe that it has the shape of a camel's hump.

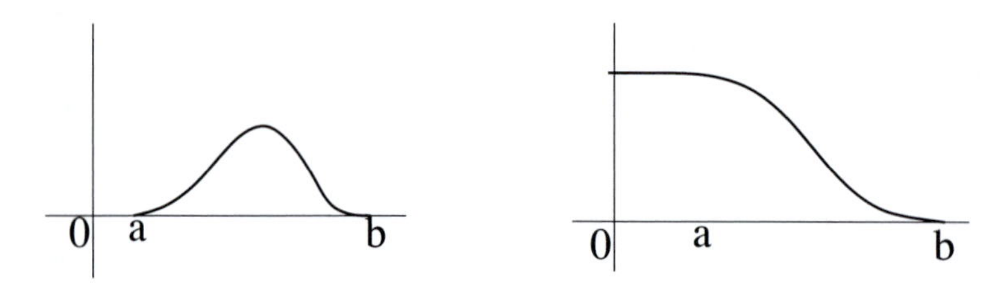

Figure 8 A camel's hump and a smooth step.

We now define

$$\gamma(x) = \frac{\int_x^\infty h(t)\,dt}{\int_{-\infty}^\infty h(t)\,dt}. \tag{1.78}$$

The function γ has the following properties:
(i) $\gamma(x) = 1$ for all $x \le a$;
(ii) monotonically decreasing in $[a, b]$;
(iii) and $\gamma(x) = 0$ for all $x \ge b$.

We call such a function a *smooth unit step-down function.* By scaling, shifting, reflecting, or combining several of these operations on this function, we can get smooth functions with other geometric properties. For instance, let us illustrate how to use this to get smooth "bump" functions on $\mathbb{R}^n$. Consider

$$\alpha(\mathbf{x}) = \gamma(\|\mathbf{x}\|) \tag{1.79}$$

The norm function is not smooth at $\mathbf{x} = 0$. However, the composite of this with γ is a smooth function all over. [This is another typical tool used in building up smooth functions out of possible nonsmooth ones. Pay attention to this.] Then $\alpha : \mathbb{R}^n \to [0, 1]$ is a smooth function which vanishes for $\|\mathbf{x}\| \ge b$ and equals 1 for $\|\mathbf{x}\| \le a$. For $n = 2$, its graph looks like a nice hill with a football ground at the top!

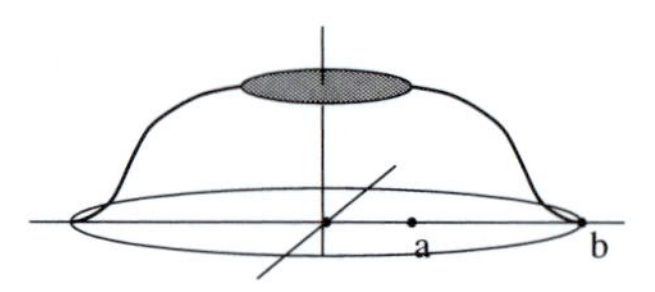

Figure 9 Hilltop with a football ground.

Remark 1.7.1 It is worth noting that the smooth functions that we construct "depend" smoothly on the initial data that we choose according to the properties that we want the functions to have. This is so because of the "algorithmic" nature of the constructions that we have discussed so far. This fact will be used in a nontrivial way in a later chapter. At this stage, you are welcome to try the first few exercise at the end of the section, which emphasize this phenomenon.

Remark 1.7.2 An important technique that allows passage from "local" to "global" in differential topology, is the so-called "partition of unity". Though the concept itself can be studied in a more general setup (of paracompact spaces), we shall restrict ourselves to the case of subspaces of Euclidean spaces.

Theorem 1.7.1 Partition of Unity: *Let X be any subspace of $\mathbb{R}^n$ and $\{U_\alpha\}_{\alpha\in\Lambda}$ be an open covering of X. Then there exists a countable family $\{\theta_j\}$ of smooth functions with compact support on $\mathbb{R}^n$ such that*
(i) $0 \le \theta_j(x) \le 1$, *for all j and $x \in X$;*
(ii) *for each $x \in X$ there exists a neighborhood N_x of x in X, such that only finite number of θ_j are nonzero on N_x;*
(iii) *for each j, $(\operatorname{supp}\theta_j)\cap X \subset U_{\alpha_j}$ for some α_j; and*
(iv) $\sum_j \theta_j(x) = 1$, *for all $x \in X$.*

Proof: Recall that U_α are open in X means that there exist V_α that are open in $\mathbb{R}^n$ and $U_\alpha = X \cap V_\alpha$. Put $W = \cup_\alpha V_\alpha$. Then $X \subset W$ and $\{V_\alpha\}$ is an open cover of the open set W. If we prove the theorem, with X replaced by W and $\{U_\alpha\}$ replaced by $\{V_\alpha\}$, we can then take the restriction of η_j to X to complete the proof of the original statement. So, instead of changing X to W, etc., we may and will assume that X itself is an open set in $\mathbb{R}^n$.

We now use the fact that every open set in $\mathbb{R}^n$ is the increasing union of a countable family of open sets whose closure is compact. There are different ways to see this; here is one such way: For positive integer i, take

$$K_i = \{\mathbf{x} \in X \;:\; \|\mathbf{x}\| < i \;\&\; d(\mathbf{x}, \mathbb{R}^n \setminus X) > 1/i\}.$$

Check that

$$X = \cup_i K_i; \quad K_i \subset \overline{K}_i \subset K_{i+1} \subset \overline{K}_{i+1}.$$

Here $\overline{K}$ denotes the closure of a subset K in $\mathbb{R}^n$. Also, each $\overline{K}_i$ being closed and bounded, is compact. We set $K_{-1} = K_0 = \emptyset$ for the sake of inductive steps that follow.

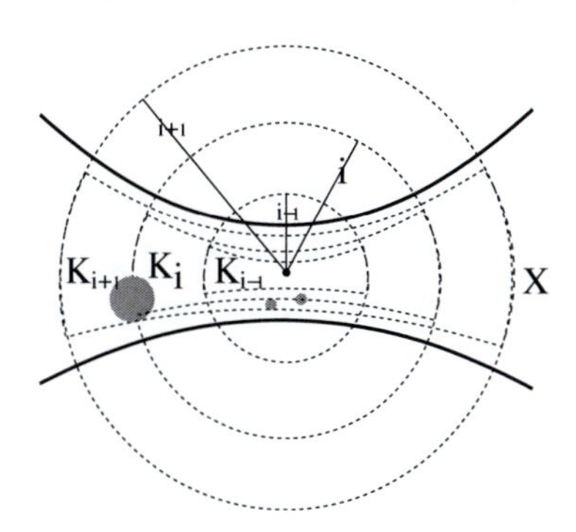

Figure 10 Nested union of compact sets.

For any ball B in $\mathbb{R}^n$, let us agree to denote by $\frac{1}{2}B$, the ball with half the radius and the same center. Let now $\mathcal{B}$ be the collection of all balls in $\mathbb{R}^n$ each of which is contained in some member of $\{U_\alpha\}$. Among these balls, for each $i \ge 1$, there will be finitely many B_k not intersecting $\bar{K}_{i-2}$ and such that their halves $\frac{1}{2}B_k$ will cover the compact set $\overline{K}_i \setminus K_{i-1}$. Together, let us index all these balls by $\{B_j\}$. (Strictly speaking we should use double indices like B_{ij} but we avoid this for simplicity.) Clearly, the family $\{B_j\}$ is an open cover for X.

Now, a crucial fact is that the family $\{B_j\}$ is *locally finite* on X, i.e., for each $x \in X$, we can find a neighborhood N_x of x such that only finitely many of $\{B_j\}$ intersect N_x.

To see this, suppose $x \in K_i$. Then take $N_x = K_i$. Clearly, N_x does not meet any of the balls chosen after the stage $i+2$.

For each j, choose some smooth bump function $\eta_j : \mathbb{R}^n \to [0,1]$ with its support equal to B_j and which equals 1 on $\frac{1}{2}B_j$. One can verify straight away that (i)-(iii) are all satisfied by η_j in place of θ_j. In particular, $\eta := \sum_j \eta_j$ makes sense and gives a smooth function on $\eta : X \to [0, \infty)$. Moreover, for each $x \in X$, since x belongs to at least one of the half balls $\frac{1}{2}B_j$, it follows that $\eta_j(x) = 1$ and hence $\eta(x) \geq 1$. Hence, for each j,

$$\theta_j := \frac{\eta_j}{\eta} \tag{1.80}$$

makes sense on X and defines a smooth map $\theta_j : X \to [0,1]$. Since each θ_j vanishes outside the ball B_j, we can extend it by zero on all of $\mathbb{R}^n \setminus \bar{B}_j$ to obtain smooth maps $\theta_j : \mathbb{R}^n \to [0,1]$. Verify that all the properties required by $\{\theta_j\}$ are satisfied. ♠

Remark 1.7.3
(i) The partition of unity is going to be extremely useful for us. Here, we shall give two immediate corollaries. The first one already tells you that there are enough smooth functions to separate disjoint closed subsets in $\mathbb{R}^n$.
(ii) The family $\{\theta_j\}$ may fail to be locally finite at all points of $\mathbb{R}^n$, viz., at the boundary points of the set X. Because of this, the function η may fail to make sense at these points. Our next corollary is an illustration of how the partition of unity can be combined with other tools in analysis such as taking convergent series to define smooth maps with other properties.

Corollary 1.7.1 Smooth Urysohn's Lemma: Let F_0, F_1 be any two disjoint closed subsets of $\mathbb{R}^n$. Then there exists a smooth function $f : \mathbb{R}^n \to [0,1]$ such that $f(x) = 0$ for all $x \in F_0$ and $f(x) = 1$ for all $x \in F_1$. If F_1 is compact then we can choose f to have compact support also.

Proof: Consider the open covering $\{\mathbb{R}^n \setminus F_0, \mathbb{R}^n \setminus F_1\}$ of $\mathbb{R}^n$ and apply the above theorem to get a partition of unity $\{\theta_j\}$ subordinate to this cover. (This just means that the family satisfies the conditions (i)–(iv) of the above theorem with X replaced by $\mathbb{R}^n$.) Define

$$A = \{j \ : \ \text{supp}\ \theta_j \cap F_1 \neq \emptyset\}; \ \ f(x) := \sum_{j \in A} \theta_j.$$

Then by condition (iii), f makes sense and is smooth. For $j \in A$, we have supp $\theta_j \subset X \setminus F_0$, which means that $\theta_j(F_0) = \{0\}$. Therefore, $f(F_0) = \{0\}$. On the other hand, given any point $x \in F_1$ since $\sum_j \theta_j(x) = 1$ and if $\theta_j(x) \neq 0$ then $j \in A$, it follows that $f(x) = 1$. If F_1 is compact, then there will be a finite open cover $\{U_1, \ldots, U_k\}$ of F_1 such that each U_i intersects only finitely many members of $\{\text{supp}\ \theta_j\}_{j=1}^{\infty}$. Therefore, it follows that the set A is finite. In particular, being a finite sum of functions with compact support f itself has compact support. ♠

Corollary 1.7.2 A subset C of $\mathbb{R}^n$ is closed if and only if it is the precise zero set of a smooth function $f : \mathbb{R}^n \to \mathbb{R}$.

Proof: Since the zero set of any continuous real valued function is closed, we have only to prove the "only if" part here.

Given an n-tuple of nonnegative integers $\alpha = (\alpha_1, \ldots, \alpha_n)$, let $|\alpha| = \sum_i \alpha_i$ and let D^α denote the partial derivative

$$D^\alpha = \frac{\partial^{|\alpha|}}{\partial x_1^{\alpha_1} \cdots \partial x_n^{\alpha_n}}.$$

Let $X = \mathbb{R}^n \setminus C$, with the open cover $\{X\}$, in the above theorem and let $\{\theta_j\}$ be the partition of unity obtained therein. Let $c_r = sup\{|D^\alpha(\theta_r)| \, : \, |\alpha| \leq r\}$. Since each θ_r is a smooth function with compact support, c_r is a finite quantity. Choose $\epsilon_r > 0$ such that $\sum_r \epsilon_r c_r < \infty$. Now consider $f = \sum_{r=1}^{\infty} \epsilon_r \theta_r$. This sum is not necessarily locally finite. However, it is convergent. Indeed, observe that for any fixed α, the series $\sum_{r \geq |\alpha|} \epsilon_r c_r$ is a majorant series for $\sum_{r \geq |\alpha|} D^\alpha(\epsilon_r \theta_r)$. By the majorant criterion, the series $\sum_r D^\alpha(\epsilon_r \theta_r)$ is uniformly convergent for each multi-index α. It follows that f is a smooth function on $\mathbb{R}^n$. If $x \in X$, then since some $\theta_r(x) \neq 0$, it follows that $f(x) \neq 0$. On the other hand, if $x \in C$ then $\theta_r(x) = 0$ for all r and hence the sum, $f(x) = 0$. ♠

We shall end this section with the following approximation theorem.

Theorem 1.7.2 Approximation Theorem: *Let C be a closed subset of an open set U in $\mathbb{R}^n$ and $f : U \to \mathbb{R}^m$ be a continuous map that is smooth restricted to C. Given a continuous function $\epsilon : U \to (0, \infty)$, there exists a smooth function $g : U \to \mathbb{R}^m$ such that*

$$g(y) = f(y), \quad \forall\ y \in C \text{ and } \|f(y) - g(y)\| < \epsilon(y), \ \forall\ y \in U. \tag{1.81}$$

Proof: By the smooth version of the Tietze Extension Theorem (see Exercise 10 below), there exists a smooth function $h : U \to \mathbb{R}^m$ such that $h = f$ on C. Since ϵ is a continuous positive function, for each $x \in U$, we can choose a neighborhood U_x of x in U so that $\epsilon'(x) = \inf\{\epsilon(y) \, : \, y \in U_x\} > 0$. By the continuity of f and h, for each $x \in C$, there exists a neighborhood $V_x \subset U_x$ of x such that

$$\|f(y) - h(y)\| < \epsilon'(x) \leq \epsilon(y), \quad y \in V_x.$$

Also by continuity of f for each $x \in U \setminus C$, there exists neighborhood $V_x \subset U_x$ of x in $U \setminus C$ such that

$$\|f(x) - f(y)\| < \epsilon'(x) < \epsilon(y), \quad y \in V_x.$$

Let us put $g_x(y) = h(y), y \in U$ for all $x \in C$ and $g_x(y) = f(x), y \in U$ (the constant map) for all $x \in U \setminus C$.

Let $\{\lambda_j\}$ be a smooth partition of unity on U subordinate to the open cover $\{V_x\}_{x \in U}$. Suppose that $\operatorname{supp} \lambda_j \subset V_{x_j}$. Define $g = \sum_j \lambda_j g_{x_j}$. Then clearly g is a smooth function on U and if $y \in C$ then

$$g(y) = \sum_j \lambda_j(y) g_{x_j}(y) = \sum_j \lambda_j(y) h(y) = f(y).$$

Finally for $y \in U$ we have,

$$\begin{aligned} \|f(y) - g(y)\| &= \|\sum_j \lambda_j(y)(f(y) - g_{x_j}(y))\| \\ &\leq \sum_j \lambda_j(y) \|f(y) - g_{x_j}(y)\| \\ &\leq \sum_j \lambda_j(y) \epsilon(y) = \epsilon(y), \end{aligned}$$

which completes the proof. ♠

Remark 1.7.4 You may have learned in your real analysis course that any continuous map $f : K \to \mathbb{R}$ on a compact subset K of $\mathbb{R}^n$ can be approximated by a polynomial function (Weierstrass' approximation theorem). The above result is clearly a big generalization of this result in a sense, though you cannot deduce Weierstrass' approximation theorem from it. Also, when K is noncompact, you cannot hope to get polynomials to approximate a given continuous function.

Exercise 1.7

1. Fix $a < b \in \mathbb{R}$. Show that there exists a smooth function $\phi : \mathbb{R} \times \mathbb{R} \times \mathbb{R} \to \mathbb{R}$ such that for each fixed $(\alpha, \beta) \in \mathbb{R} \times \mathbb{R}$, the map $\phi_{\alpha,\beta} : \mathbb{R} \to \mathbb{R}$ given by $\phi_{\alpha,\beta}(t) = \phi(\alpha, \beta, t)$ has the properties:
(i) $\phi_{\alpha,\beta}(t) = \alpha,\ t \le a,\ \phi_{\alpha,\beta}(t) = \beta,\ t \ge b$;
(ii) $\phi_{\alpha,\beta}$ is monotonic.

2. Fix $M > \epsilon > 0$.
(a) Show that there is smooth map $g : [0, M] \times [0, 1] \to [0, M]$ such that for each $\alpha \in [0, M]$, the map defined by $g_\alpha(t) = g(\alpha, t)$ has the properties:
(i) $g_\alpha(t) = \alpha$ near 0 and $g_\alpha(t)$ is some positive constant near 1;
(ii) $\int_0^1 g_\alpha(t)dt < \epsilon$.
(b) Deduce that there is a smooth map $h : [0, M] \times [0, M] \times [0, 1] \to [0, M]$ such that
(i) $h(\alpha, \beta, t) = \alpha$ near 0 and $h(\alpha, \beta, t) = \beta$ near 1 for each fixed (α, β);
(ii) $\int_0^1 h(\alpha, \beta, t)dt < \epsilon$.

3. Fix $M > 0$. Show that there is a smooth map $f : (0, M] \times (0, M] \times [0, 1] \to [0, 1]$ such that for each $(\alpha, \beta) \in (0, M] \times (0, M]$, the map $f_{\alpha,\beta} : [0, 1] \to [0, 1]$ has the properties:
(i) $f_{\alpha,\beta}(0) = 0,\ f_{\alpha,\beta}(1) = 1$;
(ii) $f'_{\alpha,\beta}(t) = \alpha$ near 0, $f'_{\alpha,\beta}(t) = \beta$ near 1;
(iii) f is strictly increasing.

4. Construct a $\mathcal{C}^\infty$ function $f : [-1, 1] \to \mathbb{R}$ such that the zero set of f is equal to the closure of $\left(\left\{\frac{1}{n} : n \neq 0 \text{ is an integer}\right\}\right)$.

5. Give an example of a smooth function $f : \mathbb{R} \to \mathbb{R}$ such that $\operatorname{supp} f = [0, 1]$ and $|f'(x)| \le 1$ for all x.

6. Let $-\infty \le \alpha < \beta \le \infty$ and $-\infty \le \gamma < \delta \le \infty$ and let $\alpha = a_0 < a_1 < \cdots < a_{2k+1} = \beta$ and $\gamma = c_0 < c_1 < \cdots < c_{2k+1} = \delta$. Given order preserving diffeomorphisms $\phi_i : [a_{2i-1}, a_{2i}] \to [c_{2i-1}, c_{2i}], i = 1, \ldots, k$, construct a diffeomorphism $\psi : [\alpha, \beta] \to [\gamma, \delta]$ such that $\psi|_{[a_{2i-1}, a_{2i}]} = \phi_i, i = 1, 2, \ldots, k$. If any one of the ϕ_i is given only on the open interval, does the conclusion hold?

7. Given $a < c < b$, show that there exists a smooth function $f : [a, b] \to \mathbb{R}$ such that f vanishes near a and b and is a constant near c and such that $\displaystyle\int_a^b f(x)dx = 1$.

8. Given $0 < a < b \in \mathbb{R}$, find a smooth function $f : \mathbb{R} \to \mathbb{R}$ such that
(i) near 0, $f \equiv a$;
(ii) near b, $f \equiv 0$; and
(iii) $-1 < f'(x) \le 0$, for $0 \le x \le b$.

9. Prove or read from a point set topology book the following simpler version of Wallman's theorem: Let $K \subset \mathbb{R}^n \times \{0\}$ be a compact set, and U be an open subset of $\mathbb{R}^{n+1}$ containing K. Then there exists $\epsilon > 0$ and an open subset V of $\mathbb{R}^n$ such that $K \subset V \times (-\epsilon, \epsilon) \subset U$.

10. **Smooth Tietze Extension Theorem:** Let C be any closed subset of $\mathbb{R}^n$ and $f : C \to \mathbb{R}$ be any smooth map. Then there exists a smooth function $\hat{f} : \mathbb{R}^n \to \mathbb{R}$ such that $\hat{f}|_C = f$.

11. Let $f : U \to \mathbb{R}$ be a smooth function that vanishes in a neighborhood of $p \in U$, where U is an open subset of $\mathbb{R}^n$. Show that there exists a smooth function $\lambda : U \to \mathbb{R}$, that vanishes in a neighborhood of p and satisfies $f = \lambda f$ on U.

12. By a *smooth homotopy,* we mean a smooth map $H : X \times \mathbb{R} \to Y$. Put $h_t(x) = h(x, t)$. Two smooth maps $f, g : X \to Y$ are said to be smoothly homotopic if there exists a smooth homotopy H such that $h_t = f, H_s = g$ for some $s, t \in \mathbb{R}$. Show that being smoothly homotopic is an equivalence relation on the set of smooth maps from X to Y.

1.8 Miscellaneous Exercises for Chapter 1

1. Given a $\mathcal{C}^\infty$ function $f : \mathbb{R} \to \mathbb{R}$ such that $f(1/n) = 0$ for all positive integers n, can you determine the values of $f(0), f'(0), f''(0), \cdots$?

2. **An example of Weierstrass' function (due to McCarthy):**[5] A continuous function $f : \mathbb{R} \to \mathbb{R}$ which is nowhere differentiable is called a *Weierstrass' function* because he was the first one to discover such functions. (This should not be confused with the Weierstrass' pee-function $\wp$.) Here is an easy example: Define

$$g(x) = \begin{cases} 1 + x, & \text{if} \quad -2 \le x \le 0; \\ 1 - x, & \text{if} \quad 0 \le x \le 2; \\ g(x - 4n), & \text{if} \quad -2 \le x - 4n \le 2, \text{ for some integer } n \neq 0. \end{cases}$$

Observe that g is a periodic function with period 4. Put $g_k(x) = g(2^{2^k} x)$ and $f(x) = \sum_0^\infty g_k(x)/2^k$. Show that f is a Weierstrass' function. [Hint: to show that f is not differentiable at x, consider the sequence $\{x + 2^{-2^k}\}$ or $\{x - 2^{-2^k}\}$ depending upon whether mod 2 you have $0 \le x \le 1$ or $1 \le x \le 2$.]

3. Examine the following functions for continuity at $(0, 0)$. The expressions below give the value of the function at $(x, y) \neq (0, 0)$. At $(0, 0)$ you are free to take any value you like.
(i) $\dfrac{x^3 y}{x^2 - y^2}$; (ii) $\dfrac{x^2 y}{x^2 + y^2}$; (iii) $xy\dfrac{x^2 - y^2}{x^2 + y^2}$;
(iv) $|[|x| - |y|]| - |x| - |y|$; (v) $\dfrac{\sin^2(x + y)}{|x| + |y|}$.

4. Suppose $f, g : \mathbb{R} \to \mathbb{R}$ are continuous functions. Show that each of the following functions on $\mathbb{R}^2$ are continuous.
(i) $(x, y) \mapsto f(x) + g(y)$; (ii) $(x, y) \mapsto f(x)g(y)$;
(iii) $(x, y) \mapsto max\{f(x), g(y)\}$; (iv) $(x, y) \mapsto min\{f(x), g(y)\}$.
If f, g are differentiable, are all the above functions differentiable?

5. Deduce from the above exercise that every polynomial function in two variables is continuous (smooth). Can you generalize this?

[5] *Amer. Math. Monthly* Vol. LX No. 10 Dec. 1953.

6. Examine each of the following functions for continuity.

(i) $f(x,y) = \begin{cases} \frac{y}{|y|}\sqrt{x^2+y^2}, & y \neq 0, \\ 0 & y = 0. \end{cases}$

(ii) $g(x,y) = \begin{cases} x\sin\frac{1}{x} + y\sin\frac{1}{y}, & x \neq 0, y \neq 0; \\ x\sin\frac{1}{x}, & x \neq 0, y = 0; \\ y\sin\frac{1}{y}, & x = 0, y \neq 0; \\ 0, & x = 0, y = 0. \end{cases}$

7. Let $f : B_r(0) \to \mathbb{R}$ be some function where $B_r(0)$ is the open ball of radius r and center 0 in $\mathbb{R}^2$. Assume that the two limits

$$\lim_{x\to 0} f(x,y); \qquad \lim_{y\to 0} f(x,y) \tag{1.82}$$

exist for all sufficiently small y and for all sufficiently small x, respectively. Assume further that the limit $\lim_{(x,y)\to(0,0)} f(x,y) = l$ also exists. Then show that the iterated limits

$$\lim_{y\to 0}[\lim_{x\to 0} f(x,y)], \qquad \lim_{x\to 0}[\lim_{y\to 0} f(x,y)] \tag{1.83}$$

both exist and are equal to l.

8. Put $f(x,y) = \frac{x-y}{x+y}$, for $(x,y) \neq (0,0)$. Show that the two iterated limits (1.83) exist but are not equal. Conclude that the limit $\lim_{(x,y)\to(0,0)} f(x,y)$ does not exist.

9. Put $f(x,y) = \frac{x^2y^2}{x^2y^2+(x-y)^2}$, $(x,y) \neq (0,0)$. Show that the iterated limits (1.83) both exist. Compute them. Show that the $\lim_{(x,y)\to(0,0)} f(x,y)$ does not exist.

10. Express the definition of $\lim_{(x,y)\to(0,0)} f(x,y)$ in terms of polar coordinates and analyze it for the following functions:

(i) $f(x,y) = \frac{x^3 - xy^2}{x^2+y^2}$; (ii) $g(x,y) = \tan^{-1}\left(\frac{|x|+|y|}{x^2+y^2}\right)$;

(iii) $h(x,y) = \frac{y^2}{x^2+y^2}$.

11. Consider the function

$$f(x,y) = \begin{cases} \frac{x^3}{x^2+y^2}, & (x,y) \neq (0,0) \\ 0, & (x,y) = (0,0). \end{cases}$$

(a) Show that f is continuous and all the directional derivatives f exist and are bounded all over $\mathbb{R}^2$.
(b) For any C^1 curve $g : \mathbb{R} \to \mathbb{R}^2$ (i.e., continuously differentiable and $g' \neq 0$) show that $f \circ g$ is a C^1- mapping. [Hint: Use the Taylor expansion for g at points $t \in \mathbb{R}$ such that $g(t) = (0,0)$.]
(c) Yet, f is not differentiable at $(0,0)$. [Hint: Use polar coordinates.]

12. Compute $D(\tau)$ at the point $-Id$, where $\tau(A) = AA^t$ as in (1.43).

13. For a fixed $A \in GL(n; \mathbb{R})$, and consider the map $\kappa(B) = ABA^{-1}$. Compute $D(\kappa)$.

14. Derive the Liebnitz rule for differentiation of the product fg of two differentiable, matrix-valued functions $f : X \to M(p \times q; \mathbb{R})$ and $g : X \to M(q \times r; \mathbb{R})$, where X is any open set of $\mathbb{R}^n$.

15. Consider $\eta : GL(n; \mathbb{R}) \to GL(n; \mathbb{R})$ given by $\eta(A) = A^{-1}$. Show that η is a $\mathcal{C}^\infty$ map and compute $D(\eta)$ at Id.

16. Compute the derivative of $A \mapsto A^2$ where $A \in M(n; \mathbb{R})$.

17. Recall that a map $f : \mathbb{C} \to \mathbb{C}$ is holomorphic if $f = u + \imath v$, where $u, v : \mathbb{R}^2 \to \mathbb{R}$ are smooth functions satisfying Cauchy-Riemann equations. Let f be holomorphic. Suppose $g : \mathbb{C} \to \mathbb{R}$ is a smooth map. Show that

$$\nabla^2(g \circ f)(z) = |f'(z)|^2 \nabla^2(g)(f(z))$$

Deduce that if g is harmonic then $g \circ f$ is harmonic.

18. For $a > 0$, consider the function $f : \mathbb{R} \times \mathbb{R} \to \mathbb{R}$ given by $f(r, s) = (a - rs)(r + s)^2$.

   (a) Discuss the extremum values of f.
   (b) Deduce that $(2\cos\theta - rs)(r + s)^2 \leq 2 + 2\cos\theta$, for all $r, s \in [0, 1]$ and $\theta \in \mathbb{R}$.
   (c) Deduce that for complex numbers $z_j, \;\; j = 1, 2$ of modulus < 1,

$$||z_1| - |z_2|| \leq \left| \frac{z_1 - z_2}{1 - z_1 \bar{z}_2} \right| \leq |z_1| + |z_2|.$$

   [Hint: Use Cosine Rule: $|z + w|^2 = |z|^2 + |w|^2 + 2\Re(z\bar{w})$.]

19. Given a homeomorphism $f : \mathbb{S}^{n-1} \to \mathbb{S}^{n-1}$, the cone over f, $C(f) : \mathbb{D}^n \to \mathbb{D}^n$ is defined by $C(f)[\mathbf{v}, r] = [f(\mathbf{v}), r]$, in the notation introduced in Remark 1.2.7.
(a) Show $C(f)$ is a homeomorphism of $\mathbb{D}^n$ to itself.
(b) If f is a diffeomorphism, can you say that $C(f)$ is a diffeomorphism?
(c) Can you give sufficient conditions on f under which this holds?
(d) Can you characterize all diffeomorphisms $f : \mathbb{S}^{n-1} \to \mathbb{S}^{n-1}$ such that $C(f)$ is a diffeomorphism?

Chapter 2

Integral Calculus

As in Chapter 1, we shall assume that the reader is familiar with line integrals, surface integrals, the standard results such as the Green theorem and the Stokes' theorem for surfaces in $\mathbb{R}^3$. In Section 2.1, we shall begin with recalling some important results from the integral calculus of several variables. We introduce the notion of measure zero sets, with the restricted aim of formulating a very important result about the largeness of the set of regular values of a smooth function. This result goes under the name Sard's theorem and is dealt with in Section 2.2. With the aim of generalizing the integration theory, we then develop the necessary algebraic preliminaries of differential forms in Sections 2.3, 2.4, and 2.5. In Section 2.6, integration on singular chains is introduced and a generalization of the Stokes' theorem is proved.

2.1 Multivariable Integration

In this section, we begin with a quick review of the Riemann integration theory for several variables, basic properties of measure-zero sets, and present the "change of variable formula" for integration.

Definition 2.1.1 By a *box* B in $\mathbb{R}^n$ we mean a product of intervals

$$B = \Pi_{i=1}^n [a_i, b_i] \tag{2.1}$$

The volume of the box B is defined to be $\mu(B) := \Pi(b_i - a_i)$.

Definition 2.1.2 Let $A \subset \mathbb{R}^n$. We say A has *measure zero in* $\mathbb{R}^n$ if for every $\epsilon > 0$, there exists a countable cover $\{B_1, \ldots, B_k, \ldots\}$ of A by boxes such that $\sum_i \mu(B_i) < \epsilon$.

Remark 2.1.1
(a) If $B \subset A \subset \mathbb{R}^n$ then A has measure zero in $\mathbb{R}^n$ implies so has B.
(b) Any countable subset of $\mathbb{R}^n$ has measure zero. More generally, a countable union of measure zero sets has measure zero.
(c) In Definition 2.1.2, we can use open boxes instead of closed ones. We can even use closed balls or open balls instead of rectangles. All these give the same notion of measure zero.
(d) If A is a compact subset of measure zero, then for every $\epsilon > 0$, there exists a finite cover of A by closed boxes $\{B_1, \ldots, B_k\}$ such that $\sum_i \mu(B_i) < \epsilon$.
(e) For any finite cover of $[a, b]$ by closed intervals $[a_i, b_i]$, we have $\sum_i (b_i - a_i) \geq b - a$. Thus,

we are happy that the interval $[a, b]$ for $a < b$ is not of measure zero in $\mathbb{R}$.
(f) The following lemma tells you that being "measure zero" is a kind of local property of sets.

Lemma 2.1.1 $A \subset \mathbb{R}^n$ is of measure zero in $\mathbb{R}^n$ iff $A \cap U$ is of measure zero for every open subset in $\mathbb{R}^n$.

Proof: Cover A by a family of open sets in $\mathbb{R}^n$. Then by the Lindelöff property[1] for subsets of $\mathbb{R}^n$, there exists a countable subcover $\{U_i\}$ for A. Now each $A \cap U_i$ is of measure zero and hence for each $\epsilon > 0$, there is a countable cover of $A \cap U_i$ such that the total measure is $< \frac{\epsilon}{2^i}$. The collection of all these open sets forms a cover for A with the property that the total measure is $< \sum_i \frac{\epsilon}{2^i} = \epsilon$. The converse is obvious from Remark 2.1.1(a). ♠

Definition 2.1.3 By a *partition* P of an interval $[a, b]$, we mean a finite set of points $a = t_0 < t_1 < \cdots < t_n = b$. By a partition of a box B as in (2.1), we mean an n-tuple $P = (P_1, \ldots, P_n)$ where P_i is a partition of $[a_i, b_i]$. The boxes $\Pi_{i=1}^n [t_{\alpha(i)}, t_{\beta(i)}]$, where $[t_{\alpha(i)}, t_{\alpha(i)+1}]$ is a sub-interval occurring in P_i are called *subboxes of the partition* P. If each subbox of a partition Q is contained is some subbox of a partition P, then we say Q *is a refinement of* P.

Definition 2.1.4 Now consider a bounded function $f : B \to \mathbb{R}$ and let P be a partition of the box B. For each subbox S of P put

$$m_S(f) = \inf\{f(x) \ : \ x \in S\}; \quad M_S(f) = \sup\{f(x) \ : \ x \in S\}.$$

The *lower sum* and the *upper sum* of f with respect to P are defined by:

$$L(f, P) := \sum_S m_S(f)\mu(S); \quad U(f, P) := \sum_S M_S(f)\mu(S),$$

where the summations are taken over all subboxes belonging to the partition P.

Remark 2.1.2
(a) If Q is a refinement of P then

$$L(f, P) \leq L(f, Q); \quad U(f, Q) \leq U(f, P).$$

(b) Given any two partitions P, P' of B, there exists a partition Q of B, which is a refinement of both P, P'.
(c) For any two partitions P, P' of B, we have $L(f, P) \leq U(f, P')$.
(d) Let now

$$L(f) := sup\{L(f, P) \ : \ P \text{ is a partition of} B\}$$

and

$$U(f) := inf\{U(f, P) \ : \ P \text{ is a partition of} B\}.$$

Then $L(f) \leq U(f)$.

Definition 2.1.5 Let $f : B \to \mathbb{R}$ be a bounded function. We say f is *(Riemann) integrable* on B if $L(f) = U(f)$. In this case, we put

$$\int_B f := L(f) = U(f)$$

and call this the integral of f over B.

---

[1]That is, every open cover has a countable subcover. Every second countable space has the Lindelöff property. In particular, every subspace of a Euclidean space has the Lindelöff property.

Remark 2.1.3
(a) The condition in the above definition is equivalent to say that for every $\epsilon > 0$, there exists a partition P of B such that $U(f,P) - L(f,P) < \epsilon$.
(b) For the constant function $f = c$, $\int_B c = c\mu(B)$.
(c) As a typical example of a bounded function that is not integrable, take $f(x_1, \ldots, x_n) = 0$ if x_1 is rational and $= 1$ if x_1 is irrational.
(d) If f, g are integrable then for any scalars α, β, $\alpha f + \beta g$ is integrable and

$$\int_B \alpha f + \beta g = \alpha \int_B f + \beta \int_B g.$$

(e) If f and g are integrable then so is fg.
(f) Given any continuous function f on a closed box B, it is not hard to see that f is integrable on B. However, we need a result which a bit stronger than this, viz., a bounded function f on a (bounded) box is integrable iff the set of points where f is discontinuous is of measure zero. This is indeed a deep theorem due to Lebesgue. Toward a quick proof of this, let us introduce the following notion:

Definition 2.1.6 Let $X \subset \mathbb{R}^n$, and $f : X \to \mathbb{R}$ be a bounded function. For each $x \in X$ and $r > 0$, introduce the notation:

$$M(x,r) = \sup\{f(y) \ : \ y \in B_r(x) \cap X\}; \quad m(x,r) = \inf\{f(y) \ : \ y \in B_r(x) \cap X\}.$$

Note that $M(x,r)$ is a decreasing function of r and $m(x,r)$ is an increasing function of r. We define the *oscillation of f* at x to be the quantity

$$\mathcal{O}(f,x) := \lim_{r \to 0}(M(x,r) - m(x,r)).$$

Remark 2.1.4 (a) For any bounded function, the oscillation $\mathcal{O}(f,x)$ is a nonnegative upper semi-continuous function: the nonnegativity follows from the fact that $M(x,r) \geq m(x,r)$. "Upper semi-continuity" means that $B_\epsilon := \{x \in B : \mathcal{O}(f,x) < \epsilon\}$ is an open subset of B for all $\epsilon > 0$. To see this, suppose $x \in B_\epsilon$. By definition of $\mathcal{O}(f,x)$, there exists $r > 0$ such that $M(x,r') - m(x,r') < \epsilon$ for all $0 < r' \leq r$. Let $y \in B_r(x)$. Then for $0 < s < r - \|x - y\|$, $B_s(y) \subset B_r(x)$ and hence $M(y,s) - m(y,s) \leq M(x,r) - m(x,r) < \epsilon$. This implies that $B_s(y) \cap B \subset B_\epsilon$.
(b) A bounded function f is continuous at a point x iff $\mathcal{O}(f,x) = 0$.

Theorem 2.1.1 (Lebesgue) *A bounded function $f : B \to \mathbb{R}$ is (Riemann) integrable iff the set of points Z at which it is discontinuous is of measure zero. In particular, every continuous function is integrable.*

Proof: Suppose Z is of measure zero. Given $\epsilon > 0$ by the above remark, first of all we have $B^\epsilon := B \setminus B_\epsilon = \{x \in B : \mathcal{O}(f,x) \geq \epsilon\}$ is a closed subset B and hence is compact. Also $B^\epsilon \subset Z$ and hence is of measure 0. Therefore there exist finitely many closed boxes $B_1, \ldots, B_k$ whose interiors cover B^ϵ and such that $\sum_i \mu(B_i) < \epsilon$. By Lebesgue covering lemma, it follows that we can find a partition P of B such that each subbox S of this partition is
(1) either contained in of the B_i
(2) or is contained in B_ϵ. Accordingly, let us denote the set of all subboxes S of P satisfying (1) by P_1 and (2) by P_2 so that $P = P_1 \cup P_2$.

Let $|f(x)| < M$ for all $x \in B$ for some $M > 0$. Then $M_S(f) - m_S(f) < 2M$ for all $S \in P$. Therefore

$$\sum_{S \in P_1} (M_S(f) - m_S(f))\mu(S) < 2M \sum_{S \in P_1} \mu(S) \leq 2M \sum_i \mu(B_i) < 2M\epsilon.$$

Now let $S \in P_2$. Then for every $x \in S$ we have $\mathcal{O}(f,x) < \epsilon$. Therefore, there exists $r > 0$ such that $M(x,r) - m(x,r) < \epsilon$. By compactness of S, this means that we can partition S so that on each subbox S' of this partition, we have $M_{S'}(f) - m_{S'}(f) < \epsilon$.

Let P' be the partition of B which is a refinement of P and when restricted to each $S \in P_2$ refines the above partition on S. Then

$$\begin{aligned} U(f,P') - L(f,P') &= \sum_{S' \in P_1' \cup P_2'} (M_{S'}(f) - m_{S'}(f))\mu(S') \\ &\leq \sum_{S \in P_1} (M_S(f) - m_S(f))\mu(S) + \sum_{S' \in S \in P_2'} (M_{S'}(f) - m_{S'}(f))\mu(S') \\ &\leq 2M\epsilon + \epsilon \sum_{S \in P_2} \mu(S) \leq \epsilon(2M + \mu(B)). \end{aligned}$$

Since $\epsilon > 0$ is arbitrary, this proves that f is integrable.

To prove the converse, note that $Z = \cup_n B^{1/n}$ and hence it is enough to prove that $B^{1/n}$ is of measure zero for each n. Given $\epsilon > 0$ let P be a partition of B such that

$$U(f,P) - L(f,P) < \epsilon/n.$$

Let P' be the set of all those boxes in P which intersect $B^{1/n}$. Then $B^{1/n} \subset \cup_{S \in P'} S$ and for $S \in P'$, we have $M_S(f) - m_s(f) \geq 1/n$. Therefore

$$\begin{aligned} \sum_{S \in P'} \mu(S) &\leq n \sum_{S \in P'} (M_S(f) - m_S(f))\mu(S) \\ &\leq n \sum_{s \in P} (M_S(f) - m_S(f))\mu(S) = n(U(f,P') - L(f,P')) < \epsilon. \end{aligned}$$

This shows that $\mu(B^{1/n}) = 0$ and completes the proof of the theorem. ♠

Remark 2.1.5

(a) Given any subset A of $\mathbb{R}^n$, recall that the boundary of A is defined by $\delta A = \bar{A} \setminus int(A)$ where $\bar{A}$ is the closure of A in $\mathbb{R}^n$ and $intA$ denotes its interior in $\mathbb{R}^n$.

(b) Recall that the characteristic function χ_A of a subset of a space X is defined to by $\chi_A(x) = 1$ if $x \in A$ and $= 0$ if $x \notin A$. It follows that χ_A is discontinuous precisely at its boundary points $x \in \delta A$. Therefore, for any closed box B, $\chi_A|_B$ is integrable over B iff $B \cap \delta A$ is of measure zero.

(c) Given a bounded function f on a bounded set A, choose a closed box B such that $A \subset B$. If $\chi_A f$ is integrable over B then we put

$$\int_A f := \int_B \chi_A f.$$

(For this to make sense, we have to verify that the right-hand side (RHS) is independent of the choice of B.)

(d) For a bounded subset $A \subset \mathbb{R}^n$ with δA having measure zero, we set

$$\mu_n(A) := \int_A 1 = \int_A \chi_A,$$

the integral of the constant function 1; $\mu_n(A)$ is called the *n-dimensional volume of A*.

(e) For any $t \in \mathbb{R}$, $\mathbb{R}^{n-1} \times \{t\} \subset \mathbb{R}^n$ has its n-dimensional volume equal to zero. So is the case with any subset of $\mathbb{R}^{n-1} \times \{t\}$.

Theorem 2.1.2 Fubini's Theorem: *Let $A \subset \mathbb{R}^n, B \subset \mathbb{R}^m$ be rectangles, $f : A \times B \to \mathbb{R}$ be an integrable function. For each $x \in A$, let $f_x : B \to \mathbb{R}$ be defined by $f_x(y) = f(x, y)$. Then the lower and upper sum functions $L : x \mapsto L(f_x)$, $U : x \mapsto U(f_x)$ are integrable functions on A and we have*

$$\int_{A\times B} f = \int_A L = \int_A U. \tag{2.2}$$

In particular, if f_x is integrable for all $x \in A$ then we have

$$\int_{A\times B} f = \int_A \left(\int_B f(x,y)dy \right) dx. \tag{2.3}$$

Remark 2.1.6
(a) The RHS of (2.3) is an iterated integral. That is the reason why we call even the RHS of (2.2) also as an iterated integral for f.
(b) Of course we can interchange A and B and get another valid statement. The change of order of integration in an iterated integral is a very useful tool that is available if f is continuous.
(c) Thus, for all continuous functions f on any rectangle, the integral can be evaluated by integrating the function with respect to one variable at a time and in whichever order we may prefer.

Theorem 2.1.3 Change of Variable Formula:
Let $f : U \to V$ be a diffeomorphism of open subsets in $\mathbb{R}^n$. Then for any continuous function α on V, with compact support, we have

$$\int_V \alpha = \int_U (\alpha \circ f)|\det\,(Df)|. \tag{2.4}$$

Proof: Consider the case when $n = 1$. Without loss of generality, we may assume that U and V are closed intervals, say $U = [a, b]$ and $V = [c, d]$. Formula (2.4) can be rewritten as:

$$\int_{[c,d]} \alpha = \int_{[a,b]} (\alpha \circ f)|f'|. \tag{2.5}$$

Suppose that f is strictly increasing. Then we know from the calculus of 1-variable that

$$\int_c^d \alpha(y)\,dy = \int_a^b \alpha(f(x))f'(x)\,dx. \tag{2.6}$$

Since $f'(x) > 0$ this formula is the same as (2.5). On the other hand, if f is strictly decreasing, then $c = f(a) > f(b) = d$ and hence

$$\int_c^d \alpha(y)\,dy = \int_{f(a)}^{f(b)} \alpha(f(x))f'(x)\,dx = \int_{f(b)}^{f(a)} \alpha(f(x))|f'(x)|\,dx \tag{2.7}$$

which is again the same as (2.4). (In the integration theory that we have developed here, we have ignored the orientations on subsets of $\mathbb{R}^n$. That is the reason why we need to introduce a correction factor of taking modulus of the determinant of Df. Later in Chapter 4, we shall restore the orientation factor as a necessity, in order to develop the theory of integration on manifolds.)

Combining this with Fubini's theorem, it follows that (2.4) holds when f is a primitive mapping (see (1.56)).

Also, we easily check that (2.4) is valid if f is a permutation of variables.

Now we claim that if (2.4) holds for $f : U \to V$ and $g : V \to W$ then it holds for $g \circ f : U \to W$. This is verified in a routine way, using the chain rule for differentiation (see (1.36)).

Now by the Primitive Mapping Theorem 1.4.7 for each $x \in U$, there exists a neighbourhood W_x on which f can be expressed as a composite of primitive mappings and a permutation. Therefore, (2.4) is valid for all α having its support inside $f(W_x)$ for some $x \in U$. Let $\{\theta_j\}$ be a smooth partition of unity subordinate to the cover $\{f(W_x)\}_{x \in U}\}$ of V. Then (2.4) is applicable to each $\theta_j\alpha$ and hence we have

$$\begin{aligned} \int_U \theta_j\alpha &= \int_{W_j} \theta_j\alpha = \int_{f(W_j)} \theta_j(\alpha_j \circ f)|\det D(f)| \\ &= \int_V \theta_j(\alpha \circ f)|\det D(f)|. \end{aligned} \tag{2.8}$$

Since α has compact support, these terms are nonzero only for a finite number of $j's$. Therefore, taking summation over j in (2.8) is allowed and this yields (2.4). ♠

Exercise 2.1

Let $\{I_i\}$ be a finite cover of an interval (a, b) by open intervals. Show that there exists a subcover $\{I_j\}_{1 \le j \le r}$ such that $I_j \cap I_k = \emptyset$, for $|j - k| > 2$. Conclude from this that the sum of the of lengths of intervals belonging to this subcover is less than $2(b - a)$.

2.2 Sard's Theorem

The behavior of a real valued smooth function of one variable changes at a point where the derivative vanishes. This fact is not simply an exception but a phenomenon that is present in the general situation also, though in a more complicated manner. Sard's theorem assures us that every smooth map takes the set of all such points where its derivative is not surjective, into a set of 'negligible size'. Going through the proof you will have an opportunity to use the surjective form of the implicit function theorem as well as the rank theorem in a nontrivial way.

Definition 2.2.1 Let X be an open subset of $\mathbb{R}^m$ or the half-space $\mathbf{H}^m$ and let $f : X \to \mathbb{R}^n$ be a smooth map. A point $x \in X$ is called a *critical point* of f if Df_x is not surjective. We shall denote the set of critical points of f by C_f. Any point $y \in \mathbb{R}^m$ such that $f(x) = y$ for some critical point x of f is called a *critical value* of f.

Theorem 2.2.1 Sard's Theorem: *Let $f : X \to \mathbb{R}^n$ be a C^∞ map. Then the image $f(C_f)$ of the set of critical points of f is of measure zero in $\mathbb{R}^n$.*

Remark 2.2.1 The proof of this will be through several steps. Recall the definition of measure zero sets in $\mathbb{R}^n$ (Definition 2.1.1). The first step is to examine the behavior of a measure zero set under a smooth map and in particular, under a diffeomorphism. This will enable us later to define the concept of measure zero sets in arbitrary manifolds.

Theorem 2.2.2 *Let $A \subset \mathbb{R}^n$ be of measure zero and $f : A \to \mathbb{R}^n$ be any smooth function. Then $f(A)$ is of measure zero.*

Proof: Given a point $x \in A$, choose a ball B around x so that $f|_{\bar{B}\cap A}$ is the restriction of a smooth function $\hat{f} : \bar{B} \to \mathbb{R}^n$. Since A can be covered by a countable union of such balls, we may as well assume that $A \subset B \subset \bar{B}$ and $f : \bar{B} \to \mathbb{R}^n$ is a smooth map. By continuity and compactness of $\bar{B}$ we can now choose $M > 0$ such that $\|Df_x\| \leq M$ for all $x \in \bar{B}$. A simple application of Theorem 1.3.3 (WMVT), now yields that

$$\|f(x) - f(y)\| \leq M\|x - y\|, \quad \forall\, x, y \in \bar{B}.$$

Therefore, if D is a disc of radius r in $\bar{B}$, it follows that $f(D)$ is contained in a disc of radius rM. Now given $\epsilon > 0$ cover A by balls D_i of radius r_i such that $\sum vol_n(D_i) < \epsilon$. It follows that $f(A)$ is contained in a countable collection of balls of total volume $< M\epsilon$. ♠

Remark 2.2.2
(i) In particular, it follows that the property of being measure zero is a diffeomorphism invariant.
(ii) Also, if $f : A \to \mathbb{R}^l$ is a smooth map where $A \subset \mathbb{R}^n$ for $n < l$, then $f(A)$ is of measure zero in $\mathbb{R}^l$. For, we can consider $\mathbb{R}^n$ as a subset of $\mathbb{R}^n \times \mathbb{R}^{l-n}$, define $\hat{f}$ on $A \times \mathbb{R}^{l-n}$ by $\hat{f}(a, b) = f(a)$ and apply the above theorem to $\hat{f}$ in place of f. Of course, we have to appeal to the fact that $A \times 0$ is of measure zero in $\mathbb{R}^l$.

We now proceed to give a proof of Morse-Sard's theorem. Suppose $X \subset \mathbb{R}^m$. The case $m < n$ is covered by Remark 2.2.2(ii), the entire set $f(X)$ being of measure zero, in this case. So, **we may and shall assume that** $m \geq n \geq 1$. Given $A \subset \mathbb{R}^k \times \mathbb{R}^l$ and $u \in \mathbb{R}^k$, define the slice of A at u by

$$A_u = \{x \in \mathbb{R}^k \ : \ (x, u) \in A\}.$$

The following is an easy consequence of Fubini's theorem.

Lemma 2.2.1 Let A be a closed subset of $\mathbb{R}^{k+l}$. If A_u is of measure zero in $\mathbb{R}^k$ for all $u \in \mathbb{R}^l$, then A is of measure zero in $\mathbb{R}^{k+l}$.

Now by the local nature of the problem, we may assume that X is an m-dimensional cube in $\mathbb{R}^m$ of side-length $\alpha > 0$ and f is defined and smooth in a neighbourhood of X. Let $C = C_f$ be the set of critical points of f. For $r \geq 1$, put C_r equal to the set of all $x \in X$ such that all partial derivatives of f of order $\leq r$ vanish at x. Then clearly

$$C \supset C_1 \supset C_2 \supset \cdots$$

Now the proof of Theorem 2.2.1 is completed by proving the following three statements:
(I) $f(C_r)$ is of measure zero if $r > \dfrac{m}{n} - 1$.
(II_m) $f(C \setminus C_1)$ is of measure zero.
(III_m) For $r \geq 1$, $f(C_r \setminus C_{r+1})$ is of measure zero.
Proof of (I): By Taylor's theorem, it follows that

$$f(x + h) = f(x) + R(x, h)$$

where $\|R(x, h)\| < M\|h\|^{r+1}$ for all $x \in C_r$ and $x + h \in X$, where M is a constant that depends on the partial derivatives of f of order $r + 1 > \frac{m}{n}$. Divide X into l^m cubes each of side α/l. If S is one of these small cubes, it follows that $f(S \cap C_r)$ is contained in a cube of side $\sqrt{n}M(\alpha\sqrt{m}/l)^{r+1}$. Thus, $f(C_r)$ is contained in the union of at most l^m such cubes which have a total volume less than or equal to

$$[\sqrt{n}M(\alpha\sqrt{m}/l)^{r+1}]^n l^m = M' l^{m-(r+1)n}.$$

Since $r+1 > \dfrac{m}{n}$, it follows that this quantity converges to zero as $l \to \infty$. This completes the proof of (I).

Before proceeding further, we observe that statement (I) yields a proof of the theorem for the case $m = n = 1$. For, here, $C = C_1$ and therefore $f(C) = f(C_1)$ is of measure zero. So, from now onwards, we assume that $m > 1$.

Inductively assume that the theorem is true for all maps $h : W \to Y$, where W is an open subset of $\mathbb{R}^k, k < m$. Under this induction hypothesis we shall prove (II_m) and (III_m). Combined with (I) this will imply the theorem for subsets $X \subset \mathbb{R}^m$.

Proof of (II_m): Let $a \in C \setminus C_1$. Then one of the first order derivatives of f at a is not zero and hence the rank of $Df_a > 0$. Hence by the rank theorem, there exists a neighbourhood V of a such that after a suitable change of co-ordinates at a as well as at $f(a)$, we may assume that f is of the form $f(x,y) = (x, g(x,y))$, for all $(x,y) \in (\mathbb{R}^p \times \mathbb{R}^{m-p}) \cap V$, where $p = rk\, Df_a$ and $g : V \to \mathbb{R}^{n-p}$ is a smooth function. Now $(x,y) \in C_f \cap V$ iff $y \in C_{g_x}$. (Here $g_x : \{x\} \times \mathbb{R}^{m-p} \cap V \to \mathbb{R}^{n-p}$ is given by $g_x(y) = g(x,y)$.) From this, it follows that for every $x \in \mathbb{R}^p$, we have

$$\{x\} \times \mathbb{R}^{n-p} \cap f(C \cap V) = \{x\} \times g_x(C_{g_x}).$$

By the induction hypothesis, each of these slices is of measure zero and hence $f(C \cap V)$ is of measure zero. Since $C \setminus C_1$ can be covered by countably many such sets, we are through.

Proof of (III_m):

For each partial differential operator P of order r and $1 \le j \le m$, put

$$U(P,j) = \left\{x \in X \ : \ \left(\frac{\partial}{\partial x_j} \circ Pf\right)(x) \neq 0\right\}.$$

Then $X \setminus C_{r+1}$ is covered by the open sets $\{U(P,j)\}$ where P and j take all possible values as above. Therefore, if we prove that $f((C_r \setminus C_{r+1}) \cap U(P,j))$ is of measure zero for each P, j, then the proof of ($\mathrm{III_m}$) will be complete.

Fix P and j, put

$$U = U(P,j), \ g = Pf : U \to \mathbb{R}, \ Z = g^{-1}(0), \quad \text{and } h = f|_Z.$$

Observe that $C_r \cap U \subset Z$. Clearly, the critical set C_h of h contains $C_r \cap U$. Therefore, it suffices to prove that $f(C_h)$ is of measure zero.

Clearly g is a smooth map, with $\dfrac{\partial g}{\partial x_j} \neq 0$ on U. Therefore by the surjective form of the implicit function theorem, around each point $z \in Z$, we can choose a coordinate neighbourhood V_z on which g takes the form $g(x,y) = x$. This means $V_z \cap Z$ looks like an open subset of $\{0\} \times \mathbb{R}^{m-1}$. Therefore, by induction hypothesis, $f(C_h \cap V_z) = h(C_h \cap V_z)$ is of measure zero. Since C_h can be covered by countably many such neighbourhoods V_z, it follows that $f(C_h)$ is of measure zero, as claimed. ♠

As an entertaining exercise, we shall give an application of Sard's theorem here to a central result in homotopy theory, which is however, not the main import of Sard's theorem.

Recall from Exercise 1.7.12, that by a smooth homotopy, we mean a smooth map $H : X \times \mathbb{R} \to Y$. Two smooth maps $f, g : X \to Y$ are said to be (smoothly) homotopic if there is a smooth homotopy H such that $H_t = f, H_s = g$ for some $t, s \in \mathbb{R}$. If f is homotopic to a constant map, then we say f is null homotopic. Given any smooth map $f : X \to \mathbb{R}^n$ we can consider $H(x,t) = tf(x)$, which defines a smooth homotopy of f with the constant map 0. We say Y is simply connected if every smooth map $f : \mathbb{S}^1 \to Y$ is null homotopic. We shall prove:

Theorem 2.2.3 *For $m > n$, every map $f : \mathbb{S}^n \to \mathbb{S}^m$ is null homotopic. In particular, $\mathbb{S}^m$ is simply connected for all $m \geq 2$.*

Proof: We first note that any map $f : X \to \mathbb{S}^m$, which is not surjective is null homotopic. This follows easily from the fact that $\mathbb{S}^m \setminus \{p\}$ is diffeomorphic to $\mathbb{R}^m$ and the observation that we have made above that every map into $\mathbb{R}^m$ is null homotopic. So, it is enough to prove that any smooth map $f : \mathbb{S}^n \to \mathbb{S}^m$ is not surjective.

This is precisely where we use Sard's theorem. Choose any point $p \in \mathbb{S}^n$, $q = f(p)$, and let $\eta : \mathbb{S}^m \setminus \{q\} \to \mathbb{R}^m$ be the stereographic projection. Then $(\eta \circ f)^{-1}(\mathbb{R}^m)$ is an open subset U of $\mathbb{S}^n \setminus \{p\}$. So, we can choose the stereographic projection (see Remark 1.6.3(i)) $\tau : \mathbb{S}^n \setminus \{p\} \to \mathbb{R}^n$ and put $V = \tau(U)$. Now put $g = \eta \circ f \circ \tau^{-1} : V \to \mathbb{R}^m$. Then g is smooth. Since $n < m$, it follows that Dg is never surjective. This means the entire set V is the critical set of g. Therefore, $g(V)$ is of measure zero in $\mathbb{R}^m$. In particular, g is not surjective. This means f is not surjective. ♠

Exercise 2.2

1. Give an example of a smooth map $f : \mathbb{R} \to \mathbb{R}$ such that the set of critical values is dense.

2. Does the exercise above contradict Sard's theorem?

3. Show that for any $\mathcal{C}^1$ function f, the critical set C_f is a closed subset.

4. Let $U \subset \mathbb{R}^3$ and $V \subset \mathbb{R}^2$ be open sets and $f : U \to V$ be a surjective $\mathcal{C}^1$ function. Does f necessarily have rank 2 at some point of U?

5. Let $f : \mathbb{C}^n \to \mathbb{C}$ be a complex analytic function. Show that the set of critical points of f is of measure zero.

2.3 Exterior Algebra

In this section, we shall develop the multilinear algebraic machinery needed to introduce the notion of differential forms.

Throughout this section, V denotes a vector space of dimension n, over the field[2] $\mathbb{R}$.

Let $p \geq 1$ be an integer and let V^p denote the Cartesian product of p copies of V :

$$V^p := \underbrace{V \times \cdots \times V}_{p \text{ copies}}.$$

Definition 2.3.1 By a *p-tensor* on V, we mean a multilinear map

$$\phi : V^p \to \mathbb{R},$$

i.e., for every $1 \leq i \leq p$ and vectors, $\mathbf{v}_1, \ldots, \mathbf{v}_n$, the map

$$\mathbf{v} \mapsto \phi(\mathbf{v}_1, \ldots, \mathbf{v}_{i-1}, \mathbf{v}, \mathbf{v}_{i+1}, \ldots, \mathbf{v}_p)$$

is linear.

[2]Indeed, most of the stuff here is valid if we replace $\mathbb{R}$ by any field. At places such as when dealing with alternating and symmetric tensors etc., we will have to assume that the field is of characteristic zero.

Remark 2.3.1

(i) For $p = 1$, a 1-tensor on V is nothing but a linear map on V. A typical example of a 2-tensor is the dot product on $\mathbb{R}^n$. By writing elements of $\mathbb{R}^n$ as column vectors and then by writing p of them side by side, we identify $(\mathbb{R}^n)^p$ with the space of $n \times p$ matrices over $\mathbb{R}$. Then for $p = n$, we have another familiar tensor, viz., the determinant.

(ii) If ϕ, ψ are p-tensors, then for any $\alpha, \beta \in \mathbb{R}$, $\alpha\phi + \beta\psi$ is also a p-tensor. Thus, the set of all p-tensors on V forms a vector space and we denote this vector space by $\mathcal{T}^p(V)$. Let us determine its dimension. For this purpose, let us consider a mechanism to produce higher order tensors, out of lower order tensors.

Definition 2.3.2 Given a p-tensor ϕ and a q-tensor ψ on V define the product $(p+q)$-tensor $\phi \otimes \psi$ on V as follows:

$$(\phi \otimes \psi)(\mathbf{u}, \mathbf{w}) = \phi(\mathbf{u})\psi(\mathbf{w}), \quad \mathbf{u} \in V^p, \mathbf{w} \in V^q. \tag{2.9}$$

We call $\phi \otimes \psi$ the tensor product of ϕ with ψ. It is not difficult to see that tensor product is an associative operation. But it is **not** commutative. (See the exercise below.)

For positive integers p, n, let us denote the set of all functions from

$$\{1, 2, \ldots, p\} \to \{1, 2, \ldots, n\}$$

by $S(p, n)$. For an element $I \in S(p, n)$, we shall write i_k for $I(k)$ and display I as

$$I = (i_1, \ldots, i_p).$$

Given a set $\{\phi_1, \phi_2, \ldots, \phi_n\}$ of 1-tensors, and $I = (i_1, \ldots, i_p) \in S(p, n)$, let us denote

$$\phi(I) := \phi_{i_1} \otimes \cdots \otimes \phi_{i_p}. \tag{2.10}$$

We can now determine the dimension of $\mathcal{T}^p(V)$:

Theorem 2.3.1 *Let $\{\phi_1, \ldots, \phi_n\} \in V^\star = \mathcal{T}^1(V)$ form a basis for $V^\star$. Then the set*

$$\{\phi(I) \ : \ I \in S(n, p)\} \tag{2.11}$$

forms a basis for $\mathcal{T}^p(V)$. Consequently, $\dim \mathcal{T}^p(V) = \#(S(p, n)) = n^p$.

Proof: Let $\{\mathbf{v}_1, \ldots, \mathbf{v}_n\}$ be the basis of V dual to $\{\phi_1, \ldots, \phi_n\}$, i.e., $\phi_i(\mathbf{v}_j) = \delta_{ij}$. We first claim that two p-tensors ϕ, ψ, are equal iff $\phi(\mathbf{v}_{j_1}, \ldots, \mathbf{v}_{j_p}) = \psi(\mathbf{v}_{j_1}, \ldots, \mathbf{v}_{j_p})$ for all $J \in S(p, n)$. Just like the similar statement for linear maps on a vector space, this follows by multilinearity of ϕ. The details are left to the reader.

Now observe that,

$$\phi(I)(\mathbf{v}_{j_1}, \ldots, \mathbf{v}_{j_p}) = \begin{cases} 1, & \text{if } (j_1, \ldots, j_p) = I, \\ 0, & \text{otherwise.} \end{cases} \tag{2.12}$$

Therefore, (2.11) defines an independent set in $\mathcal{T}^p(V)$. Next, given any p-tensor ψ, consider the p-tensor

$$\xi = \sum_{I \in S(p,n)} \psi(\mathbf{v}_{i_1}, \ldots, \mathbf{v}_{i_p})\phi(I).$$

It follows that

$$\xi(\mathbf{v}_{j_1}, \ldots, \mathbf{v}_{j_p}) = \psi(\mathbf{v}_{j_1}, \ldots, \mathbf{v}_{j_p})$$

for all $J \in S(p, n)$. Therefore, $\xi = \psi$. This proves that the set (2.11) generates $\mathcal{T}^p(V)$. Thus, we have proved that the set (2.11) is a basis for $\mathcal{T}^p(V)$. ♠

Definition 2.3.3 Let $p \geq 2$. Let S_p denote the group of permutations of $\{1, 2, \ldots, p\}$. Given $\sigma \in S_p$ and a p-tensor ϕ, we define the p-tensor ϕ^σ by

$$\phi^\sigma(\mathbf{v}_1, \ldots, \mathbf{v}_p) = \phi(\mathbf{v}_{\sigma(1)}, \mathbf{v}_{\sigma(2)}, \ldots, \mathbf{v}_{\sigma(p)}). \tag{2.13}$$

This defines a linear action of the permutation group S_p on the space $\mathcal{T}^p(V)$. For $p \geq 2$, we say a p-tensor ϕ is symmetric if $\phi^\sigma = \phi$ for all $\sigma \in S_p$. More interesting for us for the time being are the tensors that are antisymmetric or alternating, viz., those p-tensors satisfying:

$$\phi^\sigma = \text{sgn}\ (\sigma)\phi, \tag{2.14}$$

where sgn (σ) is the signature of the permutation σ ($= \pm 1$ according as σ is even or odd). Since every permutation of n letters is a composite of finitely many transpositions, it follows that a p-tensor ϕ is alternating iff

$$\phi^\tau = -\phi \tag{2.15}$$

for all transpositions τ.

Notice that any 1-tensor is automatically symmetric as well as alternating.

Let us denote by $S^p(V^*)$ and $\wedge^p(V^*)$, the subset of all symmetric tensors and the subset of alternating tensors, respectively. We can then verify that these two subsets are indeed vector subspaces of $\mathcal{T}^p(V)$. (Note that $S^1(V^*) = \wedge^1(V^*) = V^*$.) Further, let us define two linear maps Sym : $\mathcal{T}^p(V) \to S^p(V^*)$ and Alt : $\mathcal{T}^p(V) \to \wedge^p(V^*)$ as follows:

$$\text{Sym}(\phi) = \frac{1}{p!}\sum_\sigma \phi^\sigma; \quad \text{Alt}(\phi) = \frac{1}{p!}\sum_\sigma \text{sgn}\ (\sigma)\phi^\sigma, \tag{2.16}$$

where the sum is taken over $\sigma \in S_p$.

The following lemma is easy to prove:

Lemma 2.3.1
(a) For any $\phi \in \mathcal{T}^p(V)$, $\text{Alt}(\phi) = \phi$ iff $\phi \in \wedge^p(V^*)$; in particular, $\text{Alt} \circ \text{Alt} = \text{Alt}$.
(b) For any $\phi \in \mathcal{T}^p(V)$, $\text{Sym}(\phi) = \phi$ iff $\phi \in S^p(V^*)$. In particular, $\text{Sym} \circ \text{Sym} = \text{Sym}$.
(c) For $p \geq 2$, $\text{Alt} \circ \text{Sym} = \text{Sym} \circ \text{Alt} = 0$.

Let us now consider the behavior of these special tensors under the tensor product. Immediately we see that there is trouble here: $\phi \otimes \psi$ is neither symmetric nor alternating, even when both ϕ and ψ are symmetric or both are alternating. We shall not give up so easily.

Definition 2.3.4 Define *exterior product* $\wedge : \wedge^p(V^*) \times \wedge^q(V^*) \to \wedge^{p+q}(V^*)$ as follows:

$$\phi \wedge \psi = \text{Alt}\ (\phi \otimes \psi). \tag{2.17}$$

Remark 2.3.2 We can then verify easily that
(a) $(a_1\phi_1 + a_2\phi_2) \wedge \psi = a_1(\phi_1 \wedge \psi) + a_2(\phi_2 \wedge \psi)$;
(b) $\phi \wedge (b_1\psi_1 + b_2\psi_2) = b_1(\phi \wedge \psi_1) + b_2(\phi \wedge \psi_2)$. That means that "$\wedge$" is bilinear. We would like to see whether this operation is associative just like the tensor product. That needs to be verified carefully.
(c) If ϕ_1, ϕ_2 are 1-tensors then $\psi_1 \wedge \psi_2 = -\psi_2 \wedge \psi_1$. In particular, $\phi_1 \wedge \phi_1 = 0$. Can you say that $\phi \wedge \phi = 0$ for any p-tensor ϕ? Wait for a while.

Lemma 2.3.2 If Alt $(\phi) = 0$ then for all ψ, we have $\text{Alt}(\phi \otimes \psi) = 0$.

Proof: Let $\phi \in \wedge^p(V^*), \psi \in \wedge^q(V^*)$. Consider S_p as a subgroup of S_{p+q} consisting of those permutations that leave $\{p+1, \ldots, p+q\}$ pointwise fixed. Let $\{\tau_j\} \subset S_{p+q}$ be a set of right coset representatives for S_p. Then

$$\begin{aligned}
\phi \wedge \psi &= \text{Alt } (\phi \otimes \psi) \\
&= \textstyle\sum_{\sigma \in S_{p+q}} \text{sgn } (\sigma)(\phi \otimes \psi)^\sigma \\
&= \textstyle\sum_{\tau_j} \left(\sum_{\alpha \in S_p} \text{sgn } (\alpha\tau_j)(\phi \otimes \psi)^{\alpha\tau_j}\right) \\
&= \textstyle\sum_{\tau_j} \text{sgn } (\tau_j) \left(\sum_{\alpha \in S_p} \text{sgn } (\alpha)(\phi^\alpha \otimes \psi)\right)^{\tau_j} \\
&= \textstyle\sum_{\tau_j} \text{sgn } (\tau_j)(\text{Alt}(\phi) \otimes \psi)^{\tau_j} = 0
\end{aligned}$$

since Alt $(\phi) = 0$. ♠

Theorem 2.3.2 *The exterior product is associative.*

Proof: Let ϕ, ψ, ξ be any three alternating tensors. We have to prove that

$$(\phi \wedge \psi) \wedge \xi = \phi \wedge (\psi \wedge \xi).$$

Instead we shall prove that

$$(\phi \wedge \psi) \wedge \xi = \text{Alt } (\phi \otimes \psi \otimes \xi) = \phi \wedge (\psi \wedge \xi).$$

Using the associativity of the tensor product, this is the same as proving

$$(\phi \wedge \psi) \wedge \xi = \text{Alt}((\phi \otimes \psi) \otimes \xi); \;\; \text{Alt}(\phi \otimes (\psi \otimes \xi)) = \phi \wedge (\psi \wedge \xi). \tag{2.18}$$

Now

$$\begin{aligned}
& (\phi \wedge \psi) \wedge \xi - \text{Alt}((\phi \otimes \psi) \otimes \xi) \\
= \; & \text{Alt}((\phi \wedge \psi) \otimes \xi) - \text{Alt}((\phi \otimes \psi) \otimes \xi) \\
= \; & \text{Alt}[(\phi \wedge \psi) \otimes \xi - (\phi \otimes \psi) \otimes \xi] \\
= \; & \text{Alt}[(\phi \wedge \psi - \phi \otimes \psi) \otimes \xi] \\
= \; & (\phi \wedge \psi - \phi \otimes \psi) \wedge \xi = 0,
\end{aligned}$$

the last equality holds from the above lemma, because

$$\text{Alt } (\phi \wedge \psi - \phi \otimes \psi) = \phi \wedge \psi - \phi \wedge \psi = 0.$$

This proves $(\phi \wedge \psi) \wedge \xi = \text{Alt}((\phi \otimes \psi) \otimes \xi)$

Similarly, we can prove the other equality in (2.18). ♠

Remark 2.3.3 Thus, we need not put any brackets while taking the exterior product of more than two (alternating) tensors. We can now derive a basis for $\wedge^p(V^*)$ as follows: Let $s(p, n)$ denote the set of all strictly monotonic sequences $I = (i_1, i_2, \ldots, i_p)$ in $\{1, 2, \ldots, n\}$. Let $\{\phi_1, \ldots, \phi_n\}$ be a basis for $V^\star$. For $I = (i_1, \ldots, i_p) \in s(p, n)$, put

$$\phi_I := \phi_{i_1} \wedge \phi_{i_2} \wedge \cdots \wedge \phi_{i_p} = \text{Alt } (\phi(I)). \tag{2.19}$$

Theorem 2.3.3 *The set* $\{\phi_I \; : \; I \in s(n, p)\}$ *forms a basis for* $\wedge^p(V^*)$. *In particular, we have* dim $\wedge^p (V^*) = \binom{n}{p}$.

Proof: Since $\text{Alt}(\phi) = \phi$ for $\phi \in \wedge^p(V^*)$, it follows that Alt is a surjective linear map. Therefore, the image of the basis $\{\phi(J) \; : \; J \in S(p, n)\}$ of $\mathcal{T}^p(V)$, is $\{\phi_J \; : \; J \in S(p, n)\}$, a generating set for $\wedge^p(V^*)$. The given set is a subset of this set. We shall see why the extra elements are unnecessary for generating $\wedge^p(V^*)$. First of all, observe that for each sequence

$J \in S(p,n)$, $(p \leq n)$, there is a permutation $\sigma \in S_p$ such that $J \circ \sigma$ is monotonic. But Alt $(\phi(J \circ \sigma) = \text{sgn}(\sigma)\text{Alt}(\phi(J))$. Therefore, we can cut down this generating set to

$$\{\text{Alt}(\phi(J)) \; : \; J \text{ is monotonic}\}.$$

We now observe that if J is not injective, i.e., if a certain number has repeated then $\phi_J = 0$. Hence we can cut down the generating set to the required size as above.

It remains to see that the $\{\phi_I \; : \; I \in s(p,n)\}$ is independent. This can be proved by using a property similar to (2.12). (Take this as an exercise.) ♠

Remark 2.3.4

1. Using the fact $\phi_1 \wedge \phi_2 = -\phi_2 \wedge \phi_1$ for 1-tensors, we can immediately deduce that

$$\phi_I \wedge \phi_J = (-1)^{pq}\phi_J \wedge \phi_I,$$

where I, J are of length p, q, respectively. Therefore, from the bilinearity of the exterior product, it follows that this relation of *anticommutativity* should hold for all tensors:

$$\phi \wedge \psi = (-1)^{pq}\psi \wedge \phi, \tag{2.20}$$

where ϕ is a p-tensor and ψ is a q-tensor.

2. In particular, we have an affirmative answer for the question raised in Remark 2.3.2(c).

3. Take $V = \mathbb{R}^n$ and $p = n$. We get dim $\wedge^n(\mathbb{R}^{n*}) = 1$. This is the same as saying that there is only one alternating n-tensor on $\mathbb{R}^n$ up to a scalar multiple. Since the determinant function is one such, this gives the uniqueness of the determinant function as a multilinear alternating function taking the identity matrix to 1.

4. What happens to $\wedge^p(V^*)$, where p exceeds the dimension of V? Clearly, none of the sequences I is now injective and hence Alt $(\phi(I)) = 0$. Therefore, $\wedge^p(V^*) = 0$.

   For the sake of completeness we define $\wedge^0(V^*) = \mathbb{R}$.

5. We define the exterior algebra $\wedge(V^*)$ of V to be the direct sum of vector spaces,

$$\wedge(V^*) := \wedge^0(V^*) \oplus \wedge^1(V^*) \oplus \cdots \oplus \wedge^n(V^*) \oplus 0 \oplus \cdots \tag{2.21}$$

together with the exterior product

$$\wedge : \wedge^p \times \wedge^q \longrightarrow \wedge^{p+q} \tag{2.22}$$

It is a graded, anticommutative algebra. Any vector space basis for $V^\star = \mathcal{T}^1(V) = \wedge^1(V^*)$ would generate $\wedge V^*$ as an algebra. (See Theorem 2.3.3.)

Definition 2.3.5 Given a linear function $f : V \to W$ between two vector spaces, consider the linear map $f^* : \mathcal{T}^1(W) = W^* \to V^* = \mathcal{T}^1(V)$, the dual map of f. More generally, for any $\phi \in \mathcal{T}^p(W)$ we can consider $f^*(\phi) \in \mathcal{T}^p(V)$ defined by

$$f^*(\phi) = \phi \circ (f, f, \ldots, f). \tag{2.23}$$

It is easily checked that f^* is a linear map. Since

$$\sigma \circ (f, f, \ldots, f) = (f, f, \ldots, f) \circ \sigma$$

for any permutation σ of $\{1, 2, \ldots, p\}$, it follows that

$$f^*(\phi^\sigma) = (f^*(\phi))^\sigma. \tag{2.24}$$

Therefore, f^* takes alternating tensors to alternating ones. (Of course, it takes symmetric tensors to symmetric ones also.) Hence, we get an induced linear map

$$\wedge^p f : \wedge^p(W^*) \to \wedge^p(V^*). \tag{2.25}$$

Moreover, f^* clearly respects the tensor product and hence the exterior product. Therefore, (2.25) actually defines a graded algebra homomorphism

$$\wedge f : \wedge W^* \to \wedge V^*; \quad \wedge f(\phi \wedge \psi) = (\wedge f(\phi)) \wedge (\wedge f(\psi)). \tag{2.26}$$

Moreover, even at the tensor product level, if $f : V \to W, g : W \to U$ are linear maps then

$$(g \circ f)^* = f^* \circ g^*. \tag{2.27}$$

This means that

$$\wedge(g \circ f) = \wedge f \circ \wedge g. \tag{2.28}$$

Observe that $\wedge(Id) = Id$, where Id denotes the identity map of an appropriate vector space.[3]

Example 2.3.1 Let U, V be any two n-dimensional vector spaces, with bases $\{\mathbf{u}_1, \ldots, \mathbf{u}_n\}$ and $\{\mathbf{v}_1, \ldots, \mathbf{v}_n\}$, respectively. Then any linear map $A : U \to V$ corresponds, in a natural way, to an $n \times n$ matrix, which we shall denote by A itself, via,

$$A(\mathbf{u}_j) = \sum_i a_{ij}\mathbf{v}_i. \tag{2.29}$$

Now, both $\wedge^n U^*$ and $\wedge^n V^*$ are 1-dimensional vector spaces spanned by $\mathbf{u}_1^* \wedge \cdots \wedge \mathbf{u}_n^*$ and $\mathbf{v}_1^* \wedge \cdots \wedge \mathbf{v}_n^*$, respectively. Therefore,

$$(\wedge^n A)(\mathbf{v}_1^* \wedge \cdots \wedge \mathbf{v}_n^*) = \alpha(A)(\mathbf{u}_1^* \wedge \cdots \wedge \mathbf{u}_n^*), \tag{2.30}$$

where $\alpha : M(n, \mathbb{R}) \to \mathbb{R}$ is some function. We want to determine this function α.

On the space of 1-tensors, we have

$$A^*(\mathbf{v}_j^*)(\mathbf{u}_i) = (\mathbf{v}_j^* \circ A)(\mathbf{u}_i) = \mathbf{v}_j^*\left(\sum_{k=1}^n a_{ki}\mathbf{v}_k\right) = a_{ji}.$$

This in turn implies that $A^*(\mathbf{v}_j^*) = \sum_{i=1}^n a_{ji}\mathbf{u}_i^*$ (i.e., the matrix corresponding to A^* is the transpose of the matrix corresponding to A). Therefore,

$$\alpha(A)(\mathbf{u}_1^* \wedge \cdots \wedge \mathbf{u}_n^*) = (\wedge^n A)(\mathbf{v}_1^* \wedge \cdots \wedge \mathbf{v}_n^*) = A^*(\mathbf{v}_1^*) \wedge \cdots \wedge A^*(\mathbf{v}_n^*). \tag{2.31}$$

It follows from this that α is an alternating n-tensor on the space of $n \times n$ matrices, in terms of its rows. Thus, from Remark (2.3.4)(iii), it follows that $\alpha(A) = \lambda \det(A)$, since det is a nonzero alternating n-tensor.

[3] In the fancy language of categories and functors, what we have just seen above amounts to saying that the exterior algebra defines a contravariant functor from the category of finite dimensional vector spaces to the category of finitely generated, graded, anticommutative algebras.

The value of λ can now be checked by evaluating both sides on the special map $A = Id : U \to U$. Since the matrix of Id is Id, we have det $(Id) = 1$. So, $\lambda = \alpha(Id)$. On the other hand, (2.31) gives $\alpha(Id) = 1$. Therefore, $\lambda = 1$. Thus,

$$(\wedge^n A)(\mathbf{v}_1^* \wedge \cdots \wedge \mathbf{v}_n^*) = \det(A)(\mathbf{u}_1^* \wedge \cdots \wedge \mathbf{u}_n^*). \tag{2.32}$$

In particular, if $U = \mathbb{R}^n = V$, it follows that

$$(\wedge^n A)(\phi) = \det(\mathrm{A})\phi, \quad \forall\, \phi \in \wedge^n(\mathbb{R}^n). \tag{2.33}$$

Exercise 2.3

1. Consider $\pi_i : \mathbb{R}^2 \to \mathbb{R}, i = 1, 2$ the two coordinate projections. Show that $\pi_1 \otimes \pi_2 \neq \pi_2 \otimes \pi_1$.

2. Let $f : U \to V$ be an injective (surjective) linear map of finite dimensional vector spaces. Then show that $\wedge f : \wedge V \to \wedge U$ is surjective (resp., injective). (Hint: Use (2.28.)

3. Show that a given set $\{\phi_i \in V^*, i = 1, 2, \ldots, p\}$ is dependent iff $\phi_1 \wedge \cdots \wedge \phi_p = 0$.

4. For $\phi_i \in V^*$ and $\mathbf{v}_i \in V$ show that

$$(\phi_1 \wedge \cdots \wedge \phi_k)(\mathbf{v}_1, \ldots, \mathbf{v}_k) = \frac{1}{k!}\det((\phi_i(\mathbf{v}_j)).$$

5. Let $\phi \in \wedge^n V^*$ be any nonzero element. For any linear transformation $A : V \to V$ and any basis $\{\mathbf{v}_1, \ldots, \mathbf{v}_n\}$ of V, show that

$$\phi(A\mathbf{v}_1, \ldots, A\mathbf{v}_n) = (\det A)\phi(\mathbf{v}_1, \ldots, \mathbf{v}_n). \tag{2.34}$$

6. The space of symmetric tensors is also important especially in differential geometry and representation theory. Show that the dimension of $S^p(V^*)$ is equal to $\binom{p+n-1}{p}$, where $n = \dim V$.

7. Indeed, analogous to the exterior product, define a symmetric product

$$\diamondsuit : S^p(V^*) \times S^q(V^*) \to S^{p+q}(V^*)$$

and establish that it is associative. Show that the graded algebra $\oplus_{p\geq 0} S^p(V^*)$ is isomorphic to the polynomial algebra $\mathbb{R}[x_1, \ldots, x_n]$, where $n = \dim V$.

2.4 Differential Forms

Throughout this section, X will be an open subset of $\mathbf{H}^n$.

Definition 2.4.1 For $p \geq 0$, by a p-form on X, we mean a function ω, which assigns to each point $x \in X$, an alternating p-tensor $\omega(x) \in \wedge^p(\mathbb{R}^{n*})$. Two p-forms on X can be added together pointwise and also any p-form can be multiplied by a scalar function on X. Thus, the space of all p-forms on X forms a module over the ring of all scalar functions on X.

Moreover, we can take exterior product of a p-form and a q-form to get a $(p+q)$-form; this again is done pointwise. It is not difficult to see that these operations make the space of all forms on X into a graded anticommutative algebra over the ring of all real valued functions on X.

Example 2.4.1

(i) Clearly, any 0-form on X is nothing but a scalar valued function on X.

(ii) We know that $T_x(X) = \mathbb{R}^n$ for all $x \in X$. For any $\mathcal{C}^\infty$-function $f : X \to \mathbb{R}$, recall that the total derivative $Df_x : \mathbb{R}^n \to \mathbb{R}$ is a linear map, and that the assignment $x \to Df_x$ is a smooth function on X. In the new terminology, each Df_x is a 1-tensor and hence is an element of $\wedge^1(\mathbb{R}^{n*})$. Thus, Df is a 1-form on X. In this new avatar, we shall denote Df by df.

(iii) Now, suppose f is the ith-coordinate projection $\pi_i : X \to \mathbb{R}$. Then $Df_x = \pi_i$ for all x. It is customary to denote π_i by a more transparent notation x_i and we shall follow this practice. The corresponding 1-form is then dx_i. Thus, dx_i denotes the 1-form on $\mathbb{R}^n$ which assigns, to each point $x \in \mathbb{R}^n$, the 1-tensor on $\mathbb{R}^n$ which is nothing but the projection onto the ith coordinate.

(iv) Now, use the fact that $(\mathbb{R}^n)^*$ is spanned by $\{x_1, \ldots, x_n\}$, to see the following: Given any 1-form ω on X, for each $x \in X$, there exist unique scalars, $f_1(x), \ldots, f_n(x)$ such that

$$\omega(x) = \sum_{i=1}^{n} f_i(x) x_i. \tag{2.35}$$

Since $dx_i(x) = x_i$ for all $x \in X$, we have,

$$\omega = \sum_{i} f_i dx_i, \tag{2.36}$$

where $f_i : X \to \mathbb{R}$ are some scalar functions.

(v) What we saw above applies, in general, to p-forms for $p \geq 2$ also. As in (2.19), let us introduce the notation

$$x_I = x_{i_1} \wedge \cdots \wedge x_{i_p}, \text{ where } I = \{i_1 < i_2 < \cdots < i_p\}.$$

Since $\{x_1, \ldots, x_n\}$ is a basis for $(\mathbb{R}^n)^*$, it follows from (2.3.2) that $\wedge^p(\mathbb{R}^{n*})$ is spanned by $\{x_I \ : \ I \in s(n,p)\}$. We shall introduce the notation dx_I for the p-form on X, which assigns $x_I \in \wedge^p(\mathbb{R}^{n*})$ for each point $a \in X$. Then as above, given any p-form ω on X, there exist functions $f_I : X \to \mathbb{R}, \ I \in s(n,p)$, such that

$$\omega = \sum_{I \in s(n,p)} f_I dx_I. \tag{2.37}$$

The functions f_I are called the *coefficient functions* of the p-form ω.

(vi) What are the coefficient functions of the 1-form df, where $f : X \to \mathbb{R}$ is a smooth function? The answer is in the elementary calculus that you have studied. These are nothing but the partial derivatives $\dfrac{\partial f}{\partial x_i}$. Thus,

$$df = \sum_{i=1}^{n} \frac{\partial f}{\partial x_i} dx_i. \tag{2.38}$$

(vii) A very important aspect of forms is the beautiful way they transform. Consider a smooth map $\phi : X \to Y$. Then for each $x \in X$, the map $d\phi_x : \mathbb{R}^n \to \mathbb{R}^m$ is a linear map. This in turn induces a linear map

$$\wedge^p(d\phi_x) : \wedge^p(\mathbb{R}^{m*}) \to \wedge^p(\mathbb{R}^{n*}).$$

Thus, given a p-form ω on Y, we can "*pull it back*" to a p-form on X as follows:

$$\phi^*(\omega)(x) := \wedge^p(d\phi_x)(\omega)(\phi(x)). \tag{2.39}$$

It is routine to verify that for each p, ϕ^* itself is a linear map. Further, they are algebra homomorphisms also, in the sense

$$\phi^*(\omega \wedge \tau) = \phi^*(\omega) \wedge \phi^*(\tau) \tag{2.40}$$

Let us work out this one completely in a simple situation first. Let X, Y be open subsets of some Euclidean spaces and let $\phi : X \to Y$ be a smooth function. For the sake of clarity and to be conformal with standard practice, let us denote the coordinate projections on X, Y by $\{x_1, \ldots, x_n\}$ and $\{y_1, \ldots, y_m\}$, respectively. For $x \in X$, put $y = \phi(x)$. Put $\pi_i \circ \phi = \phi_i$. Consider the 1-forms dy_i on Y. We have

$$\begin{aligned} \phi^*(dy_i)(x) &= (d\phi_x)^*(dy_i) \\ &= (d\phi_x)^*(d(\pi_i)_y) \\ &= d(\pi_i)_y \circ d(\phi_x) \\ &= d(\pi_i \circ \phi)_x = d(\phi_i)(x). \end{aligned} \tag{2.41}$$

Now using the homomorphic property of ϕ^*, we can say something on p-forms. (For simplicity, we don't write the point x at which the action is taking place.)

$$\phi^*(dy_I) = \phi^*(dy_{i_1} \wedge \cdots \wedge dy_{i_p}) = d\phi_{i_1} \wedge \cdots \wedge d\phi_{i_p} =: d\phi_I. \tag{2.42}$$

It follows that

$$\phi^*\left(\sum_I f_I dy_I\right) = \sum_I (f_I \circ \phi) d\phi_I. \tag{2.43}$$

To summarize, the pullback of forms under a smooth map is completely determined in terms of the pullback of 0-forms and 1-forms.

Observe that if ϕ is a diffeomorphism then $d\phi_x$ is an isomorphism for all x and hence so is ϕ^* for each p.

Definition 2.4.2 A p-form on an open subset of $\mathbb{R}^n$ is said to be *continuous/smooth* if all its coefficient functions are continuous/smooth.

Remark 2.4.1

1. We leave the verification that the smoothness of a p-form is independent of the local parameterization chosen to the reader as an exercise.

2. The set of all smooth p-forms is a vector subspace of the space of all p-forms. This subspace will be denoted by $\Omega^p(X)$. Note that $\Omega^0(X) = \mathcal{C}^\infty(X;\mathbb{R})$, the ring of real valued $\mathcal{C}^\infty$-functions on X. Each $\Omega^p(X)$ is a module over $\Omega^0(X)$.

   Moreover, the exterior product of smooth forms turns out to be smooth. Thus,

$$\Omega^*(X) := \Omega^0(X) \oplus \Omega^1(X) \oplus \cdots \oplus \Omega^n(X), \tag{2.44}$$

   forms an anticommutative graded algebra over the ring of smooth functions $\Omega^0(X)$. Given a smooth map $f : X \to Y$, we have the algebra homomorphism $\Omega(f) : \Omega^*(Y) \to \Omega^*(X)$, with the property that

$$\Omega(g \circ f) = \Omega(f) \circ \Omega(g); \quad \Omega(Id) = Id. \tag{2.45}$$

Exercise 2.4 If $\eta : \mathbb{R} \to \mathbb{R}^2$ is the inclusion map $x \mapsto (x, 1)$ what are $\eta^*(dx_i), i = 1, 2$? Generalize your result for coordinate inclusions $\mathbb{R}^n \to \mathbb{R}^{n+k}$ and p-forms dx_I.

2.5 Exterior Differentiation

Throughout this section X, Y etc. will denote open subsets of $\mathbf{H}^N$ for some N.

We have used the concept of derivative of a function to define a 1-form associated with a smooth function:

$$\phi \mapsto d\phi \tag{2.46}$$

which now defines a function $d : \Omega^0(X) \to \Omega^1(X)$. The elementary properties of the derivative are all present in this function and can be summarized as follows:
(i) $d(\alpha\phi + \beta\psi) = \alpha d(\phi) + \beta d(\psi),\ \alpha, \beta \in \mathbb{R}$;
(ii) $d(\phi\psi) = \phi d(\psi) + \psi d(\phi)$.
(iii) If $f : X \to Y$ is a smooth map, then

$$f^* \circ d = d \circ f^*.$$

The last one is a consequence of the chain rule for differentiation.

Our aim now is to extend this operator to all over $\Omega^*(X)$, so that all these properties are preserved. That is, we would like to define $d : \Omega^p(X) \to \Omega^{p+1}(X)$ for each p which should satisfy (i), (ii), and (iii). Property (i) tells us that it is enough to define d on the basis elements dx_I. The correct extension of property (ii) means that

$$\text{(iie)} \qquad d(\phi \wedge \psi) = \phi \wedge d(\psi) + d(\phi) \wedge \psi. \tag{2.47}$$

This relates d defined on Ω^p with that defined on Ω^{p-1}, for $p \geq 2$, and so on.

Thus, it follows that we are only free to define d on Ω^1. We have reduced the entire work to the task of defining d of dx_i. Here again, we fall back on our knowledge of calculus. Recall that for any smooth function $\phi : U \to \mathbb{R}$, the derivative is a smooth map $Df : U \to (\mathbb{R}^n)^*$ and its derivative is a smooth map $D^2 f : U \to M(n;\mathbb{R})$. Thus, we can think of $D^2 f$ as a 2-tensor valued function on U. In order to get an alternating tensor out of it, we follow the good old method of "alternatising" this, viz., Alt $(D^2 f)$. But the entries of $D^2(f)(x)$ are nothing but $\dfrac{\partial^2 f}{\partial x_i \partial x_j}(x)$ and hence it follows that $D^2 f$ is a symmetric 2-tensor. Therefore, Alt $(D^2 f) = 0$. In particular, we are led to define

$$d(dx_i) = 0. \tag{2.48}$$

Combine this with (i) and (iie) to get,

$$d(df) = \sum d\left(\sum_i \frac{\partial f}{dx_i} dx_i\right) = \sum_{i,j} \frac{\partial^2 f}{\partial x_i \partial x_j} dx_j \wedge dx_i = 0, \tag{2.49}$$

the last equality is justified because the terms $\frac{\partial^2 f}{\partial x_i \partial x_j} dx_j \wedge dx_i$ and $\frac{\partial^2 f}{\partial x_j \partial x_i} dx_i \wedge dx_j$ cancel out each other.

As pointed out before, together with the definition of d on $\Omega^0(X)$, this completes the definition of d on the entire of $\Omega^*(X)$. Let us go through this carefully, once again.

To start with we have the operator d on $\Omega^0(X)$. Then we define $d(dx_i) = 0$ for all i. Next, for a general 1-form $\phi = \sum_i f_i dx_i \in \Omega^1(X)$, by property (i), we have

$$d(\phi) = \sum_i d(f_i dx_i). \tag{2.50}$$

By property (iie) we get,

$$d(f_i dx_i) = f_i d(dx_i) + d(f_i) \wedge dx_i = df_i \wedge dx_i. \tag{2.51}$$

Thus,

$$d(\phi) = \sum_i df_i \wedge dx_i. \tag{2.52}$$

Further, for $p \geq 2$, and $I \in S(p, n)$ we have,

$$d(dx_I) = \sum_{j=1}^{p} dx_{i_1} \wedge \cdots \wedge d(dx_{i_j}) \wedge \cdots \wedge dx_{i_p} = 0. \tag{2.53}$$

Finally, for any p-form, $\phi = \sum_I f_I dx_I \in \Omega^p(X)$, we have,

$$d(\phi) = \sum_I d(f_I dx_I) = \sum_I df_I \wedge dx_I. \tag{2.54}$$

That the operator d satisfies (i) and (iie) is clear from the way we have defined d. We call d the *exterior derivative.*

As a bonus, we also have

(iv) $$d^2 = 0. \tag{2.55}$$

Let us prove this. In (2.53), we have verified that $d(dx_I) = 0$ for all I. Now from (2.54),

$$d(d(\phi)) = \sum_I d(df_I \wedge dx_I) = \sum_I d(df_I) \wedge dx_I = 0 \tag{2.56}$$

since $d(df_I) = 0$ from (2.49) for each I.

Can there be another operator $\hat{d}$ say, which agrees with d on $\Omega^0(X)$ and satisfies the properties (i), (iie), and (iv)? The answer is NO, which may give you a mild surprise. Let us see why this is so. First of all, since on 0-forms $\hat{d}$ should be the same as d, we have, $\hat{d}(x_i) = dx_i$ and hence it follows that $\hat{d}dx_I = 0$ for all I. Therefore, $\hat{d}(\phi) = \sum_I \hat{d}(f_I) \wedge dx_I = \sum_I d(f_I) \wedge dx_I = d(\phi)$.

In this sense the exterior derivation d is unique on open subsets of Euclidean spaces.

We may now worry about property (iii) of the extended operator d. Not really, because, we have another surprise bonus: Suppose the statement is true for ϕ and ψ then

$$\begin{aligned} f^*d(\phi \wedge \psi) &= f^*(d(\phi) \wedge \psi + \phi \wedge d(\psi)) \\ &= f^*(d(\phi)) \wedge f^*(\psi) + f^*(\phi) \wedge f^*(d(\psi)) \\ &= d(f^*(\phi)) \wedge f^*(\psi) + f^*(\phi) \wedge d(f^*(\psi)) \\ &= d(f^*(\phi) \wedge f^*(\psi)) \\ &= d(f^*(\phi \wedge \psi)). \end{aligned}$$

Now, to begin with, (iii) is valid on $\Omega^0(X)$. Also if $\phi = d\omega$ is a 1-form then $f^*(d(\phi)) = f^*(0) = 0$. On the other hand,

$$d(f^*(\phi)) = d(d(f^*(\omega))) = 0.$$

In particular, the formula is valid for 1-forms dx_i. Using the above, it is valid for all $f_I dx_I$ in $\Omega^p(X), p \geq 1$. Finally, because of the linearity of the two sides, it is then valid for sums of such elements in $\Omega^p(X), p \geq 1$ also. Thus, property (iii) is also valid.

Example 2.5.1 Let X be an open subset of $\mathbf{H}^3$. For a smooth function $f : X \to \mathbb{R}$, we have,

$$df = \sum_{i=1}^{3} \frac{\partial f}{\partial x_i} dx_i. \tag{2.57}$$

The vector formed by the coefficient functions is nothing but the

$$\text{grad } f = \left(\frac{\partial f}{\partial x_1}, \frac{\partial f}{\partial x_2}, \frac{\partial f}{\partial x_3} \right). \tag{2.58}$$

Thus, for a 0-form f, we can say

$$df = \text{grad } f. \tag{2.59}$$

Next, given any smooth 1-form $\omega = f_1 dx_1 + f_2 dx_2 + f_3 dx_3$, we have,

$$d(\omega) = df_1 \wedge dx_1 + df_2 \wedge dx_2 + df_3 \wedge dx_3$$

Use (2.57) and the anticommutativity of $\wedge$ to see that

$$\begin{aligned} d(\omega) &= \left(\frac{\partial f_2}{\partial x_1} - \frac{\partial f_1}{\partial x_2} \right) dx_1 \wedge dx_2 \\ &+ \left(\frac{\partial f_3}{\partial x_2} - \frac{\partial f_2}{\partial x_3} \right) dx_2 \wedge dx_3 + \left(\frac{\partial f_1}{\partial x_3} - \frac{\partial f_3}{\partial x_1} \right) dx_3 \wedge dx_1 \end{aligned}$$

Choosing the correct order for basis elements of $\wedge^2(\mathbb{R}^{3^*})$, viz.,

$$\{dx_2 \wedge dx_3, dx_3 \wedge dx_1, dx_1 \wedge dx_2\} = \{\mathbf{i}, \mathbf{j}, \mathbf{k}\},$$

we see that the vector formed out of the coefficients of $d(\omega)$ is

$$\left(\frac{\partial f_3}{\partial x_2} - \frac{\partial f_2}{\partial x_3}, \frac{\partial f_1}{\partial x_3} - \frac{\partial f_3}{\partial x_1}, \frac{\partial f_2}{\partial x_1} - \frac{\partial f_1}{\partial x_2} \right) = \text{curl}\,(f_1, f_2, f_3).$$

Thus, for a 1-form ω, we may say

$$d\omega = \text{curl } \omega. \tag{2.60}$$

Let now $\tau = g_1 dx_2 \wedge dx_3 + g_2 dx_3 \wedge dx_1 + g_3 dx_1 \wedge dx_2$ be a smooth 2-form. Then

$$d(\tau) = \left(\frac{\partial g_1}{\partial x_1} + \frac{\partial g_2}{\partial x_2} + \frac{\partial g_3}{\partial x_3} \right) dx_1 \wedge dx_2 \wedge dx_3. \tag{2.61}$$

Here the coefficient function gives us the divergence, div (g_1, g_2, g_3). Thus, we may say that for a 2-form τ,

$$d\tau = \text{div } (\tau). \tag{2.62}$$

Of course, $d(\phi) = 0$ for any 3-form ϕ.
In this sense, $d \circ d = 0$ has the following familiar interpretation:

$$\text{curl(grad)} = 0 \quad \& \quad \text{div(curl)} = 0. \tag{2.63}$$

Exercise 2.5

1. Show that $d(f) = 0$ for a 0-form on $\mathbb{R}^n$ iff f is a constant.

2. Show that $d(\omega) = 0$ for a 1-form ω on $\mathbb{R}^n$ iff $\omega = d(f)$ for some 0-form f.

3. Show that $d(\tau) = 0$ for a 2-form on $\mathbb{R}^n$ iff $\tau = d(\omega)$ for a 1-form ω.

4. Generalize the above three results to the case when $\mathbb{R}^n$ is replaced by an open convex subset.

5. Which of the above results hold for $\mathbb{R}^2 \setminus \{0\}$ also?

2.6 Integration on Singular Chains

In this section, we shall once again, use the notation $I := [0,1]$, for the closed unit interval in $\mathbb{R}$. Thus, I^n will denote the n-fold Cartesian product of I which is the unit n-cube in $\mathbb{R}^n$. We shall consider topological spaces X, Y etc., which are all subspaces of Euclidean spaces.

Definition 2.6.1 For $n \geq 1$, by a *singular n-cube* in X, we mean a smooth map $\gamma : I^n \to X$. By a *singular* 0*-cube* in X we mean a point of X.

Remark 2.6.1 Here the adjective "singular" is used to indicate that the map need not be injective. Indeed, we should have used the terminology "singular smooth n-cubes" so as to allow room for continuous functions $\gamma : I^n \to X$ also as 'singular n-cubes'. For our limited purpose of integration theory, it is necessary and sufficient to restrict ourselves to smooth functions only.

Thus, a singular 1-cube γ is nothing but a smooth curve in X. Notice that the image of this curve may be a single point. Here the emphasis is on the function γ rather than its image set as a subset of X. Thus a singular 1-cube γ is a parameterization of the image curve $\gamma[I]$.

Example 2.6.1 The most important singular n-cube is the identity map of I^n to itself. We shall denote this by ξ_n. We shall also denote the inclusion map $I^n \hookrightarrow \mathbb{R}^n$ by ξ_n, when there is no confusion. Next, we have the face-cubes $F^n_{j,i} : I^n \to I^{n+1}$ for each $1 \le j \le n+1$ and $i = 0, 1$ defined by

$$F^n_{j,i}(x_1, \ldots, x_n) = (x_1, \ldots, x_{j-1}, i, x_j, \ldots, x_n).$$

For $i = 0, 1$, they are called *back* (respectively) *front j-face* of ξ_{n+1}. For instance, the two faces of ξ_1 are $\{0\}$ and $\{1\}$. The singular 2-cube ξ_2 has four singular 1-cubes as its faces. Indeed, ξ_n has $2n$ faces.

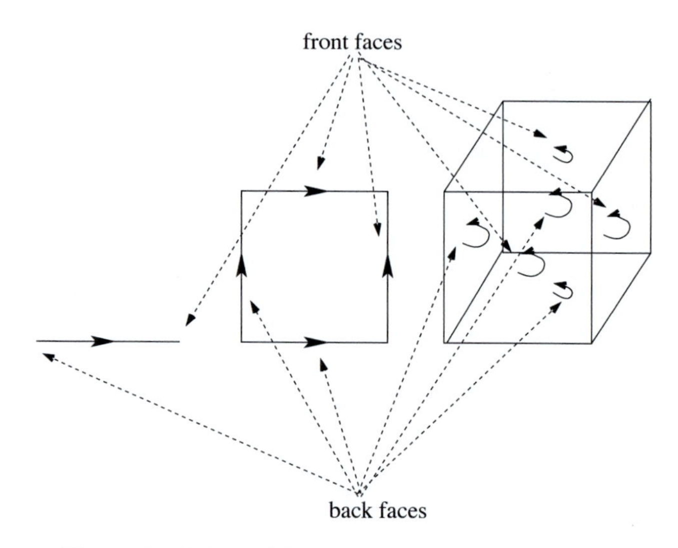

Figure 11 Oriented faces of cubes of dimension ≤ 3.

Example 2.6.2 The map $t \mapsto e^{2\pi\imath t}$ defined on $[0,1]$ is a singular 1-cube that represents the oriented circle. Likewise the spherical coordinates $[0,1] \times [0,1] \to \mathbb{R}^3$ given by

$$(t,s) \mapsto \left(\cos 2\pi t \cos \pi\frac{2s-1}{2}, \sin 2\pi t \cos \pi\frac{2s-1}{2}, \sin \pi\frac{2s-1}{2}\right) \tag{2.64}$$

is a singular 2-cube which represents the oriented 2-sphere.

Definition 2.6.2 Let $\mathfrak{s}_n(X)$ denote the set of all singular n-cubes in X. By a *n-chain* in X we mean a function $c : \mathfrak{s}_n(X) \to \mathbb{Z}$ with the property that $c(\gamma) \neq 0$ only for finitely many $\gamma \in \mathfrak{s}_n(X)$. The set of all n-chains on X will be denoted by $S_n(X)$.

Proposition 2.6.1 Pointwise addition of two members makes $S_n(X)$ into an abelian group, i.e.,

$$(c_1 + c_2)(\gamma) = c_1(\gamma) + c_2(\gamma).$$

Remark 2.6.2 The chain that sends every singular n-cube to 0 is the additive identity of the abelian group $S_n(X)$ and so, will be denoted by 0 itself. For a chain c, suppose $\{\gamma_1, \ldots, \gamma_k\}$ is the set of all singular n-cubes on which c is nonzero. It is customary to denote c by the formal sum $\sum_i n_i\gamma_i$, where $n_i = c(\gamma_i)$. This notation has a lot of advantage without loosing any rigor. Thus, a singular n-cube γ gets identified with an n-chain which takes the value 1 on γ and 0 on every other singular n-cube. Thus, the negative of γ is nothing but $-\gamma$. More generally, the negative of a chain $\sum_i n_i\gamma_i$ is nothing but $-\sum_i n_i\gamma_i$. If you know what is a free abelian group, then you will be able to see that $S_n(X)$ is actually a free abelian group with $\mathfrak{s}_n(X)$ as a basis. In particular, $S_0(X)$ is a free abelian group with X itself as a basis.

Definition 2.6.3 The boundary $\partial\gamma$ of a singular $(n+1)$-cube is

$$\partial\gamma = \sum_{i=0}^{1}\sum_{j=1}^{n+1}(-1)^{i+j}\gamma\circ F^n_{j,i}.$$

For any $(n+1)$-chain $c = \sum_i n_i\gamma_i$, we define ∂c by

$$\partial c = \sum_i n_i\partial(\gamma_i).$$

It is then clear that $\partial : S_{n+1}(X) \to S_n(X)$ is a group homomorphism. It is called the *boundary operator.*

For $n < 0$, we define $S_n(X) = 0$ and then naturally, take $\partial : S_{n+1}(X) \to S_n(X)$ to be 0 as well.

Definition 2.6.4 Given a smooth map $f : X \to Y$, we define $f_* : S_n(X) \to S_n(Y)$ by

$$f_*\left(\sum_i n_i\gamma_i\right) = \sum_i n_i f\circ\gamma_i.$$

Remark 2.6.3 It is not difficult to check that
(i) f_* is a group homomorphism;
(ii) If $f : X \to Y, g : Y \to Z$ are smooth functions then $(g\circ f)_* = g_*\circ f_*$;
(iii) $(Id)_* = Id$.

We also have one more naturality property:

Lemma 2.6.1 $\partial\circ f_* = f_*\circ\partial$.

Proof: For any smooth map $f : X \to Y$ and any $(n+1)$-chain c in X, we have to verify that $\partial\circ f_*(c) = f_*\partial(c)$. Since the singular $(n+1)$-cubes form a basis for $S_{n+1}(X)$, it suffices to verify this for singular $(n+1)$-cubes γ in place of c. But then

$$\begin{aligned} f_*\partial(\gamma) &= f_*(\textstyle\sum_i\sum_j(-1)^{i+j}\gamma\circ F^n_{j,i}) = \sum_i\sum_j(-1)^{i+j}f\circ(\gamma\circ F^n_{j,i}) \\ &= \textstyle\sum_i\sum_j(-1)^{i+j}(f\circ\gamma)\circ F^n_{j,i} = \partial(f\circ\gamma) \\ &= \partial\circ f_*(\gamma). \end{aligned}$$

This completes the proof. ♠

Remark 2.6.4 Notice that the assignments $f \rightsquigarrow f^*$ on tensors and $f \rightsquigarrow \Omega(f)$ on differential forms also have similar properties as seen in Sections 2.3, 2.4. It may interest you to know that the operator ∂ also has the property $\partial\circ\partial = 0$. This indicates a deep connection between the operator ∂ on smooth chains on the one hand and and the operator d on differential forms on the other, which leads to what is known as homology theory, which is 'dual' to the 'cohomology' that we are going to introduce in Chapter 4. It may be remarked that by the linearity and the naturality properties of ∂, in order to prove $\partial^2 = 0$, it is sufficient to prove that $\partial^2(\xi_n) = 0$. And the proof of $\partial^2(\xi_n) = 0$ is combinatorial in nature and is very straightforward. Try it.

Definition 2.6.5 Given an n-form ω on I^n, we know that there is a unique (continuous) function f such that $\omega = f dx_1 \wedge \cdots \wedge dx_n$. We define

$$\int_{I^n} \omega = \int_{I^n} f dx_1 \wedge \cdots \wedge dx_n := \int_{I^n} f(x_1, \ldots, x_n) dx_1 \cdots dx_n = \int_{I^n} f.$$

Next, if γ is a singular n-cube in X and ω is an n-form on X, then we define

$$\int_{\gamma} \omega := \int_{\xi_n} \gamma^* \omega := \int_{I^n} \gamma^* \omega.$$

And finally, we extend this definition to all singular n-chains $c = \sum_i n_i \omega_i$ by linearity:

$$\int_c \omega = \sum_i n_i \int_{\gamma_i} \omega = \sum_i n_i \int_{I^n} \gamma_i^* \omega.$$

Example 2.6.3

(i) Let $\gamma : [0,1] \to \mathbb{R}^3$ be a smooth function. Let ω be a smooth 1-form on a neighbourhood of image C of γ. Then

$$\int_C \omega = \int_{[0,1]} \gamma^*(\omega).$$

If we write $\gamma(t) = (\gamma_1(t), \gamma_2(t), \gamma_3(t))$ and $\omega = f_1 dx_1 + f_2 dx_2 + f_3 dx_3$, then

$$\gamma^*(\omega) = \left(\sum_{i=1}^{3} f_i(\gamma(t)) \frac{d\gamma_i}{dt}(t) \right) dt,$$

where dt denotes the 1-form on the interval, which assigns the identity map at each point of $[0,1]$. Thus,

$$\int_C \omega = \int_0^1 \left(\sum_{i=1}^{3} f_i(\gamma(t)) \frac{d\gamma_i}{dt}(t) \right) dt = \int_0^1 \mathbf{F} \cdot d\gamma$$

which is the familiar line integral of the vector field $\mathbf{F} = (f_1, f_2, f_3)$ on the curve γ.

(ii) Suppose $f : U \to \mathbb{R}$ is a smooth map defined in a neighbourhood U of the image of γ, and $\omega = df$. Then

$$\int_C \omega = \int_a^b \gamma^*(df) = \int_a^b (f \circ \gamma)' dt = f(\gamma(b)) - f(\gamma(a)).$$

In particular, this proves that on all closed curves γ in X, $\int_{\gamma} df = 0$. Later, we shall generalize this result.

(iii) Let now

$$\gamma(t) = (ar\cos\theta, br\sin\theta), \quad (0 \le \theta \le 2\pi, 0 \le r \le 1)$$

be a parameterization of the elliptical region for some $a, b > 0$. Then, $\partial\gamma$ is a parameterization of the ellipse and we have

$$\int_{\partial\gamma} x dy = \int_0^{2\pi} ab\cos^2\theta \, d\theta = \pi ab.$$

On the other hand, $d(x\,dy) = dx \wedge dy$ and we have

$$\begin{aligned} \gamma^*(dx \wedge dy) &= (-ar\sin\theta\,d\theta + a\cos\theta\,dr) \wedge (b\sin\theta\,dr + br\cos\theta\,d\theta) \\ &= abr\,dr \wedge d\theta. \end{aligned}$$

Therefore,

$$\int_\gamma d(x\,dy) = \int_\gamma dx \wedge dy = \int_{I\times[0,2\pi]} \gamma^*(dx \wedge dy) = ab\pi.$$

That these two integrals are the same is not a coincidence but a special case of Green's theorem.

(iv) Let now $F : [0,1]^2 \to \mathbb{R}^3$ be a parameterization of a smooth surface in $\mathbb{R}^3$. Let

$$\omega = g_1 dx_2 \wedge dx_3 + g_2 dx_3 \wedge dx_1 + g_3 dx_1 \wedge dx_2$$

be a smooth 2-form defined in a neighbourhood of the image S of F. Then,

$$\int_S \omega = \int_{[0,1]^2} F^*(\omega).$$

Write

$$F(u_1, u_2) = (f_1(u_1,u_2), f_2(u_1,u_2), f_3(u_1,u_2))$$

Then,

$$F^*(dx_i) = \sum_{j=1}^{2} \frac{\partial f_i}{\partial u_j} du_j,$$

for each i and hence,

$$F^*(dx_k \wedge dx_l) = \left(\frac{\partial f_k}{\partial u_1}\frac{\partial f_l}{\partial u_2} - \frac{\partial f_l}{\partial u_1}\frac{\partial f_k}{\partial u_2} \right) du_1 \wedge du_2.$$

Write

$$\begin{aligned} n_1(du_1 \wedge du_2) &= F^*(dx_2 \wedge dx_3), \\ n_2(du_1 \wedge du_2) &= F^*(dx_3 \wedge dx_1), \\ n_3(du_1 \wedge du_2) &= F^*(dx_1 \wedge dx_2). \end{aligned}$$

Then check that

$$\mathbf{N} := (n_1, n_2, n_3) = \frac{\partial F}{\partial u_1} \times \frac{\partial F}{\partial u_2}.$$

$\mathbf{N}$ is the fundamental vector product of the parameterized surface S. Let $\mathbf{n} = \dfrac{\mathbf{N}}{\|\mathbf{N}\|}$ denote the unit vector in the same direction as this and $dS = ||(n_1, n_2, n_3)||du_1 \wedge du_2$. Put $\mathbf{G} = (g_1 \circ F, g_2 \circ F, g_3 \circ F)$. Then,

$$F^*(\omega) = [(g_1 \circ F)n_1 + (g_2 \circ F)n_2 + (g_3 \circ F)n_3]du_1 \wedge du_2$$

and

$$\int_S \omega = \int_{[0,1]^2} F^*(\omega) = \int_{[0,1]^2} (\mathbf{G} \bullet \mathbf{n})\,dS.$$

The 2-form dS is called the *area form*. The vector (g_1, g_2, g_3) which corresponds to the given 2-form ω, represents the flux density, the integral depends only on its component normal to the surface on which integration is taken.

Put $X = [0,1] \times [0,\pi] \times [0, 2\pi]$. Let now $P : X \to \mathbb{R}^3$ be given by

$$P(r,\theta,\phi) = (r\sin\theta\cos\phi, r\sin\theta\sin\phi, r\cos\theta)$$

be spherical coordinate representation of the unit ball $\mathbb{D}^3$ in $\mathbb{R}^3$. One can easily verify that $P^*(dx\wedge dy\wedge dz) = r^2\sin\theta\, dr\wedge d\theta\wedge d\phi$ and

$$\int_P dx\wedge dy\wedge dz = \int_X P^*(dx\wedge dy\wedge dz) = \int_X r^2\sin\theta\, dr\wedge d\theta\wedge d\phi = \frac{4\pi}{3}.$$

On the other hand, take $\omega = z\, dx\wedge dy$ and check that $d(\omega) = dx\wedge dy\wedge dz$. If $F : [0,\pi]\times[0,2\pi] \to \mathbb{R}^3$ given by $F(\theta,\phi) = P(1,\theta,\phi)$ is the representation of the unit sphere, we have,

$$\int_F \omega = \int_{[0,\pi]\times[0,2\pi]} F^*(\omega) = \int_0^{\pi}\int_0^{2\pi} \cos^2\theta\sin\theta\, d\theta\wedge d\phi = \frac{4\pi}{3}.$$

Once again, the equality of the two expressions $\int_P d\omega$ and $\int_F \omega$ is not a coincidence but a very special case of the divergence theorem of Gauss.

We shall now present a sweeping generalization of these results as a reward for going through the seemingly meaningless algebrization that we have indulged in, in the preceding sections.

Theorem 2.6.1 Stokes' Theorem for Chains: *Let ω be a $(n-1)$-form on an open subset X of $\mathbb{R}^m$. Then for any singular n-chain c in X, we have,*

$$\int_c d\omega = \int_{\partial c} \omega. \tag{2.65}$$

Proof: Since both sides of the identity are linear in c, it suffices to prove (2.65) when $c = \gamma$ is a singular n-cube. But then

$$\int_\gamma d\omega = \int_{I^n} \gamma^* d(\omega) = \int_{I^n} d(\gamma^*\omega) = \int_{\xi_n} d(\gamma^*\omega)$$

whereas

$$\begin{aligned} \textstyle\int_{\partial\gamma}\omega &= \sum_{i,j}(-1)^{i+j}\int_{\gamma\circ F^n_{j,i}} \omega = \sum_{i,j}(-1)^{i+j}\int_{I^{n-1}} (\gamma\circ F^n_{j,i})^*\omega \\ &= \sum_{i,j}(-1)^{i+j}\int_{I^{n-1}} (F^n_{j,i})^*\circ\gamma^*\omega = \int_{\partial\xi_n} \gamma^*\omega. \end{aligned}$$

Now $\gamma^*\omega$ is just an $(n-1)$ form on I^n. Thus, it suffices to prove (2.65) for the singular n-cube ξ_n and an arbitrary $(n-1)$ form ω' on I^n, viz.,

$$\int_{\xi_n} d\omega' = \int_{\partial\xi_n} \omega'. \tag{2.66}$$

Let us introduce a temporary notation for the forms

$$d\mathbf{x} := dx_1\wedge\cdots\wedge dx_n;\quad d\widehat{\mathbf{x}_k} := dx_1\wedge\cdots dx_{k-1}\wedge dx_{k+1}\wedge\cdots\wedge dx_n.$$

For a fixed $\mathbf{x} = (x_1,\ldots,x_n)$ and $1 \le k \le n$ let

$$\widehat{\mathbf{x}_k} = (x_1,\ldots,x_{k-1},x_{k+1},\ldots,x_n);\quad g(\widehat{\mathbf{x}_k},t) = f(x_1,\ldots,x_{k-1},t,x_{k+1},\ldots,x_n).$$

Since each $(n-1)$-form ω is a finite sum:

$$\omega = \sum_k f_k d\widehat{\mathbf{x}_k},$$

it is enough to prove (2.66) when ω is of the form $f\, d\widehat{\mathbf{x}_k}$ for each $1 \le k \le n$. Now for such an ω we have,

$d\omega = d(f\, d\widehat{\mathbf{x}_k}) = \frac{\partial f}{\partial x_k} dx_k \wedge d\widehat{\mathbf{x}_k} = (-1)^{k-1} \frac{\partial f}{\partial x_k} d\mathbf{x}$. Therefore,

$$\begin{aligned}
\int_{\xi_n} d\omega &= (-1)^{k-1} \int_{I^n} \frac{\partial f}{\partial x_k} \\
&= (-1)^{k-1} \int_0^1 \cdots \left(\int_0^1 \frac{\partial f}{\partial x_k} dx_k \right) dx_1 \cdots \widehat{dx_k} \cdots\cdots dx_n \text{ (Fubini)} \\
&= (-1)^{k-1} \int_0^1 \cdots \int_0^1 [g(\widehat{\mathbf{x}_k}, 1) - g(\widehat{\mathbf{x}_k}, 0)] dx_1 \cdots \widehat{dx_k} \cdots dx_n \\
&= (-1)^{k-1} \int_{I^n} [g(\widehat{\mathbf{x}_k}, 1) - g(\widehat{\mathbf{x}_k}, 0)] dx_1 \cdots dx_n.
\end{aligned}$$

The last two steps are justified by the fundamental theorem of integral calculus of 1-variable and the simple fact that $\int_I 1 = 1$. On the other hand,

$$\int_{I^{n-1}} (F^n_{j,i})^*(\omega) = \begin{cases} 0, & \text{if } j \ne k, \\ \int_{I^n} g(\widehat{\mathbf{x}_k}, i) dx_1 dx_2 \cdots dx_n, & \text{if } j = k. \end{cases}$$

Therefore,

$$\begin{aligned}
\int_{\partial I^n} \omega &= \sum_{i,j} (-1)^{i+j} \int_{I^{n-1}} (F^n_{j,i})^*(\omega) \\
&= (-1)^{1+k} \int_{I^n} g(\widehat{\mathbf{x}_k}, 1) dx_1 \cdots dx_n + (-1)^k \int_{I^n} g(\widehat{\mathbf{x}_k}, 0) dx_1 \cdots dx_n.
\end{aligned}$$

The claim follows. ♠

Exercise 2.6 Verify Proposition 2.6.1.

2.7 Miscellaneous Exercises for Chapter 2

1. An element $\tau \in \wedge^p(V^*)$ is said to be *decomposable* if there exist $\phi_1, \dots, \phi_p \in V^*$ such that $\tau = \phi_1 \wedge \cdots \wedge \phi_p$. Clearly every element of $\wedge^1(\mathbb{R}^{n*})$ and $\wedge^n(\mathbb{R}^{n*})$ is decomposable. Show that every element of $\wedge^{n-1}(\mathbb{R}^{n*})$ is decomposable. In particular, it proves that every element of $\wedge^p(\mathbb{R}^{3*})$ is decomposable for all p.

2. Show that $\phi_1 \wedge \phi_2 + \phi_3 \wedge \phi_4$ is not decomposable if $\{\phi_1, \phi_2, \phi_3, \phi_4\}$ is an independent set in $\wedge^1 V^*$. (Hint: Any independent set in V^* can be completed to a basis.) Thus, in dimension ≥ 4, there are indecomposable elements.

3. Recall that given a linear map $f : \mathbb{R}^n \to \mathbb{R}$, there is a unique vector $\mathbf{w} \in \mathbb{R}^n$ such that $f(\mathbf{u}) = \mathbf{u} \bullet \mathbf{w}$, the dot product of the two vectors, for all $\mathbf{u} \in \mathbb{R}^n$. Given $n-1$ vectors $\mathbf{v}_1, \dots, \mathbf{v}_{n-1} \in V$, we define their *cross product* $C(\mathbf{v}_1, \dots, \mathbf{v}_{n-1}) =: \mathbf{w}$ by the property that $\det(\mathbf{v}_1, \dots, \mathbf{v}_{n-1}, \mathbf{u}) = \mathbf{u} \bullet \mathbf{w}$. Show that the cross product is multilinear and alternating. For $n = 3$, this is indeed the classical cross product: $C(\mathbf{v}_1, \mathbf{v}_2) = \mathbf{v}_1 \times \mathbf{v}_2$.

Chapter 3

Submanifolds of Euclidean Spaces

The basic motivation for the concept of a manifold is to be able to talk about differentiability of functions defined on such objects and then to be able to assign a meaning to the derivative of a differentiable function on such objects. In this chapter, first we introduce the basic concept of a manifold, though we shall restrict ourselves to those that are subspaces of Euclidean spaces. Next, we introduce the concept of tangent space and define the derivative of a differentiable function. We then introduce the reader to special types of maps such as immersions, submersions embeddings etc. In particular, we shall see how the concept of regularity is generalized to transversality. Finally, the concept of "perturbations" is realized in the precise form of a homotopy and some of the special types of maps above are shown to be "stable under small perturbations". This makes the study of these special maps more important.

3.1 Basic Notions

Definition 3.1.1 Let N, k, and r be fixed positive integers. Let X be a nonempty subspace of $\mathbb{R}^N$. We say X is a *k-manifold* (or *k-dimensional manifold*) of class $\mathcal{C}^r$, if for each $x \in X$ there exists an open neighborhood U_x of x in X and a $\mathcal{C}^r$-diffeomorphism $\phi : U_x \to \phi(U_x)$ onto an open subset $\phi(U_x)$ of $\mathbb{R}^k$. The integer k is called *the dimension* of X and the map ϕ is called a *chart* for X. Observe that there is no uniqueness in the choice of charts and in particular, we can choose a chart x so that $\phi(x) = 0$. Such a chart will be called a *local coordinate system at x* for X and U_x will be called a *coordinate neighborhood* of x. The inverse of ϕ will be called a *local parameterization.* Any collection of charts $\{(U_\alpha, \phi_\alpha)\}$ such that $X = \cup U_\alpha$ will be called an *atlas* for X. We also call X a *smooth manifold in $\mathbb{R}^N$* or simply a *differentiable manifold,* if we do not want to mention the class to which it belongs nor the Euclidean space where it lives.

Definition 3.1.2 A nonempty subset X of a Euclidean space is called a *0-dimensional manifold* if it is a discrete subset. If $X = \emptyset$, we call it a manifold of dimension -1.

Remark 3.1.1
(i) In the first definition above we have frozen the integer k. There is not much harm if we allow it to vary provided we assume that X is connected. For, because of the invariance of domain (see Theorem 1.6.1), it follows that $k(x)$ is locally a constant and hence a constant. However, we do not consider disjoint union of manifolds of different dimensions as a manifold.

(ii) Observe that we can allow $r = 0$ also but then we should drop out the word 'differentiable'. That will give us the definition of a *topological manifold* (for more, see Chapter 5).
(iii) Observe that any $\mathcal{C}^r$-manifold is a $\mathcal{C}^s$-manifold for $0 \le s \le r$. A deep theorem due to Whitney (see e.g., Ch.2 of [Hi]) tells us that any $\mathcal{C}^1$-manifold is $\mathcal{C}^1$-diffeomorphic to a $\mathcal{C}^\infty$-manifold. This is a partial justification for the way we use the word "smooth manifold" to mean a $\mathcal{C}^r$-manifold, where r could be any integer between 1 and ∞.
(iv) In the second definition, we want to emphasize the fact that X is not necessarily a closed subset of the Euclidean space. For instance, the set $\{1/n : n \in \mathbb{N}\}$ is a 0-dimensional manifold in $\mathbb{R}$, whereas its closure is not.

Example 3.1.1

1. Any nonempty open subset of $\mathbb{R}^k$ is a k-manifold. Indeed, any nonempty open subspace of a k-manifold is again a k-manifold.

2. All vector subspaces of $\mathbb{R}^n$ are manifolds. Suppose V is a k-dimensional subspace. Choose a basis $\{\mathbf{v}_1, \dots, \mathbf{v}_k\}$ for V and consider the map $\phi : \mathbb{R}^k \to V$ given by

$$(x_1, \dots, x_k) \mapsto \sum_{i=1}^{k} x_i \mathbf{v}_i.$$

   This is clearly a bijective linear map of $\mathbb{R}^k$ onto V. Hence it defines a global parameterization of V which makes it a k-dimensional manifold.

3. A Cartesian product of any two manifolds is again a manifold. Inductively, this will hold for a Cartesian product of any finite number of manifolds. Observe that the dimension adds up except when you take an empty product.

4. Given any open subset $U \subset \mathbb{R}^n$ and a smooth map $f : U \to \mathbb{R}$, the graph

$$\Gamma_f = \{(\mathbf{x}, f(\mathbf{x})) : \mathbf{x} \in U\}$$

   is a manifold in $\mathbb{R}^{n+1}$. It is covered by a single parameterization, viz., $\mathbf{x} \mapsto (\mathbf{x}, f(\mathbf{x}))$. This also shows that Γ_f is actually diffeomorphic to U. By combining the above examples we can produce a few more examples but not many.

5. For $n \ge 2$, consider the $(n-1)$-dimensional sphere $\mathbb{S}^{n-1}$ in $\mathbb{R}^n$ given by the following equation:

$$x_1^2 + \cdots + x_n^2 = 1. \tag{3.1}$$

   Put $U_\pm = \mathbb{S}^{n-1} \setminus \{\pm N\}$ where $N = (0,\dots,0,1)$. Let $\eta_\pm : U_\pm \to \mathbb{R}^{n-1}$ be the stereographic projections (see Example 1.6.3). It follows that $\{(U_+, \eta_+), (U_-, \eta_-)\}$ is a $\mathcal{C}^\infty$-atlas for $\mathbb{S}^{n-1}$, showing that $\mathbb{S}^{n-1}$ is a smooth manifold.

   Later on, we shall see that this is a special case of a phenomenon, viz., *a nonempty regular level set of a smooth function on an n-manifold is an $(n-1)$-manifold.*

6. You are familiar with the representation of the unit circle by $(\cos\theta, \sin\theta), 0 \le \theta \le 2\pi$. For $0 < \theta < 2\pi$ this is nothing but a parameterization of the unit circle which covers the whole circle except the point $(1, 0)$. Even the point $(1, 0)$ is covered by the map but we see that the map does not define a parameterization in the sense that we have introduced above. This can be slightly rectified by taking a larger open interval $(-\epsilon < \theta < 2\pi + \epsilon)$, and taking the same map and then restricting it to suitable smaller open intervals, depending on the point that we want to cover so as to get a

parameterization of the circle by a single map which is allowed to be not injective globally. Such maps are called "global parameterizations". In this sense, the circle is indeed a very special manifold–very few manifolds admit such parameterizations.

7. Having said that, we can generalize this aspect of the circle a little bit. Take any parameterized smooth curve C in $\mathbb{R}^2$. For definiteness assume that it does not intersect the y-axis. Now rotate the curve around the y-axis inside $\mathbb{R}^3$ to obtain a smooth surface $\rho(C)$. It is easy to give a "global parameterization" of $\rho(C)$ starting with a parameterization $\gamma : [0,1] \to \mathbb{R}^2$ of C : Observe that each point $(\gamma_1(t), \gamma_2(t), 0)$ is at a distance $|\gamma_1(t)|$ from the y-axis, and hence when rotated about the y-axis, traces the circle with center $(0, \gamma_2(t), 0)$ and radius $|\gamma_1(t)|$ in the plane $y = \gamma_2(t)$. Therefore, $\Gamma : [0,1] \times [0, 2\pi] \to \mathbb{R}^3$ given by the above formula

$$(t, \psi) \mapsto (\gamma_1(t) \cos \psi, \gamma_2(t), \gamma_1(t) \sin \psi)$$

is a parameterization of the surface $\rho(C)$. A typical example of this is when we take the curve C to be a circle disjoint from the y-axis, say,

$$\gamma(\theta) = (2 + \cos \theta, \sin \theta, 0)$$

the circle with centre $(2, 0)$ and radius 1. The surface of rotation is nothing but a *torus* given by the global parameterization

$$(\theta, \psi) \mapsto ((2 + \cos \theta) \cos \psi, \sin \theta, (2 + \cos \theta) \sin \psi)).$$

8. As a typical example of a subset of $\mathbb{R}^2$, that is not a manifold, consider X to be the union of the two coordinate axes. Observe that at each point $x \in X \setminus \{(0,0)\}$ the required condition is satisfied with $k = 1$. However, for $x = (0,0)$, no neighborhood U of $(0,0)$ in X can be even homeomorphic to $\mathbb{R}$. For, $U \setminus \{x\}$ will have at least four connected components, whereas, if we remove a point from $\mathbb{R}$, we get only two connected components.

9. Another typical example of a subspace of $\mathbb{R}$, that is not a manifold is a closed interval $[0,1]$. No neighborhood of the point 0 in the interval $[0,1]$ can be diffeomorphic (homeomorphic) to an open subset of $\mathbb{R}$. Can you see why?

Remark 3.1.2
(i) Let $X \subset \mathbb{R}^N$ be an n-manifold, $x \in X$, and let $\phi : \mathbb{R}^n \to X$ be a parameterization for X at $x \in X$. Then for all $z \in \mathbb{R}^n$, $D(\phi)_z : \mathbb{R}^n \to \mathbb{R}^N$ is a linear map of rank n and hence injective. For, if $\psi : U \to \mathbb{R}^n$ is the inverse of ϕ, then there exists $\hat{\psi} : W \to \mathbb{R}^n$, which is a smooth extension of ψ to a neighborhood W of x in $\mathbb{R}^N$. Now $\hat{\psi} \circ \phi(z) = \psi \circ \phi(z) = z$ for all $z \in \mathbb{R}^n$. Therefore, $D(\hat{\psi})_{\phi(z)} \circ D(\phi)_z = Id$.
(ii) In the situation described above, applying the injective form of implicit function theorem (1.4.5), we get a neighborhood W of $x \in \mathbb{R}^N$ and a diffeomorphism $\tau : W \to \mathbb{R}^N$ such that

$$\tau \circ \phi(y_1, \ldots, y_n) = (y_1, \ldots, y_n, 0, \ldots, 0).$$

Therefore,

$$X \cap W = \phi(\mathbb{R}^n) = \tau^{-1}(\mathbb{R}^n \times 0) = \{z \in W \ : \ \tau_i(z) = 0, \ \ i \geq n+1\}.$$

Here for each $1 \leq i \leq N$, τ_i denotes the ith-coordinate function of τ. In other words, every manifold in $\mathbb{R}^N$ is locally equal to the zero set of some "coordinate functions".

Definition 3.1.3 Let $X \subset \mathbb{R}^N$ be a manifold. A subspace Y of X is called a *submanifold* if $Y \subset \mathbb{R}^N$ is a manifold on its own.

Remark 3.1.3
(i) There is need for caution in this definition, while we are dealing with "abstract manifolds". We shall discuss this point in Chapter 5.
(ii) It is natural to expect that the dimension of a submanifold is less than or equal to that of the manifold. Indeed, we have the following stronger result that can be proved exactly as in (ii) of the above remark.

Theorem 3.1.1 *Let X be a k-dimensional submanifold of a n-dimensional manifold Y. Then for each $a \in X$, there exists a neighborhood W of a in $\mathbb{R}^N$ and a diffeomorphism $\tau : W \to \mathbb{R}^N$ such that*
(i) $\tau(a) = 0$;
(ii) $Y \cap W = \{x \in W \ : \ \tau_i(x) = 0, i \geq n+1\}$;
(iii) $X \cap W = \{x \in W \ : \ \tau_i(x) = 0, i \geq k+1\}$.

Exercise 3.1
Let $A = \{\mathbf{x} \in \mathbb{S}^{n-1} \ : \ x_1 \geq 0\}$ be the closed right-half sphere.

1. Use the stereographic projection ϕ_+ to see that $A \setminus \{(0,\ldots,1)\}$ is diffeomorphic to the closed right-half space $\{\mathbf{y} \in \mathbb{R}^{n-1} \times 0 \ : \ y_1 \geq 0\}$.
2. Given any $p \in \partial\mathbb{D}^{n-1}$ find a diffeomorphism $\alpha : \mathbb{D}^{n-1} \to A$ such that $\alpha(p) = (0,0,\ldots,1)$.
3. Conclude that if p is any point on the boundary of a closed disc $\mathbb{D}^{n-1}$ then $\mathbb{D}^{n-1} \setminus \{p\}$ is diffeomorphic to the closed upper-half space $\mathbf{H}^{n-1}$.

3.2 Manifolds with Boundary

In (7) of Example 3.1.1, we saw that even a closed interval is not a manifold according to Definition 3.1.1. Clearly, there is a need to extend the definition of a manifold to include objects such as a closed interval, a closed disc, the half-spaces in $\mathbb{R}^n$ etc. The key is to choose a correct local model such that copies of this model would cover these objects.

Recall the notation for the upper half-space in $\mathbb{R}^k$:

$$\mathbf{H}^k = \{(x_1,\ldots,x_k) \in \mathbb{R}^k \ : \ x_k \geq 0\}$$

Definition 3.2.1 Let $k \in \mathbb{N}$. Let $\emptyset \neq X \subset \mathbb{R}^N$ be such that for all $x \in X$ there exists a neighborhood U_x of x in X and a diffeomorphism $\phi : U_x \to \phi(U_x)$ where $\phi(U_x)$ is an open subset of $\mathbf{H}^k$. Then we call X a *manifold with boundary.* Terms such as chart, atlas, coordinate neighborhood, etc., are defined exactly similarly as in Definition 3.1.1.

Remark 3.2.1 Let X be a manifold with boundary. Consider the subset

$$\partial X = \{x \in X \ : \ \phi(x) \in \mathbb{R}^{k-1} \times 0 \text{ for some chart } \phi\}.$$

Suppose $x \in \partial X$. Then for every chart ψ of X at x, it follows that $\psi(x) \in \mathbb{R}^{k-1} \times 0$. (For, $\phi \circ \psi^{-1}$ defines a diffeomorphism of an open subset A of $\mathbf{H}^k$ containing the point $y = \psi(x)$ to some open subset B in $\mathbf{H}^k$ and sends the point y to $z = \phi(x) \in \mathbb{R}^{k-1} \times 0$. If y were not in

$\mathbb{R}^{k-1} \times 0$, then $\phi \circ \psi^{-1}$ would map an open neighborhood W of y in $\mathbb{R}^k$ to a neighborhood W' of z. By invariance of domain, W' must be open in $\mathbb{R}^k$. But no neighborhood of z in $\mathbb{R}^k$ is contained in $B \subset \mathbf{H}^k$!!)

It follows that

$$X \setminus \partial X = \{x \in X \ : \ \phi(x) \in \text{int}(\mathbf{H}^k) \text{ for some chart } \phi\}.$$

From this description, it follows that $X \setminus \partial X$ is open in X. Therefore, ∂X is a closed subset of X. Moreover, we have:

Theorem 3.2.1 *Let X be a k-dimensional manifold with boundary. Then ∂X is either empty or is a manifold of dimension $k-1$.*

Proof: We first observe that ∂X may be empty. For the images of all the charts may be contained in $int(\mathbf{H}^k)$. (In such a case, X will be a manifold (without boundary).

So, let ∂X be nonempty. For each $x \in \partial X$, let (U, ϕ) be a chart at x for X. Put $V = \phi^{-1}(\mathbb{R}^{k-1} \times 0)$ and $f = \phi|_V$. Then observe that $V = \partial X \cap U$ and hence is an open subset of ∂X. Moreover $f(V) = \phi(U) \cap \mathbb{R}^{k-1} \times 0$. Hence, $f : V \to f(V)$ defines a diffeomorphism of V onto an open subset of $\mathbb{R}^{k-1}$. This proves the claim. ♠

Definition 3.2.2 Let $X \subset \mathbb{R}^N$ be a n-manifold with boundary. By a *neat submanifold* $Y \subset X$ of dimension m we mean a closed subset $Y \subset X$ which is an m-manifold with boundary on its own such that
(i) $\partial Y = Y \cap \partial X$ and
(ii) at each point $y \in \partial Y$ there exists a chart (U, ϕ) for X, $\phi : U \to \mathbb{R}^{n-m} \times \mathbf{H}^m$ such that $\phi(U \cap Y) \subset 0 \times \mathbf{H}^m$.

The figure below shows a closed strip as a 2-manifold with boundary and six of its subspaces. None of the subsets except the last one is a neat submanifold, even though each one of them is a manifold with (or without) boundary.

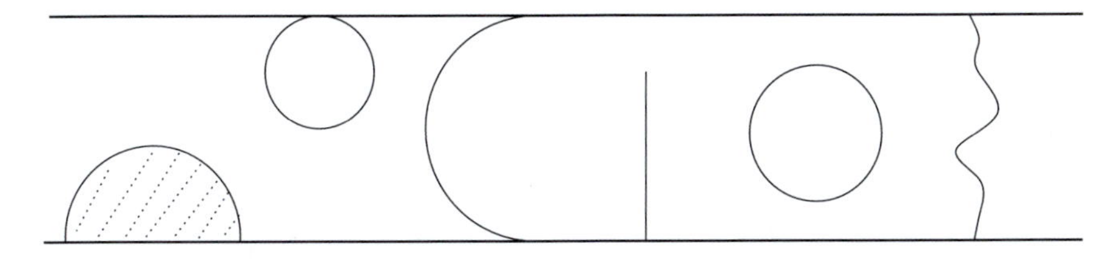

Figure 12 Only the last one is a neat submanifold.

Definition 3.2.3 A compact manifold without boundary is called a *closed manifold.* A noncompact manifold without boundary is called an *open manifold.*

Remark 3.2.2 When we say X is a closed submanifold of Y, we mean that X is a submanifold of Y and as a topological space, it is a closed subspace of Y. This should not be confused to mean that X is a closed manifold. It is true that a closed manifold, which is a submanifold of another manifold, is also a closed submanifold. However, the converse need not be true. Likewise, one should not confuse an open submanifold with an open manifold which is a submanifold of another manifold. Indeed, in this case, neither implies the other.

Example 3.2.1
(1) Clearly, the half-spaces $\mathbf{H}^k$ are themselves manifolds with boundary, $\partial \mathbf{H}^k = \mathbb{R}^{k-1} \times 0$.

More generally, if P is defined by finitely many linear inequalities inside $\mathbb{R}^n$, then it will be a topological manifold with boundary. If it is defined by only one linear inequality, then it would be a smooth manifold with boundary. In particular, all half-rays and closed intervals are 1-dimensional manifolds with boundary.
(2) The open interval $(0,1)$ is not a neat submanifold of $\mathbb{R}$ since it is not a closed subset. Nor is the closed interval $[0,1]$ since its boundary is nonempty whereas $\partial\mathbb{R} = \emptyset$. The closed interval $[-1,1]$ is a neat submanifold of the closed disc $\mathbb{D}^2$.
(3) All closed balls in $\mathbb{R}^n$ are n-dimensional manifolds with boundary. Each of them is diffeomorphic to the unit ball via a suitable translation followed by a scaling. Therefore, it is enough to see that the unit ball is a manifold with boundary. Exercises 3.1 show that $\mathbb{D}^n \setminus \{p\}$ is diffeomorphic with $\mathbf{H}^n$ where p is any point on the boundary. Thus, $\mathbb{D}^n$ can be covered by two such charts, defined on $\mathbb{D}^n \setminus \{p\}$, $\mathbb{D}^n \setminus \{q\}$, $p \neq q, p, q \in \partial\mathbb{D}^n$.

We shall give a slightly different proof of the same here, still leaving the Exercises 3.1 to you.

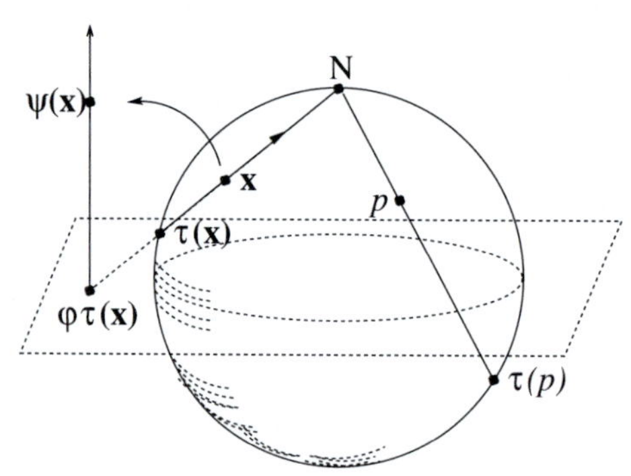

Figure 13 Extended Stereographic projection.

Consider the stereographic projection $\phi_+ : \mathbb{S}^{n-1}\setminus\{N\} \to \mathbb{R}^{n-1}$ and extend it on $\mathbb{D}^n\setminus\{N\}$ as follows. First, fix a point $\mathbf{x} \in \mathbb{D}^n$ and $\mathbf{x} \neq N$. Let $\tau(\mathbf{x}) \neq N$ denote the point of intersection of $\mathbb{S}^{n-1}$ and the line joining N and $\mathbf{x}$. Then, $\tau(\mathbf{x})$ is a smooth map. Observe that if $\mathbf{x} \in \mathbb{S}^{n-1}$, then $\tau(\mathbf{x}) = \mathbf{x}$. Moreover, each point of $\mathbb{D}^n\setminus\{N\}$ belongs to a unique line segment $[\tau(\mathbf{x}), N)$. Now put

$$\sigma(\mathbf{x}) = \frac{\|\mathbf{x} - \tau(\mathbf{x})\|^2}{\|N - \tau(\mathbf{x})\|^2 - \|\mathbf{x} - \tau(\mathbf{x})\|^2}.$$

Observe that for each $\mathbf{y} \in \mathbb{S}^{n-1} \setminus \{N\}$, σ restricted to the line segment $[\mathbf{y}, N)$ is a diffeomorphism onto $[0, \infty)$.

Finally consider $\psi(\mathbf{x}) = (\phi_+(\tau(\mathbf{x})), \sigma(\mathbf{x}))$. Then ψ is a diffeomorphism of $\mathbb{D}^n \setminus \{N\}$ with $\mathbf{H}^n$ extending the diffeomorphism ϕ_+.
(4) As a typical nonexample, let us consider the unit square I^2 in $\mathbb{R}^2$. At a first glance, we may conclude that this is a manifold with boundary. But caution is needed, particularly at the four vertices. Indeed, we shall right now prove that there is no local chart for I^2 at any of the four vertices. Say for instance (U, ϕ) is a chart at $0 = (0,0)$, i.e., $\phi : U \to \mathbf{H}^2$ is a diffeomorphism onto an open subset of $\mathbf{H}^2$. Clearly $\phi(0) \in \mathbb{R} \subset \mathbf{H}^2$. By the definition of smoothness of ϕ, we may assume that ϕ itself is defined and smooth in an open set $V \subset \mathbb{R}^2$ around 0 and $U = V \cap I^2$. It follows that $D(\phi)_0$ is an isomorphism. On the other hand we can compute the two partial derivatives of ϕ by remaining inside the boundary of the square and since this is mapped inside $\mathbb{R} \times 0$, it follows that the two partial derivatives are linearly dependent. In other words, $\{D(\phi)_0(\mathbf{e}_1), D(\phi)_0(\mathbf{e}_2)\}$ is a linearly dependent set. Hence, $D(\phi)_0$ cannot be an isomorphism. This contradiction proves the claim.

Exercise 3.2

1. Show that a manifold is locally compact, locally path connected, Hausdorff, and II-countable. If it is connected then show that it is path connected.

2. Show that any two points in a path connected manifold X can be joined by a piecewise smooth path $\omega : [0,1] \to X$ which is a 1-1 function.

3. Given $x_0 \in U_1 \subset U_2 \subset \cdots \subset U_n$ where each U_i is a manifold of dimension i, show that there exists a coordinate chart (V, ϕ) for U_n at x_0 such that
$$V \cap U_i = \{x \in V \ : \ \phi_j(x) = 0, \ i+1 \leq j \leq n\}.$$

4. Show that if X and Y are manifolds with boundary, then $X \times Y$ is a manifold with boundary provided $\partial X = \emptyset$ or $\partial Y = \emptyset$. What is $\partial(X \times Y)$? [Hint: First study the example $I \times \mathbb{S}^1$.] What goes wrong if $\partial X \neq \emptyset$ and $\partial Y \neq \emptyset$?

5. Show that if $f : X \to \mathbb{R}^n$ is a smooth map and X is a manifold then the graph of f is a manifold diffeomorphic to X.

6. Let X be a connected manifold and Y be a neat submanifold of X. If $\dim Y = \dim X$ then show that $Y = X$.

3.3 Tangent Space

Let us begin with a simple example. Let $f : \mathbb{R} \to \mathbb{R}$ be a smooth map and let us consider its graph. We know that the slope of the tangent to this curve at any point $(p, f(p))$ is equal to $f'(p)$. Therefore, it follows that the actual tangent at $(p, f(p))$ is the image of the map $t \mapsto (p, f(p)) + (1, f'(p))t$, which is the same as the graph of the map $x \mapsto f(p) - pf'(p) + f'(p)x$.

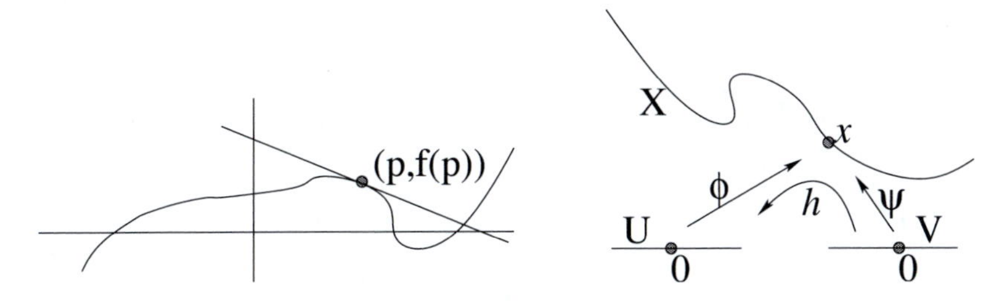

Figure 14 The tangent space is independent of the parameterization.

Observe that this is an affine linear map. If we ignore the additive constant $(p, f(p))$ for a while, then what we have is the linear map $\mathbb{R} \to \mathbb{R}^2$ given by the derivative at p of the parameterization $x \mapsto (x, f(x))$ of the graph. This is going to guide us in defining tangent space to any manifold at any of its points.

Definition 3.3.1 Let X be a manifold of dimension $k \geq 1$ in $\mathbb{R}^N$. Let $\phi : U \to X$ be a local parameterization at x. Recall that by this we mean that ϕ is the inverse of a coordinate chart, U is a neighborhood of 0 in $\mathbb{R}^k$ and $\phi(0) = x$. Consider the linear subspace $A = D(\phi)_0(\mathbb{R}^k)$ of $\mathbb{R}^N$. We shall first show that A is independent of the choice of parameterization ϕ. So, suppose ψ is another parameterization at x. Put $h = \phi^{-1} \circ \psi$, which is defined in a suitable neighborhood of 0 in $\mathbb{R}^k$, and is a diffeomorphism there. In particular, $D(h)_0 : \mathbb{R}^k \to \mathbb{R}^k$ is an isomorphism. Since $D(\psi)_0 = D(\phi)_0 \circ D(h)_0$, it follows that $D(\psi)_0(\mathbb{R}^k) = D(\phi)_0(\mathbb{R}^k) = A$.

We define the *tangent space* of X at x to be this linear subspace A and denote it by $T_x(X)$. The *geometric tangent space* at x is defined to be the affine subspace $T_x(X) + x$.

Now consider $X \times 0 \subset \mathbb{R}^N \times \mathbb{R}^N$ and the subspace

$$T(X) = \{(x, \mathbf{v}) \ : \ x \in X, \ \mathbf{v} \in T_x(X)\}.$$

$T(X)$ is called the *total tangent space* of X. Observe that the first projection $\pi : \mathbb{R}^N \times \mathbb{R}^N \to \mathbb{R}^N$ restricts to a smooth map $\pi : T(X) \to X$. We have for each $x \in X$, $\pi^{-1}(x) = \{x\} \times T_x(X) \subset \{x\} \times \mathbb{R}^N$, which we shall identify with $T_x(X)$ itself and treat it as a linear subspace of $\mathbb{R}^N$.

Now suppose Y is a l-dimensional manifold in $\mathbb{R}^M$ and $f : X \to Y$ is a smooth map. We would like to define the *derivative of* f to be a map $D(f) : T(X) \to T(Y)$. Indeed for each $x \in X$, we will define $D(f)_x : T_x(X) \to T_y(Y)$ to be a linear map, where $y = f(x)$. So, suppose that $\psi : V \to Y$ is a local parameterization for Y at y. Since f is continuous and $\psi(V)$ is open, $f^{-1}(\psi(V))$ is open. We choose a local parameterization $\phi : U \to X$ for X at x such that $\phi(U) \subset f^{-1}(\psi(V))$.

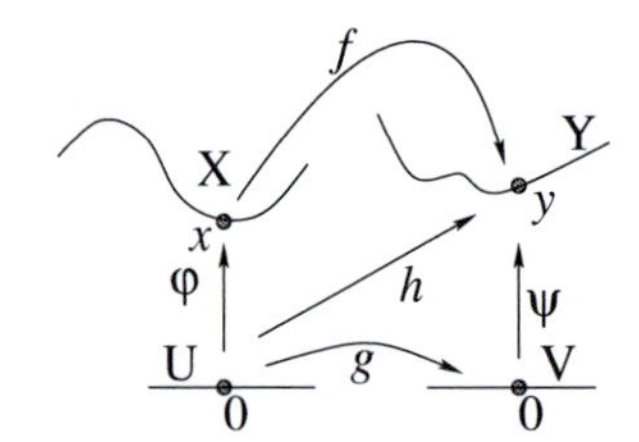

Figure 15 Using local coordinates to define the derivative.

Now consider the map $h = f \circ \phi : U \to \mathbb{R}^M$, which is smooth on U. So, $D(h)_0 : \mathbb{R}^k \to \mathbb{R}^M$ is a linear map. Since $D(\phi)_0 : \mathbb{R}^k \to T_x(X)$ is an isomorphism, it makes sense to take

$$D(f)_0 := D(h)_0 \circ (D(\phi)_0)^{-1} : T_x(X) \to \mathbb{R}^M$$

which is clearly a linear map. Put $g = \psi^{-1} \circ f \circ \phi$. Then $\psi \circ g = f \circ \phi = h$, and hence it follows that the image of $D(h)_0$ is contained in the image of $D(\psi)_0(\mathbb{R}^l) = T_y(Y)$.

It remains to prove that $D(f)_0$ as defined above is independent of the choice of ϕ. So, let ϕ_1 be another parameterization for X at x and let $h_1 = f \circ \phi_1$. We have to show that

$$D(h_1)_0 \circ (D(\phi_1)_0)^{-1} = D(h)_0 \circ (D(\phi)_0)^{-1}.$$

Let $\alpha : W \to \mathbb{R}^k$ be a smooth extension of ϕ_1^{-1} where W is open in $\mathbb{R}^N$, such that $\mathbf{x} \in W \cap X \subset \phi(\mathbb{R}^k) \cap \phi_1(\mathbb{R}^k)$. Then, it follows that $h = f \circ \phi_1 \circ \alpha \circ \phi$ and hence,

$$\begin{aligned} D(h)_0 &= D(f \circ \phi_1 \circ \alpha \circ \phi)_0 \\ &= D(h_1)_0 \circ D(\alpha)_{\mathbf{x}} \circ D(\phi)_0 = D(h_1)_0 \circ (D(\phi_1)_0)^{-1} \circ D(\phi)_0 \end{aligned}$$

which proves the claim.

It is straightforward to verify that the chain rule holds for the derivatives of functions on manifolds.

Remark 3.3.1 Here is an alternative approach to $D(f)$. Let $\hat{f}$ be any smooth extension of f in a neighborhood of $\mathbf{x}$ in $\mathbb{R}^N$. Take

$$D(f)_{\mathbf{x}} = D(\hat{f})_{\mathbf{x}}|_{T_x(X)}. \tag{3.2}$$

We have to verify that the right-hand side (RHS) of (3.2) is independent of the extension $\hat{f}$ chosen. We refer to the Remark 1.6.1 (iii). By the injective form of implicit function theorem, there exists a neighborhood W of $0 \in \mathbb{R}^N$ and a diffeomorphism $\alpha : W \to W'$ onto a neighborhood of $\mathbf{x} \in X \subset \mathbb{R}^N$ such that $U = W \cap \mathbb{R}^n \times 0$ and $\alpha|_U = \phi : U \to X$ is a parameterization of X near $\mathbf{x}$. Then $D(\hat{f} \circ \alpha)|_{\mathbb{R}^n \times 0} = D(f \circ \phi)$ for all extensions $\hat{f}$ of f. It follows that $D(\hat{f})$ restricted to $T_{\mathbf{x}}(X) = D(\phi)_0(\mathbb{R}^n)$ is independent of $\hat{f}$. Observe that (3.2) implies that the image $D(\hat{f}_{\mathbf{x}})(T_{\mathbf{x}}(X)) \subset T_{f(\mathbf{x})}(Y)$.

Definition 3.3.2 The tangent space $T(X)$ together with the projection map $\pi : T(X) \to X$ is called the *tangent bundle* on X. In this sense, for any smooth map $f : X \to Y$ of manifolds, we treat $D(f)$ as a commutative diagram of maps

$$\begin{array}{ccc} T(X) & \xrightarrow{D(f)} & T(Y) \\ \downarrow & & \downarrow \\ X & \xrightarrow{f} & Y \end{array}$$

Remark 3.3.2
(i) Observe that for each point $x \in X$, there is the vector space $\pi^{-1}(x) \subset T(X)$. There are other examples of *bundles* over a topological space. We shall meet at least one more such example during this course. A simple way to consider a bundle over X is to take $X \times \mathbb{R}^n$ with the first projection. These bundles are called trivial bundles. In fact, a bundle (E, π) over X is called *trivial bundle,* if there exists a homeomorphism $F : E \to X \times \mathbb{R}^m$ for some m such that $\pi_1 \circ F = \pi$. Often, the homeomorphism is required to satisfy additional structural conditions, depending upon the kind of bundles that we are dealing with. The situation with the tangent bundle is that the fibre $\pi^{-1}(x) = x \times T_x(X)$ over every point has a vector space structure which is very important for us. Thus the homeomorphism F above is required to be linear when restricted to each fibre: $F : \pi^{-1}(x) \to \{x\} \times \mathbb{R}^m$. It may be a good idea to read the formal definition of a vector bundle etc. from section 5.5 before reading the following examples and remarks or come back to them later.
(ii) Suppose now that X is a manifold with boundary $\partial X \neq \emptyset$. The points in the interior of X pose no problems in defining the tangent space—the old definition is applicable here verbatim. Even if $x \in \partial X$, we still define the tangent space $T_x X$ in the same manner as above. This is where Example 1.6.2 comes to our aid so that there is no ambiguity involved because of the choice of the extension of the parameterization $\phi : \mathbf{H}^n \to X$ at x to a smooth function defined in a neighborhood of 0 in $\mathbb{R}^n$.

Example 3.3.1
(1) Taking $(\mathbb{R}^n, Id)$ as a single chart for $\mathbb{R}^n$, it follows that at each point of $x \in \mathbb{R}^n$, the tangent space $T_x(\mathbb{R}^n) = \mathbb{R}^n$. Indeed, let (U, ϕ) be a local parameterization for a manifold of dimension k and let $V = \phi(U)$. Consider

$$\Psi : U \times \mathbb{R}^k \to T(X)$$

given by

$$(x, \mathbf{v}) \mapsto (\phi(x), D(\phi)_x(\mathbf{v})).$$

Clearly Ψ is a smooth map (provided ϕ is at least of class $\mathcal{C}^2$). It is also injective and the image $\pi^{-1}(V) =: T(V)$ is an open subset of $T(X)$. It is easily seen that the inverse map is also smooth and hence Ψ defines a local parameterization for $T(X)$. From this we see that $T(X)$ itself is a manifold of dimension $2k$. This discussion also shows that as a bundle $T(X)$ is *locally trivial.*

(2) Let V be any k-dimensional vector subspace of $\mathbb{R}^n$. As before we get a linear bijective mapping $\phi : \mathbb{R}^k \to V$. For every point $x \in \mathbb{R}^k$, we have $D(\phi)_0 = \phi$ and hence $T_{\mathbf{v}}(V) = V$ for every $\mathbf{v} \in V$. Indeed, if $\phi^{-1} : V \to \mathbb{R}^k$ is the inverse of ϕ then the map

$$(\mathbf{v}, \mathbf{w}) \mapsto (\mathbf{v}, \phi^{-1}(\mathbf{w}))$$

defines a diffeomorphism of $T(V)$ onto $V \times \mathbb{R}^k$ as a trivial bundle. As a subspace of $\mathbb{R}^n \times \mathbb{R}^n$, $T(V)$ is equal to $V \times V$.
(3) Consider the unit circle $\mathbb{S}^1$. We can use the global parameter $\phi : t \mapsto (\cos t, \sin t)$ defined on $\mathbb{R}$. The tangent space at a point $a = (\cos t_0, \sin t_0)$ is then the image of $D(\phi)_a =$ the line spanned by the unit vector $(-\sin t_0, \cos t_0)$, which is nothing but the line perpendicular to the position vector of the point a. Thus, we see that $T(\mathbb{S}^1) = \{(x, \imath\alpha x) \ : \ x \in \mathbb{S}^1, \ \alpha \in \mathbb{R}\}$ is diffeomorphic to $\mathbb{S}^1 \times \mathbb{R}$. Indeed, this shows that the tangent bundle of $\mathbb{S}^1$ is a trivial bundle.

Remark 3.3.3

1. A n-manifold with a trivial tangent bundle is called a *parallelizable manifold.* This is the same as having n vector fields σ_i such that at each point of the manifold, the tangent vectors $\{\sigma_i(x)\}$ are independent in T_xX. We shall see that $\mathbb{S}^3$ and $\mathbb{S}^7$ are parallelizable (see exercises below). We shall also see that $\mathbb{S}^{2n}$ is not parallelizable (see remark 7.3.2). It is a deep result that the spheres $\mathbb{S}^{2n-1}$, $n \neq 1, 2, 4$ are not parallelizable, the proof of which is beyond the scope of this book (see [Hus]). Indeed, counting maximum number of independent vector fields on some special types of manifolds itself has become a big branch of Differential Topology.

2. Let X be a submanifold of Y and $i : X \to Y$ be the inclusion map. What is $D(i)_x$ for $x \in X$? In the case, $Y = \mathbb{R}^N$, we already know that $D(i)_x : T_x(X) \to \mathbb{R}^N$ is the inclusion map. Therefore, for the inclusion maps $j_1 : X \to \mathbb{R}^N$ and $j_2 : Y \to \mathbb{R}^N$, we know that both $D(j_1)$ and $D(j_2)$ are inclusion maps. But $j_1 = j_2 \circ i$ and by chain rule, $D(j_1) = D(j_2) \circ D(i)$. Therefore, $D(i) : T_xX \to T_xY$ is also the inclusion map.

Exercise 3.3

1. Show that if $f : X \to Y$ is a diffeomorphism then $D(f) : T(X) \to T(Y)$ is a diffeomorphism. (Thus, the tangent space is a diffeomorphic invariant.)

2. Compute $D(f)$ where $f : \mathbb{S}^1 \to \mathbb{S}^1$ is given by $z \mapsto z^n$.

3. Suppose X is a manifold such that $T(X)$ is trivial. If U is an open subset of X, show that $T(U)$ is trivial.

4. Let V be a vector subspace of $\mathbb{R}^n$ of dimension k. Exhibit explicitly the tangent space of V (as a subspace of $\mathbb{R}^n \times \mathbb{R}^n$). Next if $L : V \to W$ is a linear map, where W is a vector subspace of $\mathbb{R}^m$, write down explicitly $D(L)$.

5. Let $X \subset \mathbb{R}^n$ be a closed subset and a manifold. Define $\rho : \mathbb{R}^n \to \mathbb{R}$ by

$$\rho(z) = d(z, X) := inf\,\{d(z, x) \ : \ x \in X\}$$

where d denotes the euclidean distance function. Show that
(a) there exists $x_0 \in X$ such that $\rho(z) = d(z, x_0)$ and
(b) for any $x \in X$ satisfying (a), the vector $z - x$ is perpendicular to T_xX.

6. Let X be an n-dimensional smooth manifold in $\mathbb{R}^N$. Show that for any point $x \in X$, there exist n coordinate projections $x_{i_1}, \ldots, x_{i_n}$, which when restricted to a suitable neighborhood of x in X, give a coordinate chart for X. [Hint: There is such a set of coordinate projections, which when restricted to $T_x(X)$, is independent.] Deduce that after a suitable permutation of coordinates, a neighborhood of $x \in X$ is the graph of a smooth function $f : U \to \mathbb{R}^{N-n}$ where $U \subset \mathbb{R}^n$ is open.

7. Use coordinates $(\ldots, x_j, y_j, z_j, w_j, \ldots,)$ for $\mathbb{R}^{4n}$. Check that

$$(\ldots, x_j, y_j, z_j, w_j, \ldots,) \mapsto (\ldots, -y_j, x_j, -w_j, z_j, \ldots,)$$

defines a unit vector field on the sphere $\mathbb{S}^{4n-1}$. Likewise, by merely permuting the four symbols (x, y, w, z) and choosing appropriate signs produce two more unit vector fields, so that all the three vector fields are mutually orthogonal. (In particular, this proves that $\mathbb{S}^3$ is parallelizable.)

8. Use the coordinates $(a, b, c, d, e, f, g, h) \in \mathbb{R}^8$ to write-down seven mutually orthogonal vector fields on $\mathbb{S}^7$ following a method similar to the one in the above exercise. Here are two of them.

$$(a, b, c, d, e, f, g, h) \mapsto (b, -a, d, -c, -f, e, -h, g)$$

and

$$(a, b, c, d, e, f, g, h) \mapsto (c, -d, -a, b, g, -h, -e, f).$$

Follow the same method to produce seven mutually orthogonal vector fields on $\mathbb{S}^{8n-1}$.

3.4 Special Types of Smooth Maps

Throughout this section, X, Y, etc., will denote manifolds unless specified otherwise.

Definition 3.4.1 We say $f : X \to Y$ is *immersive* (resp. *submersive*) at $x \in X$ if $(Df)_x$ is injective (resp. surjective.) If this is true at every point of X then we say f is *an immersion* (respectively, *a submersion*). Finally, we say f is an *embedding* if f is a diffeomorphism of X onto its image $f(X)$.

Remark 3.4.1 Observe that if f is immersive at a point then it is immersive at all points in a neighborhood of that point. Also, it is injective in a small neighborhood of this point. However, an immersion need not be globally injective as seen by the example $x \mapsto e^{2\pi \imath x}$. Clearly, if an immersion $f : X \to Y$ exists, then $\dim X \leq \dim Y$. Similar statements hold for submersions. Also, observe that if f is an embedding, then $f(X)$ is clearly a manifold diffeomorphic to X.

These special classes of smooth maps are very useful in the study of Differential Topology. To begin with, we shall see that they can be used to identify certain submanifolds. Before proceeding further, let us recall "proper maps".

Definition 3.4.2 A continuous function $f : A \to B$ of topological spaces is called a *proper map* if for every compact subset K of B, we have $f^{-1}(K)$ is compact.

Remark 3.4.2 It follows easily that if B is Hausdorff then every map $f : A \to B$ from a compact space A is proper. Indeed properness hypothesis on maps is a clever device introduced to take care of the vast situations wherein the domain is not compact.

Theorem 3.4.1 *Let $f : X \to Y$ be a smooth, proper, injective immersion. Then $Z = f(X)$ is a submanifold and $f : X \to Z$ is a diffeomorphism. Moreover, Z is a closed subset of Y. Conversely, if Z is a closed subset of Y, which is also a submanifold of Y then Z is the image of a smooth, proper, injective immersion.*

Proof: We shall first show that f is a homeomorphism onto Z. This is equivalent to show that $f : X \to Z$ is an open mapping. Let U be an open subset in X and suppose that $f(U)$ is not open in Z. Then there exists a sequence $y_n = f(x_n)$, say, in $Z \setminus f(U)$ such that $y_n \to y$ and $y \in f(U)$, say $y = f(x)$, for some $x \in U$. Since $K = \{y\} \cup \{y_n : n \geq 1\}$ is a compact subset of Y and f is proper, it follows that $f^{-1}(K)$ is compact. Any sequence in a compact set has a subsequence that is convergent. Hence, after passing to a subsequence, if necessary, we may assume that $x_n \to x'$ say. Since $x_n \in X \setminus U$, a closed subset, it follows that $x' \notin U$. But then $f(x') = \lim f(x_n) = \lim y_n = y = f(x)$. Since f is injective, we have, $x = x' \in U$, a contradiction. Hence, f is a homeomorphism onto Z.

(Indeed, we have just proved the following: *a continuous proper bijection of two metric spaces is a homeomorphism.*)

In order to show that Z is a submanifold, let $x \in X$ be any point and $y = f(x)$. Choose a chart (U, ψ) at y for Y and consider $g = \psi \circ f$ on $V = f^{-1}(U)$. It is enough to show that $g(V)$ is a submanifold of $\mathbb{R}^m$. Thus, we have reduced the problem to the case when $Y = \mathbb{R}^m$. Similarly, by precomposing with a parameterization at x for X, we may assume that $X = \mathbb{R}^n$.

Now by the injective form of implicit function theorem, there is a diffeomorphism $\alpha : W \to \mathbb{R}^m$ of some neighborhood W of 0 in $\mathbb{R}^m$, such that

$$\alpha \circ f(x_1, \ldots, x_n) = (x_1, \ldots, x_n, 0, \ldots, 0).$$

Put $\beta = \alpha^{-1}|_{\mathbb{R}^n \times 0}$. Then β is a parameterization for $f(X)$.

This also proves the smoothness of f^{-1} since $\beta(\mathbf{x}, 0) = f(\mathbf{x})$. Hence, f is a diffeomorphism.

To show that Z is a closed subset of Y, let $y_n = f(x_n)$ be a sequence converging to $y \in Y$. Now argue exactly as in the proof that $f : X \to Z$ is an open mapping to see that $y \in Z$.

The converse is obvious: we simply take $X = Z$ and f as the inclusion map. ♠

Example 3.4.1 As a typical counterexample consider the mapping $f : \mathbb{R} \to \mathbb{R}^2$, which traces the figure-eight curve:

$$f(t) = (\sin \pi t, \sin 2\pi t).$$

If g is the restriction of f to the open interval $(-1, 1)$, it is not difficult to see that f is an injective immersion. However, the image fails to be a manifold precisely at the point $(0, 0)$. This failure can be attributed to the fact that g is not a proper map.

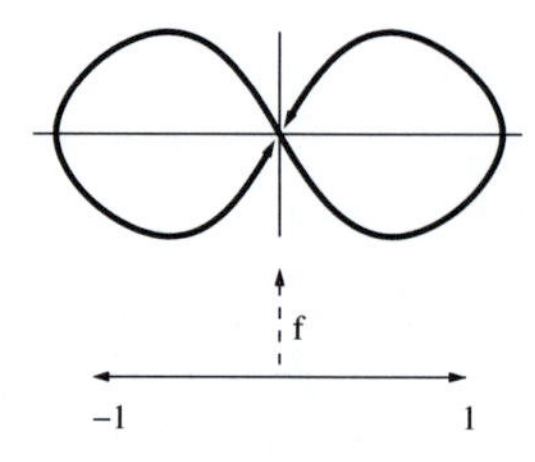

Figure 16 An injective immersion that does not yield a submanifold.

Submersive maps have a special role to play. We first make a few a definitions.

Definition 3.4.3 Let $f : X \to Y$ be a smooth map of manifolds. We say $x \in X$ is a *regular* point for f if Df_x is surjective. A point that is not regular may be referred to as *critical* or *singular.* We say y is a *regular value* if all points of $f^{-1}(y)$ are regular for f. Thus, we see that if $f^{-1}(y) = \emptyset$ then clearly y is a regular value. [This is a typical case of a phenomenon in modern mathematics wherein *a red herring need be neither red nor a herring.*]

Remark 3.4.3 We shall denote the set of regular points of f by $\mathcal{R}_f$ and the set of critical points of f by $C_f := X \setminus \mathcal{R}_f$. We shall denote the set of regular values of f by $\mathcal{V}_f$. Note that $\mathcal{V}_f \cap f(X) \subset f(\mathcal{R}_f)$ and often these two sets are not equal. It is an easy consequence of the implicit function theorem that,

$$\mathcal{R}_f \subset X, \quad f(\mathcal{R}_f) \subset Y,, \quad \mathcal{V}_f \subset Y$$

are all open.

The following theorem that may be also called *regular inverse image theorem,* is of utmost importance to us. It is a direct consequence of the surjective form of implicit function theorem.

Theorem 3.4.2 Preimage Theorem: *Let X and Y be boundaryless manifolds. Let $f : X \to Y$ be a smooth map,* $\dim X = n$, $\dim Y = m$. *Then for all regular values y of f the inverse image $W = f^{-1}(y)$ is either empty or is a closed submanifold of X of codimension m. Moreover, the tangent space to W at any point $w \in W$ is precisely the kernel of Df_w.*

Proof: Being the inverse image of a single point, clearly W is closed. In order to show that it is a submanifold, by composing f with a chart at y for Y, we may assume that $Y = \mathbb{R}^m$ and $y = 0$. Similarly, by precomposing with a parameterization for X at $x \in W$, we may assume that $X = \mathbb{R}^n$. But then the claim follows directly from the surjective form of the implicit function Theorem 1.4.4. ♠

Example 3.4.2 As we know already, this theorem can be used to produce a large number of submanifolds of Euclidean spaces. Typical examples of these are conics in $\mathbb{R}^2$, quadratic hypersurfaces in $\mathbb{R}^n$ such as the sphere, ellipsoids, hyperboloids, etc.

Let us now discuss a few important examples occurring in the space of matrices. We know that the space $M(n, \mathbb{R})$ of $n \times n$ real matrices can be thought of as the Euclidean space $\mathbb{R}^{n^2}$. A standard notation here is $\mathbb{R}^{m\times n}$, which indicates that we are considering the Euclidean space $\mathbb{R}^{mn}$ as the space of all $m \times n$ matrices with real entries instead of space of column vectors of column-size mn.

Consider the mapping $det : M(n, \mathbb{R}) \to \mathbb{R}$ the determinant mapping. It is indeed a polynomial mapping and hence is smooth. Now, the space $GL(n, \mathbb{R})$ of all invertible matrices is the inverse image of the open set $\mathbb{R} \setminus \{0\}$ under this mapping and hence is an open submanifold of $M(n, \mathbb{R})$. Also, its total tangent space is actually equal to $GL(n, \mathbb{R}) \times M(n, \mathbb{R})$. Likewise the space $S(n)$ of all symmetric matrices, being a vector subspace of dimension $n(n + 1)/2$, is a closed submanifold of $M(n, \mathbb{R})$. (Similarly the space of all antisymmetric matrices is a closed submanifold of dimension $n(n - 1)/2$.) We also know that the tangent space to any vector subspace V of the Euclidean space is $V \times V$ itself.

Now consider, the space $O(n)$ of all orthogonal matrices in $M(n, \mathbb{R})$:

$$O(n) := \{A \in M(n; \mathbb{R}) \; : \; AA^t = Id\}.$$

Thus, if $\tau : M(n, \mathbb{R}) \to S(n)$ is defined by $\tau(A) = AA^t$, then $O(n) = \tau^{-1}(Id_n)$. Since τ is easily seen to be smooth map, this leads us to the question whether Id is a regular value for τ or not. To check this, we must fix some $B \in O(n)$ and compute $D\tau_B$. This has been done in (1.43).

We have $D\tau_B(A) = AB^t + BA^t$ for all $A \in M(n, \mathbb{R})$. To show that for each fixed $B \in O(n)$ the mapping $A \mapsto AB^t + BA^t$ is surjective onto the space $S(n)$ of all symmetric matrices, given $C \in S(n)$, we consider $A = \frac{1}{2}CB$. Since $BB^t = Id$, we get $AB^t = \frac{1}{2}C$. Also, since C is symmetric, we have $BA^t = \frac{1}{2}C^t = \frac{1}{2}C$ and hence $AB^t + BA^t = C$.

It follows that $O(n)$ is a submanifold of $M(n, \mathbb{R})$ of codimension equal to $n(n+1)/2$ and hence $\dim O(n) = n^2 - n(n+1)/2 = n(n-1)/2$. Moreover, the tangent space to $O(n)$ at any point can now be identified with the kernel of $D\tau$. At $B = Id$, this is easily seen to be the space of all antisymmetric matrices.

Example 3.4.3 Consider $S = \{(A, t) \in M(n; \mathbb{R}) \times \mathbb{R} \ : \ t(\det A) = 1\}$. It is easily verified that every nonzero real number is a regular value of $f(A, t) = t(\det A)$. Therefore $S = f^{-1}(1)$ is a smooth submanifold of $M(n; \mathbb{R}) \times \mathbb{R}$. It is also easily verified that $A \mapsto (A, 1/\det, A)$ defines a diffeomorphism of $GL(n; \mathbb{R})$ with S. Thus $GL(n, \mathbb{R})$ is the zero set of a polynomial in $n^2 + 1$ variables, which makes it an *affine variety.*

Another rich source of obtaining manifolds and closely related to the preimage theorem is the "Jacobian criterion". The difference is that instead of implicit function theorem, we have to appeal to the rank theorem. Below, we state this criterion, the details of the proof of which is left to the reader as an exercise.

Theorem 3.4.3 Jacobian Criterion: *Let $f : X \to Y$ be a smooth map of manifolds. Suppose Df_x is of constant rank r throughout X. Then for any point $y \in Y$, $W = f^{-1}(y)$ is either empty or is a closed submanifold of X of codimension r. Moreover, the tangent space T_xW of $x \in W$ is the kernel of $Df_x : T_xX \to T_yY$.*

Example 3.4.4 The Rank Manifolds $\mathcal{R}_k(m, n; \mathbb{R})$: Here is an important class of submanifolds of the space of matrices, which do not occur, in general, as regular inverse image of some nice functions as such.[1] Let

$$\mathcal{R}_k(m, n; \mathbb{R}) = \{A \in \mathbb{R}^{m \times n} \ : \ \operatorname{rank} A = k\}.$$

We shall show that this is a smooth submanifold of the Euclidean space $\mathbb{R}^{m\times n}$, of codimension $(m-k)(n-k)$. Let $C_{n,k}$ denote the set of all k-subsets of the set $\{1, 2, \ldots, n\}$. For $\alpha \in C_{n,k}$, let U_α denote the open subset of $\mathcal{R}_k(m, n; \mathbb{R})$ consisting of those matrices A whose columns corresponding to the indices $\alpha_1, \ldots, \alpha_k$ are independent. It is clear that $\mathcal{R}_k(m, n; \mathbb{R}) = \cup_{\alpha \in C_{n,k}} U_\alpha$. Therefore, it is enough to show that each U_α is diffeomorphic to an open subset of $\mathbb{R}^t$, where $t = mn - (m-k)(n-k) = k(m+n-k)$.

First, consider the case when $\alpha = \{1, 2, \ldots, k\}$. Let W be the open set of all elements in $\mathbb{R}^{m\times k}$ of maximal rank k. Consider the map

$$\phi : W \times \mathbb{R}^{k\times(n-k)} \to \mathcal{R}_k(m, n; \mathbb{R})$$

given by

$$(B, C) \mapsto [B, BC].$$

The smoothness of this map is clear. It is elementary linear algebra to verify that ϕ is a bijective map onto the open set U_α of $\mathcal{R}_k(m, n; \mathbb{R})$. Being a rational function, the inverse is also a smooth map.

Now, in the general case, the map ϕ_α is nothing but ϕ as above followed by an appropriate permutation of columns; and hence all the claims made in the special case are valid.

[1]In the language of Algebraic Geometry, this phenomenon can be expressed by saying that $\mathcal{R}_k(m, n; \mathbb{R})$ is **not** a *complete intersection*, in general. See exercise 24 in 3.7.

Example 3.4.5 Steifel Manifolds $V_{k,n}$: For integers $1 \leq k \leq n$, let $V_{k,n}$ denote the subspace of $\mathbb{R}^{n\times k}$ of k-tuples $(\mathbf{v}_1, \ldots, \mathbf{v}_k)$ of vectors $\mathbf{v}_j \in \mathbb{R}^n$ such that $\langle \mathbf{v}_i, \mathbf{v}_j \rangle = \delta_{ij}$. These are called *orthonormal k-frames in* $\mathbb{R}^n$. Let $\phi : \mathcal{R}_k(n,k;\mathbb{R}) \to S(k) \subset \mathbb{R}^{k\times k}$ be defined by $\phi(A) = A^t A$. If we write $A = [\mathbf{u}_1, \ldots, \mathbf{u}_k]$ where $\mathbf{u}_j$ are columns of A then we have,

$$\phi(A) = \phi[\mathbf{u}_1, \ldots, \mathbf{u}_k] = (((\mathbf{u}_i, \mathbf{u}_j))).$$

One can write down the matrix form for $D(\phi)$, using the double-indices $\{(i,j), 1 \leq i \leq m,\ 1 \leq j \leq n\}$. Check that the columns corresponding to (i,j) for $1 \leq i \leq j \leq k$ are independent, whereas other columns are repetitions. [Write down the matrix of $D\phi$ for some small values of k and n.] Therefore, it follows that the rank of $D\phi$ is equal to $k(k+1)/2$ everywhere and hence, by the Jacobian Criterion, $V_{k,n} = \phi^{-1}(Id_k)$ is a submanifold of $\mathcal{R}_k(n,k;\mathbb{R})$ of codimension $= k(k+1)/2$. It is clear that $V_{k,n}$ is a closed subspace of $\mathbb{S}^{n-1} \times \cdots \times \mathbb{S}^{n-1}$ (k-copies) and hence is compact. Also note that $V_{1,n} = \mathbb{S}^{n-1}$ and $V_{n,n} = O(n)$, the space of the orthogonal group and $V_{n-1,n} = SO(n)$, the space of special orthogonal group. (See exercise 2 at the end of the section.)

We now consider an extension of Theorem 3.4.2 to the case when X may have boundary. The importance of such a result being more or less obvious, will, in any case, be demonstrated in a later section. We begin with:

Lemma 3.4.1 Let S be a manifold without boundary and $\pi : S \to \mathbb{R}$ be a smooth map such that 0 is a regular value of π. Then $\{s \in S \ :\ \pi(s) \geq 0\}$ is either empty or is a manifold of the same dimension as S with boundary equal to $\{s \in S \ :\ \pi(s) = 0\}$.

Proof: The set $\{s \in S \ :\ \pi(s) > 0\}$ is an open subset of S and hence is a submanifold. (What happens if it is empty?) So, let $s \in S$ be such that $\pi(s) = 0$. By the surjective form of implicit function theorem, we may as well assume that π is a coordinate projection in a neighborhood of $s \in S$ and then the result is obvious. ♠

Theorem 3.4.4 Extended Preimage Theorem: *Let X, Y be smooth manifolds, $\partial Y = \emptyset$, $f : X \to Y$ be a smooth map and $y \in Y$ be a regular value for both f and $f|_{\partial X}$. Then $W := f^{-1}(y)$ is either empty or is a neat submanifold of X of codimension* $= \dim Y$. *Also, for all $x \in W$, we have, $T_x W = ker\,(Df_x)$.*

Proof: Clearly W is a closed subset of X. As in the boundaryless case, from the Implicit Function Theorem 1.4.3, it follows that $W \cap \text{int}(X)$ and $W \cap \partial X$ are both submanifolds, the former being a manifold without boundary. So, we have to analyze the situation at a point $x \in W \cap \partial X$.

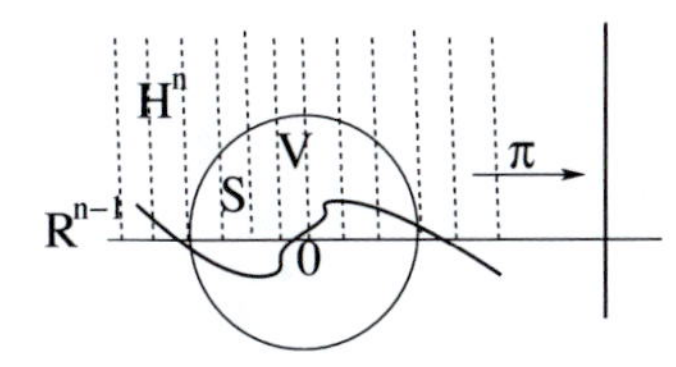

Figure 17 A neat submanifold produced by preimage.

By choosing a parameterized neighborhood of x in X, etc., we reduce the entire situation to the case of a map $f : \mathbf{H}^n \to \mathbb{R}^k$, $x = 0, f(0) = 0$ and $0 \in \mathbb{R}^k$ is a regular value of both f and ∂f. The smoothness of f means that f has a smooth extension $\tilde{f}$ in a neighborhood U of 0 in $\mathbb{R}^n$. Also by definition $Df = D\tilde{f}$ and hence $D\tilde{f}_0$ is surjective. From the result that we have proved in the boundaryless case, it follows that for a smaller neighborhood V of 0 in

$\mathbb{R}^n$, we have, $S = \tilde{f}^{-1}(0) \cap V$ is a codimension k submanifold. Clearly, $\mathbf{H}^n \cap S = f^{-1}(0) \cap V$ and hence, we must show that $\mathbf{H}^n \cap S$ is a manifold with boundary $\partial \mathbf{H}^n \cap S$. For this, we consider the last coordinate projection $\pi : V \to \mathbb{R}$ and claim that 0 is a regular value for $\pi|_S$. Since $\mathbf{H}^n \cap S = \{x \in S \ : \ \pi(x) \geq 0\}$, by appealing to the lemma above, this will complete the proof.

To show that $D(\pi|_S) : T_0(S) \to \mathbb{R}$ is a nonzero map, assume the contrary. This means that $T_0(S) \subset \mathbb{R}^{n-1}$. But $T_0(S) = Ker\, D\tilde{f}_0$ and $D\tilde{f}_0|_{\mathbb{R}^{n-1}} = Df_0|_{\mathbb{R}^{n-1}} = D(\partial f)_0$. Therefore, $Ker\, D\tilde{f}_0 = Ker\, D(\partial f)_0$. On the other hand, both maps are surjective and hence by the rank-nullity theorem, nullity of $D\tilde{f}_0$ is $n-k$ whereas that of $D(\partial f)_0$ is $n-1-k$, which is absurd. ♠

Remark 3.4.4
(i) In Figure 17, the curve S and the x-axis intersect in a very special way. In layman's language, one may say that the two curves "cross" each other at the point of intersection. Now consider another situation in which S is the graph of $y = x^3$. Here also the curve S crosses the x-axis. But there is a difference, viz., the curve appears to "linger on" a little, before crossing over. Mathematically, this can be described in a very neat way. The tangent to the graph $y = x^3$ coincides with the tangent to the x-axis, whereas in the situation of Figure 17 the two tangent lines are different. This leads us to the geometric concept called "transversality" which is going to play a big role in the rest of the development of the subject and we propose to take it up in the next section.
(ii) Discussion of regular values of maps between manifolds of the same dimension deserves some special attention. We end this section with a result that we come across in many situations.

Theorem 3.4.5 Stack-Record Theorem: *Let X be compact and $f : X \to Y$ be a smooth map of manifolds of the same dimension and without boundary.*
(i) *If $y \in Y$ is a regular value, then $f^{-1}(y)$ is a finite set $= \{x_1, \ldots, x_k\}$ and there exist open neighborhoods U_i of x_i and V of y such that $f : U_i \to V$ is a diffeomorphism for $1 \leq i \leq k$ and $f^{-1}(V) = \amalg_{i=1}^k U_i$, a disjoint union.*
(ii) *On the space of regular values of f, the number of points $\#(f^{-1}(y))$ is locally a constant function.*
(iii) *If f is a submersion and Y is connected, then f is a covering map.*

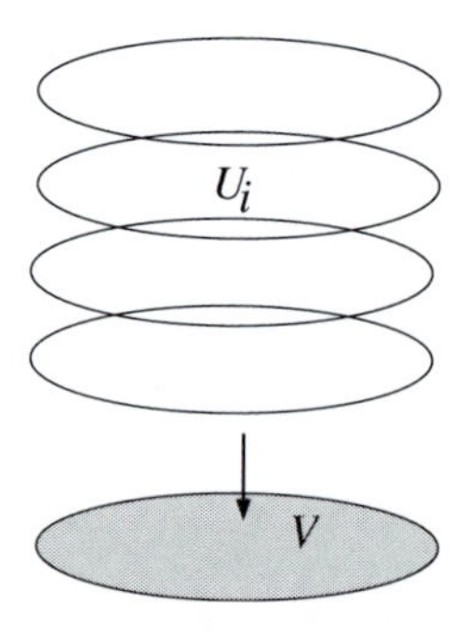

Figure 18 The records are stacked one over other.

Proof: (i) By the inverse function theorem, for every point $x \in X$ such that $f(x) = y$, there is an open set U_x in X such that $x \in U_x$, $f : U_x \to f(U_x)$ is a diffeomorphism and $f(U_x)$ is open in Y. In particular, this implies that $f^{-1}(y)$ is an isolated subset of the compact space X and hence is a finite set, say $f^{-1}(y) = \{x_1, \ldots, x_k\}$. Therefore, there exists $W_i \subset U_{x_i}$

such that $W_i \cap W_j = \emptyset, i \neq j$. Put $V' = \cap_{i=1}^k f(W_i)$. We claim that there exists an open neighborhood V of y contained in V', such that we have $f^{-1}(V) \subset \cup_{i=1}^k W_i = W$. By putting $U_i = W_i \cap f^{-1}(V')$, (i) follows.

Suppose the claim is not true. Then there will be a sequence $z_n \in X \setminus W$ such that $f(z_n) \to y$. Passing to a subsequence we may assume that $z_n \to z$. Then $f(z) = y$ whereas $z \notin W$ and hence, cannot be equal to any of the $x_1, \ldots, x_k$; a contradiction. (Observe that this takes care of the case when y is not a value at all.)

(ii) Follows from (i). However, first notice that it also follows that the set of regular values is an open set.

(iii) If f is a submersion, then $f(X)$ is open. Since X is compact, $f(X)$ is closed as well. Therefore, $f(X) = Y$ and every point of Y is a regular value. The conclusion (i) for every point of Y means that f is a covering map. ♠

Exercise 3.4

1. Show that the following subsets of $M(n, \mathbb{R})$ are submanifolds; determine their dimension and the tangent space at Id_n.
   (i) $T = \{A = ((a_{ij})) \;:\; det\, A \neq 0, \;\&\; a_{ij} = 0, i < j\}$.
   (ii) $SL(n, \mathbb{R}) = \{A \;:\; det\, A = 1\}$.
   (iii) $SO(n) = \{A \in M(n, \mathbb{R}) \;:\; AA^t = Id_n, \;\&\; det\, A = 1\}$.
   (iv) $O_{p,q}(\mathbb{R}) = \{A \in M(n, \mathbb{R}) \;:\; AJ_{p,q}A^t = J_{p,q}\}$, where
   $$J_{p,q} = \begin{bmatrix} I_p & 0 \\ 0 & -I_q \end{bmatrix}, \quad 1 \leq p \leq n-1, p+q = n.$$
   (v) $SO_{p,q} = \{A \in O_{p,q} \;:\; det\, A = 1\}$.
2. Show that $V_{n-1,n}$ is diffeomorphic to $SO(n)$.
3. Identify the group of all rigid motions of $\mathbb{R}^n$ as a subspace of $M(n+1, \mathbb{R})$ and show that it is a manifold. Similarly identify the group of all affine isomorphisms of $\mathbb{R}^n$ with a subspace of $M(n+1, \mathbb{R})$ and show that it is a manifold. What are the dimensions?
4. Under the identification $\mathbb{C} = \mathbb{R}^2$ show that the group $U(n)$ of all $n \times n$ unitary matrices (i.e., those which satisfy $AA^* = Id$) form a manifold. What is the tangent space at Id and what is the dimension?
5. Show that $GL(n, \mathbb{C})$ is connected. Show that $GL(n, \mathbb{R})$ has precisely two components.

3.5 Transversality

Definition 3.5.1 Let $f : X \to Y$ be a smooth map of smooth manifolds and let Z be a submanifold of Y. We say f is *transversal* to Z, (written $f \pitchfork Z$) if for each point $x \in X$ such that $y = f(x) \in Z$, we have,

$$Im(Df)_x + T_y(Z) = T_y(Y).$$

Remark 3.5.1
(i) It may be noted that, no condition is imposed at points $x \in X$ such that $f(x) \notin Z$. Thus, it follows that if $f(X) \cap Z = \emptyset$, then f is transversal to Z in a vacuous way.
(ii) Consider the case when $Z = \{z\}$ is a singleton. Then the above condition says that f is

a submersion at x. This is true for each $x \in f^{-1}\{z\}$ iff z is a regular value for f. Therefore, we see that the notion of transversality is a generalization of the notion of regularity. This will become more evident in the proof of the following theorem.

Theorem 3.5.1 Transversal Inverse Image Theorem: *Let $f : X \to Y$ be a smooth map and $Z \subset Y$ be a submanifold of codimension k, where Z and Y are boundaryless. Suppose that $f \pitchfork Z$ and $\partial f \pitchfork Z$ where $\partial f := f|_{\partial X}$. Then $f^{-1}(Z) =: W$ is a submanifold of X of codimension $= k$ such that $\partial W = \partial X \cap W$. Moreover, we have $T_x(W) = (df_x)^{-1}(T_{f(x)}(Z))$, for every $x \in W$.*

Proof: We shall prove this theorem by reducing it to the case of the Preimage Theorem 3.4.2, as follows:

Let $x \in W$ be any point. Choose a coordinate neighborhood (U, ϕ) at $y = f(x)$ such that, $Z \cap U = \{y : \phi_1(y) = \cdots = \phi_k(y) = 0\}$, where $\phi = (\phi_1, \ldots, \phi_n)$. Let p denote the projection to the first k factors in the Euclidean space, so that we can rewrite $Z \cap U = (p \circ \phi)^{-1}(0)$. Observe that $p \circ \phi$ is a submersion. Hence, by the Preimage Theorem 3.4.2, we have, $T_y(Z) = T_y(Z \cap U) = Ker\, d(p \circ \phi)_y$. Now consider the map $g : V \to \mathbb{R}^k$ given by $g = p \circ \phi \circ f$, where $V = f^{-1}(U)$. Then clearly $W \cap V = g^{-1}(0)$. Thus, if we prove that 0 is a regular value of g, then from the Preimage Theorem, the conclusion of the above theorem follows.

Therefore, to see that Dg is surjective, let $\mathbf{v} \in \mathbb{R}^k$ be any vector. Since $d(p \circ \phi)$ is surjective, there exists $\mathbf{u} \in T_y(Y)$ such that $d(p \circ \phi)(\mathbf{u}) = \mathbf{v}$. Now from the hypothesis of the theorem, we can write $\mathbf{u} = (df)_x(\mathbf{w}_1) + \mathbf{u}_2$, where $\mathbf{u}_2 \in T_y(Z)$. Therefore, $\mathbf{v} = d(p \circ \phi)(\mathbf{u}) = (dg_x)(\mathbf{w}_1)$, since $d(p \circ \phi)(\mathbf{u}_2) = 0$.

Finally, for the last part, we have,

$$T_x W = (dg_x)^{-1}(0) = (df_x)^{-1}(d(p \circ \phi)_y)^{-1}(0) = (df_x)^{-1} T_y Z,$$

as required. This completes the proof. ♠

Remark 3.5.2 It should be remarked that, in the above theorem, the hypothesis on f restricted to V is indeed equivalent to say that g is a submersion. It is in this strong sense that we say that transversality is a generalization of the regularity. Thus, taking transversal inverse images of already known manifolds is another sure way of constructing new manifolds.

Example 3.5.1 Let us now examine a few simple examples of transversal maps. Consider $Y = \mathbb{R}^2$ and let Z be the real axis. Consider the map $f : \mathbb{R} \to \mathbb{R}^2$ given by $t \mapsto (t, at + b.)$ For $a = 0 = b$, f maps $\mathbb{R}$ inside Z. At any point $t \in \mathbb{R}$, we then have $df_t = (1, 0)$ and hence the image of df is contained in the tangent space to Z. Thus, f is not transversal. However, for $a \neq 0$, the only point $t \in \mathbb{R}$ for which $f(t) \in Z$ is $t = -b/a$, and at this point $df = (1, a)$. Therefore, the image of df is the line $y = ax$ which, together with the tangent space to Z, spans $\mathbb{R}^2$. We conclude that, f is transversal to Z iff $a \neq 0$ or $b \neq 0$. Now consider the map $g(t) = (t, p(t))$, where $p(t)$ is some polynomial function of some higher degree. The points common to the image of g and Z are given by the roots of the polynomial $p(t)$. At such a root t_0, the derivative is given by $df = (1, p'(t_0))$. Therefore, the image of df at such a point is not contained in Z iff $p'(t_0) \neq 0$, which is the same as saying that t_0 is not a multiple root of $p(t)$. Thus, g is transversal to Z iff $p(t)$ has no multiple real roots.

As a typical counterexample, take $p(t) = t^3$. As observed in Remark 3.4.4, the graph of p crosses the x-axis at the origin. Yet it is **not** a "point of transversal intersection".

Exercise 3.5

1. Consider the hyperboloid $x^2+y^2-z^2=1$ and the sphere $x^2+y^2+z^2=a^2$. Determine the values of a for which the intersection is transversal.

2. Let X, Z be submanifolds of Y. Say X is transversal to Z if $\iota_X : X \hookrightarrow Y$ is transversal to Z. Show that this is equivalent to say that for every $y \in X \cap Z$, we have, $T_y(X \cap Z) = T_y(X) \cap T_y(Z)$. In particular, this implies that the inclusion map $\iota_Y : Z \to Y$ is transversal to X. Thus, the symmetric terminology, "X, Z *intersect transversally*" for this property is justified.

3. Given smooth maps $f : X \to Y$ and $g : Y \to Z$ and a submanifold W of Z, such that g is transversal to W, show that $g \circ f$ is transversal to W iff f is transversal to $g^{-1}(W)$.

4. Give an example to show that the intersection of two submanifolds can still be a manifold even though they are not intersecting transversally. In such a case, what can be said about the dimension of the intersection? Illustrate with examples.

3.6 Homotopy and Stability

One of the important aspects of the algebraic and differential topology is to study properties that remain invariant under a deformation or a homotopy. Recall that by a homotopy we mean a continuous map $F : X \times I \to Y$, where X, Y are any two topological spaces and I denotes the unit interval. We write, for each $t \in I$, $f_t(x) = F(x,t), x \in X$ and say that f_0 is homotopic to f_1. In fact by a reparameterization, we can see that f_t and f_s are also homotopic for any $t, s \in I$. A property that remains invariant under "small" deformations is called a *stable property.* More precisely, suppose $\mathbf{P}$ is a property of a map $f : X \to Y$. Then $\mathbf{P}$ is said to be *stable* if for all homotopies $F : X \times I \to Y$ of f (i.e., $f_0 = f$,) f_t possess the property $\mathbf{P}$, for all $0 \leq t \leq \epsilon$, for some $\epsilon > 0$. Thus, the study of stability property can be termed as a "local behavior" under homotopies (deformations).

While dealing with smooth maps of manifolds $f : X \to Y$, it is not unreasonable to demand that the homotopy F of f is also a smooth map. Later on, we shall see that this is indeed not too much of a restriction. So, from now on, we shall assume that a homotopy F is a smooth map also. The following theorem tells you that many interesting properties that we studied so far are all stable.

Definition 3.6.1 For any homotopy $F : X \times I \to Y$, the *track* of F is defined to be the map $\tilde{F} : X \times I \to Y \times I$ given by $\tilde{F}(x,t) = (F(x,t), t)$.

Lemma 3.6.1 Given any smooth map $F : X \times I \to Y$ and a point $q = (x,s) \in X \times I$, we have $rank\,(D\tilde{F})_q = rank\,(Df_s)_x + 1$.

Proof: First consider the case when $X = \mathbb{R}^m$ and $Y = \mathbb{R}^n$. Then $D(\tilde{F})_q$ is represented by a $(n+1) \times (m+1)$ matrix of the form the

$$D(\tilde{F})_q = \begin{bmatrix} A_s & \star \\ 0 & 1 \end{bmatrix}$$

where $A_s = (Df_s)_x,\ f_s(x) = F(x,s)$. Hence, the conclusion.

In the general case, take a coordinate neighborhood (U, ϕ) at $y = F(x,s)$ and a parameterization $(\mathbb{R}^n, \psi)$ at x for X such that $F(\psi \times Id)(\mathbb{R}^n \times I) \subset U$ and consider $G = \phi \circ F \circ (\psi \times Id) : \mathbb{R}^m \times I \to \mathbb{R}^n$ in place of F. ♠

Theorem 3.6.1 *Let $f : X \to Y$ be a smooth map of manifolds where $\partial X = \emptyset$ and X is compact. Then all of the following properties of f are stable properties:*
(a) Immersion.
(b) Submersion.
(c) Local diffeomorphism.
(d) Transversal to a given submanifold $Z \subset Y$.
(e) Embedding.
(f) Diffeomorphism.

Proof: Let $F : X \times I \to Y$ be a homotopy of f and $\tilde{F}$ be its track.
(a) Let f be an immersion. This means that at each point $x \in X$, the rank of Df_x is equal to the dimension of X. Then by the lemma above, $D(\tilde{F})_{(x,0)}$ is injective, i.e, $\tilde{F}$ is an immersion at $(x, 0)$ for all $x \in X$. By the injective form of implicit function theorem 1.4.5, there exist neighborhood U_x of x and $\epsilon_x > 0$ such that $\tilde{F}|_{U_x \times [0,\epsilon_x)}$ is an immersion. Since X is compact, there exist finitely many points x_i such that $X = \cup_i U_{x_i}$. Take $\epsilon = min\{\epsilon_{x_i}\}$. It follows that $\tilde{F}|_{X\times[0,\epsilon)}$ is an immersion. But then from the above lemma, we conclude that f_s is also an immersion for all $0 \le s \le \epsilon$, as required.
(b) Simply replace the word "immersion" by "submersion" everywhere in the proof of (a).
(c) This is a special case of (a) or (b) when $\dim X = \dim Y$.
(d) Let $x \in X$ and $f(x) = y \in Z$. Let $\dim Z = k$. Choose a coordinate neighborhood (U, ϕ) at y for Y such that $Z \cap U$ is given by vanishing of $\phi_1, \ldots, \phi_k$. Consider the composition $H = \pi \circ \phi \circ F$ in a neighborhood of $(x, 0)$ say, in $V \times [0, \alpha)$. Here $\pi : \mathbb{R}^n \to \mathbb{R}^{n-k}$ is the projection to the last $n - k$ coordinates space. As in the proof of Theorem 3.5.1, to say that $f|_V$ is transversal to Z is equivalent to say that $H|_{V\times 0}$ is a submersion. Now by the surjective form of implicit function theorem 1.4.4, conclude that $H|_{W\times[0,\beta)}$ is a submersion, for a neighborhood W of x and for $\beta > 0$. This means that $f_t|_W$ is transversal to Z for all $t \in [0, \beta)$. Now, use the compactness of X to get a finite cover and choose ϵ to be the minimum of the corresponding positive numbers β_i.
(e) Since X is compact, any continuous map to a Hausdorff space Y is proper. Thus, in view of (a), and Theorem 3.4.1, the only thing that needs to be proved is injectivity of f_t for sufficiently small t. If this is not true, then for all n, we can find $x_n \neq x'_n \in X$ and $0 \le t_n \le 1/n$ such that $F(x_n, t_n) = F(x'_n, t_n)$. By the compactness of X, passing to subsequences if needed, we can assume that $x_n \to x_0$, and $x'_n \to x'_0$. Then $F(x_n, t_n) \to F(x_0, 0)$, and also $F(x'_n, t_n) \to F(x'_0, 0)$. This means that $f(x_0) = F(x_0, 0) = F(x'_0, 0) = f(x'_0)$. Since f is injective this means that $x_0 = x'_0$. On the other hand, by the injective form of implicit function theorem, it follows that there exists a neighborhood $U \times [0, \epsilon)$ of $(x_0, 0)$ such that $\tilde{F}$ is injective in this neighborhood. Then for large n, we have $(x_n, t_n), (x'_n, t_n) \in U \times [0, \epsilon)$ and we have, $F(x_n, t_n) = F(x'_n, t_n)$. This means that $\tilde{F}(x_n, t_n) = \tilde{F}(x'_n, t_n)$ and hence $x_n = x'_n$, for all sufficiently large n, which is a contradiction.
(f) Observe that since f is a diffeomorphism, $\dim X = \dim Y$. Again in view of (e), it remains to prove that f_t is also surjective. Since, connected components are mapped inside connected components, we may as well assume that both X and Y are connected. It now follows that for small t, f_t being a submersion, is an open mapping. Hence, $f_t(X)$ is an open subset of Y. It is also closed because X is compact. Hence, $f_t(X) = Y$, as required. ♠

Exercise 3.6 Let $f : X \times Y \to Z$ be a smooth map and let $F : X \times Y \to Z \times Y$, be defined by $F(x, y) = (f(x, y), y)$. For any point $y \in Y$, let $f^y(x) = f(x, y)$. Show that for any point $(x, y) \in X \times Y$, we have rank $DF_{(x,y)} =$ rank $D(f^y)_x + n$, where $n = \dim Y$. (Compare Lemma 3.6.1.)

3.7 Miscellaneous Exercises for Chapter 3

Throughout these exercises, X, Y etc. denote manifolds in $\mathbb{R}^N$ or $\mathbf{H}^N$.

1. Give an explicit atlas for $\mathbb{S}^1 \times \mathbb{S}^1$. What is the minimum number of charts needed?

2. For what values of r, does the equation $z^2 - x^2 - y^2 = r^2$ define a manifold?

3. For positive real numbers a, b consider the set $S(a,b)$ of points $P \in \mathbb{R}^3$, which are at a distance b from the circle $x^2 + y^2 = a$. Show that for $0 < b < a$, $S(a,b)$ is a 2-dimensional manifold and identify this manifold. What happens when $a = b$ and $b > a$?

4. The graph of a function $f : X \to Y$ is defined to be

$$\Gamma(f) = \{(x,y) \in X \times Y \ : \ y = f(x)\}.$$

   Show that if X, Y are smooth manifolds and f is smooth then $\Gamma(f)$ is a submanifold of $X \times Y$ and is diffeomorphic to X.

5. Let $f : X \to Y$ be a continuous map of differentiable manifolds such that $\Gamma(f)$ is a smooth submanifold of $X \times Y$. Does this imply that f is differentiable? [Compare Exercise 3.3.6 and Example 3.1.1.4. Hint: Example 1.6.1.3.]

6. Let V be a vector subspace of $\mathbb{R}^n$. Show that the tangent space $T_x(V) = V$ for all $x \in V$.

7. Exhibit a basis for the tangent space of the paraboloid $x^2 + y^2 - z^2 = 1$ at the point $(1,0,0)$.

8. For any two manifolds X and Y without boundary, show that

$$T_{(x,y)}(X \times Y) = T_xX \times T_yY.$$

   Compute dp where $p : X \times Y \to Y$ is the projection.

9. For smooth maps $f : X \to X', g : Y \to Y'$, show that $d(f \times g) = df \times dg$.

10. Compute $d(\eta)$, where $\eta : X \to X \times X$ is the diagonal map $x \mapsto (x,x)$.

11. Prove that the tangent space to the graph of a smooth map $f : X \to Y$ at $(x, f(x))$ is the graph of the tangent map

$$df_x : T_x(X) \to T_{(x,f(x))}(X \times Y).$$

12. Let $c : I \to X$ be a smooth curve where X is a smooth manifold. The *velocity vector* to c at $t = s$ is defined to be the vector $dc_s(1) \in T_{c(s)}(X)$. We denote this simply by $\frac{dc}{dt}\big|_s$. If $X = \mathbb{R}^n$ and $c(t) = (c_1(t), \ldots, c_n(t))$ then check that

$$dc_s(1) = \left.\frac{dc}{dt}\right|_s = (c_1'(s), \ldots, c_n'(s))$$

   Prove that every vector in $T_x(X)$ is the velocity vector of some curve in X passing through x. (This exercise gives an alternate description of the tangent space.)

13. If $f : \mathbb{R} \to \mathbb{R}$ is a local diffeomorphism, then show that f is a diffeomorphism onto an open interval.

14. Show that the map $z \mapsto 2z^3 - 3z^2$ restricts to a local diffeomorphism of $\mathbb{C} \setminus \{0,1\}$ onto $\mathbb{C}$. Now, use the following fact that there is a complex differentiable map that is a covering projection $\lambda : \mathbb{D}^2 \to \mathbb{C} \setminus \{0,1\}$ (the modular function), to construct a surjective holomorphic mapping $B_1(0) \to \mathbb{C}$, which is a local diffeomorphism of the open unit disc onto $\mathbb{C}$.

15. Give an example of a surjective local diffeomorphism $\mathbb{R}^2 \to \mathbb{S}^2$. [Hint: Consider $z \mapsto 2z^3 - 3z^2/z + 1$.]

16. Show that any injective local diffeomorphism $f : X \to Y$ is a diffeomorphism onto an open subset, where X, Y are smooth manifolds.

17. Let $f : X \to Y, g : Y \to Z$ be smooth maps. If f, g are embeddings then so is $g \circ f$. If $g, g \circ f$ are embeddings then so is f.

18. Let $f : X \to Y$ be a submersion. If X is compact and Y is connected, show that f is surjective.

19. Let $p(x_1, \ldots, x_n)$ be a homogeneous polynomial of degree k with real coefficients. For any $r \neq 0$, show that $p(X) = r$ defines a submanifold of $\mathbb{R}^n$ of codimension 1. Show that if $r_1 r_2 > 0$ then $p(X) = r_1$ and $p(X) = r_2$ define diffeomorphic submanifolds. Give an example to show that the condition $r_1 r_2 > 0$ cannot be dropped.

20. Let $P_1, \ldots, P_{n+1} \in \mathbb{R}^n$ be any $n+1$ points that do not lie in a hyperplane. Consider the mapping $\phi = (\phi_1, \ldots, \phi_{n+1})$ given by $\phi_i(\mathbf{x}) = \frac{1}{\|\mathbf{x} - P_i\|}, i = 1, \ldots, n+1$. Show that $\phi : \mathbb{R}^n \setminus \{P_1, \ldots, P_{n+1}\} \to \mathbb{R}^{n+1}$ is an embedding.

21. Let U be a nonempty open subset of $\mathbb{R}^n$ and $\phi(x) = d(x, \mathbb{R}^n \setminus U)$. Show that

$$\psi(x) = \left(x, \frac{1}{\phi(x)}\right)$$

defines a proper embedding of U in $\mathbb{R}^{n+1}$.

22. Show that every rational function $f(z) = P(z)/Q(z)$ (where P, Q are polynomials in one variable) extends to a smooth map of $\hat{f} : \hat{\mathbb{C}} \to \hat{\mathbb{C}}$, where $\hat{\mathbb{C}} = \mathbb{S}^2$ is the extended complex plane (which is also called the *Riemann sphere*).

23. **Fundamental Theorem of Algebra** Let $p(z) = z^n + a_1 z^{n-1} + \cdots + a_n$ be a non-constant polynomial with complex coefficients. Show that p has a root by making the following simple observations:
(i) p is smooth.
(ii) p extends smoothly to a map $\hat{\mathbb{C}} = \mathbb{S}^2 \to \mathbb{S}^2$.
(iii) p is a local diffeomorphism at z_0 iff $p'(z_0) \neq 0$.
(iv) There are only finitely many critical points of p.
(v) $\#(p^{-1}(w))$ is finite for all $w \in \mathbb{S}^2$.
(vi) Property (ii) of Theorem 3.4.5 holds for p.
(vii) $\mathbb{C} \setminus S$ is connected where S is a finite set.
(viii) A locally constant function is a constant on a connected set.
(ix) $p : \mathbb{C} \to \mathbb{C}$ defines a surjective mapping.

24. In Example 3.4.4, we saw that the space $\mathcal{R}_1(2, 2; \mathbb{R})$ of all 2×2 real matrices of rank 1 is a 3-dimensional submanifold of $\mathbb{R}^{2\times 2}$. Display it as the regular inverse image of a smooth map $\mathbb{R}^{2\times 2} \setminus \{0\} \to \mathbb{R}$. Generalize this to the case of $\mathcal{R}_{n-1}(n, n; \mathbb{R})$. [Indeed, among all $\mathcal{R}_k(m, n; \mathbb{R})$, this is as far as you can go– other $\mathcal{R}_r(m, n; \mathbb{R})$ are not even intersection of an appropriate number of hypersurfaces.]

25. Show that every closed subset of $\mathbb{R}^n$ is the intersection of $\mathbb{R}^n \times 0$ and another submanifold Y of $\mathbb{R}^{n+1}$. [Hint: Compare Corollary 1.7.2.]

26. For a linear map $f : \mathbb{R}^n \to \mathbb{R}^n$ show that the graph of f is transversal to the diagonal Δ iff $+1$ is not an eigenvalue of f.

27. Let $f : X \to X$ be a smooth map with x as a fixed point. We say x is *Lefschetz type* if $+1$ is not an eigenvalue of $df_x : T_xX \to T_xX$. If all fixed points of f are of Lefschetz type then we say f is a Lefschetz map. If X is compact, show that a Lefschetz map f has only finitely many fixed points.

28. Let $f : X \to Z, g : Y \to Z$ be continuous maps of topological spaces. The fibered product denoted by $X \times_Z Y$ is a subspace of the product $X \times Y$ defined by

$$X \times_Z Y = \{(x, y) \in X \times Y \ : \ f(x) = g(y)\}.$$

Suppose now that X, Y, Z are smooth manifolds and f, g are smooth. We say f is transversal to g (and write $f \pitchfork g$) if $df_x(T_xX) + dg_y(T_y(Y)) = T_z(Z)$ whenever $f(x) = g(y) = z$.
(i) Show that $f \pitchfork g$ iff $(f \times g) \pitchfork \Delta_Z$ in $Z \times Z$.
(ii) Deduce that $X \times_Z Y$ is a submanifold of $X \times Y$ if $f \pitchfork g$.
(iii) If f is a submersion, show that the projection $p_2 : X \times_Z Y \to Y$ is a submersion. This submersion is called the *pullback* of f via g.

$$\begin{array}{ccc} X \times_Z Y & \xrightarrow{p_1} & X \\ {\scriptstyle p_2}\downarrow & & \downarrow{\scriptstyle f} \\ Y & \xrightarrow{g} & Z \end{array}$$

29. Show that $S0(n)$ is connected for all $n \geq 1$ and $O(n)$ has precisely two components.

30. Show that $U(n)$ is connected for all $n \geq 1$.

31. Show that $V_{k,n}, \ \ 1 \leq k < n$ are connected.

Chapter 4

Integration on Manifolds

In this chapter, we extend the notion of integration discussed in Chapter 2, to the class of smooth manifolds. First, we introduce the notion of orientation on manifolds and differential forms on manifolds. Via partition of unity, we then define integration on oriented manifolds. Stokes' theorem comes almost as a free reward. As an illustration of the fact that the integration on manifolds can be effectively used to bring out the geometric and the topological behaviour of the manifolds, we then introduce De Rham cohomology groups and compute them for spheres and give a few applications.

4.1 Orientation on Manifolds

In your calculus course, you have met with the concept of orienting a curve by putting an arrow or even indicating orientations on surfaces by placing curved arrows, etc. In this section, we shall consolidate the notion of orientation, which is very fundamental especially in the theory of integration on manifolds.

Definition 4.1.1 Let V be a vector space of dimension $n \geq 1$. Consider the set $\mathcal{B}$ of all ordered bases for V. On this set let us introduce an equivalence relation as follows. We know that any two ordered bases correspond under an invertible matrix, i.e., if α, β are any two ordered bases, written columnwise, then there is a unique invertible $n \times n$ matrix M, such that $M\alpha = \beta$. We shall define $\alpha \sim \beta$ if $\det M > 0$. It is straightforward to verify that "$\sim$" is an equivalence relation. We shall call an equivalence class of an ordered basis an *orientation* for V. Since for any invertible matrix M either $\det M > 0$ or < 0, it follows easily that there are precisely two equivalence classes of ordered bases. It is also clear that, if one class α is represented by $(\mathbf{v}_1, \ldots, \mathbf{v}_n)$ then $(-\mathbf{v}_1, \mathbf{v}_2, \ldots, \mathbf{v}_n)$ represents the other class. For this reason, we may refer to the other orientation as the "opposite orientation" or the "negative" of the first orientation and indicate this by writing $-\alpha$. This does not necessarily mean that α is *the positive* orientation and the other one is the *'negative'*. As such we do not have any bias for one orientation from the other and treat both the orientations with equal dignity.

Finally, it remains to define orientation on the vector space (0). Since on every nonzero vector space there are precisely two orientations, we shall declare the same to be true on the (0) space as well. These two orientations will be denoted by $\pm$, by convention. We shall denote by E_n the standard orientation on $\mathbb{R}^n$, i.e.,

$$E_n = [e_1, e_2, \ldots, e_n].$$

Remark 4.1.1 Observe that if $f : V \to W$ is a linear isomorphism then f takes an ordered basis of V to an ordered basis of W. Moreover, it is easily verified that if two of the ordered bases of V are equivalent then the corresponding bases of W are also so. Thus, f induces a function $f_\star$ on the set of orientations of V to that of W. This function is clearly a bijection.

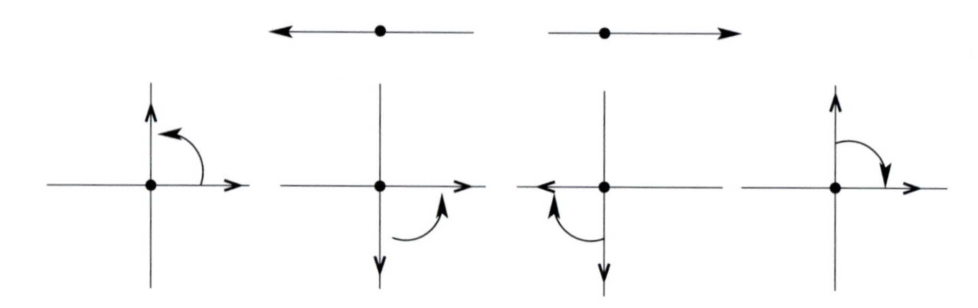

Figure 19 Orientations on $\mathbb{R}$ and $\mathbb{R}^2$.

Example 4.1.1 Consider the following orientation on $\mathbb{R}^3$:

(i)	$\{\mathbf{e}_1, \mathbf{e}_2, \mathbf{e}_3\}$,	(ii)	$\{-\mathbf{e}_1, -\mathbf{e}_2, -\mathbf{e}_3\}$,	(iii)	$\{\mathbf{e}_1, \mathbf{e}_3, \mathbf{e}_2\}$,
(iv)	$\{-\mathbf{e}_3, \mathbf{e}_2, \mathbf{e}_1\}$,	(v)	$\{3\mathbf{e}_1, \mathbf{e}_2, 5\mathbf{e}_3\}$,	(vi)	$\{\mathbf{e}_1+\mathbf{e}_2, \mathbf{e}_2+\mathbf{e}_3, \mathbf{e}_3+\mathbf{e}_1\}$.

Since there are three negative signs in (ii) as compared to (i) and no order change, it follows that the determinant of the transformation matrix is actually -1. So, (i) and (ii) are in different orientation classes. Arguing similarly, we can check that (i), (iv), (v), and (vi) are in one class whereas, (ii) and (iii) in another.

Definition 4.1.2 Let X be a manifold of dimension $n \geq 1$. By a *preorientation* α on X we mean a choice α_x of an orientation on T_xX for each $x \in X$. A preorientation α of X is said to be *smooth* or *continuous* if it satisfies the following smoothness condition: to each $x \in X$, there exists a parameterization (U, ϕ) for X at x such that $d\phi_y : \mathbb{R}^n \to T_{\phi(y)}(X)$ carries the orientation class E_n to $\alpha_{\phi(y)}$ for all $y \in U \subset \mathbb{R}^n$, i.e., $d\phi_y(E_n) = \alpha_{\phi(y)}$. (Observe that we could have replaced E_n by $-E_n$.) A preorientation which is smooth will be called an *orientation.* If there exists an orientation of X then we say X is *orientable.* In such a case, X together with a fixed orientation will be called an *oriented manifold.*

Finally, let X be a manifold of dimension 0. Then by an orientation on X, we mean a function $\epsilon : X \to \{+, -\}$. Thus, an orientation on X simply assigns a sign $\pm$ for each point of X. Obviously, there is no need to put any smoothness condition on ϵ, since every function on a discrete space is continuous.

Lemma 4.1.1 Let X be a connected manifold and ω, θ be any two orientations. Suppose there exists a point $x \in X$ such that $\omega_x = \theta_x$. Then $\omega = \theta$.

Proof: As usual, we take $A = \{x \in X \ : \ \omega_x = \theta_x\}$. The hypothesis says that A is nonempty. We shall show that A is both open and closed in X and conclude that $A = X$. So consider $y \in A$. Let $\phi : U \to \mathbb{R}^n$ and $\psi : U \to \mathbb{R}^n$ be charts for X at y such that $d\phi_x(\omega_x) = E_n = d\psi_x(\theta_x)$ for all $x \in U$. Then, since $\omega_y = \theta_y$, we have, $d(\psi \circ \phi^{-1})_0(E_n) = d\psi_y(\omega_y) = d\psi_y(\theta_y) = E_n$. Therefore, $d(\psi \circ \phi^{-1})_z$ has positive determinant at $z = 0$ and hence this must be the case for all z in a neighbourhood W of 0. This means $d(\psi \circ \phi^{-1})_z(E_n) = E_n$ for all $z \in W$. Taking $V = \phi^{-1}(W)$, this is the same as saying $\omega_x = \theta_x$ for all $x \in V$. Therefore, $V \subset A$. This proves the openness of A. Exactly the same way, one can show that $X \setminus A$ is open. This completes the proof of the lemma. ♠

As an immediate consequence we have:

Theorem 4.1.1 *On a connected manifold there can be at most two orientations.*

Remark 4.1.2
(i) Clearly every open subset of $\mathbb{R}^n$ is orientable. Indeed we need to fix one basis for $\mathbb{R}^n$ and take the class of the same basis for all $T_xU = \mathbb{R}^n$. In fact, a similar reason tells you that if X is an orientable manifold then every open subset of X is also so. Observe that this is also true for any subset Y of X, which is the closure of an open subset.
(ii) Of course there need not be any orientation on a given manifold. The simplest example is the Möbius band. We leave it to you to figure out why this surface is not orientable. At this stage we do not want to give a formal proof of this fact.
(iii) Observe that if ω is an orientation on X, U is a connected open subset of X and $\phi : U \to \mathbb{R}^n$ is a diffeomorphism onto an open set, then $d(\phi)_x$ preserves orientation for all $x \in U$ or it reverses orientation for all $x \in U$. The argument that goes into proving this has occurred in the proof of the above lemma.

Definition 4.1.3 A diffeomorphism $\phi : X \to Y$ of oriented manifolds is said to be *orientation preserving* if at each point $x \in X$, we have $D\phi : T_xX \to T_{f(x)}Y$ mapping the orientation of T_xX to that of $T_{f(x)}Y$.

Remark 4.1.3 It is a straightforward exercise to verify that if X is connected then any diffeomorphism ϕ preserves orientation iff it does so at a single point. [See Remark 4.1.2.(iii).] Therefore, every diffeomorphism of connected oriented manifolds, either preserves orientations or reverses them. Also, a composite of two orientation preserving diffeomorphisms is again orientation preserving. It is also true that composite of two orientation reversing diffeomorphisms is orientation preserving. On $\mathbb{R}^n$ itself, translations and rotations are easily seen to be orientation preserving, whereas the reflection in any hyperplane is orientation reversing. A holomorphic function defined in a region in $\mathbb{C}$ is orientation preserving at all points z such that $f'(z) \neq 0$.

Let us now give a criterion for orientability of a manifold.

Theorem 4.1.2 *An n-dimensional manifold is orientable iff it has an atlas $\{(U_\alpha, \phi_\alpha)\}$ such that for each pair of indices α, β, $\phi_\beta \circ \phi_\alpha^{-1}$ is an orientation preserving diffeomorphism of open subsets of $\mathbb{R}^n$.*

Proof: Assume that X is orientable and θ is an orientation on X. Let $\{(U_\alpha, \phi_\alpha)\}$ be an atlas that guarantees the local smoothness of θ. This just means that $D(\phi_\alpha)_y(\theta_y) = E_n$ for all $y \in U_\alpha$ and for all α. Therefore, whenever $U_\alpha \cap U_\beta \neq \emptyset$, for all $z \in \phi_\alpha(U_\alpha \cap U_\beta)$, we have, $D(\phi_\beta \circ \phi_\alpha^{-1})_z(E_n) = E_n$.

Conversely, given such an atlas, we define an orientation ω on X as follows: Let $x \in X$ be such that $x \in U_\alpha$. Among the two possible orientations on T_xX, take ω_x to be that one such that $D(\phi_\alpha)_x(\omega_x) = E_n$. From the condition given on the atlas, it follows that ω_x is well defined, i.e., it does not depend upon the choice of α. The smoothness of ω follows by the very definition, since for all points y in the neighbourhood U_α we have, $\omega_y = D(\phi_\alpha)_y(\omega_x) = E_n$. This completes the proof of the theorem. ♠

We shall now take up the task of getting an orientation on the boundary of an oriented manifold. In this, we shall be guided by the standard practice of choosing orientation on the boundary of a bounded plane region, viz., when you trace the boundary curve forward, the region should be on your left. So, to begin with we fix an orientation on $\mathbb{R}^n$ say, the standard one E_n, given by the ordered basis $(\mathbf{e}_1, \ldots, \mathbf{e}_n)$. This then gives an orientation on $\mathbf{H}^n$.

Definition 4.1.4 The *induced orientation* on $\mathbb{R}^{n-1} \times \{0\}$ from the standard orientation E_n of $\mathbf{H}^n$ is defined and denoted by

$$\partial E_n := (-1)^n E_{n-1} \tag{4.1}$$

Remark 4.1.4

1. First observe that if $\{\mathbf{v}_1, \ldots, \mathbf{v}_{n-1}\}$ is an ordered basis of $\mathbb{R}^{n-1}$ representing ∂E_n, then $\{-\mathbf{e}_n, \mathbf{v}_1, \ldots, \mathbf{v}_{n-1}\}$ represents the orientation E_n on $\mathbb{R}^n$. We express this by simply writing
$$-\mathbf{e}_n \partial E_n = E_n.$$

2. Now, if $h : U \to V$ is a diffeomorphism of two open subsets in $\mathbf{H}^n$, then for any $x \in U \cap \mathbb{R}^{n-1}$ we have,
$$Dh_x(\mathbb{R}^{n-1}) \subseteq \mathbb{R}^{n-1} \text{ and } Dh_x(\mathbf{e}_n) = \lambda \mathbf{e}_n + \mathbf{v}; \;\; \lambda > 0, \; \mathbf{v} \in \mathbb{R}^{n-1}. \tag{4.2}$$

Definition 4.1.5 Let X be a manifold, $x \in \partial X$. A vector $\mathbf{v} \in T_x(X)$ is called *inward* (resp. *outward*) if there exists a local parameterization $\phi : \mathbf{H}^n \to U$ at x for X such that $d\phi_0(e_n) = \mathbf{v}$ (resp. $= -\mathbf{v}$). From the remark above, it follows that this definition is independent of the choice of parameterization.

Remark 4.1.5 It is now clear that while adopting the Definition 4.1.4, we have followed the standard convention of orienting the boundary of a planar region by the rule that the region lies to your left while you trace the curve in the chosen direction.

Theorem 4.1.3 *Let X be an orientable manifold. Then ∂X is orientable. Indeed, if ω is an orientation on X, there is a unique orientation $\partial\omega$ such that at each point $x \in \partial X$, we have, $\omega_x = [\eta_x][\partial\omega_x]$, where η_x is a nonzero outward vector to ∂X at x.*

Proof: We shall verify the local constancy of $\partial\omega$, which will automatically prove that it is well defined also. So, let $\phi : \mathbf{H} \to U$ denote a local parameterization around x, such that for all $z \in \partial\mathbf{H}$,
$$\begin{aligned} \omega_{\phi(z)} &= (d\phi)_z(E_n) = (d\phi)_z(-\mathbf{e}_n \partial E_n) \\ &= \eta_{\phi(z)}(d\phi)_z(\partial E_n) = (-1)^n \eta_{\phi(z)}(d\phi)_z(E_{n-1}). \end{aligned}$$
Taking $x = \phi(z)$ and $\partial\omega_x = (-1)^n (d\phi)_z(E_{n-1})$ we are done. ♠

Remark 4.1.6 It is not difficult to prove the above theorem directly by using the criterion for orientability in 4.1.2. We leave this to you as an exercise.

Definition 4.1.6 The orientation $\partial\omega$ is called the *induced orientation* on the boundary of X from the given orientation ω on X.

Remark 4.1.7
(i) Let X and Y be manifolds, at least one of them being boundaryless. If both are oriented with orientations ω and θ respectively, then the product $X \times Y$ is oriented with the *product orientation* defined by
$$(\omega\theta)_{(x,y)} := \omega_x \theta_y.$$
Recall that $T_{(x,y)}(X \times Y) = T_x(X) \times T_y(Y)$ and hence an ordered basis for this product vector space is obtained by simply taking an ordered basis for the first factor followed by an ordered basis for the second.
(ii) Consider the unit interval I with the standard orientation E_1. On the two points $0, 1 \in \partial I$, what are the induced orientations? If we follow the outward normal rule, the point 1 acquires the $+$ sign and the point 0 acquires the sign $-$. Next consider cylinder $I \times \mathbb{S}^1$. Fixing an orientation on $\mathbb{S}^1$ and one on I say the standard ones, we get an orientation on $I \times \mathbb{S}^1$. The boundary of this consists of two copies of the circle, viz., $\{0\} \times \mathbb{S}^1$ and $\{1\} \times \mathbb{S}^1$.

What are the induced orientations on them? If we denote the standard orientations on I and $\mathbb{S}^1$ by E_1 and τ, we see that on $\{0\} \times \mathbb{S}^1$, we have $\partial(\eta\tau) = \partial(\eta)(\tau) = -\tau$ and hence the induced orientation on $\{0\} \times \mathbb{S}^1$ is $-\tau$. It is easily seen that the induced orientation on $\{1\} \times \mathbb{S}^1$ is τ itself.

More generally, for any oriented manifold X without boundary, the oriented boundary of $I \times X$ should be expressed as

$$\partial[I \times X] = [\{1\} \times X] - [\{0\} \times X].$$

Figure 20 Boundary orientations on the plane region.

Figure 20 shows the induced orientations on the boundary components of a disc with three wholes in $\mathbb{R}^2$.

Exercise 4.1

1. Let X be a connected nonorientable manifold.

   (i) Let U_α be an atlas for X of open subsets diffeomorphic to $\mathbb{D}^n$. Show that there exists a smooth loop $\gamma : [0,1] \to X$ with the following properties:
   (a) There exist a partition $0 = t_0 < t_1 < \cdots < t_n = 1$ such that $\gamma([t_{i-1}, t_i]) \subset U_{\alpha_i}$.
   (b) If we choose the constant orientation on each of U_{α_i} such that at $\gamma(t_i)$ the two orientations from $U_{\alpha_{i-1}}$ and U_{α_i} agree for $1 \le i \le n-1$, then the orientations on $\gamma(t_1) = \gamma(t_n)$ coming from U_{α_1} and U_{α_n} do **not** agree.
   Such a loop is called an *orientation reversing loop.*

   (ii) Show that dim $X > 1$.

   (iii) Improve the result in (i) by getting an immersed loop first. This means that the loop may have some self-intersections.

   (iv) Extract an embedded loop out of (iii), which is orientation reversing.

2. Conversely, show that if there is an orientation reversing loop in X then X is nonorientable.

3. Show that if a loop is contained in a single chart then it is orientation preserving.

4. Show that any null-homotopic loop is orientation preserving. Deduce that a simply connected manifold is orientable.

5. Obtain an embedding of the Möbius band inside the solid torus in $\mathbb{R}^3$ which itself is obtained by rotating the disc

$$\{(x,y),0) \in \mathbb{R}^3 \;:\; (x-2)^2 + y^2 \le 1\}$$

around the y-axis. Use this description to explicitly write down an diffeomorphism $\phi : \mathbb{M} \to \mathbb{M}$ which is isotopic to identity.

4.2 Differential Forms on Manifolds

Throughout this section, X will denote a manifold of dimension n, with or without boundary. Recall that $T_x(X)$ denotes the tangent space to X at $x \in X$, $T(X) : \cup_{x\in X} T_x(X)$, the total tangent space, and $\pi : T(X) \to X$, the projection map. If X is an open subset of $\mathbb{R}^n$ or $\mathbf{H}^n$, then all the tangent spaces $T_x(X) = \{x\} \times \mathbb{R}^n$ can be identified with $0 \times \mathbb{R}^n$ in a natural way. In Chapter 2, we have extensively discussed the differential forms on X in this special case. The situation with a submanifold of Euclidean spaces is only a little bit more complicated and all those notions and results which are "local" in nature, generalize in a natural way. Moreover, whatever we do here is then applicable to the so-called "abstract smooth manifolds" (introduced in Chapter 5) as well, thanks to the embedding theorems that we shall prove therein.

Definition 4.2.1 For $p \geq 0$, by a p-form on X, we mean a function ω that assigns to each point $x \in X$, an alternating p-tensor $\omega(x) \in \wedge^p(T_x(X)^*)$. Two p-forms on X can be added together pointwise and also any p-form can be multiplied by a scalar-function on X. Thus, the space of all p-forms on X forms a module over the ring of all scalar functions on X.

Moreover, we can take the wedge product of a p-form and a q-form to get a $(p+q)$-form; this again is done pointwise. It is not difficult to see that these operations make the space of all forms on X into a graded anticommutative algebra over the ring of all real valued functions on X.

Remark 4.2.1 As before, a very important aspect of forms is the beautiful way they transform. Suppose $\phi : X \to Y$ is a smooth map. Then for each $x \in X$, the map $d\phi_x : T_x(X) \to T_y(Y), (\ y = \phi(x))$, is a linear map. This in turn induces a linear map

$$\wedge^p(d\phi_x) : \wedge^p(T_y(Y)^*) \to \wedge^p(T_x(X)^*).$$

Thus, given a p-form ω on Y, we can *pull it back* to a p-form on X as follows:

$$\phi^*(\omega)(x) := \wedge^p(d\phi_x)(\omega)(\phi(x)). \tag{4.3}$$

It is routine to verify that for each p, ϕ^* itself is a linear map. Further, they are algebra homomorphisms also, in the sense

$$\phi^*(\omega \wedge \tau) = \phi^*(\omega) \wedge \phi^*(\tau). \tag{4.4}$$

Definition 4.2.2 A p-form on an open subset of $\mathbb{R}^n$ is said to be *smooth* if all its coefficient functions are smooth. A p-form ω on a manifold X is said to be smooth at $x \in X$ if for some local parametrization $\phi : U \to X$ at x, (where U is an open subset of $\mathbb{R}^n$,) the p-form $f^*(\omega)$ is smooth.

Remark 4.2.2

1. We leave the verification that the smoothness of a p-form is independent of the local parameterization chosen to the reader as an exercise.

2. The set of all smooth p-forms is a vector subspace of the space of all p-forms. This subspace will be denoted by $\Omega^p(X)$. Note that $\Omega^0(X) = \mathcal{C}^\infty(X;\mathbb{R})$, the ring of real valued $\mathcal{C}^\infty$-functions on X. Each $\Omega^p(X)$ is a module over $\Omega^0(X)$.

   Moreover, the wedge product of smooth forms turns out to be smooth. Thus,

$$\Omega^*(X) := \Omega^0(X) \oplus \Omega^1(X) \oplus \cdots \oplus \Omega^n(X), \tag{4.5}$$

   forms an anticommutative graded algebra over the ring of smooth functions $\Omega^0(X)$. Given a smooth map $f : X \to Y$, we have the algebra homomorphism $\Omega(f) : \Omega^*(Y) \to \Omega^*(X)$, with the property that

$$\Omega(g \circ f) = \Omega(f) \circ \Omega(g); \quad \Omega(Id) = Id. \tag{4.6}$$

   Thus, the assignment $X \rightsquigarrow \Omega^*(X)$ is a contravariant functor from the category of smooth manifolds to the category of graded anticommutative algebras.

3. Recall from Section 2.5, that for an open subset U of $\mathbb{R}^n$, the exterior differentiation $d : \Omega(U) \to \Omega(U)$ satisfied the properties (i), (ii), (iii), and (iv), which are verified pointwise and hence all these are valid when we replace U by a manifold X.

   For future reference, here is the summary of the properties of the exterior differentiation operator that we have already established.

Theorem 4.2.1 *The exterior differentiation operator $d : \Omega^*(X) \to \Omega^*(X)$ is the unique operator satisfying the following conditions:*
(E1) *d is homogeneous of degree 1, i.e., $d(\Omega^p(X)) \subset \Omega^{p+1}(X)$, for all $0 \le p \le n-1$, and $d(\Omega^n(X)) = (0)$.*
(E2) *Let $\pi : T(\mathbb{R}) = \mathbb{R} \times \mathbb{R} \to \mathbb{R}$ be the projection to the second factor. For any smooth map $f : X \to \mathbb{R}$ let $Tf : TX \to T\mathbb{R}$ be the map induced on the tangent space. Then $\pi \circ Tf : TX \to \mathbb{R}$ is a smooth map which when restricted to each fibre is a linear map. We can identify it with df which assigns to each point $x \in X$, the 1-tensor $df_x : T_xX \to \mathbb{R}$. Thus $\pi \circ Tf = df$.*
(E3) *$d(\omega + \tau) = d(\omega) + d(\tau)$.*
(E4) *$d(\omega \wedge \tau) = d(\omega) \wedge \tau + \omega \wedge d(\tau)$.*
(E5) *$d \circ d = 0$. (Co-cycle condition).*
(E6) *For any smooth map $f : X \to Y$, $f^* \circ d = d \circ f^*$.*

Exercise 4.2

1. Verify various claims made in the section, especially those in Remark 4.2.2 (1).

2. Let X be a connected manifold with an oriented atlas $\{U_\alpha\}$. We say a smooth n-form ϕ on X is an orientation form on X if $\phi(\mathbf{v}_1, \ldots, \mathbf{v}_n) > 0$ whenever, $[\mathbf{v}_1, \ldots, \mathbf{v}_n]$ is the orientation on X at a point. Show that there is such a form ϕ on X. Obviously, such a form is nonzero at every point of X. Conversely, show that given any nowhere vanishing smooth n-form ϕ on X, there is an orientation on X with respect to which ϕ is an orientation form.

4.3 Integration on Manifolds

We begin with recalling Theorem 2.1.3.

Theorem 4.3.1 Change of Variable Formula: *Let $f : U \to V$ be a diffeomorphism of open subsets in $\mathbb{R}^n$. Then for any integrable function α on V, we have*

$$\int_V \alpha = \int_U (\alpha \circ f)|\det(df)| \tag{4.7}$$

If f is an orientation preserving diffeomorphism then $\det df$ is positive everywhere and hence there is no need to take the modulus in the expression for the integrand on the right-hand side (RHS) of (4.7) and we get

$$\int_U \alpha = \int_V (\alpha \circ f)\det(df) \tag{4.8}$$

Just as in Definition 2.6.5, given any n-form ω on an open set U there is a unique function α such that $\omega = \alpha\, dx_1 \wedge \cdots \wedge dx_n$. So, if α is integrable, we can and shall define

$$\int_U \omega := \int_U \alpha. \tag{4.9}$$

We can now rewrite (4.8) in the form

$$\int_U \omega = \int_V f^*\omega. \tag{4.10}$$

[The existence of the integral the RHS of (4.9) can be guaranteed if we assume that α is continuous and has compact support.] Consider now the simple line integrals $\int_a^b f(t)\, dt$ and $\int_b^a f(t)\, dt$. By convention, one is the negative of the other. The domain of integration, viz., the interval $[a, b]$ is considered with different orientations in the two integrals. A diffeomorphism of the interval that takes one orientation to the other, is given by $\phi(t) = a + b - t$; the determinant of the Jacobian of this diffeomorphism is -1 everywhere. Thus, the change of variable formula explains why the two integrals are negatives of each other. The notation $\int_a^b$ is too special for the case of intervals and is not conveniently generalizable. This problem is already resolved in the definition of the integration of forms on chains as well as the definition of forms adopted above.

The advantage of (4.10) over (4.7) is not just in elegance. Only this way we shall be able to formulate the notion of integration on a manifold. The price we have to pay is that we are now restricted to taking diffeomorphisms that preserve orientations. Naturally, this makes sense only on manifolds that are oriented. However, (4.10) can be used for orientation reversing ones as well, by changing the orientation of the domain on one of the sides.

Let then X be an oriented n-dimensional manifold. Recall (from 4.1.2) that this means that:

(i) there is an assignment $\Theta : x \mapsto \Theta_x$, where for each $x \in X$, Θ_x is an orientation (an equivalence class of an ordered basis) of $T_x(X)$;

(ii) Θ is "smooth" or "locally consistent" in the sense that there is an atlas $\{(U_j, \phi_j)\}$ of parameterizations for X such that for each $y \in U_j$, $D(\phi_j)_y : \mathbb{R}^n \to T_{\phi(y)}X$ takes the standard orientation of $\mathbb{R}^n$ to the orientation $\Theta_{\phi_j(y)}$. Such parameterizations are said to be orientation preserving.

Let now ω be a smooth n-form on X with compact support, i.e., with the closure of the set of all points where $\omega \neq 0$ being compact. We shall define the integral of ω on X. This will take several steps:

Step I Let the support of ω be contained in an open set W of X, which can be covered by a single parameterization. If $\phi : U \to W$ is an orientation preserving parameterization, then we define

$$\int_X \omega := \int_U \phi^*(\omega). \tag{4.11}$$

The most important thing to observe now is that the RHS above is independent of ϕ chosen. For, if $\psi : V \to W$ is another such parameterization, we put $\xi = \psi^{-1} \circ \phi$ and see that $\xi : U \to V$ is an orientation preserving diffeomorphism and hence

$$\int_V \psi^*(\omega) = \int_U \xi^*(\psi^*(\omega)) = \int_U (\psi \circ \psi^{-1} \circ \phi)^*(\omega) = \int_U \phi^*(\omega).$$

We also observe that this integral has the usual linearity properties on forms that satisfy the assumption above.

Step II: In order to remove the assumption that we made in Step I, we now invoke the technique of partition of unity. Let $\{\alpha_i\}$ be a smooth partition of unity subordinate to an atlas $\{(U_j, \phi_j)\}$ as above. Then for each i, $\alpha_i\omega$ satisfies the assumption in Step I and hence $\int_X \alpha\omega$ makes sense. Since the support of ω is compact, only finitely many of $\alpha_i\omega$ are nonzero and hence we may take

$$\int_X \omega := \sum_i \int_X \alpha_i \omega \tag{4.12}$$

provided we verify that the RHS of (4.12) is independent of the partition of unity chosen.

Step III Once again, assume first that support of ω satisfies the assumption in Step I. Then all $\alpha_i\omega$ also satisfy this and we have $\sum_i \alpha_i\omega = \omega$. Therefore,

$$\int_U \phi^*(\omega) = \sum_i \int_U \phi^*(\alpha_i\omega).$$

This means that the two definitions of $\int_X \omega$ as in (4.11) and (4.12) coincide, in this special case.

Step IV Now suppose $\{\beta_i\}$ is another smooth partition of unity subordinate to the cover $\{U_i\}$. Then for each fixed j, we apply Step III to $\beta_j\omega$ to see that

$$\int_X \beta_j\omega = \sum_i \int_X \alpha_i\beta_j\omega.$$

Similarly, we apply it to each $\alpha_i\omega$ to see that

$$\int_X \alpha_i\omega = \sum_j \int_X \beta_j\alpha_i\omega.$$

(Observe that all sums involved are finite even though the indexing sets may be infinite.) Therefore, upon taking the sum over both i, j, we obtain that

$$\sum_j \int_X \beta_j\omega = \sum_{i,j} \int_X \alpha_i\beta_j\omega = \sum_i \int_X \alpha_i\omega.$$

This proves that the RHS of (4.12) is independent of the partition chosen.

Step V We have more or less subsumed that the dimension n of X is ≥ 1. However, this is not the case. Indeed, the case when $n = 0$ is much easier to handle and let us do it separately here. X is now a discrete space. An orientation on X means a function $\Theta : X \to \{-1, 1\}$. A 0-form on X is nothing but a function $f : X \to \mathbb{R}$. To say f has compact support means that $f(x) = 0$ except for finitely many $x \in X$. Now we define

$$\int_X f = \sum_x \Theta(x) f(x). \tag{4.13}$$

For a bijection $f : X \to Y$ of 0-dimensional oriented manifolds, we define $\det f : X \to \mathbb{R}$ by

$$(\det f)(x) = \Theta(x)\Theta(f(x)). \tag{4.14}$$

Then to say that f preserves orientation is the same as saying that $\det(f)(x) = 1$ for all $x \in X$.

This completes the definition of $\displaystyle\int_X \omega$ for a smooth compactly supported n-form ω. It has all the usual properties of the integral. We shall summarize them in the following theorem, for future reference.

Theorem 4.3.2 (Properties of the Integral)
(i) $\displaystyle\int_X a\omega + b\tau = a\int_X \omega + b\int_X \tau.$
(ii) *If* $X_1 \cap X_2 = \emptyset$, *then* $\displaystyle\int_{X_1 \cup X_2} \omega = \int_{X_1} \omega + \int_{X_2} \omega.$
(iii) *If* $-X$ *denotes the manifold with the orientation opposite to that of* X, *then*

$$\int_{-X} \omega = -\int_X \omega.$$

(iv) *If* $f : X \to Y$ *is an orientation preserving diffeomorphism then*

$$\int_X f^*\omega = \int_Y \omega.$$

Remark 4.3.1 Now suppose that Y is a smooth submanifold of X of dimension k. Then the restriction of a smooth k-form τ on X to Y makes sense. It is just $\iota^*(\tau)$ where $\iota : Y \to X$ is the inclusion map. For the sake of notational simplicity, we write τ or $\tau|_Y$ for $\iota^*(\tau)$. If this restriction is compactly supported on Y, and Y is oriented, then $\int_Y \tau$ makes sense. Special cases of this will lead you to some familiar integrals. We have already seen enough examples of this.

We end this section with the final version of Stokes' theorem. In reality, it is the general form of the Fundamental Theorem of Integral Calculus. Given a continuously differentiable function $f : [a, b] \to \mathbb{R}$ we know that

$$\int_a^b f'(t)dt = f(b) - f(a), \quad a < b. \tag{4.15}$$

We first observe that in terms of our new terminology, the left-hand side (LHS) of the above equation can be written as $\displaystyle\int_{[a,b]} df$, with the standard orientation on the interval $[a, b]$. Even the RHS can also be written as an integral, viz., integral of the 0-form f on the

0-dimensional oriented manifold $\{a, b\}$, which is treated as the boundary of the oriented 1-manifold $[a, b]$. If g is a continuously differentiable function on $[0, \infty)$ with compact support then,

$$\int_0^\infty g'(t)dt = -g(0) \tag{4.16}$$

Once again the RHS of this formula can be treated as an integral of the 0-form on the boundary 0 of the ray $[0, \infty)$, which receives the negative orientation sign. Thus, we see that both (4.15), (4.16) can be written as

$$\int_X df = \int_{\partial X} f \tag{4.17}$$

where X is a 1-dimensional oriented manifold, ∂X is the oriented boundary, and f is a smooth 0-form on X with compact support.

Our aim is to generalize (4.17) to all dimensions. As a first step we have:

Lemma 4.3.1 Let ω be a smooth $(n-1)$-form on $X = \mathbb{R}^n$ or $X = \mathbf{H}^n$ with compact support. Then

$$\int_X d\omega = \int_{\partial X} \omega, \tag{4.18}$$

where X has standard orientation and ∂X has the induced boundary orientation.

Proof: First consider the case when $X = \mathbb{R}^n$. Then $\partial X = \emptyset$ and hence RHS has to be interpreted to be $= 0$. On the other hand, since ω has compact support, there exists $R > 0$ such that ω vanishes identically on the boundary and outside of the box $B = [-R, R]^n$. Therefore, LHS$=\int_B d\omega$.

Write $I_p := (1, 2, \dots, p-1, p+1, \dots, n)$ so that

$$dx_{I_p} = dx_1 \wedge \cdots \wedge \widehat{dx_p} \wedge \cdots \wedge dx_n$$

where the "hat" indicates that the corresponding term is missing. Write $\omega = \sum_p f_p dx_{I_p}$, where f_p are some smooth functions each vanishing outside of B and on ∂B. Then

$$d(\omega) = \sum_p (-1)^{p-1} \frac{\partial f_p}{\partial x_p} dx_1 \wedge \cdots \wedge dx_n.$$

Now for each p and for each fixed $(x_1, \dots, \widehat{x_p}, \dots, x_n)$,

$$\begin{aligned} & \int_{-R}^{R} \frac{\partial f_p}{\partial x_p}(x_1, \dots, \widehat{x_p}, \dots, x_n) dx_p \\ = \;& f_p(x_1, \dots, x_{p-1}, R, \dots, x_{p+1}, \dots, x_n) - f_p(x_1, \dots, x_{p-1}, -R, x_{p+1}, \dots, x_n) \\ = \;& 0 - 0 = 0. \end{aligned}$$

Therefore, by Fubini's theorem, we have

$$\int_B d(\omega) = \sum_p (-1)^{p-1} \int_{-R}^{R} \cdots \int_{-R}^{R} \frac{\partial f_p}{\partial x_p} dx_1 dx_2 \cdots dx_n = 0.$$

Thus, the lemma is proved in this case.

Next, consider the case when $X = \mathbf{H}^n$. Here, the modifications that we have to make in the above argument are the following:
(i) The box B should be replaced by the half-box $H = [-R, R]^{n-1} \times [0, R]$.
(ii) The $(n-1)$-form ω vanishes outside H and on all faces of it except perhaps on the face $[-R, R]^{n-1} \times 0$.
(iii) On $\mathbb{R}^{n-1} \times \{0\}$, we have, $dx_n = 0$, and hence $dx_{I_p} = 0$ for all $p \neq n$. Therefore, ω restricted to $\mathbb{R}^{n-1} \times \{0\}$ is equal to $f_n dx_{I_n}$.

Now as before, $\int_{-R}^{R} \frac{\partial f_p}{\partial x_p} dx_p = 0$ for all $p < n$. However, for $p = n$ we have to consider

$$\int_0^R \frac{\partial f_n}{\partial x_n} dx_n = f_n(x_1, \ldots, x_{n-1}, R) - f_n(x_1, \ldots, x_{n-1}, 0) = -f_n(x_1, \ldots, x_{n-1}, 0).$$

Now the orientation on $\mathbb{R}^{n-1} \times 0$ as the boundary of $\mathbf{H}^n$ is equal to $(-1)^n$ times the standard one.

Therefore,

$$\int_{\partial \mathbf{H}^n} \omega = (-1)^n \int_{\mathbb{R}^{n-1} \times 0} \omega = (-1)^n \int_{\mathbb{R}^{n-1} \times 0} f_n \, dx_1 dx_2 \cdots dx_{n-1}. \tag{4.19}$$

Therefore,

$$\begin{aligned} \int_{\mathbf{H}^n} d\omega &= \int_{\mathbf{H}^n} \left(\sum_p (-1)^{p-1} \frac{\partial f_p}{\partial x_p} \right) dx_1 \wedge \cdots \wedge dx_n \\ &= (-1)^{n-1} \int_{\mathbf{H}^n} \frac{\partial f_n}{\partial x_n} dx_1 \wedge \cdots \wedge dx_n \\ &= (-1)^{n-1} \int_{\mathbb{R}^{n-1} \times 0} (-f_n(x_1, \ldots, x_{n-1}, 0)) dx_1 \cdots dx_{n-1} \\ &= (-1)^n \int_{\mathbb{R}^{n-1} \times 0} f_n \, dx_1 dx_2 \cdots dx_{n-1}. \end{aligned}$$

Combining this with (4.19), completes the proof of the lemma. ♠

Theorem 4.3.3 Stokes' Theorem: *Let X be a compact oriented n-manifold and let ∂X be oriented with the induced orientation. Then for any smooth $(n-1)$-form ω on X we have*

$$\int_X d\omega = \int_{\partial X} \omega. \tag{4.20}$$

Proof: Since the two sides are linear in ω, using partition of unity technique, we may assume that ω has its support contained in a parameterized open set W of X. Pulling back both sides the statement is reduced to the statement of the above lemma. ♠

Corollary 4.3.1 Let X be an oriented closed n-manifold (compact and with empty boundary). Then for any $(n-1)$-form ω on X, we have $\int_X d\omega = 0$.

Apart from the obvious classical applications implied via Green's theorem and Gauss's theorem, etc., the general form of Stokes' theorem has many applications. We give one such application below as an exercise. For others, you will have to wait a bit.

Exercise 4.3 Let $f, g : X \to Y$ be smooth maps where X and Y are oriented closed n-manifolds. Suppose f is homotopic to g. Then for any smooth n form ω on Y we have

$$\int_X f^* \omega = \int_X g^* \omega.$$

4.4 De Rham Cohomology

In Section 2.4 we introduced the functorial graded algebra $\Omega^*(X)$ of smooth differential forms over the ring $\mathcal{C}^\infty(X;\mathbb{R})$, for a smooth manifold. In Section 2.5, we introduced the functorial differential operator $d : \Omega^*(X) \to \Omega^*(X)$ with the property that $d^2 = 0$. $\Omega^*(X)$ together with the operator d is called the *De Rham complex* of X.

That $\partial^2 = 0$ is the same as saying that $ker\,[d : \Omega^p(X) \to \Omega^{p+1}(X)]$ contains the image of $d : \Omega^{p-1}(X) \to \Omega^p(X)$. For $X = \mathbb{R}^3$, we have seen that this is the same as saying $\text{curl} \circ \text{grad} = 0$ and $\text{div} \circ \text{curl} = 0$. Classically, if you had a differential equation such as

$$pdx + qdy = 0,$$

solving this equation means that we have to find a function f such that $df = pdx + qdy$. In the terminology of the De Rham complex, this just means that the 1-form $\omega = pdx + qdy$ is in the image of $d : \Omega^0 \to \Omega^1$. Of course, you know that it is necessary to assume that $\frac{\partial p}{\partial y} = \frac{\partial q}{\partial x}$ in order to solve this equation and then how to solve it in $\mathbb{R}^2$. Such an equation was called an *exact equation.*

Now suppose you consider

$$\omega(x,y) = \frac{x\,dy - y\,dx}{x^2 + y^2}$$

on $\mathbb{R}^2 \setminus \{0\}$. You can easily check that $d\omega = 0$. Integrate this form on any circle C_r of radius $r > 0$ and center 0 and see that

$$\int_{C_r} \frac{x\,dy - y\,dx}{x^2 + y^2} = 2\pi. \tag{4.21}$$

On the other hand, if $\omega = df$ for some f then by Stokes' theorem, we know that $\int_{C_r} \omega = 0$. Therefore, we conclude that if U is any domain contained in $\mathbb{R}^2 \setminus \{0\}$ and contains some circle C_r then there cannot be any smooth function f such that $df = \omega$.

But now, consider a domain such as the right-half space $G = \{(x,y) \in \mathbb{R}^2 \,:\, x > 0\}$. Then we can take $f(x,y) = arctan\, y/x$ and see that $df = \omega$.

Thus, it seems that the integration theory of forms on spaces can reveal to us certain topological properties of the space. In this case U had a hole in it and $\text{Ker}\, d$ was bigger than image of d, whereas G does not have any holes and $\text{Ker}\, d$ is equal to image of d.

In this section, we shall strive to make this observation into a meaningful mathematical result.

We make a small beginning with the following theorem.

Theorem 4.4.1 *Let ω be a smooth 1-form on $\mathbb{S}^1$. Then $\omega = df$ for some f iff $\int_{\mathbb{S}^1} \omega = 0$.*

Proof: We have already seen the "only if" part in example 2.6.3 (ii). Now suppose the integral vanishes. Let $h : \mathbb{R} \to \mathbb{S}^1$ be the map $\theta \mapsto (\cos\theta, \sin\theta)$. Define $g : \mathbb{R} \to \mathbb{R}$ by the formula,

$$g(x) := \int_0^x h^*(\omega).$$

Then for any fixed $x \in \mathbb{R}$, the restriction $h : [x, 2\pi + x] \to \mathbb{S}^1$ is a parameterization of $\mathbb{S}^1$ and therefore,

$$g(2\pi + x) - g(x) = \int_x^{2\pi + x} h^*(\omega) = \int_{\mathbb{S}^1} \omega = 0.$$

Therefore, g factors through h to define a map $f : \mathbb{S}^1 \to \mathbb{R}$ by the formula $f(\cos\theta, \sin\theta) = g(\theta)$. Your knowledge of 1-variable calculus is enough to tell you that g is smooth and $dg = h^*(\omega)$. Then, so is f and $df = \omega$. ♠

Definition 4.4.1 Let ω be a differential p-form. We say, ω *is closed* if $d(\omega) = 0$. We say ω is exact if there exists a $(p-1)$-form τ such that $d(\tau) = \omega$.

Remark 4.4.1 Thus, from (2.55), all exact forms are closed forms. On an n-dimensional manifold, all n-forms are closed. By convention, the only exact 0-form is the constant function 0. Of course, all constant functions are closed 0-forms. On the other hand, on a connected manifold, the only closed 0-forms are constant functions. (Can you prove this?) The question of whether a closed form is exact or not is surprisingly related to the topological properties of the manifold X. To lay down the foundations for a systematic study of this question is the aim of this section.

Observe that the set $Z^p(X)$ of all closed p-forms constitutes a linear subspace of $\Omega^p(X)$ and contains the linear subspace $B^p(X)$ of all exact p-forms. The quotient space

$$H^p_{DR}(X) := Z^p(X)/B^p(X) \tag{4.22}$$

is called the pth *De Rham Cohomology group of* X. We shall drop the lower suffix $_{DR}$ for the time being and use a simpler notation,

$$H^p(X) = H^p_{DR}(X).$$

The direct sum

$$H^*(X) := H^0(X) \oplus H^1(X) \oplus \cdots \oplus H^n(X) \tag{4.23}$$

is called the *total De Rham cohomology group* of X. Observe that if $f : X \to Y$ is a smooth map of manifolds, then from Theorem 4.2.1 (f), it follows that the pullback of a closed (resp. exact) p-form on Y is a closed (resp. exact) p-form on X, i.e., $f^*(Z^p(Y)) \subset Z^p(X)$ and $f^*(B^p(X)) \subset B^p(Y)$. (It may be noted here that, this property is expressed by saying that the map $\Omega(f) : \Omega^*(Y) \to \Omega^*(X)$ is a homomorphism of chain groups.) Therefore, we obtain a linear map on the cohomology groups, induced by $\Omega(f)$. Let us denote this by $H(f) : H^p(Y) \to H^p(X)$. The property (2.45) also continues to hold for these induced maps, viz.,

$$H(g \circ f) = H(f) \circ H(g); \quad H(Id) = Id. \tag{4.24}$$

Example 4.4.1 We remarked earlier that on a connected manifold the only closed 0-forms are constants. Let us prove this. Fix any point $x \in X$ and for any other point $y \in X$, choose an embedded smooth arc $\gamma : [0,1] \to X$ joining x and y. Let now $f : X \to \mathbb{R}$ be a smooth 0-form on X such that $df = 0$. Then the pullback 0-form on $[0,1]$ viz., $g = \gamma^*(f) = f \circ \gamma$, is such that $dg = \gamma^*(df) = 0$. This means $\left(\frac{dg}{dt}\right) dt = 0$ at all points of $[0,1]$ and hence $\frac{dg}{dt} = 0$. This means $g = c$, a constant on $[0,1]$. In particular, it means that $f(x) = g(0) = g(1) = f(y)$. Therefore, f is a constant.

In terms of cohomology, this just means that $H^0(X) = \mathbb{R}$. More generally, it is not hard to see that $H^0(X)$ is a vector space over $\mathbb{R}$ of dimension equal to the number of connected components of X.

Also observe that if $X = \{\star\}$ is the singleton space, then there are no nonzero p-forms on X for $p > 0$ and hence $H^p(X) = 0$ for $p > 0$. Thus,

$$H^{(p)}(\star) \approx \begin{cases} \mathbb{R}, & p = 0, \\ 0, & p > 0. \end{cases} \tag{4.25}$$

In a sense, the size of this space gives you an idea of how "many" closed p-forms are not exact.

Remark 4.4.2 Connectivity of manifolds, which is the same as path-connectivity can be rephrased as follows. Any two point-maps in X are homotopic in X. This should ring a familiar bell in us. We may anticipate something more general about the behavior of homotopic maps vis-à-vis differential forms. And this anticipation turns out to be true. As a first step we have:

Lemma 4.4.1 Let X be any manifold and ω be any 1-form on X. Then $\omega = df$ for some smooth function f iff for every closed piecewise smooth path $\gamma : [a, b] \to X$, we have $\displaystyle\int_\gamma \omega := \int_a^b \gamma^*(\omega) = 0.$

Proof: Assume $\omega = df$. Then by the Fundamental Theorem of Integral Calculus,

$$\int_a^b \gamma^*(df) = \int_a^b (f \circ \gamma)'(t)\, dt = f(\gamma(b)) - f(\gamma(a)) = 0,$$

since γ is a closed path.

To prove the converse, we fix $p \in X$ and for any $q \in X$ choose a smooth path γ from p to q and define $f(q) = \int_\gamma \omega$. If τ is another path from p to q, then $\tau \cdot \gamma^{-1}$ is a closed piecewise smooth path and by hypothesis, $\int_{\tau\cdot\gamma^{-1}} \omega = 0$. This implies that $\int_\tau \omega = \int_\gamma \omega$. Thus, $f(q)$ is independent of the path chosen.

We claim that $df = \omega$. So, let $q \in X$ be fixed. Choose a coordinate neighbourhood U around q. Fix a smooth path γ from p to q. Then for every point $x \in U$ we can take any path α lying within U, and write

$$f(x) = \int_{\gamma * \alpha} \omega = \int_\gamma \omega + \int_\alpha \omega.$$

Using some coordinates for U we may therefore workout as if we are inside $\mathbb{R}^n$. So, we can first of all write $\omega = \sum_i g_i dx_i$. In these coordinates, we may assume $q = 0$. Therefore,

$$\begin{aligned} \frac{\partial f}{\partial x_i}(0) &= \lim_{h\to 0} \frac{f(0, \ldots, h, \ldots, 0) - f(0, \ldots, 0, \ldots, 0)}{h} \\ &= \lim_{h\to 0} \frac{1}{h} \int_0^h g_i dx_i = g_i(0). \end{aligned}$$

(The last equality is due to the mean value theorem of integral calculus.) Since this is true for all $i = 1, 2, \ldots, n$, this means $df(q) = \omega(q)$. Therefore, $df = \omega$. ♠

Theorem 4.4.2 *If X is a simply connected manifold, then every closed 1-form in X is exact.*

Proof: Recall that if X is simply connected then every smooth loop $\gamma : \mathbb{S}^1 \to X$ has a smooth extension $\hat{\gamma} : \mathbb{D}^2 \to X$. Now if ω is a closed 1-form on X then

$$\int_\gamma \omega = \int_{\mathbb{S}^1} \gamma^*(\omega) = \int_{\mathbb{S}^1} \hat{\gamma}^*(\omega) = \int_{D^2} d(\hat{\gamma}^*(\omega))$$

by Stokes' theorem. But $d(\hat{\gamma}^*(\omega)) = \hat{\gamma}^*(d\omega) = 0$. Therefore, $\displaystyle\int_\gamma \omega = 0$ for every smooth loop. Therefore, from the above lemma, ω is exact. ♠

Indeed, we have the best that we can wish, viz., homotopic maps inducing same homomorphisms at the cohomology.

Theorem 4.4.3 *Let $h : \mathbb{R} \times X \to Y$ be a smooth map and put $h_r(x) = h(r, x)$. Then for any two points $t, s \in \mathbb{R}$, we have $h_t^* = h_s^* : H^p(X) \to H^p(Y)$.*

A more fundamental result from which we can deduce the above theorem is:

Lemma 4.4.2 (Poincaré) For any fixed $t \in \mathbb{R}$, let $\iota(x) = (t, x)$ and let $\pi : \mathbb{R} \times X \to X$ be the projection to the second factor. Then, ι^* and π^* are inverses of each other on cohomology groups. In particular, $H^p(\mathbb{R} \times X) \approx H^p(X)$.

The proof of the theorem is immediate from this lemma: Take $\iota_r(x) = (r, x)$ for $r = t, s$. Then

$$h_t^* = (h \circ \iota_t)^* = \iota_t^* \circ h^* = \iota_s^* \circ h^* = (h \circ \iota_s)^* = h_s^*.$$

Moreover, taking $X = \mathbb{R}^n$ and applying the second part of the lemma iteratively, we obtain the following:

Corollary 4.4.1 $H^*(\mathbb{R}^n) \approx H^*(\star)$, where $\star$ denotes a pointspace.

A more general result that is also immediate is:

Theorem 4.4.4 Homotopy Invariance: *A smooth homotopy equivalence $f : X \to Y$ of manifolds, induces an isomorphism $H(f) : H^*(Y) \to H^*(X)$ of the De Rham cohomology algebras.*

Proof: If $g : Y \to X$ is a smooth map that is a homotopy inverse to f then, we have $f \circ g \simeq Id_Y$ and $g \circ f \simeq Id_X$. Therefore, $H(f \circ g)$ and $H(g \circ f)$ are identity homomorphisms of appropriate cohomology algebras. Hence, $H(g)$ is the inverse of $H(f)$. ♠

Proof of Poincaré Lemma:

In order to prove the lemma, since $\iota^* \circ \pi^* = (\pi \circ \iota)^* = (Id_X)^*$, we need to prove the other equality, viz., $\pi^* \circ \iota^* = (Id_{\mathbb{R} \times X})^*$ on $H^*(\mathbb{R} \times X)$.

For this purpose, we introduce the *homotopy operator*

$$P : \Omega^p(\mathbb{R} \times X) \to \Omega^{p-1}(\mathbb{R} \times X)$$

satisfying the following property:

$$dP\omega + Pd\omega = \omega - \pi^* \iota^* \omega. \tag{4.26}$$

Assume that we have defined such an operator. Consider an element of $H^p(\mathbb{R} \times X)$ represented by a closed p-form ω on $\mathbb{R} \times X$. Then by (4.26), it follows that $\omega - \pi^* \circ \iota^*(\omega) = dP\omega$ is an exact form and therefore represents the 0 element of $H^{p-1}(X)$. This will complete the proof of the lemma.

As in the case of the operators d and $\int$, the task of defining an operator on the forms on manifolds becomes easy (and indeed possible at all) if we demand that the operator behaves well under transformations. Here, let us demand that for any diffeomorphism $\phi : X \to Y$ we have,

$$P \circ (\phi \times Id)^* = (\phi \times Id)^* \circ P. \tag{4.27}$$

where $Id : \mathbb{R} \to \mathbb{R}$ is the identity map.

Having made this demand, we now proceed from local to global. So, let X be an open subset of the Euclidean space. Then every p-form on $\mathbb{R} \times X$ can be written in the form:

$$\omega = \sum_{I \in s(p-1,n)} f_I \, dt \wedge dx_I + \sum_{J \in s(p,n)} g_J \, dx_J$$

where f_I, g_J are smooth functions on $\mathbb{R} \times X$. (See Remark 2.3.3.) Now define

$$P(\omega)(t,x) = \sum_{I \in s(p-1,n)} \left(\int_0^t f_I(s,x)ds \right) dx_I. \tag{4.28}$$

Our task is to check that P, first of all, satisfies (4.26), at least in this special case. By additivity of either side, it is enough to check this for forms of the type $f\, dt \wedge dx_I$ and $g\, dx_J$. We leave this to the reader as a routine exercise.

The next task is to prove (4.27).

Once again, we need to verify this for forms of the type $f\, dt \wedge dy_I$ and $g\, dy_J$. On the latter, i.e., on $g\, dy_J$ both sides vanish. Checking this on the former is left to the reader as an exercise.

Finally, let X be any manifold. Cover it with an atlas $\{(\phi_i, U_i)\}$. For any p-form ω on $\mathbb{R} \times X$, define $P(\omega)$ to be that (unique) $(p-1)$-form, which when pulled back onto $\mathbb{R} \times U_i$ via $Id \times \phi_i$ gives the differential $(p-1)$-form $P((Id \times \phi^*)(\omega))$.

It remains to prove that P satisfies (4.26), in the general case. This follows, by using partition of unity, since both sides of (4.26) are linear.

This completes the construction of the operator P and thereby the proofs of the lemma, the theorem, and the corollary. ♠

We shall now take up the task of determining the cohomology groups of the spheres $\mathbb{S}^n$. We plan to attack this using induction on n. To begin with from Example 4.4.1, we know that $H^0(\mathbb{S}^0) \approx \mathbb{R}^2$ and $H^p(\mathbb{S}^0) = 0$ for all $p \geq 1$. We also know that $H^0(\mathbb{S}^n) \approx \mathbb{R}$ for all n. However, our induction begins with $n = 1$.

Lemma 4.4.3 $H^1(\mathbb{S}^1) \approx \mathbb{R}$.

Proof: Consider the linear map $\eta : \Omega^1(\mathbb{S}^1) \to \mathbb{R}$ given by

$$\eta(\omega) = \int_{\mathbb{S}^1} \omega.$$

By (4.21), it follows that η is not the zero map. By Theorem 4.4.1, it follows that η induces a linear map $\bar{\eta} : H^1(\mathbb{S}^1) \to \mathbb{R}$, which is injective. Therefore, $\bar{\eta}$ is an isomorphism. ♠

We now use the standard topological approach. Write $\mathbb{S}^n = U_1 \cup U_2$, where, U_1, U_2 are obtained by deleting north pole and south pole respectively, from $\mathbb{S}^n$. Use the stereographic projection to see that both U_i are diffeomorphic to $\mathbb{R}^n$ and $U_1 \cap U_2$ is diffeomorphic to $\mathbb{R} \times \mathbb{S}^{n-1}$. By Lemma 4.4.2, it follows that $H^p(U_1 \cap U_2) \approx H^p(\mathbb{S}^{n-1})$. Our inductive argument then needs only the following key step:

Lemma 4.4.4 For $p \geq 1$, $H^p(U_1 \cap U_2)$ is isomorphic to $H^{p+1}(U_1 \cup U_2)$.

Proof: Let us directly define an isomorphism

$$\beta : H^{p+1}(U_1 \cup U_2) \to H^p(U_1 \cap U_2) \tag{4.29}$$

as follows: Given a closed $(p+1)$-form ω on $U_1 \cup U_2$, since U_i are diffeomorphic to $\mathbb{R}^n$, the closed forms $\omega|_{U_i}$ are also closed and hence both are exact. So, there exist p-forms ν_i on U_i such that $d(\nu_i) = \omega|_{U_i}$. Consider the p-form $\nu_1 - \nu_2$ on $U_1 \cap U_2$. This is a closed form, for $d(\nu_1 - \nu_2) = d(\nu_1) - d(\nu_2) = \omega - \omega = 0$ on $U_1 \cap U_2$. Put $\beta(\omega) = [\nu_1 - \nu_2]$. If we have chosen some other p-forms ν_i' in place of ν_i, then it follows that $\nu_i - \nu_i'$ is a closed form on U_i and hence is an exact form. Therefore, $(\nu_1 - \nu_2) - (\nu_1' - \nu_2') = (\nu_1 - \nu_1') - (\nu_2 - \nu_2')$ is also exact on $U_1 \cap U_2$. This means that in the cohomology group, $[\nu_1 - \nu_2] = [\nu_1' - \nu_2']$. Therefore, β is independent of the choice of ν_i's.

Using this, it is now easy to verify that $\beta : Z^{p+1}(U_1 \cup U_2) \to H^p(U_1 \cap U_2)$ is a linear map such that $\beta \circ d = 0$. Therefore, β factors through the quotient space and defines a linear map that we shall again denote by $\beta : H^{p+1}(U_1 \cup U_2) \to H^p(U_1 \cap U_2)$.

Suppose $\beta[\omega] = 0$. This means $\beta(\omega) = 0$ and hence there exist a $(p-1)$-form θ on $U_1 \cap U_2$ such that $\nu_1 - \nu_2 = d(\theta)$. Put

$$V_1 = \{(x_0, \dots, x_n) \in \mathbb{S}^n : x_n > -1/2\};\ V_2 = \{(x_0, \dots, x_n) \in \mathbb{S}^n : x_n < 1/2\};\ W = V_1 \cap V_2.$$

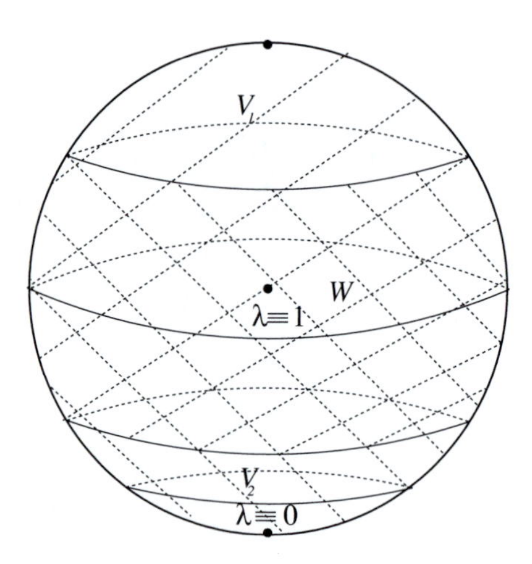

Figure 21 Patching up forms.

Choose a smooth function λ on V_2 that is $\equiv 1$ in $V_1 \cap V_2$ and $\equiv 0$ in a small neighbourhood of the south pole. Then the $(p-1)$-form $\lambda\theta$ is defined on V_2 and is equal to θ on $V_1 \cap V_2$. Let ν be the p-form defined by

$$\nu = \begin{cases} \nu_1, & \text{on } V_1, \\ \nu_2 + d(\lambda\theta), & \text{on } V_2. \end{cases}$$

This is well defined since, on the intersection $V_1 \cap V_2$ we have, $\nu_1 = \nu_2 + d(\theta) = \nu_2 + d(\lambda\theta)$. Now clearly, $d(\nu) = \omega$ on the whole of $V_1 \cup V_2 = \mathbb{S}^n$. Therefore, $[\omega] = 0$. This proves β is injective.

Now let $\{\rho_1, \rho_2\}$ be a smooth partition of unity on $\mathbb{S}^n$ such that supp $\rho_i \subset V_i,\ i = 1, 2$. Given a closed p-form ν on $U_1 \cap U_2$, the forms $\rho_i \nu$ make sense on $V_j, j \neq i$. Consider the $(p+1)$-form ω defined by:

$$\omega = \begin{cases} d(\rho_1 \nu), & \text{on } V_2, \\ -d(\rho_2 \nu), & \text{on } V_1. \end{cases}$$

On $U_1 \cap U_2$, we have $d(\rho_1\nu) + d(\rho_2\nu) = d((\rho_1 + \rho_2)\nu) = d(\nu) = 0$, which means that $d(\rho_1\nu) = -d(\rho_2\nu)$. Therefore, ω is well-defined. It is easily verified that ω is a closed form and $\beta(\omega) = [\rho_1\nu + \rho_2\nu] = [\nu]$. Therefore, $\beta[\omega] = [\nu]$. This proves surjectivity of β and thereby the lemma. ♠

Theorem 4.4.5 *For $n \geq 1$ we have,*

$$H^p(\mathbb{S}^n) = \begin{cases} \mathbb{R}, & p = 0, n, \\ 0, & \text{otherwise.} \end{cases} \tag{4.30}$$

Proof: Since all $\mathbb{S}^n, n \geq 1$ are connected $H^0(\mathbb{S}^n) \approx \mathbb{R}$, by Example 4.4.1. We now use a result in algebraic topology that we proved in Theorem 2.2.3 viz., that all $\mathbb{S}^n, n \geq 2$ are simply connected. By Theorem 4.4.1, it follows that $H^1(\mathbb{S}^k) = (0)$ for all $k > 1$. By iterated application of Lemma 4.4.4 it follows that $H^{p+1}(\mathbb{S}^{k+p}) = (0)$ for $k > 1$ and $H^n(\mathbb{S}^n) \approx H^1(\mathbb{S}^1) \approx \mathbb{R}$. ♠

Here are some immediate applications of the computation of the De Rham cohomology of $\mathbb{D}^n$ and $\mathbb{S}^n$. We begin with an elementary lemma which often used in algebraic topology.

Lemma 4.4.5 The following two statements are equivalent.
(a) There is no smooth map $r : \mathbb{D}^n \to \mathbb{S}^{n-1}$ such that $r(x) = x$ for all $x \in \mathbb{S}^{n-1}$.
(b) Every smooth map $f : \mathbb{D}^n \to \mathbb{D}^n$ has a fixed point.

Proof: (A map as in (a) is called a *smooth retraction.*) Suppose there is a smooth map $f : \mathbb{D}^n \to \mathbb{D}^n$, such that $f(x) \neq x$ for any $x \in \mathbb{D}^n$. Let L_x denote the line passing through x and $f(x)$. Then $L_x \cap \mathbb{S}^{n-1}$ has precisely two points. Let $g(x) \in L_x \cap \mathbb{S}^{n-1}$ be the point so that $\{g(x), x, f(x)\}$ are located in that order on L_x. (It is given by an explicit root of a quadratic equation.) It follows that $g : \mathbb{D}^n \to \mathbb{S}^{n-1}$ is smooth and $g(x) = x$ for all $x \in \mathbb{S}^{n-1}$. This proves (a) $\implies$ (b).

To see (b) $\implies$ (a), suppose $r : \mathbb{D}^n \to \mathbb{S}^{n-1}$ is a smooth map such that $r(x) = x$ for all $x \in \mathbb{S}^{n-1}$. Then the map $-\eta \circ r$, where $\eta : \mathbb{S}^{n-1} \to \mathbb{D}^n$ is the inclusion map, is a smooth map $\mathbb{D}^n \to \mathbb{D}^n$ having no fixed points.

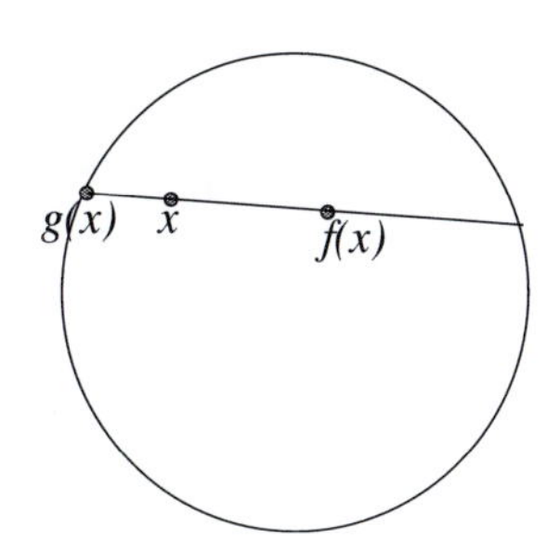

Figure 22 "No fixed points" yields retraction.

Theorem 4.4.6 (**Brouwer**) *Every (smooth) map $f : \mathbb{D}^n \to \mathbb{D}^n$ has a fixed point.*

Consider the case $n \geq 2$. Suppose we have such a map, then by the above lemma, we have a smooth retraction $r : \mathbb{D}^n \to \mathbb{S}^{n-1}$, i.e., $r \circ \eta = Id_{\mathbb{S}^{n-1}}$. Therefore, on the $(n-1)^{th}$ cohomology groups, we have

$$Id = Id^* = (r \circ \eta)^* = \eta^* \circ r^* = 0,$$

because $H^{n-1}(\mathbb{D}^n) = (0)$. This is a contradiction to the fact that $H^{n-1}(\mathbb{S}^{n-1}) = \mathbb{R}$. ♠

Remark 4.4.3 Using the approximation theorem 1.7.2, first show that every continuous map $f : \mathbb{D}^n \to \mathbb{D}^n$ can be approximated by a smooth map $g : \mathbb{D}^n \to \mathbb{D}^n$. Now if f has no fixed points, put $\epsilon = \min\{d(x, f(x)) : x \in \mathbb{D}^n\}$ and choose g such that $d(f(x), g(x)) < \epsilon/2, x \in \mathbb{D}^n$, to conclude that g cannot have a fixed point. This contradicts the above result and thereby establishes the continuous versions of (a) and (b).

Exercise 4.4 Verify all the details left to you in the proof of the Poincaré lemma.

4.5 Miscellaneous Exercises for Chapter 4

1. Show that an n-form ω on $\mathbb{S}^n$ is exact iff $\int_{\mathbb{S}^n} \omega = 0$.

2. Show that a closed $(n-1)$-form η on $\mathbb{R}^n \setminus B^n$ is exact iff $\int_{\partial B^n} \eta = 0$.

3. Let X be an orientable smooth n manifold. Say X satisfies (CI)-property if for every compactly supported n-form on ω, there is a compactly supported $(n-1)$-form η such that $d(\eta) = \omega$ iff $\int_X \omega = 0$.
   (a) Show that $\mathbb{R}^n$ satisfies (CI).
   (b) Let $X = U_1 \cup U_2$, where U_i are connected open and $U_1 \cap U_2$ is connected. Show that X satisfies (CI) if U_1 and U_2 satisfy (CI).
   (c) Show that every connected orientable manifold X satisfies (CI).

4. Let X be a connected closed manifold. Show that $H^n(X) = \mathbb{R}$ or (0) according as X is orientable or not.

Chapter 5

Abstract Manifolds

We are familiar with the concept of a differential manifold as a subspace of a Euclidean space. We shall now study this concept in an abstract setup. This will enable us to see that the notion of a "manifold" is independent of any particular embedding. It also gives us more natural examples of differential manifolds such as projective spaces, without the tedious method of realizing them as submanifolds of some Euclidean space. All the notions developed so far for manifolds do make sense in this general setup as well. Finally, we shall see that every abstract differential manifold can be embedded in some Euclidean space, coming full circle. Thus, after all, one could have stuck to submanifolds of Euclidean spaces only.

We introduce abstract topological manifolds and abstract smooth manifolds in Section 5.1 and 5.2, respectively. In Section 5.3, we introduce the fundamental gluing lemma which will be used several times in what follows. In Section 5.4, a simple proof of classification of 1-dimensional manifolds is proved. Sections 5.5 and 5.6 deal with tangent space. In Section 5.7, we shall first show that every manifold can be embedded in a suitable Euclidean space. We shall then prove some general results that bring down the dimension of the ambient Euclidean space. (These are the so-called easy Whitney embedding theorems.)

5.1 Topological Manifolds

In this section, let us get familiar with the general concept of a topological manifold. For simplicity, we shall first consider only manifolds "without boundary".

Definition 5.1.1 Let X be a topological space. By a (n-dimensional) *chart* for X we mean a pair (U, ψ) consisting of an open neighborhood U of x and a homeomorphism $\psi : U \to \mathbb{R}^n$ onto an open subset of $\mathbb{R}^n$. By an *atlas* $\{(U_j, \psi_j)\}$ for X, we mean a collection of charts for X such that $X = \cup_j U_j$. If there is an atlas for X, we say X is *locally Euclidean.*

A chart (U, ψ) is called a *chart at* $x_0 \in X$ if $\psi(x_0) = 0$.

Let $n \geq 1$ be an integer. We say X is a *topological manifold* of dimension n if:
(i) X is locally Euclidean, i.e., there is an atlas consisting of n-dimensional charts,
(ii) II-countable, it has a countable base for its topology and
(iii) a Hausdorff space.

Any countable discrete space is called a 0-*dimensional manifold.*

Remark 5.1.1

1. Observe that once a chart (U, ψ) exists at a point $x_0 \in X$, then we can choose a chart (V, ϕ) at x_0 such that $\phi(U) = \mathbb{R}^n$. For, by composing with a translation, we can assume that $\psi(x_0) = 0$ and then we can choose $r > 0$ such that the open ball $B_r(0) \subset \psi(U)$ and put $V = \psi^{-1}(B_r(0))$, and $\phi = f \circ \psi$ where $f : B_r(0) \to \mathbb{R}^n$ is the diffeomorphism given by $\mathbf{x} \mapsto \frac{\mathbf{x}}{r^2 - \|\mathbf{x}\|^2}$.

2. For an atlas, it is necessary to assume that the integer n is the same for all the charts. Of course, if X is connected, such an assumption is not necessary—it is a consequence of topological *Invariance of Domain.* (See Theorem 5.1.1, below.) It should be noted that any known proof of the purely set-topological invariance of domain is very hard as compared to the differential topological one. However, the differential topological version of this is easy as seen in Theorem 1.6.1.

3. For a topological space that is locally Euclidean, the II-countability condition is equivalent to many others, such as metrizability or paracompactness. We find II-countability the most suitable for our purpose.

Example 5.1.1

1. Clearly, differential manifolds inside Euclidean spaces that you have studied earlier are topological manifolds in the above sense.

2. Let X be the union of the two axes in $\mathbb{R}^2$. If U is any connected neighbourhood of $(0,0)$ in X then $U \setminus \{(0,0)\}$ has four components. It follows that X cannot have any chart covering $(0,0)$ and hence fails to be a topological manifold.

3. Let X be the set of all real numbers together with one extra point that we shall denote by $\tilde{0}$. We shall make X into a topological space as follows: Let $\mathcal{T}$ be the collection of all subsets A of X of the form $A = B \cup C$ where B is either empty or an open subset of $\mathbb{R}$ with the usual topology such that $(C \cap \mathbb{R}) \cup \{0\}$ is a neighborhood of 0 in $\mathbb{R}$. We leave it to you to verify that $\mathcal{T}$ forms a topology on X in which $\mathbb{R}$ is a subspace. Since $\tilde{0}$ also has neighborhoods that are homeomorphic to an interval, it follows that X has an atlas. It is easily seen that X has a countable base also. But however, observe that X fails to be a Hausdorff space, since neighborhoods of 0 and $\tilde{0}$ cannot be disjoint.

4. Likewise, one can also give examples of spaces that are Hausdorff and has an atlas but not II-countable. The typical example is the so-called *long line:* Consider the set Ω of all countable ordinals and put $X = \Omega \times [0, 1)$. Define a total order $\ll$ on X as follows:
$$(\alpha, t) \ll (\beta, s) \quad \text{if } \alpha < \beta \text{ or } \alpha = \beta \text{ and } t < s.$$
With the order topology induced by this order, X is a connected, Hausdorff space, having a smooth structure, locally diffeomorphic to $\mathbb{R}$. But X does not have a countable base. (For more details, See [J].)

5. Another type of nonexample is obtained by taking the disjoint union of manifolds of different dimensions. Thus, the subspace of $\mathbb{R}^2$, consisting of the x-axis together with the point $(0, 1)$, is not a manifold.

6. **The real projective space.** Let us now consider some examples of manifolds that do not occur naturally as subspaces of any Euclidean space but as quotients of subspaces of Euclidean spaces. The foremost one is the n-dimensional real projective space $\mathbb{P}^n$. This is the quotient space of the unit sphere $\mathbb{S}^n$ by the antipodal action, viz., each

element x of $\mathbb{S}^n$ is identified with its antipode $-x$. By the definition of quotient topology, if $q : \mathbb{S}^n \to \mathbb{P}^n$ denotes the quotient map then a subset U of $\mathbb{P}^n$ is open iff its inverse image $q^{-1}(U)$ is open in $\mathbb{S}^n$. We first observe that the quotient map q is both an open mapping as well as a closed mapping. This follows easily from the fact that for any subset $F \subset \mathbb{S}^n$, $F \cup (-F)$ is open (closed) if F is open (closed, respectively). From this, many of the topological properties of $\mathbb{S}^n$ pass onto the quotient space $\mathbb{P}^n$. For instance, using the openness of q we can easily conclude that $\mathbb{P}^n$ is II-countable. Indeed given any base $\mathcal{B}$ for the topology of $\mathbb{S}^n$, it follows that $\{q(U) \ : \ U \in \mathcal{B}\}$ is a base for $\mathbb{P}^n$. Since $\mathbb{S}^n$ is compact, it follows that $\mathbb{P}^n$ is also so.

We may represent points of $\mathbb{P}^n$ by the symbols $[x]$, where $[x] = q(x),\ x \in \mathbb{S}^n$. Given $x \in \mathbb{S}^n$ consider V to be the set of all points in $\mathbb{S}^n$, that are at a distance less than $\sqrt{2}$ from x. Then check that $U = q(V)$ is a neighborhood of $[x]$ in $\mathbb{P}^n$ and q itself restricts to a homeomorphism from V to U. Since V is anyway homeomorphic to an open subset of $\mathbb{R}^n$, this proves the existence of an n-dimensional atlas for $\mathbb{P}^n$.

To see that $\mathbb{P}^n$ is Hausdorff, let $[x] \neq [y] \in \mathbb{P}^n$ be two points. Clearly, in $\mathbb{S}^n$, we can choose $\epsilon > 0$ such that $B_\epsilon(\pm x) \cap B_\epsilon(\pm y) = \emptyset$. It then follows that $q(B_\epsilon(x))$ and $q(B_\epsilon(y))$ are disjoint neighborhoods of $[x]$ and $[y]$ in $\mathbb{P}^n$.

Definition 5.1.2 A topological space X is called a *manifold with boundary* if it is a II-countable, Hausdorff space, such that each point of x has an open neighborhood U_x and a homeomorphism $\phi : U_x \to \mathbf{H}^n$ onto an open subset of $\mathbf{H}^n$.

Denote by $\operatorname{int} X$ the set of all those points in x having a neighborhood U_x, that is homeomorphic to an open subset of

$$\operatorname{int} \mathbf{H}^n = \{(x_1, \ldots, x_n) \in \mathbb{R}^n \ : \ x_n > 0\}.$$

Clearly this forms an open subset of X and is a topological n-manifold in the old sense. Can you see why this is nonempty? The complement of this set in X is denoted by ∂X and is called boundary of X. Clearly it is a closed subset of X.

Remark 5.1.2 It may happen that ∂X is empty which means precisely that X is a manifold. The points of ∂X are characterized by the following property. There is a neighborhood U_x of x and a homeomorphism $\phi : U_x \to \mathbf{H}^n$ such that the nth-coordinate of $\phi(x)$ vanishes, i.e., $\phi_n(x) = 0$. This is a simple consequence of the following profound result:

Theorem 5.1.1 Topological Invariance of Domain: *Let A and B be any two subspaces of $\mathbb{R}^n$ that are homeomorphic to each other. If one of them is open in $\mathbb{R}^n$ then the other one is also open.*

We shall not prove this here. We shall not use this theorem either. Usually, a first course in algebraic topology offers a proof of this theorem. For a purely point-set topological proof of this theorem, you may read [H-W].

Remark 5.1.3 It follows that $\hat{U} = \phi^{-1}(\mathbb{R}^{n-1} \times 0)$ is a neighborhood of x in ∂X if we take ϕ as given in Remark 5.1.2. Also, then ϕ itself restricts to a homeomorphism $\phi : \hat{U} \to (\mathbb{R}^{n-1} \times \{0\}) \cap \phi(U)$. As a consequence, it follows that ∂X, if nonempty, is itself a topological $(n-1)$-dimensional manifold (without boundary).

Exercise 5.1

1. Give 5 examples of manifolds of different dimension. Give examples of Hausdorff, II-countable spaces that are not manifolds. Can you think of some manifolds other than projective spaces, that "do not naturally occur" as subspaces of Euclidean spaces?

2. Consider an equivalence relation on $\mathbb{R}^{n+1} \setminus \{0\} : \mathbf{x} \sim \mathbf{y}$ if there exists $\lambda \neq 0$ such that $\mathbf{y} = \lambda \mathbf{x}$. Show that the quotient space $(\mathbb{R}^{n+1} \setminus \{0\})/\sim$ is homeomorphic to $\mathbb{P}^n$.

3. Prove that every locally Euclidean, II-countable, Hausdorff space is paracompact or read it from some book, say, e.g., [Du].

5.2 Abstract Differential Manifolds

We would now like to introduce the notion of an abstract differential manifold. To begin with, we shall concentrate only on manifolds without boundary. Most of the concepts that we are going to introduce apply to manifolds with boundary as well and can be obtained routinely. However, when extra care needs to be taken for manifolds with boundary, we shall take care of them.

The "new" objects that we are going to introduce should be such that we should be able to talk about the differentiability of maps defined on such objects. So, first of all, these objects should be 'locally' Euclidean. That is precisely what we have done in the previous section. On the other hand, consider a differential manifold X as a subspace of a Euclidean space as introduced in Ch. 3. If $f : X \to \mathbb{R}^k$ is any map then, f is differentiable iff for each parameterization $\phi : \mathbb{R}^n \to X$ of an open set of X, the composite $f \circ \phi$ is differentiable. Can we then simply turn the table around and say that, in the case of an abstract topological manifold, a map $f : X \to \mathbb{R}$ is differentiable at a point $x \in X$ iff $f \circ \phi$ is differentiable at $0 = \phi^{-1}(x)$, where, ϕ is a parameterization of X around x? Clearly, such a definition will depend heavily on the parameterization of X near x that we may choose and can run into serious difficulties. This leads us precisely to the notion of "compatibility of charts" and "differential structures".

Definition 5.2.1 Let X be an n-dimensional topological manifold (with or without boundary) and $\Psi = \{(U_i, \psi_i)\}$ be an atlas for X. For each pair of indices i, j, such that $U_i \cap U_j \neq \emptyset$, consider the homeomorphisms $\psi_{ij} : \psi_i(U_i \cap U_j) \to \psi_j(U_i \cap U_j)$ defined by $\psi_{ij} := \psi_j \circ \psi_i^{-1}$. These are called the *transition functions* associated to the atlas Ψ.

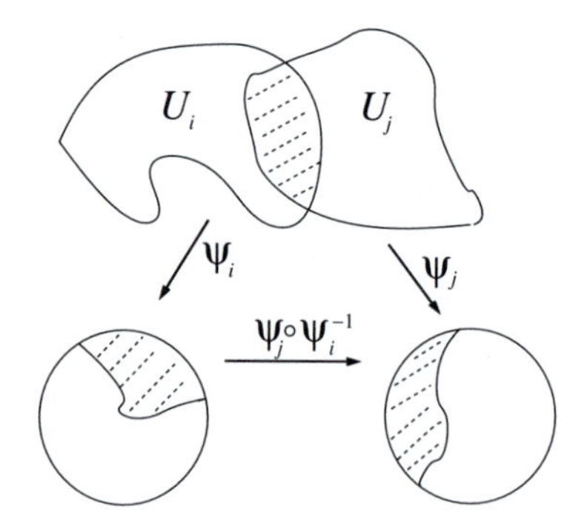

Figure 23 Smooth transition functions.

Now fix a positive integer r. We say the atlas Ψ is of class $\mathcal{C}^r$ if all its transition functions are at least of class $\mathcal{C}^r$.

Let $\Psi = \{(U_i, \psi_i)\}$, $\Theta = \{(V_j, \theta_j)\}$ be two $\mathcal{C}^r$ atlases for X. We say Ψ is $\mathcal{C}^r$*-equivalent* to Θ if for each pair of indices i, j, such that $U_i \cap V_j \neq \emptyset$, the map

$$\theta_j \circ \psi_i^{-1} : \psi(U_i \cap V_j) \to \theta_j(U_i \cap V_j)$$

is of class C^r. Check that this is an equivalence relation on the family of all C^r-atlases on X.

It follows easily that if Ψ and Θ are two equivalent C^r atlases, then their union $\{(U_i, \psi_i)\} \cup \{(V_j, \theta_j)\}$ is also a C^r atlas on X in the same equivalence class. From this it follows that, given any C^r atlas on X, there exists a unique maximal C^r atlas in the same equivalence class. Indeed, it is nothing but the the union of all members of the equivalence class.

Definition 5.2.2 Let $0 \leq r \leq \infty$ be an integer or equal to ∞. By a C^r *structure* on a topological manifold X, we mean an equivalence class Φ of C^r atlas on it. The ordered pair (X, Φ) where X is a topological manifold and Φ is a C^r structure on it will be called a C^r *manifold.* If $r \geq 1$, it will be called a *smooth manifold.*

Often, in practice, we simply take one single atlas to represent the entire C^r structure and of course the maximal atlas in the class is a natural choice. Also, we shall dispense away with this elaborate notation (X, Φ), and simply say X is a smooth manifold, whenever there is no confusion. This is similar to the usual practice followed elsewhere such as "X is a topological space" or 'G is a group", etc.

Definition 5.2.3 Let (X, Φ) and (Y, Ψ) be two smooth manifolds. A continuous map $f : X \to Y$ is said to be *smooth at* $x \in X$ if for some $(U, \phi) \in \Phi$ and $(V, \psi) \in \Psi$ such that $x \in U$ and $f(U) \subset V$, we have $\psi \circ f \circ \phi^{-1}$ is smooth at $\phi(x)$. [Using the chain rule for differentiation, it follows easily that this condition is independent of the choice of ψ and ϕ.] If f is smooth at all points of X, then we say f is *smooth on* X. A bijection f such that f and f^{-1} are both smooth will be called a *diffeomorphism.*

It is also clear that a diffeomorphism is a homeomorphism. Hence, if two manifolds are nonhomeomorphic, they cannot be diffeomorphic. However, there are examples (difficult) of homeomorphic manifolds, that are not diffeomorphic. [Milnor gave examples of nondiffeomorphic smooth structures on the sphere $\mathbb{S}^7$.]

Definition 5.2.4 Let X be an n-dimensional manifold, and $Y \subset X$ a subspace. We say Y is a k-*dimensional submanifold* if for each $y \in Y$ there is a chart (U, ϕ) for X at y such that $\phi(y) = 0$ and

$$Y \cap U = \{x \in U \ : \ \phi_1(x) = \cdots = \phi_{n-k}(x) = 0\},$$

where $\phi = (\phi_1, \ldots, \phi_n)$. Letting $\psi = (\phi_{n-k+1}, \ldots, \phi_n)$, it follows easily that the collection $\{(Y \cap U, \psi|_{Y \cap U})\}$ forms an atlas for Y and hence Y itself is a k-dimensional manifold on its own. The maximal atlas containing this atlas will be called the *smooth structure induced from that of* X *on* Y.

Remark 5.2.1

1. It follows that the inclusion map is now an embedding of Y in X. Conversely, if Y is a subspace and a k-dimensional manifold on its own and if $\iota : Y \hookrightarrow X$ is an immersion, then by applying the injective form of the implicit function theorem, we see that Y is a submanifold.

2. Theorem 3.4.1 is valid in the abstract setup as well. The proof is identical.

3. As in the case of manifolds in Euclidean spaces and unlike the topological manifolds, the abstract differential manifolds with boundary do **not** cause any problem. In the definition 5.2.1, by allowing the charts to take values in the upper-half space $\mathbf{H}^n$ as in the definition 5.1.2, we get a differential n-manifold with boundary. The boundary

of such a manifold precisely consists of those points that are mapped inside $\mathbb{R}^{n-1} \times 0$, by any chart. Remark 3.2.1 and Theorem 3.2.1 etc. are valid ditto in this situation as well.

4. Replacing open subsets of $\mathbb{R}^n$ by open subsets of $\mathbb{C}^n$ and requiring that the transition functions to be complex differentiable, one gets the notion of complex n-dimensional manifolds. However, the analogy almost ends there. Due to the "rich" analytical aspects of complex differentiability, complex manifolds need to be studied on their own. On the other hand, due to the nonavailability of the partition of unity, many of the techniques available in differential topology are not available to complex manifold theory. On the positive side, since every complex n-dimensional manifold is a real $2n$-dimensional smooth manifold, all the results applicable to real smooth manifolds are available to complex manifolds after treating them as real smooth manifolds.

5. Having pointed out the importance of existence of smooth partition of unity, we shall now see that what we have proved for subsets of $\mathbb{R}^n$ in Theorem 1.7.1, is also available for abstract manifolds. And this result will be needed immediately in section 5.7 to prove that every abstract manifold can be embedded in a Euclidean space.

Lemma 5.2.1 Let X be a topological manifold. Then there exists a nested sequence of open subsets $\{W_i\}$ in X such that
(i) $\overline{W}_i$ is compact for each i;
(ii) $\overline{W}_i \subset W_{i+1}$, for each i;
(iii) $X = \cup_i W_i$.

Proof: Let $\phi_i : U_\alpha \to B^n$ be homeomorphisms such that $X = \cup_\alpha \phi_\alpha^{-1}(B^n_{1/2})$. Since X is II-countable, it is Lindelöff, i.e., every open cover of X has a countable subcover. Therefore we may assume that the family $\{U_\alpha\}$ is countable and index them by natural numbers: $\{U_i\}_{i\in\mathbb{N}}$. Put $V_i = \phi_i^{-1}(B^n_{1/2})$. Then each V_i is an open subset of X. The closure of V_i is compact being homeomorphic to the closed ball $\mathbb{D}^n_{1/2}$.

Put $W_1 = V_1$. Inductively having defined W_k, satisfying (i) and (ii), there are finitely many members of $\{V_i\}$, that cover $\overline{W}_k$. Let W_{k+1} be the union of all these members and V_{k+1}. Check the property (iii). ♠

We now have the following version of Theorem 1.7.1, that is obtained by merely replacing $\mathbb{R}^n$ by an abstract smooth manifold Y. Even the proof is similar. You are welcome to write down the details of the proof by yourself and read what is given here later.

Theorem 5.2.1 Partition of Unity on Abstract Manifolds: *Let X be any subspace of a smooth manifold Y and $\{U_\alpha\}_{\alpha\in\Lambda}$ be an open covering of X. Then there exists a countable family $\{\theta_j\}$ of smooth real valued functions on Y with compact support such that*
(i) $0 \le \theta_j(x) \le 1$, *for all j and $x \in X$;*
(ii) *for each $x \in X$ there exists a neighborhood N_x of x in X, such that only finitely many of θ_j are nonzero on N_x;*
(iii) *for each j, $(\operatorname{supp}\theta_j) \cap X \subset U_{\alpha_j}$ for some α_j; and*
(iv) $\sum_j \theta_j(x) = 1$, *for all $x \in X$.*

Proof: As before, we may replace X by a neighborhood of X in Y and assume that X itself is a smooth n-manifold. By the previous lemma, X is the increasing union of a countable family of open sets $\{K_i\}_{i\ge1}$ whose closure is compact. We set $K_0 = \emptyset$.

We shall now construct a countable family $\{B_{ij}\}$ of open sets in X with diffeomorphisms $\phi_{ij} : B_{ij} \to B^n$ such that $\{B_{ij}\}$ is a covering of X, that is a locally finite open refinement of $\{U_\alpha\}$. After that, since each B_{ij} is diffeomorphic to an open ball, we can construct a bump function on each of them and proceed exactly as in the proof of Theorem 1.7.1.

Inductively, suppose $\{B_{ij}\}$ have been constructed for $i \le k$ so as to cover $\overline{K}_k$. For each point $x \in \overline{K}_{k+1} \setminus K_k$, we can choose a neighborhood W_x contained in some member of $\{U_\alpha\}$ and not intersecting $\overline{K}_{k-1}$. We can further assume that there is a diffeomorphism $\psi_x : W_x \to B^n$ such that $\psi_x(x) = 0$. Since $\{\psi_x^{-1}(B^n_{1/2})\}$ is an open cover of $\overline{K}_{k+1} \setminus K_k$ that is compact, finitely many of these will cover it and we shall name them $B_{(k+1),j}$'s. It then follows that

$$\overline{K}_{k+1} \subset \cup_{ij}\{B_{ij} \; : \; i \le k+1\}.$$

Inductively, the construction of the family $\{B_{ij}\}$ is over. Clearly, it is an open refinement of the family $\{U_\alpha\}$ and covers X. To see that the family $\{B_{ij}\}$ is *locally finite*, given $x \in X$, suppose $x \in K_k$. Then, take $N_x = K_k$. Clearly, N_x does not meet any of the B_{ij} for $i \ge k+2$ and the family $\{B_{ij} \; : \; i \le k+2\}$ is finite. ♠

Remark 5.2.2 Naturally, we now have all other consequences of existence of the partition of unity such as the smooth Urysohn's lemma, the smooth Tietze's extension theorem, the Approximation theorem 1.7.2, etc., for any smooth manifold.

The fundamental problem in differential topology is to classify all smooth manifolds up to diffeomorphism. This problem is known to be unsolvable. However, there are plenty of restricted versions of this problem that are useful, solvable, and some of them are still not solved. Let us now study some examples and nonexamples of smooth manifolds.

Example 5.2.1
(i) The simplest example is perhaps $\mathbb{R}^n$ itself. Begin with the atlas $\{(\mathbb{R}^n, Id)\}$ and take a maximal atlas Ψ containing it. This definitely gives us a $\mathcal{C}^\infty$ structure on $\mathbb{R}^n$, called the *standard smooth structure* on $\mathbb{R}^n$. Let us denote this by $(\mathbb{R}^n, \mathcal{S})$. Indeed for any open subset U of $\mathbb{R}^n$, there is a unique smooth structure on U, that contains the inclusion map $U \hookrightarrow \mathbb{R}^n$.
(ii) The above game can be played starting with an arbitrary homeomorphism $h : \mathbb{R}^n \to \mathbb{R}^n$, as well, instead of the identity map. Clearly, $\{h\}$ is an atlas and there is a unique maximal atlas containing $\{h\}$, that gives a smooth structure on $\mathbb{R}^n$. Let us denote this by Φ_h. Observe that $h : (\mathbb{R}^n, \Phi_h) \to (\mathbb{R}^n, \mathcal{S})$ is a diffeomorphism. Thus, up to a diffeomorphism, we have not obtained any new structure on $\mathbb{R}^n$.
(iii) More generally, suppose X, Y are topological spaces and $f : X \to Y$ is a homeomorphism. If Y is a smooth manifold, then we can give a smooth structure to X also, by "pulling back" the smooth structure on Y via f, so that f is a diffeomorphism, viz., if $\{(U_i, \psi_i)\}$ is a smooth atlas for Y then take $V_i = f^{-1}(U_i)$ and $\phi_i = f \circ \psi_i$. Then $\{(V_i, \phi_i)\}$ is an atlas for X. (Verify this.) This fact should not be confused to mean that any two smooth manifolds that are homeomorphic are diffeomorphic. Now verify that, if $X = Y = \mathbb{R}^n$, then the pull back of the standard structure on $\mathbb{R}^n$ via f is Φ_f that we have defined in a slightly different way in (ii) above.
(iv) Given a nonempty open subset $A \subset X$ of a smooth manifold, by restricting each chart of X to U, we get a smooth structure on U. This is called the subspace structure. We shall always take the subspace structure on an open subset of a given manifold, unless it is mentioned otherwise. Clearly, the dimension of A is equal to that of X. The openness of A in X is crucial for this. For an arbitrary subset B, this process need not give a manifold structure. However, for a function defined on B, we have the notion of differentiability in the usual sense: $f : B \to Y$ is $\mathcal{C}^r$ at $x \in B$ if f has an extension that is $\mathcal{C}^r$ in a neighborhood of x in X.
(v) It is easy to construct a homeomorphism of the unit square $[0, 1] \times [0, 1]$ in $\mathbb{R}^2$ onto the unit disc. Then using the method described in (ii) we can think of the unit square as a smooth manifold. However, recall that the unit square is not a smooth submanifold of $\mathbb{R}^2$. (See Example 3.2.1(3).) Likewise the boundary of the unit square in $\mathbb{R}^2$ is **not** a smooth

submanifold of $\mathbb{R}^2$. However, since it is homeomorphic to $\mathbb{S}^1$, we can give it a smooth structure so that it is diffeomorphic to $\mathbb{S}^1$. In other words, here we have two different smooth structures i.e., inequivalent atlases, on the underlying topological manifold $\mathbb{S}^1$. (It is possible to have many more: merely keep taking regular n-gons in $\mathbb{R}^2$.) We shall soon see that all these inequivalent structures are diffeomorphic to each other, i.e., there is only one smooth manifold up to diffeomorphism with its underlying topological space homeomorphic to $\mathbb{S}^1$.
(vi) It is known that $\mathbb{R}^n$, $n \leq 3$ has only one smooth structure, up to diffeomorphism. Of course the proof is not easy. Soon, we shall see a proof of this for $n = 1$. Indeed, we shall classify all 1-dimensional manifolds. A somewhat more technical result is the classification of 2-dimensional manifolds, which will be studied in Chapter 8. In dimension 3, the problem already becomes formidable ([Moi]).
(vii) It was believed for a long time that each $\mathbb{R}^n$ has only one (viz., the standard) smooth structure, up to diffeomorphism. However, it is one of the shocking discoveries of the 1980's that $\mathbb{R}^4$ admits (uncountably) many nondiffeomorphic smooth structures. Donaldson was awarded the Fields medal for this outstanding discovery [Do].
(viii) **$\mathbb{P}^n$ as an abstract $\mathcal{C}^\infty$ manifold.** Let us write down at least one atlas explicitly. We will use the notation in Example 5.1.1.6. Indeed, we need to solve Exercise 5.1.2 now. Observe that $\mathbb{S}^n \subset \mathbb{R}^{n+1} \setminus \{0\}$ and the equivalence relation defined in the exercise restricts to the antipodal relation on $\mathbb{S}^n$. Moreover, observe that given any nonzero vector in $\mathbb{R}^n$, there is a unit vector in that direction. This shows that we can represent $\mathbb{P}^n$ as "the space of all lines" in $\mathbb{R}^{n+1}$, passing through 0. Let us denote the line represented by $(x_0, \ldots, x_n)$ by $[x_0, \ldots, x_n]$. For $0 \leq i \leq n$, put $U_i = \{[x] = [x_0, \ldots, x_n] \ : \ x_i \neq 0\}$. Define $\psi_i : U_i \to \mathbb{R}^n$ by the formula

$$[x] = [(x_0, \ldots, x_n)] \mapsto \left(\frac{x_0}{x_i}, \ldots, \frac{x_{i-1}}{x_i}, \frac{x_{i+1}}{x_i}, \ldots, \frac{x_n}{x_i}\right).$$

To verify that each ψ_i is a homeomorphism, we write down its inverse map, viz., $(t_0, t_1, \ldots, t_{n-1}) \mapsto [s_0, s_1, \ldots, s_i, \ldots, s_n]$ where,

$$s_j = \begin{cases} t_j, & j < i, \\ 1, & j = i, \\ t_{j-1}, & j \geq i+1. \end{cases}$$

Clearly, $\cup_i U_i = \mathbb{P}^n$. Therefore, in order to show this is a smooth atlas, we have only to check that the transition functions are all smooth. Let $i < j$. Then, observe that

$$\psi_i(U_i \cap U_j) = \{(t_0, \ldots, t_{n-1}) \in \mathbb{R}^n \ : \ t_{j-1} \neq 0.\}$$

and

$$\psi_j \circ \psi_i^{-1}(t_0, \ldots, t_{n-1}) = \left(\frac{t_0}{t_{j-1}}, \ldots, \frac{1}{t_{j-1}}, \ldots, \frac{t_{n-1}}{t_{j-1}}\right). \tag{5.1}$$

A similar formula holds for the case $i > j$ also. Hence, all the transition functions are smooth.
(ix) **The complex analytic manifold $\mathbb{C}P^n$:** In a similar fashion to that of real projective space, we can define the complex projective space $\mathbb{C}P^n$ as the quotient space of $\mathbb{C}^{n+1} \setminus \{0\}$ by the equivalence relation

$$(z_1, \ldots, z_{n+1}) \sim (\lambda z_1, \ldots, \lambda z_{n+1}), \quad \lambda \in \mathbb{C} \setminus \{0\}.$$

Of course, it will be a "n-dimensional complex manifold" and hence, a $2n$-dimensional real $\mathcal{C}^\infty$ manifold. The details are left to the reader.

Remark 5.2.3 Caution has to be taken while considering the Cartesian product $X \times Y$, when X and Y both have nonempty boundary. The natural way would be to take the product atlas $\{\phi_i \times \psi_j\}$ where, $\{\phi_i\}, \{\psi_j\}$ are, respectively, atlases for X and Y. This is perfect under the assumption that $\partial X = \emptyset$ or $\partial Y = \emptyset$. Therefore, in general, $(\text{int}\, X) \times (\text{int}\, Y) \subset X \times Y$ carries the product structure. However, if $x \in \partial X, y \in \partial Y$ then at $(x, y) \in X \times Y$ the product of charts fail to form a chart for $X \times Y$. Therefore, one needs to "modify" this. Such a process is called "smoothing the corners" in which one has to establish that there is a "unique" smooth structure on $X \times Y$ that extends the product structure on $(\text{int}\, X) \times (\text{int}\, Y)$. Here, we shall not discuss this any further. On the other hand, it makes perfect sense to talk about differentiability of functions defined on $X \times Y$ with reference to the product atlas, without bothering about whether it defines a manifold or not.

Exercise 5.2 Write down details of the Example 5.2.1.(ix).

5.3 Gluing Lemma

We shall now give a general construction that allows us to obtain new manifolds out of the old ones. As a bonus, we shall then be able to classify all 1-dimensional manifolds, in the next section. A more general version of this is employed in defining the tangent bundle of an abstract smooth manifold. Several important constructions are just special cases of this gluing lemma. Especially, we shall see some of them in Chapter 8.

Definition 5.3.1 By a *gluing data* (M, ϕ), we mean a smooth manifold M and a diffeomorphism $\phi : U \to V$ where U, V are open subsets of M such that $\overline{U} \cap \overline{V} = \emptyset$. Let R be the equivalence relation on M whose equivalence classes are the singletons outside $U \cup V$, and the pairs $\{u, \phi(u)\}, u \in U$. We shall denote the set of equivalence classes M/R with the quotient topology by M_ϕ and the quotient map by $q_\phi : M \to M_\phi$.

Remark 5.3.1 There is a category of gluing data and morphisms between gluing data: suppose $(M, \phi), (M', \phi')$ are two gluing data. By a morphism

$$\alpha : (M, \phi) \to (M', \phi')$$

we mean a smooth map $\alpha : M \to M'$ such that $\phi' \circ \alpha = \alpha \circ \phi$ for $x \in U$. Clearly then α induces a smooth map $\hat{\alpha} : M_\phi \to M'_{\phi'}$. The word "canonical" in the following lemma should be understood in this sense.

Recall that for a subset A of a topological space X, the *boundary* $\delta(A)$ is defined to be the set of all points $x \in X$ such that every neighborhood U of x in X intersects both A and $X \setminus A$. This should not be confused with the boundary of a manifold, though there are instances in which the two concepts coincide. In the following lemma, we need to use this concept.

Lemma 5.3.1 The Gluing Lemma: Let (M, ϕ) be a gluing data and $q_\phi : M \to M_\phi =: N$ be the corresponding quotient map. Then
(i) N has a "canonical" smooth structure such that the quotient map $q : M \to M_\phi$ is a local diffeomorphism. In particular, a map $f : M_\phi \to \mathbb{R}$ is smooth iff $f \circ q$ is smooth.
(ii) N is not Hausdorff iff there is a sequence $\{u_n\}$ in U converging to some point $p \in \delta(U)$ and such that $\{\phi(u_n)\}$ converges to a point in $\delta(V)$, (where $\delta(U) = \bar{U} \setminus U$, and $\delta(V) = \bar{V} \setminus V$ are the sets of boudary points of U, V respectively in X).

(iii) Suppose (M', ϕ') is another gluing data, i.e., $\phi : U' \to V'$ is another diffeomorphism where U', V' are open subsets of M' with $\overline{U'} \cap \overline{V'} = \emptyset$. Then M_ϕ is diffeomorphic to $M'_{\phi'}$, if there exists a diffeomorphism $\alpha : M \to M'$ such that $\alpha \circ \phi = \phi' \circ \alpha$ on U.

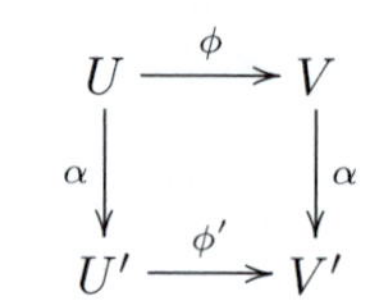

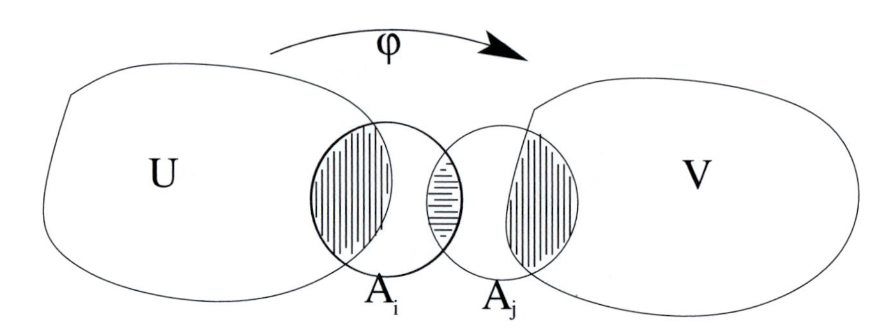

Figure 24 Verifying Hausdorffness for gluing-up.

Proof: (i) Observe that the quotient map $q = q_\phi : M \to M_\phi$ is a local homeomorphism. Indeed, for any open set G of M, that is contained in $M \setminus U$ or in $M \setminus V$, we have $q : G \to q(G)$ is a homeomorphism. Therefore, we can choose an atlas $\{(A_j, \psi_j)\}$ for M such that each A_j is contained in $M \setminus U$ or $M \setminus V$. (Here, the hypothesis $\overline{U} \cap \overline{V} = \emptyset$ is used.) We claim that $\{(q(A_j), \tau_j)\}$, where $\tau_j = \psi_j \circ q^{-1})\}$, is a (smooth) atlas for M_ϕ. For, given i, j, let us say $A_i \subset M \setminus V, A_j \subset M \setminus U$. Then the intersection of $q(A_i)$ with $q(A_j)$ can be written as a disjoint union of two open sets:

$$q(A_i) \cap q(A_j) = q(A_i \cap A_j) \coprod q(\phi(A_i \cap U) \cap A_j).$$

And the transition function $\tau_j \circ \tau_i^{-1}$ is equal to $\psi_j \circ \psi_i^{-1}$ on the first set and $\psi_j \circ \phi \circ \psi_i^{-1}$ on the second set.

(ii) Given two points $x, y \in N$, if one of them does not belong to the boundary of $q(U) = q(V)$ in N, we can easily get open sets around x and y that are disjoint. Therefore, N is not Hausdorff iff there are points $q(x), q(y)$ on the boundary of $q(U) = q(V)$, that cannot be separated by open sets. Let $\{U_n\}, \{V_n\}$ be a fundamental systems of neighborhoods at $x, y \in M$, respectively. Then $q(x), q(y)$ cannot be separated in N iff for each n, we have $q(U_n) \cap q(V_n) \neq \emptyset$. This is the same as saying that for each $n \geq 1$, there is $x_n \in U_n \cap U$ such that $\phi(x_n) \in V_n \cap V$. But then we see that $x_n \to x$ and $\phi(x_n) \to y$. The converse is easy to see. (The best way to understand this is to consider the situation when you are gluing two open intervals along two open subintervals (see the Figure 25).

(iii) Clearly, the map α factors down to give a continuous bijection $\overline{\alpha} : M_\phi \to M_\psi$:

$$\begin{array}{ccc} M & \xrightarrow{\alpha} & M \\ {\scriptstyle q}\downarrow & & \downarrow{\scriptstyle q'} \\ M_\phi & \xrightarrow{\overline{\alpha}} & M_\psi. \end{array}$$

The smoothness of $\overline{\alpha}$ and its inverse follows from the last part of (i). ♠

Example 5.3.1

1. In the lemma above, take

$$M = (-1/2, 1),\ U = (-1/2, 0),\ V = (1/2, 1)$$

and

$$\tau(t) = t + 1;\ \ \lambda(t) = 1/2 - t.$$

Can you recognize M_τ and M_λ? How do they differ?

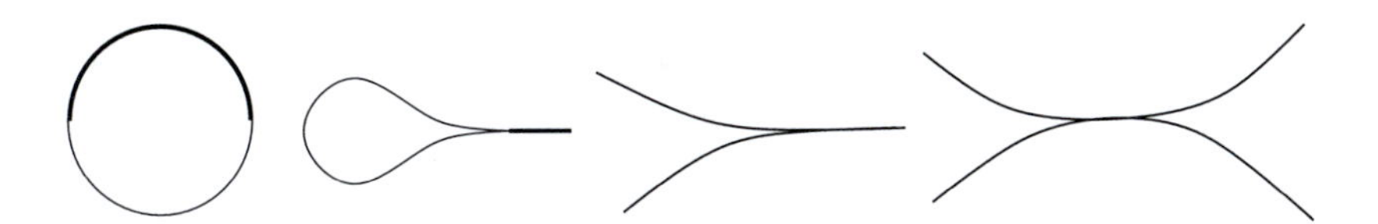

Figure 25 gluing one or two intervals.

Observe that the map $M \to \mathbb{S}^1$ given by $t \mapsto e^{2\pi\imath t}$ factors down to define a diffeomorphism $M_\tau \to \mathbb{S}^1$. However, M_λ is not a Hausdorff space since λ does not satisfy condition (ii) in Lemma 5.3.1.

2. **The cylinder and the Möbius band:** Put

$$M = (-1/2, 1) \times [0, 1];\ \ U = (-1/2, 0) \times [0, 1];\ \ V = (1/2, 1) \times [0, 1].$$

Take $\phi : U \to V$ given by $\phi(t, s) = (t + 1, s)$. As in the above example obtain a diffeomorphism of M_ϕ with $\mathbb{S}^1 \times [0, 1]$.

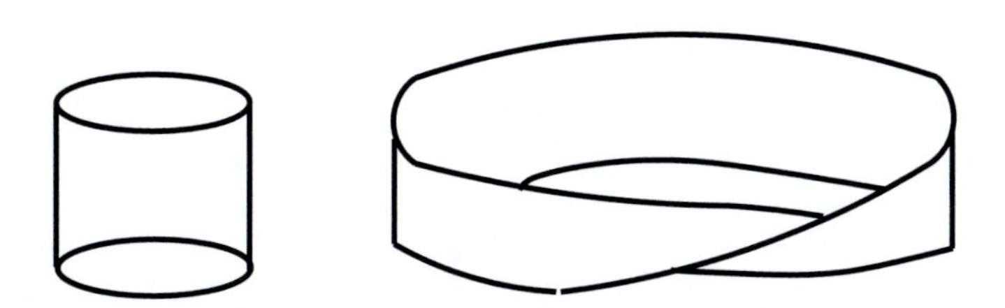

Figure 26 Cylinder and the Möbius band.

Now take $\psi : (t, s) = (t + 1, 1 - s)$. The resulting manifold M_ψ is called the *Möbius band.* We shall denote it by $\mathbb{M}$. The difference in the construction of the cylinder and the Möbius band is in the fact that the gluing map ϕ used for the cylinder is orientation preserving whereas ψ used for the Möbius band is orientation reversing. Indeed, you can easily show that the central circle in $\mathbb{M}$ is orientation reversing and hence $\mathbb{M}$ is nonorientable (see Exercise 4.1.1). Later, we shall give yet another proof of this fact (see Theorem 7.4.2).

3. In the above example begin with $M = (-1/2, 1) \times \mathbb{S}^1$ (instead of $(-1/2, 1) \times I$) and make corresponding changes everywhere else. Also take ψ to be $(t, v) \mapsto (1 + t, \bar{v})$. Exactly as in example 1 and 2, you will get a diffeomorphism from M_ϕ to $\mathbb{S}^1 \times \mathbb{S}^1$. This surface is popularly called the **torus**, that resembles the surface of a well-made

medu-vada or a donut. On the other and, M_ψ is not such a surface at all. One cannot even visualize it in the 3-dimensional Euclidean space. An "approximate picture" of this object is shown in the Figure 27 that is indeed an immersion of this surface. It is called the **Klein bottle.**[1] We shall denote it by $\mathbf{K}$.

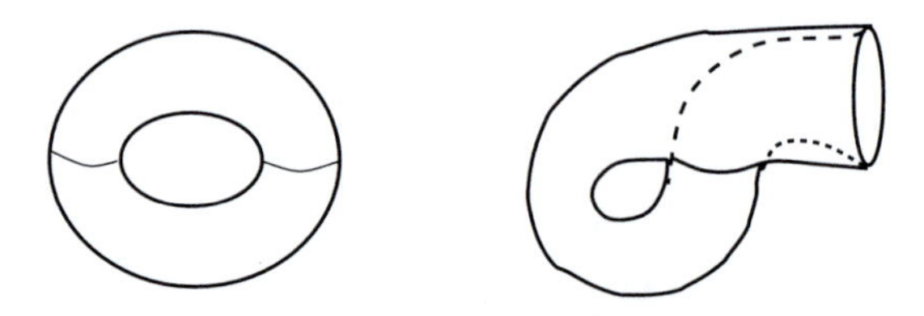

Figure 27 The torus and the Klein bottle.

Let $q : M \to \mathbf{K}$ be the quotient map. The image of the two lines $(-1/2, 1) \times \{v, \bar{v}\}$ in $\mathbf{K}$ is an embedded loop, that separates $\mathbf{K}$ into two Möbius bands. Define $\tau : M \to \mathbf{K}$ by

$$\tau(t,v) = \begin{cases} q(2t, v), & -1/4 < t < 1/2; \\ q(2t-1, v), & 1/4 < t < 1; \\ q(2t+1, v), & -1/2 < t < 0. \end{cases}$$

Then τ factors through $M \to \mathbb{S}^1 \times \mathbb{S}^1$ to give $\hat{\tau} : \mathbb{S}^1 \times \mathbb{S}^1 \to \mathbf{K}$, that is a two-to-one smooth map. (Indeed it is a 2-sheeted covering projection, if you know what a covering projection means.)

Exercise 5.3

(a) Obtain a generalization of the gluing Lemma 5.3.1 for the following gluing data: Let M be a smooth manifold. Instead of considering one single pair of open sets $\{U, V\}$, let us have a family $\phi_\alpha : U_\alpha \to V_\alpha, \alpha \in \Lambda$ of homeomorphisms (diffeomorphisms) where U_α, V_α are open subsets of M such that $\overline{U}_\alpha \cap \overline{V}_\alpha = \emptyset, \ \forall \ \alpha \in \Lambda$. Carry out the rest of the details exactly as before.

(b) Given a smooth atlas $\{(U_i, \phi_i)\}$ on X, put M equal to disjoint union of $\phi_i(U_i)$ and for each pair of indices i, j such that $U_i \cap U_j \neq \emptyset$, put $\psi_{ij} : \phi_i(U_i \cap U_j) \to \phi_j(U_i \cap U_j)$ to be the diffeomorphisms $\psi_{ij} = \phi_j \circ \phi_i^{-1}$. Use this as gluing data to obtain a space N. Consider the map $\Theta : M \to X$ defined by $\Theta|_{\phi_i(U_i)} = \phi_i^{-1}$. Show that Θ factors through the quotient map $q : M \to N$ to define a diffeomorphism $\theta : N \to X$.

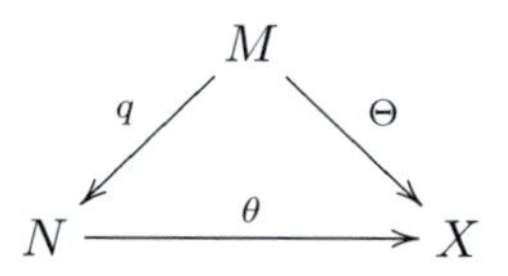

[1] The Klein bottle was first described in 1882 by the German mathematician Felix Klein. It was originally named the Kleinsche Fläche meaning "Klein surface"; however, this was incorrectly interpreted as Kleinsche Flasche meaning "Klein bottle", that ultimately led to the adoption of this term in the German language as well.–Wikipedia

5.4 Classification of 1-dimensional Manifolds

In every classification problem, we must, first of all, have plenty of examples that are "likely" to represent all possible types of objects that we want to classify. Only after that, we can make a probable list of representatives that are mutually of different types. The final step is to show that every object that we wanted to classify belongs to (precisely) one of the types mentioned in the list.

In order to classify all manifolds, clearly, it suffices to consider only connected ones. For, any manifold is locally connected and hence its connected components are open as well as closed. Therefore, any manifold is the disjoint union of its connected components, even as a topological space.

What are the examples of 1-dimensional connected manifolds that we have?

We observe that any two closed intervals are diffeomorphic via an affine linear map. This diffeomorphism can then be used to get diffeomorphisms between any two finite open intervals or between any two half-open intervals as well. Moreover, $x \mapsto \tan x$ defines a diffeomorphism of the interval $(-\pi/2, \pi/2) \to \mathbb{R}$. Thus, as far as subsets of $\mathbb{R}$ are concerned, we have three different classes of connected 1-dimensional manifolds:

(i) open intervals (ii) half-open intervals (iii) closed intervals.

As soon as we go to subspaces of $\mathbb{R}^2$, we get "other" types: circles, ellipses, parabolas, and many more smooth curves. If we have one-to-one parameterization of any of these curves then clearly they will be diffeomorphic to an interval. This is the case with a parabola for instance. One can also see easily that any two circles are diffeomorphic to each other. Indeed, placing a small circle inside an ellipse and then projecting radially from the center of the circle produces a diffeomorphism of the circle with the ellipse. Write down an explicit formula by yourself:

Do we get any other types of 1-dimensional manifolds, if we look inside higher dimensional Euclidean spaces? The answer is: **NO.**

Theorem 5.4.1 *Let X be a connected 1-dimensional, (Hausdorff and II-countable) abstract smooth manifold. Then X is diffeomorphic to one of the following:*
(i) $(0,1)$; (ii) $[0,1)$; (iii) $[0,1]$ (iv) $\mathbb{S}^1$.

Our main tool in the proof of this theorem is the basic gluing Lemma 5.3.1 that we have introduced in Section 5.3.

Let us take a look at the Examples 5.3.1. Since they become crucial, we shall recast them in a little general setup.

Example 5.4.1 Let $M = (a,b)\coprod(c,d)$, $p \in (a,b), q = b + c - p \in (c,d), U = (p,b), V = (c,q)$. Let $\phi : U \to V$ be the diffeomorphism given by $\phi(x) = c+x-p$. Then the identification space M_ϕ is diffeomorphic to an open interval. If we replace one or both open intervals $(a,b), (c,d)$ by the half-closed intervals $[a,b), (c,d]$, respectively, then M_ϕ is diffeomorphic to a half-closed or a closed interval accordingly. To see this, consider the map $f : M \to (a+c-p, d)$ defined by $f(x) = c+x-p, x \in (a,b)$ and $f(x) = x$ for $x \in (c,d)$. Clearly, this map factors through $q : M \to M_\phi$ to define a smooth map $\hat{f} : M_\phi \to (a+c-p, d)$.

Example 5.4.2 Consider $M = (0,1), U = (0,1/4), V = (3/4,1)$ and $\phi(t) = t + 3/4$. Then M_ϕ is diffeomorphic to the unit circle $\mathbb{S}^1$.

To see this, consider the map $\Theta : M \to \mathbb{S}^1$ given by $t \mapsto e^{8\pi\imath t/3}$, which factors through the quotient space to define a map $\hat{\Theta} : M_\phi \to \mathbb{S}^1$. We can easily check that $\hat{\Theta}$ is a diffeomorphism.

We can now slightly generalize these facts in the following two lemmas. The ease with which we can prove the next result is specific to 1-dimensional manifolds.

Lemma 5.4.1 Let $a < b < c$ and $d < e < f$ be real numbers, M be the disjoint union of the two intervals $M = (a,c)\coprod(d,f)$. Let $\phi : (b,c) \to (d,e)$ be a diffeomorphism with $\phi(b) = d$ and $\phi(c) = e$. Then the identification space M_ϕ is diffeomorphic to an open interval.

Proof: Choose b', c' so that $b < b' < c' < c$. Put $\phi_1 = \phi|_{[b',c']}$. Extend the diffeomorphism $\phi_1 : [b',c'] \to [\phi(b'),\phi(c')]$ to a diffeomorphism $\hat{\phi_1} : \mathbb{R} \to \mathbb{R}$. Put $g = \hat{\phi_1}(a)$. We claim that M_ϕ is diffeomorphic to the open interval (g,f).

Put $M' = (a,c')\coprod(\phi(b'),f) \subset M$. Then observe that the quotient map $q : M \to M_\phi$ is surjective restricted to M' and hence M_ϕ is the same as $q(M') = M'_{\phi_1}$. Now let $\psi : M' \to (g,f)$ be defined by

$$\psi(t) = \begin{cases} \hat{\phi_1}(t), & \text{if } t \in (a,c']; \\ t, & \text{if } t \in [\phi(b'),f). \end{cases}$$

Then ψ factors through q to give a continuous bijection $\bar{\psi} : M_\phi \to (g,f)$. On the other hand $\bar{\psi} \circ q$ restricted to (a,c) is $\hat{\phi}$ and restricted to (d,f) is the identity map. This suffices to conclude that $\bar{\psi}$ and its inverse are smooth. ♠

Lemma 5.4.2 Let $a < b < e < f$ and let $\phi : (a,b) \to (e,f)$ be an order preserving diffeomorphism. Then the identification space $(a,f)_\phi$ is diffeomorphic to the circle.

Proof: Choose $a < a' < b' < b$ and put $e' = \phi(a'), f' = \phi(b'), \phi_1 = \phi|_{(a',b')}$. As in the previous lemma, observe that the subset (a',f') surjects onto $(a,f)_\phi$ and hence $(a',f')_{\phi_1} = (a,f)_\phi$.

From the exercise 6 in 1.7, there is a diffeomorphism $\alpha : (a',f') \to (0,1)$ such that the following diagram is commutative:

$$\begin{array}{ccc} (a',b') & \xrightarrow{\phi} & (e',f') \\ \downarrow \alpha & & \downarrow \alpha \\ (0,1/4) & \xrightarrow{\tau} & (3/4,1) \end{array}$$

where $\tau(t) = t + 3/4$. Therefore, the gluing Lemma 5.3.1(iii) implies that α defines a diffeomorphism of $(a',f')_{\phi_1}$ with $(0,1)_\tau$. As seen in Example 5.3.1, the map $t \mapsto e^{8\pi \imath t/3}$ defines a diffeomorphism of $(0,1)_\tau$ with the unit circle $\mathbb{S}^1$. ♠

We now come to another result specific to 1-dimension. It is the key to the classification of 1-dimensional manifolds.

Lemma 5.4.3 A key-lemma: Let X be a connected 1-dimensional manifold having an atlas consisting of two members, U_1, U_2, such that $U_1 \neq U_1 \cap U_2 \neq U_2$. Then
(i) $U_1 \cap U_2$ is nonempty and has at most two components.
(ii) If $U_1 \cap U_2$ has one component, then X is diffeomorphic to an interval.
(iii) If $U_1 \cap U_2$ has two components, then X is diffeomorphic to $\mathbb{S}^1$.

Proof: That $U_1 \cap U_2 \neq \emptyset$ follows from the connectedness of X. Observe that this just means that X is obtained by gluing two (disjoint) intervals I_1, I_2 (diffeomorphic to U_1, U_2 respectively), along some nonempty proper open subsets $J_1 \subset I_1, J_2 \subset I_2$ via a diffeomorphism $\phi : J_1 \to J_2$. We make the following observations, each one being obvious or follows easily from the previous ones.
(a) Each I_i is an open interval or an half-open interval, i.e., cannot be a closed interval.
(b) Each component of J_1, J_2 is an interval. None of them can be a closed interval.
(c) At least one of the end points of each component C of J_i belongs to $I_i, i = 1,2$.
(d) If e is an end point of a component C of J_1, then $\lim_{x\to e}\phi(x)$ does not exist in I_2 (by

(ii) of gluing Lemma 5.3.1). Here the limit is taken where the variable x remains inside C.
(e) Number of component of J_i cannot exceed the number of end points of I_i that do not belong to I_i. This follows from (d).
(f) If both I_i are open intervals, then J_i can have at most two components each. If one of the I_i is an half-open interval then J_i are connected. This proves (i).
To prove (ii), first consider the case wherein I_1, I_2 are both open intervals. It follows that J_1 and J_2, being diffeomorphic $U_1 \cap U_2$ are also open intervals. As seen in (i), both end points of J_1 cannot be in the interior of I_1 and similarly both end points of J_2 cannot be in the interior of I_2. Therefore, the situation is precisely as in the Lemma 5.4.1. Even the diffeomorphism ϕ has to be order preserving. So, Lemma 5.4.1 is applicable.

The case when I_1 or/and I_2 is not an open interval is handled similarly, once we show that J_1 and J_2 are open intervals. As such the other possibility is that both are half-open intervals. The diffeomorphism will then take the boundary point of J_1 to that of J_2 and we will have non-Hausdorffness at the other end. Therefore, J_1 and J_2 are open intervals. The rest of the argument is the same as above, except that the conclusion in this case is that X is diffeomorphic to a half-open interval or a closed interval according as one of I_1, I_2 or both are half-open intervals. This proves (ii).
(iii) Now suppose J_1 and hence J_2 has two components. Say A_1, A_2 and B_1, B_2 are the components of J_1, J_2, respectively. Let $\phi(A_j) = B_j$. Then first of all we note that both I_i are open intervals. Then the identification space X can be obtained in two steps: first, we glue along $\phi_1 = \phi|_{A_1}$ to obtain a space M_{ϕ_1} and then perform another gluing along $\phi_2 = \phi|_{A_2}$. Of course, M_{ϕ_1} is also a Hausdorff space, for otherwise X will not be Hausdorff. By the first case discussed in (ii) it follows that M_{ϕ_1} is an interval. Therefore, we are in the situation of Lemma 5.4.2 and conclude that X is diffeomorphic to $\mathbb{S}^1$. ♠

Proof of the theorem 5.4.1: Let $\{U_\alpha\}$ be an atlas for X. By II-countability, there exists a countable subcover for $\{U_j\}$ for X. Indeed, we have:
Step I There exists a countable family $\{U_j\}$:
(i) Each U_j is diffeomorphic to an interval;
(ii) $U_k \not\subset \cup_{j\leq k-1} U_j =: W_{k-1}$;
(iii) $W_{k-1} \cap U_k \neq \emptyset, k \geq 1$.
To construct such a family, we start off with any one member from the countable family $\{U_j\}$, call it U_1. Having picked up $U_0, U_1, \ldots, U_{k-1}$ so as to satisfy the above requirement, we check whether $W_{k-1} := \cup_{j\leq k-1} U_j$ is the whole space X. If so, we stop. Otherwise, it means that there are members in $\{U_j\}$ not contained in W_{k-1}. If none of them intersect W_{k-1}, it would mean that X is disconnected. Therefore there exists a member that we label U_k with the required property.

We remark that in addition if X is compact, then the family $\{U_j\}$ can be chosen to be finite also. Moreover, if ∂X is nonempty, we can start with U_1 as a half-open interval.
Step II If $W_n \neq X$, then W_n is diffeomorphic to an interval.
We shall prove this by induction on n. This is true for $k = 0$ since $W_0 = U_0$. So assume that W_k is diffeomorphic to an interval and $W_{k+1} \neq X$. Apply the Lemma 5.4.3, we conclude that W_{k+1} is diffeomorphic to an open interval, a half-open interval, a closed interval, or $\mathbb{S}^1$. The latter two cases are ruled out, because, then W_{k+1} is compact and hence will have to be the whole of X.
Step III If the sequence $\{U_j\}$ stops at $j = m$ say, then from the previous step W_{m-1} is an open interval or a half-open interval. Therefore, the key Lemma 5.4.3 applied to $X = W_m = W_{m-1} \cup W_m$ yields that X is an interval or a circle.
Step IV Consider the case when the sequence $\{U_j\}$ is infinite. First consider the simpler case when $U_0 = W_0$ is diffeomorphic to a half-closed interval. We shall fix a diffeomorphism $f_0 : W_0 \to [0, 1)$. We can then get a diffeomorphism $f_1 : W_1 = W_0 \cup U_1 \to [0, 2)$ such that $f_1|_{W_0} = f_0$. Inductively, having found a diffeomorphism $f_k : W_k \to [0, k + 1)$ such that

$f_k|_{W_{k-1}} = f_{k-1}$, it is easily seen that we can find a diffeomorphism $f_{k+1} : W_{k+1} \to [0, k+2)$ such that $f_{k+1}|_{W_k} = f_k$. Now define $f : X \to [0, \infty)$ by $f = \lim_{k\to\infty} f_k$. This is the same as saying $f(x) = f_{k+2}(x)$ whenever $x \in U_k$. It follows that f is a diffeomorphism.

Now consider the case when $\partial X = \emptyset$. In this case, it follows that at each stage W_k is diffeomorphic an open interval. Hence the gluing of the next interval may occur on either side. If there is a U_i such that all other U_j occur on only one of the two sides, then we could have re-labeled them so that this U_i is the 0th one and proceeded as in the above case. If this is not the case, then there are infinitely many U_j occurring on either side of U_0 and hence we can relabel them by positive and negative integers, redefine $W_k = \cup_{j=-k}^{k} U_j$ and obtain diffeomorphisms $f_k : W_k \to (-k, k)$ such that $f_{k+2}|_{W_k} = f_k$. Put $f = \lim_{k\to\infty} f_k$ and check that $f : X \to (-\infty, \infty)$ is a diffeomorphism. ♠

Here is an interesting application of the classification of 1-dimensional manifolds. (Compare Lemma 4.4.5 and Theorem 4.4.6.)

Theorem 5.4.2 *Let X be a compact manifold with nonempty boundary. Then there exists no map $f : X \to \partial X$ such that $f|_{\partial X} = Id_{\partial X}$.*

Proof: [The result is true even for continuous maps. However, we shall give the proof here for smooth maps only. For continuous maps see Exercise 6.1.1.] Let $f : X \to \partial X$ be such a smooth map. By Sard's theorem, there exists $x \in \partial X$, which is a regular value for f (as well as for ∂X, which is obvious in this case). By the Extended Preimage Theorem 3.4.4, it follows that $L = f^{-1}(x)$ is a 1-dimensional neat submanifold of X. Recall that neatness means that the boundary of L is precisely equal to

$$L \cap \partial X = (f|_{\partial X})^{-1}(x) = \{x\}.$$

On the other hand, since X is compact L is also compact. By the classification theorem, L is the union of finitely many components, each one of them diffeomorphic to a closed interval or a circle. Hence, ∂L should have an even number of points. This is a contradiction. ♠

Remark 5.4.1 As a corollary, we obtain another proof of **Brouwer's Fixed Point Theorem** 4.4.6.

Exercise 5.4 Let W be a smooth 1-dimensional manifold, $W = A \cup B$ with $A \cap B = \{w\}$. Suppose both A and B are diffeomorphic to a closed interval. Show that W is also diffeomorphic to a closed interval, without appealing to the classification theorem.

5.5 Tangent Space and Tangent Bundle

It is time you have gone through the exercise 5.3. The theme of this exercise is that we can think of a smooth n-manifold as a union of a collection of open subsets of $\mathbb{R}^n$ that are glued to each other along some diffeomorphisms. This view helps us to define the tangent space of an abstract smooth manifold also, since we know that the tangent space for an open subset U of $\mathbb{R}^n$ is actually diffeomorphic to $U \times \mathbb{R}^n$. The following definition should be understood from this point of view.

Definition 5.5.1 (**Tangent space to an abstract smooth manifold**) Let (X, Φ) be a smooth n-dimensional manifold, where $\Phi = \{(U_i, \phi_i)\}_{i\in\Lambda}$ is a maximal smooth atlas. Consider the space

$$F = \{(x, i, v) \in X \times \Lambda \times \mathbb{R}^n \ : \ x \in U_i\}.$$

Give discrete topology to Λ, take the product topology on $X \times \Lambda \times \mathbb{R}^n$ and let F have the subspace topology. [Observe that F is homeomorphic to the disjoint union $\cup_{i \in \Lambda} U_i \times \mathbb{R}^n$. Now define an equivalence relation on F by $(x, i, v) \sim (y, j, u)$ iff

$$x = y, \text{ and } D(\phi_j \circ \phi_i^{-1})_{\phi_i(x)}(v) = u.$$

The verification that this is an equivalence relation is straightforward. Let TX be the space of equivalence classes with the quotient topology and $q : F \to TX$ be the quotient map.

We call TX the *total tangent space* to X. If $\pi : TX \to X$ denotes the map induced by the projection onto the first factor $F \to X$, then (TX, π) is called the *tangent bundle* of X. The subspace $\pi^{-1}(x)$ is called the *tangent space* of X at x and is denoted by $T_x(X)$.

Remark 5.5.1

(i) The first thing to observe is that in the definition of TX we could have used any atlas $\Phi' \subset \Phi$, in place of the maximal atlas Φ. For if $\Lambda' \subset \Lambda$ such that $\{U_i : i \in \Lambda'\}$ is a cover of X, and F' is defined similarly, then $q : F' \to TX$ is surjective. This observation is going to be quite useful in practice. For example, we can now immediately see that TX is second countable, by choosing a countable Λ' as above. We leave it to you to verify that TX is Hausdorff also. However, this becomes clearer, after remark (iii) below.
(ii) The image of all points $(x, i, 0)$ forms a subspace of TX diffeomorphic to X. We shall merely identify this with X itself. Under this identification, we have $\pi(x) = x$ and hence X is called the 0-section of π. Further, in $X \times \Lambda \times \mathbb{R}^n$, for each fixed (x, i), we have the subspace $\{(x, i)\} \times \mathbb{R}^n$, that can be identified with $\mathbb{R}^n$ and hence can be given the linear structure. Moreover, $q : \{(x, i)\} \times \mathbb{R}^n \to T_x(X)$ is a bijection. Verify that it is actually a homeomorphism. Since the equivalence relation respects the linearity, it follows that $T_x(X)$ is a vector space in such a way that $q : \{(x, i)\} \times \mathbb{R}^n \to T_x(X)$ is an isomorphism.
(iii) For any open subset U of X, we shall denote by $\tilde{U}$ the inverse image of U under π. [Some authors use the notation TU for $T^{-1}U$. This notation may run into difficulty when U is not an open set but a submanifold of X of lower dimension.] Once again, $q : U_i \times \{i\} \times \mathbb{R}^n \to \tilde{U}_i$ is a homeomorphism. Shifting the labeling from the domain to the map, we shall denote this homeomorphism by $q_i : U_i \times \mathbb{R}^n \to \tilde{U}_i$. Put $\xi_i = (\phi_i \times Id) \circ q_i^{-1} : \tilde{U}_i \to \mathbb{R}^n \times \mathbb{R}^n$. Since $\{\tilde{U}_i : i \in \Lambda\}$ is an open cover of TX, it follows that TX is a topological manifold of dimension $2n$, with $\{(\tilde{U}_i, \xi_i) : i \in \Lambda\}$ as an atlas. (Of course, you have to verify now that TX is Hausdorff to complete this step.)
(iv) Finally, let us look at the transition functions associated to the above atlas of TX. For any two indices i, j such that $U_i \cap U_j \neq \emptyset$, we have, the transition mappings

$$\xi_j \circ \xi_i^{-1} : \phi_i(U_i \cap U_j) \times \mathbb{R}^n \to \phi_j(U_i \cap U_j) \times \mathbb{R}^n$$

given by,

$$(y, w) \mapsto (\phi_j \circ \phi_i^{-1}(y), D(\phi_j \phi_i^{-1})_y(w)).$$

[The expression in the first slot is obvious. The expression in the second slot is precisely due to the manner in which we have introduced the equivalence relation in F.] It follows that if Ψ is a $\mathcal{C}^{r+1}$ atlas for X then the above maps define a $\mathcal{C}^r$ atlas for TX. In this manner we get TX as a $2n$-dimensional $\mathcal{C}^r$-manifold. Observe that with this smooth structure, the projection map $\pi : TX \to X$ is a $\mathcal{C}^r$ mapping and is referred to as the *projection* map of the tangent bundle. Finally, it should be noted that each transition function is a linear map when restricted to the second slot.
(v) The abstract definition of the tangent space coincides with our old geometric definition for a smooth submanifold X of $\mathbb{R}^N$ given in the previous chapter. To see this, consider the map $\tau : TX \to \mathbb{R}^N \times \mathbb{R}^N$ defined by $[x, i, v] \mapsto (x, D(\phi_i^{-1})_{\phi_i(x)}(v))$. Verify that it is well defined. The surjectivity of this map is obvious. Proving the injectivity is similar to proving

the well-definedness of the map. That the inverse map is also of class $\mathcal{C}^r$ can be verified by choosing any chart at a point.
(vi) Tangent space of a manifold with boundary is also defined exactly in a similar fashion. Note that the dimension of the tangent space T_xX is the same as the dimension of X even at points of ∂X. On the other hand, since ∂X is a manifold of one dimension lower, $T_x\partial X$ is a codimension 1 subspace of T_xX.

Definition 5.5.2 For a smooth map $f : X \to Y$ of manifolds, the induced map $Tf : TX \to TY$ is defined by the formula

$$[x, i, v] \mapsto [y, j, u],$$

where, $(U_i, \phi_i), (V_j, \phi_j)$ are charts for X and Y, respectively at x, y such that $f(U_i) \subset V_j$, $y = f(x)$ and $u = D(\phi_j \circ f \circ \phi_i^{-1})(v)$.

Remark 5.5.2
(i) Once again, the usual chain rule for differentiation can be employed to see that the above definition of Tf is independent of the choice of charts at x and $f(x)$. Clearly, Tf restricted to $\pi^{-1}(x)$ for any $x \in X$, is a linear map. We also have the obvious chain rule $T(g \circ f) = T(g) \circ T(f)$. Once we have verified this, we can use the notation Df itself for Tf and call it the *derivative* of f.
(ii) Observe that we can define the smoothness of any function defined on any non empty open set U of X as well, by simply treating U as a submanifold with the smooth structure obtained by restricting the atlas for X to U. In particular, we then ask the question: Is a chart smooth with respect to the smooth structure to which it belongs? To be precise, consider the homeomorphism $\phi_i : U_i \to \mathbb{R}^n$, that belongs to the atlas Φ. Treating U_i as a manifold with the induced structure and $\mathbb{R}^n$ with its standard structure, is ϕ_i smooth? To answer this, we first observe that the singleton $\{(U_i, \phi_i)\}$ is an atlas for U_i and hence any map $g : U_i \to \mathbb{R}^m$ is smooth iff $g \circ \phi_i^{-1}$ is smooth. In particular, we can take $g = \phi_i$ and it follows that ϕ_i is smooth.
(iii) Closely related is the question: How does the derivative of ϕ_i look like? The answer is $D(\phi_i)[x, i, v] = (\phi_i(x), v)$.
(iv) All the notions and results about immersion, submersion, etc., that you have learned for differential manifolds in Euclidean spaces are valid for abstract manifolds as well.

The concept of tangent bundle leads to a more general and very powerful concept, viz., the vector bundle. We shall take this opportunity just to introduce this concept here. For further studies the reader may consult [Hus] or [St].

Definition 5.5.3 Let B be a topological space. By a *real vector bundle of rank k over B* we mean an ordered triple $\xi = (E, p, B)$, where E is a topological space $p : E \to B$ is a continuous map such that for each $b \in B$, the fiber $p^{-1}(b)$ is a k-dimensional $\mathbb{R}$-vector space satisfying the following local triviality condition(LTC):
To each point $b \in B$ there is an open neighborhood U of b and a homeomorphism $\phi : p^{-1}(U) \to U \times \mathbb{R}^k$ such that
(i) $\pi_1 \circ \phi = p$ and
(ii) $\pi_2 \circ \phi : p^{-1}(b) \to \mathbb{R}^k$ is an isomorphism of vector spaces.

Here $\pi_1 : U \times \mathbb{R}^k \to U$ and $\pi_2 : U \times \mathbb{R}^k \to \mathbb{R}^k$ are projection maps.

E is called the *total space* of ξ and B is called the base of ξ.

If $\xi = (E_i, p_i, B_i),\ i = 1, 2$ are two vector bundles, a *morphism* $\xi_1 \to \xi_2$ *of vector bundles* consists of a pair $(f, \bar{f})$ of continuous functions such that the diagram

$$\begin{array}{ccc} E_1 & \xrightarrow{f} & E_2 \\ \pi_1 \downarrow & & \downarrow \pi_2 \\ B_1 & \xrightarrow{\bar{f}} & B_2 \end{array}$$

is commutative and such that $f|_{p_1^{-1}(b)}$ is $\mathbb{R}$-linear. If both f and $\bar{f}$ are homeomorphisms also, then we say $(f, \bar{f})$ is a vector bundle isomorphism. In this situation we say that the two bundles are *isomorphic*.

Often while dealing with vector bundles over a fixed base space B, we require a bundle morphism $(f, \bar{f}) : (E_1, p_1, B) \to (E_2, p_2, B)$ to be such that $\bar{f} = Id_B$.

Definition 5.5.4 Let $\xi = (E, p, B)$ be a vector bundle. By a *section of* ξ we mean a continuous (smooth) map $\sigma : B \to E$ such that $p \circ \sigma = Id_B$. A section σ is said to be nowhere zero, if $\sigma(b) \neq 0$ for each $b \in B$.

Remark 5.5.3

1. A simple example of a section is the zero section, that assigns to each $b \in B$ the 0-vector in $p^{-1}(b)$. (Use (LTC) to see that the zero section is continuous.)

2. The simplest example of a vector bundle of rank k over B is $B \times \mathbb{R}^k$. These are called *trivial vector bundles.* In fact any vector bundle isomorphic to a product bundle is called a trivial vector bundle. We shall denote this by $\Theta^k := B \times \mathbb{R}^k$, the base space of the bundle being understood by the context. It is easy to see that the trivial bundle has lots of sections. Indeed, if $\sigma : B \to B \times \mathbb{R}^k$ is a section then it is of the form,

$$\sigma(b) = (b, f(b))$$

   where $f : B \to \mathbb{R}^k$ is continuous. Thus, the set of sections of Θ^k is equal to $\mathcal{C}(B, \mathbb{R}^k)$.

3. Using the vector space structure of each fibre, given two sections, s_1 and s_2 of a vector bundle, we can add and scale them:

$$(s_1 + s_2)(x) = s_1(x) + s_2(x); \quad (rs_1)(x) = rs_1(x).$$

   Using local triviality, one can verify that if s_1, s_2 are continuous, then so are rs_1 and $s_1 + s_2$. Thus, the set of all sections of a vector bundle forms a vector space. Indeed, it forms a module over the ring $\mathcal{C}(X; \mathbb{R})$ of continuous real valued functions on X. Using partition of unity, we can produce lots of sections.

4. In all the above remarks, for a smooth bundle over a smooth manifold, you can replace "continuous" by "smooth" everywhere.

Example 5.5.1

1. A simple example of a nontrivial vector bundle is the infinite Möbius band M : Consider the quotient space of $\mathbb{R} \times \mathbb{R}$ by the equivalence relation $(t, s) \sim (t + 1, -s)$. The first projection gives rise to a map $p : M \to \mathbb{S}^1$, which we claim is a nontrivial real vector bundle of rank 1 over $\mathbb{S}^1$. It is easy to see that complement of the 0-section in the total space of this bundle is connected. Therefore, the bundle cannot be the trivial bundle $\mathbb{S}^1 \times \mathbb{R}$. Indeed, the total space of this bundle is not even homeomorphic to $\mathbb{S}^1 \times \mathbb{R}$ but to see that needs a little bit more topological arguments.

2. The tangent bundle $\tau(X) := (TX, p, X)$ of any smooth submanifold $X \in \mathbb{R}^N$ is a typical example of a vector bundle of rank n, where $n = dim\, X$. It satisfies the additional smoothness conditions, viz.,
   (i) both the total space and the base space are smooth manifolds;
   (ii) the projection map p is smooth and
   (iii) the homeomorphisms $\phi : p^{-1}(U) \to U \times \mathbb{R}^n$ are actually diffeomorphisms.
   Over the base space B which is a smooth manifold, a vector bundle that satisfies these additional smoothness conditions will be called a *smooth vector bundle.*

3. On a manifold X embedded in $\mathbb{R}^N$, we get another vector bundle, viz., the normal bundle, $\nu(X)$, that is also a smooth vector bundle (see section 6.1).

4. Let $B = \mathbb{P}^n$ be the n-dimensional real projective space. The canonical line bundle $\gamma_n^1 = (E, p, \mathbb{P}^n)$ is defined as follows: Recall that $\mathbb{P}^n$ can be defined as the quotient space of $\mathbb{S}^n$ by the antipodal action. Put

$$E = \{([x], \mathbf{v}) \in \mathbb{P}^n \times \mathbb{R}^{n+1} \ : \ \mathbf{v} = \lambda x\}.$$

   That is, over each point $[x] \in \mathbb{P}^n$ we are taking the entire line spanned by the vector $x \in \mathbb{S}^n$ in $\mathbb{R}^{n+1}$. Let $p : E \to \mathbb{P}^n$ be the projection to the first factor. The verification that this data forms a line bundle is easy.

Definition 5.5.5 By a continuous/smooth *vector field* on a smooth manifold X, we mean a continuous/smooth section of the tangent bundle. By a *parallelizable manifold,* we mean a smooth manifold X whose tangent bundle is trivial.

Remark 5.5.4 As before, a vector field on a parallelizable manifold X^n corresponds to a smooth map $X \to \mathbb{R}^n$. Indeed, a manifold is parallelizable iff there exists n smooth vector fields $\{\sigma_1, \ldots, \sigma_n\}$ such that for each $p \in X$, we have

$$\{\sigma_1(p), \ldots, \sigma_n(p)\}$$

is linearly independent in $T_p(X)$.

Exercise 5.5

1. Show that if $f : X \to Y$ is a diffeomorphism, then $Tf : TX \to TY$ is a diffeomorphism. Thus the tangent space is a diffeomorphic invariant of a manifold.

2. Show that a map $f : X \to Y$ is smooth, iff for every open set U of a Euclidean space and every smooth map $g : U \to X$ the composite $f \circ g$ is smooth.

3. Show that $f : X \to Y_1 \times Y_2, \ \ f = (f_1, f_2)$ is smooth iff each f_i is smooth.

4. If X and Y are smooth manifolds (without boundary), show that $X_1 \times X_2$ is a smooth manifold in such a way that the projection maps $\pi_i : X_1 \times X_2 \to X_i$ are smooth. Also show that $T(X \times Y)$ is diffeomorphic to $T(X) \times T(Y)$.

5. Show that the diagonal map $x \mapsto (x, x)$ is smooth for any smooth manifold X.

6. Show that if $f_i : X_i \to Y_i$ are smooth, then so is $f_1 \times f_2$.

5.6 Tangents as Operators

Let V be a finite dimensional vector space and V^* be the space of all linear functionals on V. Elements of V^*, by definition, operate on V. This role can be easily interchanged meaningfully as follows: Fix a vector $\mathbf{v} \in V$ and consider the assignment $f \mapsto f(\mathbf{v})$ for $f \in V^*$. This assignment defines a linear functional on V^*, that vanishes iff $\mathbf{v}$ is the zero vector. Thus, we may think of elements of V as operating on the vector space V^*. Using this idea, we shall now give an operator-theoretic interpretation of a tangent to a manifold at a point.

Given an abstract smooth manifold X, let $\mathcal{C}^\infty(X)$ denote the ring of all smooth real valued functions on X. (The ring structure on $\mathcal{C}^\infty(X)$ comes from that of $\mathbb{R}$ via pointwise addition and pointwise multiplication.) Fix a point $p \in X$ and define an equivalence relation $\sim_p$ on $\mathcal{C}^\infty(X)$ as follows: $f \sim_p g$ iff there exists a neighborhood U of p on which $f = g$. Check that this is actually an equivalence relation and denote the space of equivalence classes by $\mathcal{C}_p^\infty(X)$. Also, we shall denote by $[f]_p$ the equivalence class of f. Observe that the ring structure of $\mathcal{C}^\infty(X)$ quotients down to define a ring structure on $\mathcal{C}_p^\infty(X)$. This ring is called the *ring of germs of smooth functions* on X defined at p. The phrase "on X" in this nomenclature can further be justified, because, if we have a smooth function f defined in a neighborhood of p, using a bump function, we can extend it to a smooth function $\hat{f}$ on the whole of the manifold X and the germ $[\hat{f}]_p$ of $\hat{f}$ at p does not depend on the actual extension but only on $[f]_p$.

Let us now consider the situation when X is an open subset of $\mathbb{R}^n$. Fix a vector $\mathbf{v} \in T_p(X) = \mathbb{R}^n$. To each smooth function $f : X \to \mathbb{R}$, we have an assignment $f \mapsto Df_p(\mathbf{v})$, the directional derivative of f along $\mathbf{v}$. Let us denote this by $\mathbf{v}(f) = Df_p(\mathbf{v})$. Then
(i) $\mathbf{v}(\alpha f + \beta g) = \alpha\mathbf{v}(f) + \beta\mathbf{v}(g)$;
(ii) $\mathbf{v}(fg) = f(p)\mathbf{v}(g) + g(p)\mathbf{v}(f)$; and
(iii) If $f = g$ in a neighborhood of p then $\mathbf{v}(f) = \mathbf{v}(g)$.
All this clearly make sense, when X is any smooth manifold because of the chain rule: If $\phi : X \to Y$ is a diffeomorphism of open sets in $\mathbb{R}^n$, with $\phi(p) = q$, $D(\phi)_p(\mathbf{v}) = \mathbf{u}$ we have $\mathbf{v}(f) = \mathbf{u}(f \circ \phi)$. This proves the invariance of the properties (i), (ii), and (iii) under local coordinate changes. It allows us to adopt these properties as axioms for a tangent at a point for abstract manifolds as well. Property (i) is nothing but linearity. Property (ii) is called *derivation.* Property (iii) seems to have been added as an afterthought. This corresponds to the infinitesimal behavior of the derivative without which the entire machinery will be meaningless. To sum up, we have the following definition.

Definition 5.6.1 Let X be a manifold and $p \in X$. Let $D_p(X)$ be the space of all functions $\tau : \mathcal{C}^\infty(X) \to \mathbb{R}$ satisfying:
(A) $\tau(\alpha f + \beta g) = \alpha\tau(f) + \beta\tau(g),\ \alpha, \beta \in \mathbb{R}, f, g \in \mathcal{C}^\infty(X)$.
(B) $\tau(fg) = f(p)\tau(g) + g(p)\tau(f);\ f, g \in \mathcal{C}^\infty(X)$.
(C) If $f = g$ in a neighborhood of p then $\tau(f) = \tau(g)$.

Observe that $\tau_1, \tau_2 \in D_p(X), \alpha_1, \alpha_2 \in \mathbb{R}, \phi \in \mathcal{C}^\infty(X)$ then $\alpha_1\tau_1 + \alpha_2\tau_2 \in D_p(X)$ and $\phi\tau_j \in D_p(X)$. Verify that, in this way, $D_p(X)$ is a $\mathcal{C}^\infty(X)$-module. Property (C) implies that we can consider $\tau \in D_p(X)$ as a function $\mathcal{C}_p^\infty \to \mathbb{R}$ satisfying (A) and (B). It then follows that $D_p(X)$ is a $\mathcal{C}_p^\infty(X)$-module. Define $\Theta : T_p(X) \to D_p(X)$ by the formula:

$$\Theta(\mathbf{v})(f) = Df_p(\mathbf{v}). \tag{5.2}$$

Then Θ is a linear map.

Once again, by the chain rule, we obtain:

Theorem 5.6.1 *Θ is functorial in the sense that if $\phi : U \to \mathbb{R}^n$ is a diffeomorphism such that $\phi(p) = 0$, then we have the following commutative diagram:*

$$\begin{array}{ccc} T_p(X) & \xrightarrow{\Theta} & D_p(X) \\ {\scriptstyle d\phi_0}\downarrow & & \downarrow{\scriptstyle \phi_*} \\ T_0(\mathbb{R}^n) & \xrightarrow{\Theta} & D_0(\mathbb{R}^n) \end{array}$$

Theorem 5.6.2 *$\Theta : T_p(X) \to D_p(X)$ is an isomorphism of vector spaces.*

Proof: Because of the above naturality property, it is enough to prove this for $X = \mathbb{R}^n$ and $p = 0$. Now let $\mathbf{e}_i$ denote the standard basis from $T_0(\mathbb{R}^n) = \mathbb{R}^n$. Then $\Theta(\mathbf{e}_i)(f) = \frac{\partial f}{\partial x_i}(0)$, the partial derivative with respect to x_i. We shall denote $\Theta(\mathbf{e}_i)$ by ∂_i and prove that $\{\partial_1, \ldots, \partial_n\}$ is a basis for $D_0(\mathbb{R}^n)$.

Given a smooth function f on $\mathbb{R}^n$, from (1.30) we have the identity:

$$f(x) = f(0) + \sum_{i=1}^{n} x_i g_i(x)$$

where $g_i(0) = \partial_i(f)$. Let now $\tau \in D_0(\mathbb{R}^n)$. Using property (A) and (B) first verify that for any constant function c, we have $\tau(c) = 0$. Therefore,

$$\tau(f) = \sum_i \tau(x_i) g_i(0) = \sum_i \tau(x_i)\partial_i(f) = \left(\sum_i \tau(x_i)\partial_i\right)(f).$$

This proves that $\{\partial_1, \ldots, \partial_n\}$ generates $D_0(\mathbb{R}^n)$. To see that these vectors form an independent set, we simply evaluate them on the coordinate functions x_j's and see that $\left(\frac{\partial}{\partial x_i}\right)_p (x_j) = \delta_{ij}$. ♠

Remark 5.6.1 Thus, we can now identify the tangent space $T_p(X)$ with the space $D_p(X)$ of "local derivations" at p. Let us now consider global derivations.

Definition 5.6.2 By a *derivation on the ring* $\mathcal{C}^\infty(X)$, we mean a $\mathbb{R}$-linear mapping $\tau : \mathcal{C}^\infty(X) \to \mathcal{C}^\infty(X)$ such that

$$\tau(fg) = f\tau(g) + g\tau(f). \tag{5.3}$$

Remark 5.6.2

1. Let us denote by $D(X)$ the space of all derivations on X. Clearly the sum of any two elements of $D(X)$ is an element of $D(X)$ and we can multiply a derivation with a smooth function to obtain another. This makes $D(X)$ into a $\mathcal{C}^\infty(X)$-module.

2. Fix a point $p \in X$ and consider the map $E_p : D(X) \to D_p(X)$ defined as follows: Given a germ $[f]_p$ of a smooth function f defined in a neighborhood of p, consider some extension $\hat{f} : X \to \mathbb{R}$ of f and define

$$E_p(\tau)([f]_p) = [\tau(\hat{f})]_p.$$

In order to see that E_p is well-defined, we must check that if $f_1, f_2 : X \to \mathbb{R}$ are smooth functions such that $[f_1]_p = [f_2]_p$ then $[\tau f_1]_p = [\tau f_2]_p$. This is the same as showing that if $f \equiv 0$ in a neighborhood of p then $\tau(f) \equiv 0$ in a neighborhood of p.

To see this, choose a smooth function ρ on X such that $f = \rho f$ and $\rho(q) = 0$ in a neighborhood of p. (See Exercise 1.7.11.) Then

$$\tau(f) = \tau(\rho f) = \tau(\rho)f + \tau(f)\rho.$$

It follows that $\tau(f)$ vanishes in a neighborhood of p. Therefore, E_p is well-defined. Clearly for each τ, $E_p(\tau)$ satisfies (A) and (B). Any way we have verified (C) just now. Therefore, $E_p(\tau) \in D_p$. We can now onwards denote $E_p(\tau)$ by τ_p.

3. Thus, each element $\tau \in D(X)$ gives rise to an element of $D_p(X)$ for each $p \in X$. Is then an element $\tau \in D(X)$ just a collection of elements of $D_p(X)$ one for each $p \in X$? As in the case of a vector field, we may anticipate some "smoothness" condition on these operators, as p varies over X. To examine this, we can restrict ourselves to a coordinate neighborhood and work on an open subset U of $\mathbb{R}^n$. Now suppose $\tau \in D(U)$ so that $E_p(\tau) = \tau_p \in D_p(U)$ for each $p \in U$. Using the structure of $D_p(U)$ we can write

$$\tau_p(f) = \sum_{i=1}^{n} \sigma_i(p)\frac{\partial f}{\partial x_i}(p) \tag{5.4}$$

where $\sigma_i(p) \in \mathbb{R}$. Now, the fact that $\tau(f)$ is a smooth map for any smooth map f tells us that each of the functions σ_i must be smooth. Conversely, one can directly check that, given a smooth function $\sigma : U \to \mathbb{R}^n$, (5.4) gives a derivation τ on U.

4. Let $\mathcal{T}(X)$ denote the space of all smooth vector fields on X. This is also a $\mathcal{C}^\infty(X)$-module in an obvious fashion. The following theorem establishes a canonical isomorphism of these two modules.

Theorem 5.6.3 *The $\mathcal{C}^\infty(X)$-module $\mathcal{T}(X)$ of all smooth vector fields on X is isomorphic to the module $D(X)$ of all derivations on X.*

Proof: Once again, we define $\tilde{\Theta} : \mathcal{T}(X) \to D(X)$ as follows: For a smooth vector field σ on X, let $\tau_p = \tilde{\Theta}(\sigma)_p \in D_p(X)$ be defined by

$$\tau_p(f) = \Theta(\sigma(p))(f).$$

From the remark above, it follows that $f \mapsto \tau_p(f)$ is smooth. Hence the above definition makes sense. Clearly, $\tilde{\Theta}$ is a $\mathcal{C}^\infty(X)$-module homomorphism. We directly define the inverse of this as follows: Given $\tau \in D(X)$, define $\tau_p \in D_p(X)$ by the formula

$$\tau_p(f) = \tau(f)(p).$$

Clearly, τ_p satisfies (A) and (B) of Definition 5.6.1. Property (C) is checked as in the above remark. It follows that $\tau_p \in D_p(X)$. Hence, $\tau_p = \Theta(\mathbf{v})$ for a unique $\mathbf{v} \in T_p(X)$. From the remark above, it follows that the assignment $p \mapsto \Theta^{-1}(\tau_p)$ defines a smooth vector field σ. The verification that this gives the inverse map of $\tilde{\Theta}$ is left to the reader. ♠

Definition 5.6.3 Given $\tau, \tau' \in D(X) = \mathcal{T}(X)$ we define their *Lie bracket* $[\tau, \tau']$ by the formula:

$$[\tau, \tau'] = \tau \circ \tau' - \tau' \circ \tau.$$

Observe that, a composite of two endomorphisms is an endomorphism and hence $[\,,\,]$ is an endomorphism, being the difference of two endomorphisms. But, the composite of two derivations is not a derivation. However, the bracket $[\tau, \sigma]$ of two derivations is a derivation and you have to verify this.

Definition 5.6.4 Let $\phi : X \to Y$ be a smooth map. Let $\tau : X \to TX, \sigma : Y \to TY$ be any two smooth vector fields. We say τ is *ϕ-related to* σ and write $\tau \sim_\phi \sigma$ if

$$T\phi \circ \tau = \sigma \circ \phi. \tag{5.5}$$

Given σ, in general, there may not be any τ that is ϕ-related to σ and vice versa. However, if ϕ is a diffeomorphism, then the identity (5.5) defines τ in terms of σ and vice versa. We then say $\tau =: \phi^*(\sigma)$ is the *pullback* of σ or $\sigma = \phi_*(\tau)$ is the *pushout* of τ.

$$\begin{array}{ccc} TX & \xrightarrow{T\phi} & TY \\ \uparrow{\scriptstyle\tau} & & \uparrow{\scriptstyle\sigma} \\ X & \xrightarrow{\phi} & Y \end{array}$$

Definition 5.6.5 Similarly, $\alpha \in D(X), \beta \in D(Y)$ are said to be *ϕ-related* for a smooth map $\phi : X \to Y$ if for every $x \in X$ and a smooth map defined in a neighborhood of $f(x)$ in Y we have

$$\beta(f) = \alpha(f \circ \phi). \tag{5.6}$$

Equivalently, $\alpha \sim_\phi \beta$ if the following diagram

$$\begin{array}{ccc} \mathcal{C}^\infty(X) & \xrightarrow{\alpha} & \mathcal{C}^\infty(X) \\ \uparrow{\scriptstyle\phi^*} & & \uparrow{\scriptstyle\phi^*} \\ \mathcal{C}^\infty(Y) & \xrightarrow{\beta} & \mathcal{C}^\infty(Y) \end{array}$$

is commutative. Once again, if ϕ is a diffeomorphism, then (5.6) defines α in terms of β and vice versa.

Theorem 5.6.4 *The isomorphisms $\hat{\Theta}_X : \mathcal{T}(X) \to D(X)$, respect the relation $\sim_\phi$ for any smooth ϕ.*

Proof: Let $\phi : X \to Y$ be a smooth map, $\tau \in \mathcal{T}(X), \sigma \in \mathcal{T}(Y)$, be any. Put $\alpha = \hat{\Theta}(\tau), \beta = \hat{\Theta}(\sigma)$. Suppose $\tau \sim_\phi \sigma$. We have to show that $\alpha \sim_\phi \beta$. So, let $x \in X$ be any and f be a smooth function defined in a neighborhood of $y = \phi(x)$ in Y. We have to prove (5.6).

$$\begin{aligned} \beta(f) \circ \phi(y) &= \beta(f)(y) \\ &= \hat{\Theta}(\sigma)(f)(y) = \Theta(\sigma(y))(f) \\ &= Df_y(\sigma(y)) = Df_y(D\phi_x(\tau(x))) \\ &= D(f \circ \phi)_x(\tau(x)) = \Theta(\tau(x))(f \circ \phi) \\ &= \hat{\Theta}(\tau)(f \circ \phi)(x) = \alpha(f \circ \phi)(x) = \alpha(f \circ \phi). \end{aligned}$$

We leave the converse part to the reader as an exercise.

Corollary 5.6.1 For any smooth map $\phi : X \to Y$ of manifolds and respective vector fields, we have,

$$\tau_i \sim_\phi \sigma_i, i = 1, 2 \Longrightarrow [\tau_1, \tau_2] \sim_\phi [\sigma_1, \sigma_2].$$

Proof: We should think of τ_i and σ_i as elements of $D(X)$ and $D(Y)$ respectively. But then the hypothesis is equivalent to say that the following diagram is commutative:

$$\begin{array}{ccc} \mathcal{C}^\infty(X) & \xrightarrow{\tau_i} & \mathcal{C}^\infty(X) \\ {\scriptstyle\phi^*}\uparrow & & \uparrow{\scriptstyle\phi^*} \\ \mathcal{C}^\infty(Y) & \xrightarrow{\sigma_i} & \mathcal{C}^\infty(Y) \end{array}$$

Thus, if $\tau_i \sim_\phi \sigma_i$ then $\tau_1 \circ \tau_2 \sim_\phi \sigma_1 \circ \sigma_2$ and $\tau_2 \circ \tau_1 \sim_\phi \sigma_2 \circ \sigma_1$. By the linearity, $[\tau_1, \tau_2] = \tau_1 \circ \tau_2 - \tau_2 \circ \tau_1 \sim_\phi \sigma_1 \circ \sigma_2 - \sigma_2 \circ \sigma_1 = [\sigma_1, \sigma_2]$. ♠

Exercise 5.6

1. Let σ, τ be two vector fields on an open subset U of $\mathbb{R}^n$. Express $[\sigma, \tau]$ in terms of coordinate functions.

2. Show that $[\,,\,]$ has the following properties:
   (i) $[\,,\,]$ is antisymmetric, i.e., $[\mathbf{v}, \mathbf{u}] = -[\mathbf{u}, \mathbf{v}]$ for all $\mathbf{u}, \mathbf{v} \in \mathfrak{g}$.
   (ii) $[\,,\,]$ is bilinear, i.e., $[a_1\mathbf{u}_1 + a_2\mathbf{u}_2, \mathbf{v}] = a_1[\mathbf{u}_1, \mathbf{v}] + a_2[\mathbf{u}_2, \mathbf{v}]$.
   (iii) the Jacobi identity:

$$[[\tau_1, \tau_2], \tau_3] + [[\tau_2, \tau_3], \tau_1] + [[\tau_3, \tau_1], \tau_2] = 0 \tag{5.7}$$

   (A vector space with a binary operation like the above is called a *Lie algebra*. When X happens to be a Lie group, a particular subalgebra of $(\mathcal{T}(X), [\,,\,])$ becomes very important. (See Chapter 9 for more details.)

5.7 Whitney Embedding Theorems

In this section, we shall prove some of the celebrated results of Whitney about approximating functions by immersions and embeddings into Euclidean spaces. As a warm-up let us begin with:

Theorem 5.7.1 *For any compact smooth manifold X, there exists a smooth neat embedding $g : X \to \mathbf{H}^N$.*

Proof: First, consider the case when $\partial X = \emptyset$. At each point $x \in X$, choose a chart ϕ^x such that $\phi^x(x) = 0$. Put $U_x = (\phi^x)^{-1}(\mathbb{R}^n)$. Clearly $\{(\phi^x)^{-1}(\mathbb{D}^n), x \in X\}$ itself is an open cover of X. Since X is compact it follows that there exists a finite atlas $\{(U_i, \phi_i)\}_{1 \le i \le k}$ such that

$$U_i = \phi_i^{-1}(\mathbb{R}^n) \text{ and } \left\{\phi_i^{-1}(\mathbb{D}^n) \;:\; i = 1, 2, \ldots, k\right\} \text{ covers } X. \tag{5.8}$$

Let $\lambda : \mathbb{R}^n \to \mathbb{R}$ be a $\mathcal{C}^\infty$ bump function, such that $\lambda(\mathbb{D}^n) = \{1\}$ and $\lambda(\mathbb{R}^n \setminus 2\mathbb{D}^n) = \{0\}$. Define $\lambda_i : X \to \mathbb{R}$ by the formula

$$\lambda_i(x) = \begin{cases} \lambda \circ \phi_i(x), & x \in U_i \\ 0, & x \in X \setminus U_i. \end{cases}$$

Check that each λ_i is a smooth function on X. Now define $f_i : X \to \mathbb{R}^n$ by the formula:

$$f_i(x) = \begin{cases} \lambda_i(x)\phi_i(x), & x \in U_i \\ 0, & x \in X \setminus U_i. \end{cases}$$

Once again, observe that each f_i is smooth and restricted to $\phi_i^{-1}(\mathbb{D}^n)$, is an embedding. We put $g_i = (f_i, \lambda_i) : X \to \mathbb{R}^{n+1}$ and put $g = (g_1, \ldots, g_k) : X \to \mathbb{R}^{k(n+1)}$. It remains to show that g is an embedding. Clearly g is a smooth map. If $x \in \phi_i^{-1}(\mathbb{D}^n)$, then since f_i is immersive at x, it follows that g is immersive at x. Hence, g is an immersion. To see that g is injective, let us say $x \ne y \in X$. We may assume that $\lambda_i(y) = 1$ for some i. If $\lambda_i(x) = 1$, then both x, y belong to U_i and hence $\phi_i(x) \ne \phi_i(y)$ and hence $f_i(x) = \phi_i(x) \ne \phi_i(y) = f_i(y)$.

It follows that $g(x) \neq g(y)$. Now assume that $\lambda_i(x) \neq 1$. But then $g_i(x) \neq g_i(y)$ and hence $g(x) \neq g(y)$.

Since X is compact, the map g is closed. Therefore, g is a homeomorphism onto its image. By the injective form of the implicit function theorem, it follows that g^{-1} is also smooth. Therefore, g is an embedding. Put $\hat{g}(x) = (g(x), 1)$ and take $N = k(n+1)+1$. This takes care of the case $\partial X = \emptyset$.

In case $\partial X \neq \emptyset$, we can first find a smooth function $\alpha : X \to [0, \infty)$ such that $\alpha^{-1}(0) = \partial X$ as follows: Choose an atlas $\{(U_i, \phi_i)\}$, that is locally finite. Define $\alpha_i(x) = \pi_n \circ \phi_i$. Define $\alpha = \sum_i \alpha_i$. This makes sense and gives a smooth map because of local finiteness of U_i. Now $\alpha(x) = 0$ iff $\alpha_i(x) = 0$ for all i such that $x \in U_i$. This is equivalent to say that $x \in \partial X$.

Now consider the map $\bar{g}(x) = (g(x), \alpha(x))$, where g is constructed as before. It follows that $\bar{g}(\partial X) = g(X) \cap \mathbb{R}^{N-1} \times 0$ and $\bar{g}(X) \subset \mathbf{H}^N$. Therefore, $\bar{g} : (X, \partial X) \to (\mathbf{H}^N, \partial \mathbf{H}^N)$ is a neat embedding. ♠

Remark 5.7.1 The proof of the above theorem gives you a glimpse of certain techniques frequently used in differential topology: have a certain result locally and then use partition of unity to glue it up to obtain a global one. Here the end result is somewhat crude in the sense that it is not at all economical in the choice of the dimension of the ambient Euclidean space. Moreover it cannot be easily generalized to the case of noncompact manifolds. There is yet another related question: Given a continuous map $f : X \to \mathbb{R}^N$, can we find embeddings (immersions) $g : X \to \mathbb{R}^N$ very close to f? Whitney's idea is to address all these questions simultaneously. The key is to bring in a measure theoretic argument. We need to make a definition and recall some results from Section 2.2.

Definition 5.7.1 Let X be a smooth manifold of dimension n. A subset A of X is said to be of *measure-zero in* X if there exists an atlas $\{(U_i, \phi_i) \ : \ i \in \Lambda\}$ for X such that $\phi_i(U_i \cap A)$ is of measure zero in $\mathbb{R}^n$ for every $i \in \Lambda$.

Remark 5.7.2 By Theorem 2.2.2, it follows that the above definition is independent of the atlas. Check that this concept has the usual properties of measure zero sets in the Euclidean spaces. Also verify that for any manifold X, $X \times 0$ is of measure zero in $X \times \mathbb{R}$. As an easy consequence, we deduce the following:

Theorem 5.7.2 Mini Sard's Theorem: *Let $f : X \to Y$ be a smooth map of manifolds with $\dim X < \dim Y$. Then $f(X)$ is of measure zero in Y.*

Proof: First, consider the case when X and Y are open subsets of $\mathbb{R}^n, \mathbb{R}^m$ respectively, with $n < m$. We can then treat f as defined on an open set U in $\mathbb{R}^m$ and identify X with $U \cap \mathbb{R}^n \times (0, 0, \ldots, 0)$, that is of measure zero in $\mathbb{R}^m$. Therefore by Theorem 2.2.2, it follows that $f(X)$ is of measure zero in Y. In the general case, using coordinate charts on either side, $f(X)$ can be covered by countably many measure zero sets. Hence, the conclusion. ♠

Remark 5.7.3 In any case, this is an easy consequence of the result that we prove next. The idea of giving this proof is that, for our purpose in this section, we do not need the stronger form of Sard's theorem and the above mini Sard's theorem suffices. However, we shall use the stronger form later. We begin by recalling an important result from Chapter 2 for ready reference. (See Theorem 2.2.1).

Theorem 5.7.3 *Let $f : U \to \mathbb{R}^m$ be a C^∞ function. Then the set of critical values of f is of Lebesgue measure zero in $\mathbb{R}^m$.*

Remark 5.7.4 Recall that $x \in U$ is called a *critical point* of f if Df_x is not surjective. A point $y \in \mathbb{R}^m$ is called a *critical value* of f if $f^{-1}(y)$ consists of at least one critical point of f. Notice that there is no difficulty in extending the definition of critical points, etc., for maps between manifolds, since being a critical point is invariant under diffeomorphisms. We can now extend the above result to manifolds.

Theorem 5.7.4 Morse-Sard Theorem: *Let X, Y be any smooth manifolds, and $f : X \to Y$ be a $\mathcal{C}^\infty$ map. Then the image $f(C_f)$ of the set of critical points of f is of measure zero in Y.*

Proof: We choose an atlas for $\{(U_i, \phi_i,)\}$ for Y. To show that $f(C_f) \subset Y$ is of measure-zero is the same as to show that $\phi_i(f(C_f) \cap U_i)$ is of measure zero in $\mathbb{R}^n$, where n is the dimension of Y. Replacing X by $f^{-1}(U_i)$ and f by $\phi_i \circ f$, we have reduced the task to the case of a smooth map $f : X \to \mathbb{R}^n$. We now cover X by a countable family of parameterized neighborhoods $\{(V_j, \psi_j)\}$. It then suffices to prove that $f(V_j \cap C_f))$ is of measure zero for each j. Put $g_j := f \circ \psi_j : \mathbb{R}^m \to \mathbb{R}^n$. Then the critical set $C_f \cap V_j$ of f is precisely equal to $\psi_j(C_{g_j})$. Therefore, $f(V_j \cap C_f) = g_j(C_{g_j})$, the image of the critical set of $g_j : \mathbb{R}^m \to \mathbb{R}^n, m < n$. By Theorem 5.7.3, each $g_j(C_{g_j})$ is of measure. This completes the proof of the theorem. ♠

We can now move toward Whitney's results, with a "local extendability lemma".

Lemma 5.7.1 Let $m \geq 2n$. Let $\alpha, \beta : \mathbb{R}^n \to \mathbb{R}^m$ be two smooth maps and $C \subset \mathbb{R}^n$ be a compact subset such that
(a) $\alpha \equiv \beta$ on $\mathbb{R}^n \setminus 2\mathbb{D}^n$;
(b) $\beta|_C$ is an immersion;
(c) $\sup \|\alpha(x) - \beta(x)\| < \epsilon$, for some $\epsilon > 0$.
Then there exists a smooth map $\gamma : \mathbb{R}^n \to \mathbb{R}^m$ satisfying:
(i) $\gamma|_{\mathbb{R}^n \setminus 2\mathbb{D}^n} = \alpha$.
(ii) $\sup \|\gamma(x) - \alpha(x)\| < \epsilon$.
(iii) γ is an immersion on $C \cup \mathbb{D}^n$.

Proof: Choose a smooth bump function $\phi : \mathbb{R}^n \to [0, 1]$ such that $\phi \equiv 1$ on $\mathbb{D}^n$ and $\operatorname{supp} \phi \subset 2\mathbb{D}^n$. For $L \in \mathbb{R}^{m \times n}$, put

$$\gamma_L(x) = \beta(x) + \phi(x)L(x).$$

We shall show that there are "plenty" of choices of L for which $\gamma = \gamma_L$ will satisfy our requirements. To begin with note that (i) is satisfied for all L.
Step 1 Since C is compact, there exists $\epsilon_0 > 0$ such that

$$\sup_{x \in C} \|\alpha(x) - \beta(x)\| < \epsilon_0 < \epsilon.$$

Then for all $L \in \mathbb{R}^{m \times n}$ with $\|L\| < (\epsilon - \epsilon_0)/2$, we have

$$\sup \|\gamma_L(x) - \alpha(x)\| \leq \sup \|\gamma_L(x) - \beta(x)\| + \sup \|\beta(x) - \alpha(x)\| < \epsilon.$$

This ensures condition (ii).
Step 2 Clearly the map $\lambda : \mathbb{R}^n \times \mathbb{R}^{m \times n} \to \mathbb{R}^{m \times n}$ given by

$$(x, L) \mapsto D\gamma_L(x)$$

is continuous and $\lambda(C \times \{0\})$ is contained in the open set of all linear maps L of rank n. Therefore, there exists $\delta > 0$ such that for all $\|L\| < \delta$ and for all $x \in C$, we have $D\gamma_L(x)$ is of rank n.

Step 3 We want to choose L in such a way that for all $x \in \mathbb{D}^n$, $D\gamma_L(x)$ is of rank n. Since $\phi \equiv 1$ on $\mathbb{D}^n$, we have

$$D\gamma_L(x) = D\beta(x) + L$$

Therefore, we do not want L to be of the form $\ell - D\beta(x)$ where ℓ is of rank $< n$.

Recall that $\mathcal{R}_k(m, n; \mathbb{R})$ of rank k matrices is a smooth submanifold of codimension $(m-k)(n-k)$ in $\mathbb{R}^{m \times n}$. (See Example 3.4.4.) Since $(m-k)(n-k) \geq m - n + 1$, it follows that the dimension of $\mathbb{R}^n \times \mathcal{R}_k(m, n; \mathbb{R})$ is less than $mn + 2n - m - 1$ for all $k < n$. Since $m \geq 2n$, by mini Sard's Theorem 5.7.2, the mapping $\tau : \mathbb{R}^n \times \mathcal{R}_k(m, n; \mathbb{R}) \to \mathbb{R}^{m \times n}$ given by $(x, \ell) \mapsto \ell - D\beta(x)$ has its image of measure zero in $\mathbb{R}^{m \times n}$.

Thus, if we choose $L \in \mathbb{R}^{m \times n} \setminus \tau\left(\cup_{k<n} \mathbb{R}^n \times \mathcal{R}_k(m, n; \mathbb{R})\right)$ such that $\|L\| < \min\{\delta, \frac{\epsilon - \epsilon_0}{2}\}$, we are through. ♠

Theorem 5.7.5 *Let X be a smooth manifold of dimension n, $\partial X = \emptyset$, and let $f : X \to \mathbb{R}^m$ be a continuous function, $m \geq 2n$. Suppose that $K \subset X$ is a compact set such that $f|_K$ is smooth and is an immersion on an open set containing K. Given $\epsilon > 0$ there exist a smooth function $G : X \to \mathbb{R}^m$ such that*
(a) $G|_K = f$;
(b) $\sup \|G(x) - f(x)\| < \epsilon$;
(c) *G is immersive on X.*

Proof: Note that the approximate Theorem 1.7.2 though stated for continuous functions on open subsets of $\mathbb{R}^n$ is valid for continuous functions on any manifold. (See remark 5.2.2.) Even the proof is valid verbatim. Thus, there exists a smooth map $\hat{f} : X \to \mathbb{R}^m$ such that $\sup \|f(x) - \hat{f}(x)\| < \epsilon/2$ and $\hat{f} = f$ on a compact neighborhood of K. So, if we prove (a),(b),(c) with f replaced by $\hat{f}$ and ϵ replaced by $\epsilon/2$ then we are through. It follows that we may as well assume that f itself is smooth, which we shall do.

Choose a locally finite, countable atlas $\{(U_i, \phi_i)\}$ such that $X \setminus K = \cup_i \phi_i^{-1}(\mathbb{D})$. Put $X_k = K \cup_{i \leq k} \phi_i^{-1}(\mathbb{D}^n)$. Then clearly each X_k is campact and $X = \cup_k X_k$. We shall first construct a family of smooth functions $g_i : X \to \mathbb{R}^m$ such that
(i) $g_1|_K = f$; $g_{k+1}|_{X_k} = g_k, k \geq 1$;
(ii) g_k is immersive on X_k.
(iii) $\sup_{x \in X_k} \|f(x) - g_k(x)\| \leq \epsilon$.

We do this by induction, using Lemma 5.7.1. Put $\alpha = \beta = f \circ \phi_1^{-1} : \mathbb{R}^n \to \mathbb{R}^m$, and $C = \emptyset$. Let γ be the map obtained in the lemma and put $g_1 = \gamma \circ \phi_1$ on U_1 and $= f$ on $X \setminus \phi_1^{-1}(2\mathbb{D}^n)$. It is straightforward to check that g_1 satisfies (i), (ii), and (iii).

Inductively, suppose we have defined g_k. Put $\alpha = f \circ \phi_{k+1}^{-1}, \beta = \mathfrak{g}_k \circ \phi_{k+1}^{-1}$ and $C = \phi_{k+1}(X_k \cap \phi_{k+1}^{-1}(2\mathbb{D}^n))$. Apply Lemma 5.7.1 to get γ and put $g_{k+1} = \gamma \circ \phi_{k+1}$ on U_{k+1} and $= g_k$ on $X \setminus \phi_{k+1}^{-1}(2\mathbb{D}^n)$. Once again verifying the conditions (i), (ii), and (iii) is just routine.

Given $x \in X = \cup X_k$ we choose k such that $x \in X_k$. We then define $G(x) = g_k(x)$ whenever $x \in X_k$. By property (i) if $x \in X_l, l > k$ then it follows that $g_k(x) = g_l(x)$ and hence G is well-defined. Also $G(x) = g_1(x) = f(x)$ on K. Observe that since $\{U_i\}$ is locally finite, given any $x \in X$ there exist a neighborhood W of x and r such that $U_i \cap W = \emptyset$ for $i > r$. This then implies that $f_r(x) = f_{r+1}(x) = \cdots = G(x)$ on W. Hence, G is smooth. This same property also allows us to conclude that $\|G(x) - f(x)\| = \|f_r(x) - f(x)\| < \epsilon$. Therefore, (b) is satisfied. Condition (c) follows from (ii) and the fact that $X = \cup_k X_k$. ♠

Theorem 5.7.6 *Let $m \geq 2n + 1$. Let X be a smooth manifold of dimension n and $f : X \to \mathbb{R}^m$ be a continuous function. Suppose K is a compact subset of X on which f is an embedding. Then there exists an injective immersion $g : X \to \mathbb{R}^m$ such that $g|_K = f$ and $\sup \|g(x) - f(x)\| < \epsilon$. Moreover if f is a proper mapping, then g is an embedding.*

Proof: By the previous theorem, there exists an immersion $\hat{f} : X \to \mathbb{R}^m$ such that $\hat{f}|_K = f$ and $\sup \|f(x) - \hat{f}(x)\| < \epsilon_2$. Therefore, without loss of generality we may as well assume that f itself is an immersion. This means that f is locally injective. Therefore, there exists an atlas $\{(U_i, \phi_i)\}$ for $X \setminus K$, that is locally finite, countable, $X = K \cup (\cup_i V_i)$ where $V_i = \phi_i^{-1}(\mathbb{D}^n)$, and such that on each $W_i := \phi_i^{-1}(2\mathbb{D}^n)$, f is injective. Put $X_0 = K$ and $X_{k+1} = X_k \cup \overline{V}_{k+1}$ so that each X_k is compact and $X = \cup_k X_k$. We shall construct a family $g_k : X \to \mathbb{R}^m$ of smooth functions such that
(i) $g_0 = f; g_{k+1} = g_k$ on $X \setminus W_{k+1}$.
(ii) $\sup \|f(x) - g(x)\| < \epsilon$.
(iii) g_k is an embedding on X_k.
(iv) g_k is injective on each W_i.
We shall then put $g(x) = \lim_{k\to\infty} g_k(x)$. It then easily follows that g is as required. To see the last part, we have only to observe that if f is proper then g is also proper because of the property $\sup \|f(x) - g(x)\| < \epsilon$.

Taking $g_1 = f$, inductively, suppose we have defined g_k as required. Choose a bump function $\psi : X \to [0, 1]$ such that $\operatorname{supp} \psi = \overline{W}_{k+1}$ and $\psi \equiv 1$ on $\overline{V}_{k+1}$. For any $\mathbf{v} \in \mathbb{R}^m$, put

$$g_{k+1}(x) = \begin{cases} g_k(x), & x \in X \setminus \overline{W}_{k+1}; \\ g_k(x) + \psi(x)\mathbf{v}, & x \in U_{k+1}. \end{cases}$$

The idea is to choose $\mathbf{v} \in \mathbb{R}^m$ suitably, so that g_{k+1} will become an embedding on X_{k+1}. By continuity of the map $(x, \mathbf{v}) \mapsto Dg_{k+1}(x)$ and the fact that it is an immersion on the compact set X_k, as in the proof of the previous theorem, we can ensure that g_{k+1} is an immersion and $\sup \|g_{k+1}(x) - f(x)\| < \epsilon$ provided $\|\mathbf{v}\|$ is sufficiently small.

We now want to ensure that g_{k+1} is injective on X_{k+1} as well as on each W_i. Consider the open set

$$\Omega = \{(x, y) \in X \times Y \ : \ \psi(x) \neq \psi(y)\}$$

and the smooth map $\lambda : U \to \mathbb{R}^m$ given by

$$\lambda(x, y) = -\frac{g_k(x) - g_k(y)}{\psi(x) - \psi(y)}.$$

Since the dimension U is $2n < m$ it follows that image of λ is of measure zero in $\mathbb{R}^m$. Hence, we can choose $\mathbf{v}$ of arbitrary small length and not in the image of λ. With such a choice of $\mathbf{v}$, we shall now claim that g_{k+1} is as required.

Suppose $x \neq y$ and $g_{k+1}(x) = g_{k+1}(y)$. If $\psi(x) \neq \psi(y)$ then it follows that $\mathbf{v} = \lambda(x, y)$, which is a contradiction. Therefore, $\psi(x) = \psi(y)$. This implies that $g_k(x) = g_k(y)$. Therefore, x, y do not belong to same W_i nor both belong to X_k. It remains to see why both of them do not belong to X_{k+1}. If so, one of them say x must be in $X_{k+1} \setminus X_k \subset \overline{U}_k$. But this implies $\psi(x) = 1$ and therefore $\psi(y) = 1$. But then both are in W_{k+1}, which is a contradiction. ♠

Remark 5.7.5

1. Theorem 5.7.6 immediately tells you that all compact n-manifolds can be embedded in $\mathbb{R}^{2n+1}$. For noncompact manifolds, the missing link is that we have to find just one smooth proper map $f : X \to \mathbb{R}^{2n+1}$. Note that if $\alpha : X \to \mathbb{R}$ is a proper map then so is the map $x \mapsto (\alpha(x), 0, 0 \ldots, 0)$. If $\{\theta_i\}$ is a countable, smooth partition of unity, we can easily verify that $\alpha(x) = \sum_j j\theta_j(x)$ is a smooth proper map. Thus, we conclude that every n-manifold can be embedded in $\mathbb{R}^{2n+1}$.

2. In all these discussions we have assumed that X is a manifold without boundary just

for simplicity. All the arguments go through even if $\partial X \neq \emptyset$. The only difference is that some of the members of any atlas we take may have their range the upper-half space $\mathbf{H}^n$ instead of $\mathbb{R}^n$. As in the last part of the proof of Theorem 5.7.1, we can obtain a neat embedding of $\eta : X \to \mathbf{H}^{2n+2}$. Finally, using measure theoretic arguments as in the above proof, we can find a vector $\mathbf{u} \in \mathbb{S}^{2n+1}$ so that if $\tau : \mathbf{H}^{2n+2} \to \mathbf{H}^{2n+1}$ is the projection along $\mathbf{u}$ then $\tau \circ f : X \to \mathbf{H}^{2n+1}$ is a neat embedding.

3. Using the heavy tools from algebraic topology, Whitney has proved that every n-manifold can be embedded in $\mathbb{R}^{2n}$ for $n > 0$ and can be immersed in $\mathbb{R}^{2n-1}$, $n > 1$. But the approximate version of this result is false in general. The simplest example is obtained by the figure eight curve:

$$\gamma(\cos t, \sin t) = (\cos t, \sin 2t), \quad 0 \leq t \leq 2\pi \tag{5.9}$$

defines an immersion of $\mathbb{S}^1$ in $\mathbb{R}^2$. If $\epsilon < 1/2$, then it is not very difficult to see that no ϵ-approximation to γ can be an embedding. (Exercise: Write down details.)

4. It is possible to avoid/post-pone the definition of tangent space for an abstract smooth manifold as done in section 5, if needed, due to time constraint or otherwise. For this, one has to define immersions in terms of local parameterizations and then directly prove the embedding theorems of the last section. After that, we can of course take the geometric tangent space as defined in chapter 3. However, this is not recommended in the long run. The alternative definition treated in section 6 is used only in the last chapter in this book and is a must from the geometric viewpoint.

5.8 Miscellaneous Exercises for Chapter 5

Throughout this set of exercises, X, Y etc. denote smooth manifolds.

1. Let $S(X)$ denote the set of all points $(x, v) \in T(X)$ such that $\|v\| = 1$ where $X \subset \mathbb{R}^N$. Show that $S(X)$ is a manifold and compute its dimension. ($S(X)$ is called the *unit tangent bundle of* X.)

2. Consider the space $\mathbb{R}^{n+1} \setminus \{0\}$. Define an equivalence relation on this set by saying

$$(x_0, \ldots, x_n) \sim (y_1, \ldots, y_n)$$

iff $(x_0, \ldots, x_n) = r(y_1, \ldots, y_n)$ for a nonzero real number r. Let Y denote the quotient space.
(i) Show that Y is the same as $\mathbb{P}^n$.
(ii) Let $x = (x_0, x_2, \ldots, x_n)$ and $U_i = \{[x] \in \mathbb{P}^n \ : \ x_i \neq 0\}$. Show that each $U_i, i = 0, \ldots, n$ is an open subset of $\mathbb{P}^n$ homeomorphic to $\mathbb{R}^n$ and $\mathbb{P}^n = \cup_i U_i$.
(iii) Define $d([x],[y]) = cos^{-1}\left(\dfrac{\langle x, y\rangle}{\|x\|\|y\|}\right)$, $[x],[y] \in \mathbb{P}^n$. Verify that d is a metric on $\mathbb{P}^n$ inducing quotient topology from $\mathbb{S}^n$.

3. The two-sphere $\mathbb{S}^2$, the projective space $\mathbb{P}^2$, the torus $\mathbb{S}^1 \times \mathbb{S}^1$, the Klein's bottle, etc., can all be obtained by the above construction in Lemma 5.3.1, where U, V are diffeomorphic to $\mathbb{S}^1 \times (0,1)$. In each case, think of the correct choice for M and the diffeomorphism $\phi : U \to V$. Indeed all oriented surfaces can be obtained by this construction, taking M as the disjoint union of two copies of a disc with a number of holes and U, V as the unions of small boundary annuli. Learn the details!

4. Show that if $f : X \to Y$ is a diffeomorphism, then $Tf : TX \to TY$ is a diffeomorphism. Thus, the tangent space is a diffeomorphic invariant of a manifold.

5. Show that a map $f : X \to Y$ is smooth, iff for every open set U of a Euclidean space and every smooth map $g : U \to X$ the composite $f \circ g$ is smooth.

6. Verify that all the results proved in Section 2.4 are valid for abstract manifolds also.

7. **Whitney's Hill** There exists a $\mathcal{C}^1$ map $f : \mathbb{R}^2 \to \mathbb{R}$ and a "path" γ in $\mathbb{R}^2$ such that γ is contained in the critical set of f and $f|_\gamma$ is not a constant (see [W]). The graph of f is called *Whitney's hill.* The graph of γ then defines a path on this hill that is "flat" at every point and yet keeps going up and down. Can this path be smooth?

8. Show that every complex manifold is orientable and has a canonical orientation.

9. **Orientation Double Cover:** For any linear isomorphism $\alpha : \mathbb{R}^n \to \mathbb{R}^n$, define $sgn(\alpha)$ to be ± 1 according as α preserves (or reverses) orientation. Let X be a connected manifold. Choose a family $\psi_j : \mathbb{R}^n \to X, j \in \Lambda$ of local parameterizations that cover the whole of X. (ψ_j may be the inverses of functions belonging to an atlas.) On the space $Z = \mathbb{R}^n \times \Lambda \times \{-1, 1\}$, introduce an equivalence relation as follows:

$$(x, j, \epsilon_1) \sim (y, k, \epsilon_2) \text{ iff } \psi_j(x) = \psi_k(y) \text{ and } sgn(d(\psi_k^{-1} \circ \psi_j)_x) = \epsilon_1\epsilon_2.$$

Denote the quotient space by $\tilde{X}$. Prove that
(i) the assignment $(x, j, \epsilon) \mapsto \psi_j(x)$ defines a double covering $p : \tilde{X} \to X$ and hence $\tilde{X}$ is a smooth manifold of the same dimension as X;
(ii) $\tilde{X}$ is orientable;
(iii) $\tilde{X}$ has one or two connected components according as X is not orientable or orientable.
(iv) The map $(x, j, \epsilon) \mapsto (x, j, -\epsilon)$ defines a diffeomorphism $\tau : \tilde{X} \to \tilde{X}$ such that $p \circ \tau = p$, and $\tau \circ \tau = Id$.
(v) τ is orientation preserving iff X is oriented.

10. **Grassmannian Manifolds:** Let $G_{k,n}$ denote the set of all k-dimensional vector subspaces of $\mathbb{R}^n$ where $1 \le k \le n-1$. (The cases $k = 0, n$ are disinteresting.) Consider the space $M(n,k) := M(n,k;\mathbb{R})$ of all $n \times k$ real matrices. Let $U(n,k) =: U$ be the open set of $M(n,k)$ consisting of those matrices of maximal rank (=k). Let $C_{k,n}$ denote the set of all k-subsets of $\{1, 2, \dots, n\}$. Finally, let $\Theta : M(n,k) \to G_{k,n}$ be defined by $A \mapsto$ column space of A. We give the quotient topology to $G_{k,n}$ via Θ, i.e., a subset V of $G_{k,n}$ is open iff $\Theta^{-1}(V)$ is open in U.
(i) Show that $\Theta : U \to G_{k,n}$ is surjective.
(ii) For each $A \in C_{k,n}$, let U_A be the subset of those elements of U such that the rows corresponding to the indices in A are independent. Show that $\{U_A \ : \ A \in C_{k,n}\}$ forms an open cover for U.
(iii) The family $\{\Theta(U_A) \ : \ A \in C_{k,n}\}$ forms an open cover for $G_{k,n}$.
(iv) For each $A \in C_{k,n}$, let $\eta = \eta(A)$ denote the strictly increasing function $\{1, \dots, n-k\} \to \{1, 2, \dots, n\} \setminus A$. Given $L \in M(n-k, k)$, let L_i denote its ith row. Let $L' := \phi_A(L) \in U_A$ denote the element whose j^{th} row L'_j is defined as follows: if $j = \eta(i)$ then $L'_j = L_i$. Otherwise $j = q_i$ for some i, where $A = \{q_1 < q_2 < \cdots < q_k\}$ in which case, we take $L'_j = (0, 0, \dots, 1, 0, \dots, 0)$ with 1 exactly in the ith place. Show that $\phi_A : M(n-k, k) \to U_A$ is an embedding and $\psi_A := \Theta \circ \phi_A : M(n-k, k) \to \Theta(U_A)$ is a homeomorphism.
(v) Verify that $\{(\Theta(U_A), \psi_A^{-1}) \ : \ A \in C_{n,k}\}$ forms a smooth atlas for $G_{n,k}$ making it

into a smooth manifold of dimension $k(n-k)$. It is called the *Grassmannian manifold of type* (k,n).
(vi) Show that $\Theta : U \to G_{n,k}$ is a smooth map.

11. Consider the Steifel manifold $V_{k,n} \subset U$ as given in Example 3.4.5.
(i) Show that $\Theta(V_{k,n}) = G_{n,k}$ and hence $G_{k,n}$ is compact.
(ii) Given an element $F = (\mathbf{v}_1, \ldots, \mathbf{v}_k) \in V_{k,n}$, complete it to a basis $\{\mathbf{v}_1, \ldots, \mathbf{v}_k, \mathbf{u}_1, \ldots, \mathbf{u}_{n-k}\}$ of $\mathbb{R}^n$. Show that the set U of all $F' = (\mathbf{v}_1; \ldots, \mathbf{v}'_k) \in V_{k,n}$ such that $\{\mathbf{v}'_1 \ldots, \mathbf{v}'_k, \mathbf{u}_1, \ldots, \mathbf{u}_{n-k}\}$ is a basis for $\mathbb{R}^n$ is an open subset of $V_{k,n}$ containing F.
(iii) Apply the Gram-Schmidt process to the sequence $\{\mathbf{v}'_1, \ldots, \mathbf{v}'_k, \mathbf{u}_1, \ldots, \mathbf{u}_n\}$ to obtain the orthonormal basis

$$\{\mathbf{v}'_1, \ldots, \mathbf{v}'_k, \mathbf{v}'_{k+1}, \ldots, \mathbf{v}'_n\}.$$

Define a mapping $\hat{S} : U \to V_{n-k,n}$ by the formula

$$\hat{S}(F') = \{\mathbf{v}'_{k+1}, \ldots, \mathbf{v}'_n\}.$$

Show that $\hat{S}$ is smooth.
(iv) Consider the mapping $S : G_{k,n} \to G_{n-k,n}$, which maps any k-dimensional subspace to its orthogonal complement. Show that S is a diffeomorphism. S is canonical in the sense that it does not depend upon the choice of the basis. (However, it does depend upon the choice of the metric.)
(v) Observe that $G_{1,n} = \mathbb{P}^{n-1} = S(G_{n-1,n})$.

12. Consider the space S of all symmetric $n \times n$ real matrices A of rank k such that $A^2 = A$. Define $\eta : S \to G_{k,n}$ mapping A to the column space of A. Show that η is a diffeomorphism. Hence, we get an embedding of $G_{k,n}$ in $Sym(n) \cong \mathbb{R}^{n(n+1)/2}$.

13. Let M be a smooth n-submanifold of some $\mathbb{R}^N$. Then there is a canonical map, called the *Gauss map*, of M to $G_{n,N}$ sending $p \in M$ to its tangent space $g(p)$.
(i) Show that g is smooth.
This map is clearly of great geometrical interest; its nondegeneracy is obviously related to how *curved* M is in $\mathbb{R}^N$. (See the Gauss-Bonnet Theorem in Chapter 7.)
(ii) Gauss map could be defined by taking the normal spaces instead of the tangent spaces, and use this to describe the Gauss map of $\mathbb{S}^n \subset \mathbb{R}^{n+1}$.

14. Show that $G_{k,n}$ is connected.

Chapter 6

Isotopy

In this chapter, we introduce a "nonsingular" version of homotopy, viz., *isotopy.* In Section 6.1, we discuss the normal bundle and tubular neighborhoods laying down the foundation for homotopical aspects of manifolds. In particular, we prove the existence of "collar neighborhoods" for the boundary a manifold. In Section 6.2, we shall show how vector fields help us to construct isotopies. Isotopies together with collar neighborhoods form an essential part of the tool-kit in differential topology. In section 6.3, we obtain a few ready-to-use results which go a long way in building up new manifolds out of the old. In Section 6.4, we shall see a little bit of 'smoothing theory'.

6.1 Normal Bundle and Tubular Neighborhoods

In this section, we shall assume that all manifolds are submanifolds of some Euclidean space and use the Euclidean metric. Most of these concepts could be generalized to the situation while dealing with manifolds endowed with a "Riemannian metric", which of course, we are not discussing here.

Definition 6.1.1 Let $X \subset \mathbb{R}^N$ be a smooth manifold. For each point $x \in X$ define the space of normals to X at x to be

$$N_x(X) = \{\mathbf{v} \in \mathbb{R}^N \ : \ \mathbf{v} \perp T_xX\}.$$

The *total normal space* $N(X)$ of X in $\mathbb{R}^N$ is defined by:

$$N(X) := \{(x, \mathbf{v}) \in X \times \mathbb{R}^N \ : \ \mathbf{v} \perp T_xX\}.$$

We denote the restriction of the first projection $X \times \mathbb{R}^N \to X$ to $N(X)$ by $\pi : N(X) \to X$. Clearly it is a smooth map. $N(X)$ together with the map π is called the *normal bundle* of X in $\mathbb{R}^N$.

Theorem 6.1.1 *Given any submanifold $X \subset \mathbb{R}^N$, $N(X)$ is a submanifold of $\mathbb{R}^N \times \mathbb{R}^N$; the dimension of $N(X)$ is equal to N. For each point $(x, \mathbf{v}) \in N(X)$, we have $T_{(x,\mathbf{v})}(N(X)) = T_x(X) \times N_x(X)$. Moreover, the projection map $\pi : N(X) \to X$ is a submersion.*

Proof: Given $x \in X$ choose a diffeomorphism $\tau : \mathbb{R}^N \to V$, where V is a neighborhood of x in $\mathbb{R}^N$ such that $\tau(\mathbb{R}^n \times 0) = V \cap X = U$. We shall show that $\pi^{-1}(U)$ is diffeomorphic to $\mathbb{R}^N$. As x varies over X, this will give an atlas for $N(X)$.

Let $\{\mathbf{e}_1, \ldots, \mathbf{e}_N\}$ be the standard basis for $\mathbb{R}^N$. For $t \in \mathbb{R}^n$, put $\mathbf{v}_i(t) = (d\tau)_{(t,0)}(\mathbf{e}_i)$. Apply Gram-Schmidt's process to $\{\mathbf{v}_1(t), \ldots, \mathbf{v}_N(t)\}$ to obtain an orthonormal basis

$\{\mathbf{u}_1(t), \ldots, \mathbf{u}_N(t)\}$ for $\mathbb{R}^N$. Note that if $t \in \mathbb{R}^n, x = \tau(t,0) \in X$, then $\{\mathbf{v}_1(t), \ldots, \mathbf{v}_n(t)\}$ is a basis for T_xX. It follows that $\{\mathbf{u}_1(t), \ldots, \mathbf{u}_n(t)\}$ is an orthonormal basis for $T_x(X)$ and $\{\mathbf{u}_{n+1}(t), \ldots, \mathbf{u}_N(t)\}$ is an orthonormal basis $N_x(X)$. Since Gram-Schmidt's process is a smooth operation, it follows that $\mathbf{u}_i(t)$ are smooth functions of $t \in \mathbb{R}^n$. For $(t,s) \in \mathbb{R}^n \times \mathbb{R}^{N-n}$, we define

$$\Theta(t,s) = (\tau(t,0), \sum_{j=1}^{N-n} s_j \mathbf{u}_{j+n}(t)).$$

It is straight forward to verify that Θ is a diffeomorphism of $\mathbb{R}^N$ onto $\pi^{-1}(U) = N(U) \subset N(X)$. This proves the first part of the theorem.

Also, it follows that $T_{(x,\mathbf{v})}(N(X)) = d(\Theta)_{(t,s)}(\mathbb{R}^N) = T_x(X) \times N_x(X)$. The above argument also shows that X is a submanifold of $N(X)$. In particular, $T(X)$ is a submanifold of $T(N(X))$. Since π is the restriction of the projection, $\pi(x,0) = x$, it follows that $(d\pi)_x : T_x(N(X)) \to T_x(X)$ also has the property $d\pi_x(\mathbf{v}) = \mathbf{v}$ for all $\mathbf{v} \in T_x(X) \subset T_x(N(X))$. Hence, π is a submersion.

Indeed, for each fixed $t = (t_1, \ldots, t_n, 0, \ldots, 0)$ the mapping

$$s = (s_1, \ldots, s_{N-n}) \mapsto \Theta(t,s)$$

is a linear isomorphism of the vector spaces $0 \times \mathbb{R}^{N-n} \to N_xX$, where $x = \tau(t,0)$. Putting $\psi = \tau|_{\mathbb{R}^n \times 0} : \mathbb{R}^n \to U$, and $\phi = (\psi \times Id) \circ \Theta^{-1} : N(U) \to U \times \mathbb{R}^{N-n}$, it follows that the projection map $\pi : N(X) \to X$ satisfies the requirements as in Definition 5.5.3 for a vector bundle of rank $N - n$ over X. ♠

Example 6.1.1 Consider the unit sphere $\mathbb{S}^{n-1}$ in $\mathbb{R}^n$. Its normal bundle $N(\mathbb{S}^{n-1})$ is a submanifold of $\mathbb{R}^n \times \mathbb{R}^n$. Consider the mapping $\alpha : N(\mathbb{S}^{n-1}) \to \mathbb{R}^n$ given by

$$(x,y) \mapsto x + y.$$

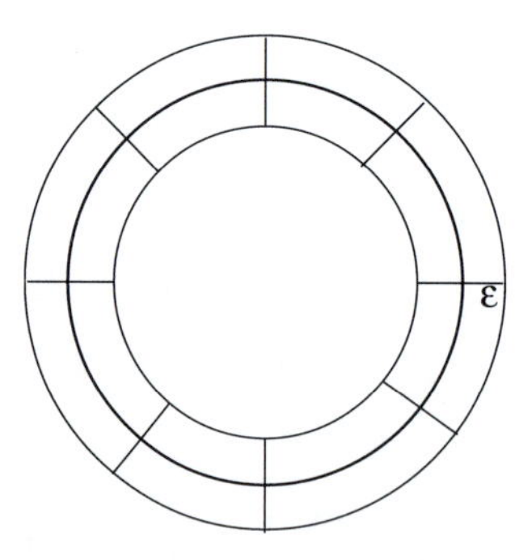

Figure 28 The ϵ-neighborhood of the unit circle.

Restricted to $\mathbb{S}^{n-1} \times 0$, it is a diffeomorphism with which we are used to identify $\mathbb{S}^{n-1} \times 0$ with $\mathbb{S}^{n-1}$. On the other hand, restricted to each normal line $N_x(\mathbb{S}^{n-1})$, it is an *affine linear embedding* onto the line spanned by x itself in $\mathbb{R}^n$. Thus, we see that if we restrict α to only those vectors of length $< \epsilon$ for some $0 < \epsilon \leq 1$, then α itself is one-to-one.

It can be easily seen that this is actually an embedding onto the annular region,

$$\{x \in \mathbb{R}^n \ : \ \epsilon < \|x\| < 1 + \epsilon\}$$

which is an open neighborhood of $\mathbb{S}^{n-1}$ in $\mathbb{R}^n$. We shall soon conceptualize this property and prove it for all proper submanifolds of $\mathbb{R}^N$.

On the other hand, consider the mapping $\beta : N(X) \to \mathbb{S}^{n-1} \times \mathbb{R}$ given by

$$(x, v) \mapsto (x, \langle x, v \rangle).$$

This is clearly a smooth map. Verify that this is a diffeomorphism by showing that the map $\gamma : \mathbb{S}^{n-1} \times \mathbb{R} \to N(X)$ defined by

$$(x, r) \mapsto (x, rx)$$

is the inverse of β. The diffeomorphism is seen to preserve the first factor and restricted to each normal line, it is linear also. Such a diffeomorphism is called an *equivalence* or *isomorphism of the vector bundles.* A vector bundle over X, which is isomorphic to $X \times \mathbb{R}^n$ for some n, is called a *trivial bundle.* Thus, what we have seen above amounts to saying that the normal bundle of $\mathbb{S}^{n-1}$ in $\mathbb{R}^n$ is trivial, i.e.,

$$N(\mathbb{S}^{n-1}; \mathbb{R}^n) \approx \Theta^1_{\mathbb{S}^{n-1}} \approx \mathbb{S}^{n-1} \times \mathbb{R}.$$

(Compare Example 3.3.1(3).)

Definition 6.1.2 Given a submanifold $X \subset \mathbb{R}^N$, and a continuous function $\epsilon : X \to (0, \infty)$, introduce the notation,

$$\mathcal{N}_\epsilon(X) = \{(x, \mathbf{v}) \in N(X) \ : \ \|\mathbf{v}\| < \epsilon(x)\}$$

Clearly, $\mathcal{N}_\epsilon(X)$ is a neighborhood of $X \times 0$ in $N(X)$.

Remark 6.1.1 Whatever we did so far in this section is valid for manifolds with or without boundary. But in the following theorem, we actually have to assume that X has no boundary.

Theorem 6.1.2 Tubular Neighborhood Theorem: *Let $X \subset \mathbb{R}^N$ be a proper submanifold without boundary and let U be a neighborhood of X in $\mathbb{R}^N$. Then there exists a continuous function $\epsilon : X \to (0, \infty)$ and a diffeomorphism $\phi : \mathcal{N}_\epsilon(X) \to V$ onto an open neighborhood V of X in U such that $\phi(x, 0) = x$ for all $x \in X$. In particular, X is a strong deformation retract of V. Moreover, if X is compact, then we can choose a constant $\epsilon > 0$, such that $V = V_{\epsilon'} = \cup_{x \in X} B_{\epsilon'}(x)$.*

Proof: We consider the linear mapping $\alpha : \mathbb{R}^N \times \mathbb{R}^N \to \mathbb{R}^N$ given by

$$(\mathbf{u}, \mathbf{v}) \mapsto \mathbf{u} + \mathbf{v}.$$

First we claim that $\alpha : X \to \mathbb{R}^N$ is a submersion. In Theorem 6.1.1, we have seen that for all points $(x, \mathbf{v}) \in N(X)$, the tangent space to $N(X)$ is given by

$$T_{(x,\mathbf{v})} N(X) = T_x X \oplus N_x(X).$$

Since α is a linear map, it follows that $D\alpha = \alpha$ for all points in $\mathbb{R}^N \times \mathbb{R}^N$. Therefore, it follows that

$$D\alpha(T_{(x,\mathbf{v})} N(X)) = \mathbb{R}^N.$$

This proves that α is a submersion. Since dimension of $N(X)$ and dimension of $\mathbb{R}^N$ are the same, by the inverse function theorem, this means that $\alpha : X \to \mathbb{R}^N$ is a local diffeomorphism.

In particular, for each $x \in X$ we can find $\delta(x) > 0$ such that on the open set $\mathrm{Box}(x, \delta(x)) = \{(y, \mathbf{v}) \in N(M) \ : \ \|x - y\| < \delta(x), \|\mathbf{v}\| < \delta(x)\}$, the map α is injective and the image is contained in U.

(If X is compact, we can immediately conclude that for suitable a choice of $\epsilon > 0$, α is injective on V_ϵ as well. In the noncompact case, we have to work harder.)

Let us put $\tau'(x)$ to be the supremum of all such $\delta(x) \leq 1$ for which α is injective on $\text{Box}(x, \delta(x))$ and such that $\alpha(\text{Box}(x, \delta)) \subset U$. As seen above, α is a local embedding implies that τ is positive. We shall show that τ is actually uniformly continuous in x, i.e.,

$$|\tau(x) - \tau(x')| \leq \|x - x'\|. \tag{6.1}$$

We may assume that $\tau(x) \geq \tau(x')$. We then have to prove that

$$\tau(x) - \tau(x') \leq \|x - x'\|.$$

If this were not true, then $\tau(x) > \tau(x') + \|x - x'\| > \|x - x'\|$. This clearly implies that $\text{Box}(x', \tau(x) - \|x - x'\|)$ is contained in $\text{Box}(x, \tau(x))$ on which α is injective. Therefore, $\tau(x') \geq \tau(x) - \|x - x'\|$, which is absurd.

Put $\epsilon(x) = \tau(x)/2$ and take $\phi = \alpha|_{\mathcal{N}_\epsilon(X)}$. We claim that ϕ is a diffeomorphism onto an open subset $V = \phi(\mathcal{N}_\epsilon(X))$. Since the dimension of the domain and codomain are the same and α is a submersion, all that we need to prove is that ϕ is injective.

Now suppose $(x, \mathbf{v}), (x', \mathbf{v}') \in N_\epsilon(X)$ are such that $\phi(x, \mathbf{v}) = \phi(x', \mathbf{v}') = z$. Once again, we may assume that $\tau(x) \geq \tau(x')$. But then

$$\|x - x'\| \leq \|z - x\| + \|z - x'\| = \|\mathbf{v}\| + \|\mathbf{v}'\| \leq \epsilon(x) + \epsilon(x') < \tau(x).$$

This implies that $(x, \mathbf{v}), (x, \mathbf{v}') \in \text{Box}(x, \tau(x))$ on which $\alpha = \phi$ is injective. Therefore $(x, \mathbf{v}) = (x', \mathbf{v}')$.

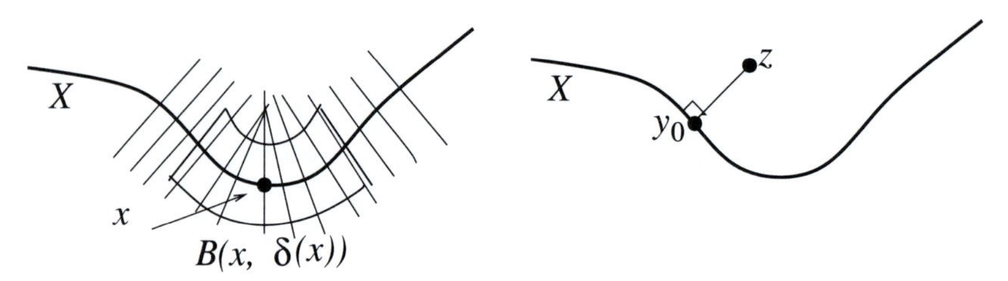

Figure 29 The normal bundle.

Now consider the case when X is compact. Take ϵ' as the constant function $min\{\epsilon(x) \ : \ x \in X\}$. Clearly, $V_{\epsilon'}$ contains

$$V = \alpha(\mathcal{N}_\epsilon(X)) = \alpha\left(\cup_{x \in X}\{x\} \times B_{\epsilon'}(x) \cap N_x(X)\right).$$

On the other hand, suppose $z \in V_{\epsilon'}$. This means that $z \in B_{\epsilon'}(x)$ for some $x \in X$. Since X is a closed subset of $\mathbb{R}^N$ (being a proper submanifold) $X \cap B_{\epsilon'}(x)$ is a closed subset of this open ball. If $r = d(z, X)$, the distance between z and X, then it follows that there is $y_0 \in X \cap B_{\epsilon'}(x)$ such that $r = d(z, y_0)$. Moreover, it follows that y_0 is a critical point of the function $f(y) = \|y - z\|^2$. This means that the derivative of the function f at the point $y = y_0$ restricted to $T_{y_0}X$ must be the zero map. This is the same as saying that the gradient vector $2(y - z)$ at y_0 is perpendicular to $T_{y_0}X$, which is the same as saying that $z - y_0 \in N_{y_0}(X)$. Clearly, $r < \epsilon'$ and hence $(y_0, z - y_0) \in \mathcal{N}_{\epsilon'}(X)$. Also, $\phi(y_0, z - y_0) = z$ and hence $z \in V$. ♠

Definition 6.1.3 Let X be a submanifold of a manifold Y. By a tubular neighborhood of X in Y we mean a pair (V, π) where V is a neighborhood of X in Y and $\pi : V \to X$ is a submersion and a strong deformation retraction.

Remark 6.1.2
(i) Recall that π is a strong deformation retraction means that there is homotopy $H : V \times I \to V$ such that

$$H(y,0) = y,\ H(y,1) = \pi(y),\ y \in V; \text{ and } H(x,t) = x,\ \forall\ x \in X, t \in I.$$

In particular, this implies that $\pi(x) = x,\ \forall\ x \in X$. The above theorem, not only establishes the existence of a tubular neighborhoods in case $Y = \mathbb{R}^n$, but also gives recipe to construct one. At least in case X is compact, for sufficiently small ϵ, an ϵ-neighborhood will do. Note that once some $\epsilon > 0$ is chosen to satisfy the above property, the same holds for all $0 < \epsilon' < \epsilon$ also. Moreover, it is easily seen that $\mathcal{N}_\epsilon(X)$ is diffeomorphic to $\mathcal{N}_{\epsilon'}(X)$ with the diffeomorphism being equal to the identity map on $X \times 0$.
(ii) Consider the situation when $X \subset Y$ is a neat submanifold with $\partial X = \partial Y \cap X \neq \emptyset$. Note that the normal bundle of X in Y makes perfect sense and is a vector bundle of rank equal to the codimension of X in Y. This is so because the normals to X at a point $x \in \partial X$ are the same as normals to ∂X in ∂Y at that point. We then take the union of some appropriate tubular neighborhood of $\operatorname{int} X$ in $\operatorname{int} Y$ with a tubular neighborhood of ∂X in ∂Y to obtain a tubular neighborhood of X in Y.
(iii) Finally, consider a situation of a nonneat submanifold such as $[0,1] \times \{0\} \subset \mathbb{R}^2$. The arguments in the above theorem break down because the local diffeomorphism does not guarantee open mapping due to the presence of boundary points of the submanifold in the interior of the ambient manifold. The situation can be remedied to some extent by allowing all normals to ∂X at the boundary points of X. By modifying the map π near the boundary points, it is possible to obtain a map $\pi : V \to X$ satisfying the conditions in Definition 6.1.3. The simplest situation of this type is depicted in Figure 30.

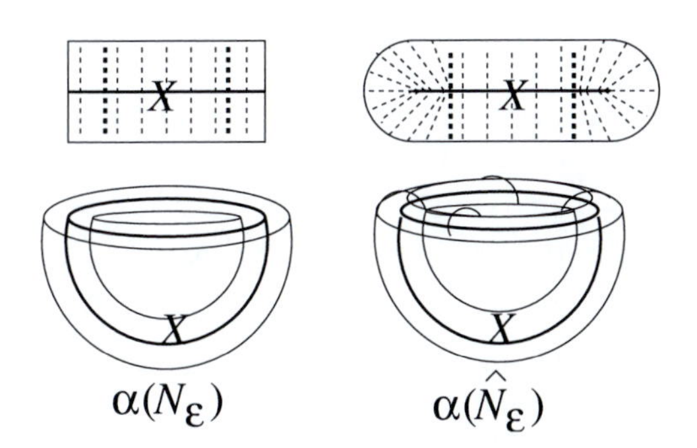

Figure 30 Tubular neighborhoods.

(iv) Suppose $X \subset Y \subset \mathbb{R}^N$. One can talk about the normal bundle and the tubular neighborhoods of X in Y extending all the results we have proved for the case $X \subset \mathbb{R}^N$. We can also deal with the situation when Y itself is an abstract manifld. However, this leads us to consider the concept of inner products on tangent spaces of Y which vary smoothly, familiarly known as Riemannian metric, which we shall not be dealing with.

As an application of the existence of tubular neighborhoods, we can now generalize the approximation Theorem 1.7.2, which says that smooth functions form a dense subspace of the space of all functions.

Theorem 6.1.3 Smooth Approximation: *Let X and Y be manifolds and let $C \subset X$ be a closed subset. Given $f : X \to Y$ a continuous function that is smooth on C, and a positive function $\epsilon : X \to (0, \infty)$, there exists a smooth map $g : X \to Y$ such that $\|f(x) - g(x)\| < \epsilon(x),\ x \in X$ and such that $g|_C = f$.*

Proof: We may assume that both X and Y are proper submanifolds of $\mathbb{R}^N$ and f is defined on an open set U containing X. Choose a continuous function $\delta : X \to (0, \infty)$ such that V_δ is a tubular neighborhood of Y in $\mathbb{R}^N$ with the submersion $\pi : V_\delta(Y) \to Y$. Observe that $\pi_Y = Id_Y$. Let $r(x) = \min \{\delta(x)/2, \epsilon(x)/2\}$. By Theorem 1.7.2, there a smooth function $p : \mathbb{R}^N \to \mathbb{R}^N$ such that $\|f(x) - p(x)\| < r(x)$ for all $x \in X$ and $p|_C = f$. Put $g = \pi \circ p$. Then for $x \in C$ we have, $g(x) = \pi \circ p(x) = \pi \circ f(x) = f(x)$. Moreover,

$$\|f(x) - g(x)\| \leq \|f(x) - p(x)\| + \|p(x) - \pi \circ p(x)\| \leq 2r(x) \leq \epsilon(x),$$

for all $x \in X$. ♠

Exercise 6.1

1. Use Theorem 6.1.3 to deduce Theorem 5.4.2 for continuous maps. Can you replace the compactness hypothesis on X by a weaker hypothesis? Can you remove the compactness hypothesis altogether?

2. Prove the continuous version of Brouwer's fixed point theorem.

3. Let $f : \mathbb{D}^n \to \mathbb{D}^n$ be any continuous (smooth) map such that $f|_{\mathbb{S}^{n-1}} = Id$. Then show that f is surjective.

6.2 Orientation on Normal Bundle

Based on this experience we have gained in the special case of the boundary of a manifold, we shall now study how to orient an arbitrary submanifold of an oriented manifold.

Definition 6.2.1 Let $X \subset \mathbb{R}^N$ be a smooth manifold and $N(X)$ denote its normal bundle. By an orientation τ of $N(X)$ we mean a choice of orientation class τ_x of $N_x(X)$ for each $x \in X$ such that τ is locally constant in the following sense. Recall that if (U, ϕ) is a local coordinate chart for X then using Gram-Schmidt process we have defined a diffeomorphism $\hat{\phi} : N(U) \to \phi(U) \times \mathbb{R}^{N-n}$. We demand that for all $x \in U$ the orientation class τ_x of $N_x(X)$ is mapped onto the same orientation class of $\mathbb{R}^{N-n}$, under $\hat{\phi}_x$.

Remark 6.2.1
(i) Observe that the above definition is similar to the definition of orientation on the manifold X itself. Indeed, in this sense an orientation on X is nothing but an orientation on the tangent bundle $T(X)$.
(ii) The above definition is available to the normal bundle of any submanifold Y inside a given manifold also.
(iii) Given a manifold $X \subset \mathbb{R}^N$ and an orientation τ on $N(X)$, we define the associated orientation ω on X by the rule

$$\tau_x \omega_x = E_n.$$

(Use Theorem 4.1.2 to verify that ω is smooth.)
(iv) More generally, consider the situation where Z is a submanifold of Y. Let us agree to use the notation $[V]$ to denote an oriented vector space, i.e., a vector space V with a specific orientation. Consider the identity

$$[N_z(Z;Y)][T_z Z] = [T_z Y]. \tag{6.2}$$

This identity can be used to define any one of the three objects whenever the other two are known. Moreover, if two of them are smooth then the third one is also smooth. As an example, suppose now that Y is oriented. Then Z will be orientable iff $N(Z;Y)$ is orientable; moreover, fixing an orientation on Z is the same as fixing one on $N(Z;Y)$.

Theorem 6.2.1 *Let X, Y, Z be smooth manifolds, $Z \subset Y$ be a closed submanifold with Z and Y boundaryless. Let $f : X \to Y$ a smooth map such that $f \pitchfork Z$ and $\partial f \pitchfork Z$. Assume that X, Y, Z are all orientable. Then $W = f^{-1}(Z)$ is also orientable.*

Proof: Let $x \in W$ and $z = f(x)$. The transversality condition says that

$$df_x(T_x(X)) + T_z Z = T_z Y. \tag{6.3}$$

Write $T_x(X) = T_x W \oplus N_x W$. Observe that $df_x(T_x W) = T_z Z$. It follows that

$$df_x(N_x W) \oplus T_z Z = T_z Y. \tag{6.4}$$

Using the orientations on Y and Z to define the orientation classes on $df_x(N_x W)$. Similar to the proof of the above theorem, the smoothness of this assignment can be seen. Now use the fact that df_x is injective on $N_x W$ and hence $df_x : N_x W \to df_x(N_x W)$ is an isomorphism. Use this to pull back the orientation class onto $N_x W$. Since df_x is smooth in x, it follows that this gives a smooth orientation on $N(W; X)$, the normal bundle of W in X. Since X is orientable, it follows that W is also orientable. ♠

Remark 6.2.2
(i) It may be noted that according to our convention, the induced orientation on W is given by the following two formulae:

$$df_x[N_x W][T_z Z] = [T_z Y]; \quad [N_x W][T_x W] = [T_x X]. \tag{6.5}$$

(ii) Once W is oriented in the above manner, its boundary gets an induced orientation, say α, by the formula:

$$\eta_x \alpha_x = [T_x W], \tag{6.6}$$

where η is the outward normal to ∂W. On the other hand, we can consider the induced orientation, say β on ∂W directly via the map $\partial f : \partial X \to Y$ given by the formula

$$d(\partial f)_x[N_x(\partial W; \partial X)][T_z Z] = [T_z Y]; \quad [N_x(\partial W; \partial X)]\beta_x = [T_x \partial X] \tag{6.7}$$

It is an interesting exercise to figure out the exact relation between α and β. Please try this and only after you have given enough time to it read the following solution. However, we are not going to use this result in the rest of the course.

Since the outward normal η_x for ∂X is also the outward normal to ∂W, by the rule for induced orientation on the boundary we have,

$$\eta_x[T_x \partial X] = [T_x X]. \tag{6.8}$$

Therefore,

$$\eta_x[N_x(\partial W; \partial X)]\beta_x = \eta_x[T_x \partial X] = [T_x X]. \tag{6.9}$$

On the other hand,

$$[T_x X] = [N_x W][T_x W] = [N_x W]\eta_x \alpha_x. \tag{6.10}$$

Therefore,

$$\eta_x[N_x(\partial W; \partial X)]\beta_x = [N_x W]\eta_x \alpha_x = (-1)^k \eta_x [N_x W]\alpha_x,$$

where $k = \dim N_x(W)$. Notice that at any point $x \in \partial W$, the normals to ∂W inside ∂X are the same as normals to W inside X. It follows that $\beta_x = (-1)^k \alpha_x$. Since this true for all points $x \in W$ and $k = codim\ Z$, we have,

$$\beta = (-1)^{codim\ Z} \alpha.$$

Exercise 6.2 Extend the concept of orientation on any vector bundle.

6.3 Vector Fields and Isotopies

Definition 6.3.1 Let X be a smooth manifold. By a vector field on X we mean a smooth map $\sigma : X \to TX$ such that $\pi \circ \sigma = Id_X$.

Remark 6.3.1
(i) Thus, a vector field assigns to each point $x \in X$, a tangent vector to X at x in a smooth fashion. Consider the case when X is an open subset of $\mathbb{R}^n$. Then $TX = X \times \mathbb{R}^n$ and π is the first projection. Therefore, we can write $\sigma(x) = (x, \tau(x))$ where $\tau : X \to \mathbb{R}^n$ is a smooth map. This establishes a one-to-one correspondence between vector fields on X and smooth maps $X \to \mathbb{R}^n$. This happens, more generally, whenever the tangent bundle of X is trivial. A typical case is the unit circle $\mathbb{S}^1$.
(ii) In general, the assignment $x \mapsto (x, 0)$ defines a vector field which is the zero-field. Given any two vector fields, we can add them pointwise to obtain another vector field. We can also multiply a vector field by a scalar. Thus, the set of vector fields on a given manifold forms a vector space. Further, we can also multiply a vector field on X by a smooth function $X \to \mathbb{R}$. This operation makes the set of all vector fields into a module over the ring $\mathcal{C}^\infty(X; \mathbb{R})$ of smooth real valued functions.
(iii) Using partition of unity, we can always patch up vector fields defined locally to obtain plenty of nonzero vector fields. However, having a vector field that does not vanish anywhere is another matter. As we shall see later the existence of a vector field that does not vanish anywhere implies a certain topological behavior of the manifold.
(iv) Vector fields arise in a natural way from many physical situations, which we shall not discuss here. The most common one is when we have to deal with a homotopy $h_t : X \to X$ of the identity map. Then, to each point $x \in X$, we have the smooth curve $c : t \mapsto h_t(x)$ which starts at x. We could simply take the tangent vector to this curve at x, to obtain a vector field on X. In this section, we want to show that the situation is reversible to a large extent.

We shall need the following fundamental result from the theory of ordinary differential equations (See for instance, Theorems 1 and 2 and Remark 1 in Section 2.3 of [Pe].)

Theorem 6.3.1 *Let $U \subset \mathbb{R}^n, V \subset \mathbb{R}^m$ be open subsets, $x_0 \in U, \mathbf{v}_0 \in V$ and let $\sigma \in \mathcal{C}^r(U \times V, \mathbb{R}^n)$. Then there exist positive real numbers δ, r such that for all $\mathbf{x} \in B_r(\mathbf{x}_0) \subset U$ and $\mathbf{v} \in B_r(\mathbf{v}_0) \subset V$, the initial value problem*

$$\frac{\partial \mathbf{f}}{\partial t}(\mathbf{x}, \mathbf{v}, t) = \sigma(f(\mathbf{x}, \mathbf{v}, t), \mathbf{v}); \quad f(\mathbf{x}, \mathbf{v}, 0) = \mathbf{x}$$

has a unique $\mathcal{C}^r$ solution $f : G \to U$, where $G = B_r(\mathbf{x}_0) \times B_r(\mathbf{v}_0) \times [-\delta, \delta]$.

Definition 6.3.2 Given a vector field σ on X, by *the initial value problem associated to σ* [abbreviated to IVP(σ)], we mean

$$\left.\frac{\partial f}{\partial t}\right|_{(x,t)} = \sigma(f(x,t)); \quad f(x,0) = x. \tag{6.11}$$

For a function $\epsilon : X \to (0, \infty)$ set up the notation

$$X_\epsilon := \{(x,t) \in X \times \mathbb{R} \; : \; |t| < \epsilon(x)\}.$$

Lemma 6.3.1 Let X be any smooth manifold and $\sigma : X \to TX$ be a smooth vector field. Then there exists a smooth map $\epsilon : X \to (0, \infty)$ and a smooth map $f : X_\epsilon \to X$ satisfying the IVP(σ) (6.11) for all points in $(x,t) \in X_\epsilon$.

Proof: Via a local coordinate system at a point $x \in X$, the differential equation (6.11) corresponds to an ordinary differential equation (initial value problem) defined in a neighborhood of $0 \in \mathbb{R}^n$. Therefore, there exists a positive number δ and a neighborhood U of 0 on which the equation has a unique solution $U \times (-\delta, \delta) \to U$. Translated back to the manifold, this means that we have a neighborhood V_x of x and $\delta > 0$ and a smooth map $f : V_x \times (-\delta, \delta) \to V_x$ satisfying (6.11). Let $\{V_\alpha\}$ be a locally finite open refinement of the open cover $\{V_x\}$ of X. Let $\{\theta_\alpha\}$ be a smooth partition of unity subordinate to $\{V_\alpha\}$ and let δ_α denote the corresponding positive numbers, so that we have a smooth function $f_\alpha : V_\alpha \times (-\delta_\alpha, \delta_\alpha) \to X$ satisfying (6.11). Put $\epsilon = \sum_\alpha \delta_\alpha \theta_\alpha$. By the uniqueness of f_α's, in each neighborhood, it follows that $f_\alpha = f_\beta$ on $V_\alpha \cap V_\beta \times (-a, a)$, where $a = min\{\delta_\alpha, \delta_\beta\}$. (Details are left to the reader as an exercise.) Therefore, $\{f_\alpha\}$ patch-up to a well-defined map $f : X_\epsilon \to X$ as required. ♠

Definition 6.3.3 For each fixed $x \in X$, the curve $t \mapsto f(x,t)$ has the property that $f(x, 0) = x$ and the tangent to this curve at any point $f(x,t)$ is equal to $\sigma(f(x,t))$, for all values of t for which the curve is defined. They are called *solution curves* or *flow curves* of the field σ, emanating from x. If the map ϵ in the above lemma can be chosen so that the values keep away from 0 i.e., there exists $r > 0$ such that $\epsilon(x) \geq r$ for all $x \in X$, then we say σ *admits global solutions on* X.

The reason for introducing the last terminology is clear from the following lemma.

Lemma 6.3.2 Suppose the vector field σ admits a global solution. Then a solution f is defined on $X \times \mathbb{R}$ and satisfies the property

$$f(x, t + s) = f(f(x, t), s). \tag{6.12}$$

In particular, the maps $f_t : X \to X$ defined by $f_t(x) = f(x, t)$ are diffeomorphisms that satisfy $f_{t+s} = f_s \circ f_t$, for all $t, s \in \mathbb{R}$.

Proof: Suppose $f : X \times (-r, r) \to X$ is a global solution of σ where $r > 0$. If $t, s \in (-r, r)$ are such that $t + s \in (-r, r)$, then the curves $t \mapsto f(f(x, s), t)$ and $t \mapsto f(x, s + t)$ are both solution curves for σ emanating from the point $f(x, s)$. By the uniqueness of the solution, it follows that $f(f(x, s), t) = f(x, s + t)$. Now given any $t \in \mathbb{R}$ write $t = m\frac{r}{2} + t'$ where $|t'| < r/2$ and m is some integer. Put $\phi = f_{r/2}$. Define $g : X \times \mathbb{R} \to X$ by the formula $g(x, t) = f(\phi^m(x), t')$. Verify first that $g = f$ wherever f is already defined, g is a solution for σ and $g_{t+s} = g_t \circ g_s$. Therefore, we can rename g as f. ♠

Certainly, if X is compact then the smooth function ϵ keeps away from 0 and hence σ admits global solutions. Even if X is not compact but the vector field σ has compact support K, we have the same conclusion. For then, around points where $\sigma = 0$, we can choose $\delta = \infty$, K itself can be covered by finitely many neighborhoods where solutions are defined for some $\delta > 0$. Therefore, we obtain:

Theorem 6.3.2 *Let X be any smooth manifold and $\sigma : X \to TX$ be a smooth vector field that vanishes outside a compact subset of X. Then there exists a smooth 1-parameter family of diffeomorphisms $h_t : X \to X$, for all $t \in \mathbb{R}$ with the following properties.*
(i) $\dfrac{dh_t(x)}{dt} = \sigma(h_t(x))$, *for all* $x \in X,\ t \in \mathbb{R}$;
(ii) $h_0(x) = x$, *for all* $x \in X$;
(iii) $h_t \circ h_s = h_{t+s}$;
(iv) $h_t(x) = x$ *for all* $t \in \mathbb{R}$ *and for* x *outside a compact set.*
Moreover, the family h_t satisfying (i) *and* (ii) *is unique.*

Definition 6.3.4 Because of the uniqueness and the property (iii) in the theorem, the 1-parameter family of diffeomorphism is called *the 1-parameter group of diffeomorphism generated by the vector field* σ.

Remark 6.3.2
(i) Without some compactness conditions, the theorem is not valid as seen by the following example. Take $X = (0, 1)$ and $\sigma = \frac{d}{dx}$. Then the solution for IVP(σ) has to be necessarily, $f(x, t) = x + t$. But, there is no $\epsilon > 0$ such that $f(x, t) \in (0, 1)\ \forall\ x \in (0, 1)$ and all $|t| < \epsilon$.
(ii) The theorem is valid for manifolds with boundary as well, except that we need to assume that the vector field σ is *never* pointing outward along the boundary points. For then the solution curves starting at a boundary point will lie inside the manifold for a small period. So, the conclusion also should be modified by restricting solutions for $t \in [0, \epsilon)$. As an important corollary we have:

Theorem 6.3.3 Collar Neighborhood Theorem: *Let X be a manifold with its boundary ∂X compact. Let U be a neighborhood of ∂X. Then there exists an embedding $\phi : \partial X \times [0, 1) \to X$ such that $\phi(x, 0) = x$ for all $x \in \partial X$.*

Proof: Let η be a vector field on X, which is pointing strictly inward at each point of the boundary. (Use partition of unity to see the existence of such a vector field.) Choose an open set V such that $\partial X \subset V \subset \bar{V} \subset U$ and such that $\bar{V}$ is compact. Then by multiplying by a suitable cut-off function, we may assume that η vanishes outside $\bar{V}$. It follows that there is an $\epsilon > 0$ and a map $h : \partial X \times [0, \epsilon) \to X$ such that
(i) $h(x, 0) = x,\ \forall\ x \in \partial X$;
(ii) $\dfrac{\partial h}{\partial t}(x) = \eta(x),\ \forall\ x \in \partial X,\ t \in [0, \epsilon)$.
This then implies that h is a local diffeomorphism at $(x, 0)$, for each $x \in \partial X$. Moreover $h|_{\partial X \times 0}$ is injective. By stability property (see Theorem 3.6.1), it follows that there exists $\delta > 0$ such that $h : \partial X \times [0, \delta] \to X$ is an injective immersion. Choosing δ smaller if necessary we may assume $h(\partial X \times [0, \delta]) \subset V$. Now we just define $\phi(x, t) = h(x, t\delta)$. ♠

Definition 6.3.5 Let X, Y be smooth manifolds, $h : X \times I \to Y$ be a smooth homotopy. Put $h_t(x) = h(x, t)$. We say h is an *isotopy,* if each h_t is an embedding. In that case, we say h_0 and h_1 are *isotopic* to each other. Clearly, this is an equivalence relation among all diffeomorphisms $X \to Y$. (Take this as an exercise.) Furthermore, when $X = Y$ and each h_t is a diffeomorphism with $h_0 = Id_X$, we call h_t a *diffeotopy,* or an *ambient isotopy.* Of course, a diffeotopy is an isotopy. An isotopy is said to have *compact support* if it is the identity map outside a compact set. Two embeddings $f, g : X \to Y$ are said to be ambient isotopic if there exists an ambient isotopy $h : Y \times I \to Y$ such that $h(f(x), 1) = g(x)$.

Remark 6.3.3 Clearly, if two maps are ambient isotopic to each other, then they are isotopic. We may always assume that an isotopy is defined for all $t \in \mathbb{R}$, instead of just on the interval I. For example, let μ be a smooth function such that $\mu(t) = 0$ for $t \leq 0$, $\mu(t) = 1$ for all $t \geq 1$ and $\mu'(t) \geq 0$, for all t. Then consider $F(x, t) = h(x, \mu(t))$. Thus, whenever an isotopy is given we may at will assume that it is defined for all t. This will help us technically in several ways. For instance, we can now allow isotopies defined on manifolds with boundary as well.

Example 6.3.1

1. Any two linear isomorphisms : $\mathbb{R}^n \to \mathbb{R}^n$ are ambient isotopic to each other iff the product of their determinant is positive: Consider a smooth map $\lambda : [0, 1] \to GL(n, \mathbb{R})$ such that $\lambda(0) = Id$. Then the map defined by $H(x, t) = \lambda(t)(x)$ is a diffeotopy of $Id_{\mathbb{R}^n}$

and $\lambda(1)$. Now given linear isomorphisms, $f, g \in GL(n, \mathbb{R})$ we can find λ as above with $\lambda(1) = g \circ f^{-1}$ iff $\det g \circ f > 0$. ($GL(n, \mathbb{R})$ has precisely two path components consisting of those matrices with $\pm$ determinant. See Exercise 3.4.5.) Check that $H(f(x), t)$ is the required diffeotopy from f to g.

2. Let $f : \mathbb{R}^n \to \mathbb{R}^n$ be any diffeomorphism (at least C^2). Then it is diffeotopic to the linear isomorphism Df_0. To see this, first consider a diffeomorphism $g : \mathbb{R}^n \to \mathbb{R}^n$ such that $g(0) = 0$ and $Dg_0 = Id$. Define

$$h(x,t) = \begin{cases} \dfrac{g(tx)}{t}, & t \neq 0 \\ x, & t = 0. \end{cases}$$

Verify that h is a diffeotopy of g. Now given any diffeomorphism f, consider $\phi(x,t) = x - tf(0)$. Then $\phi(f(x), t)$ defines a diffeotopy of f with $f - f(0)$. Therefore, we may assume $f(0) = 0$. Put $g = f \circ (Df_0)^{-1}$. Then $g(0) = 0$ and $Dg_0 = Id$. Therefore, by the earlier case, we have the diffeotopy h_t. But now $h(Df_0(x), 1) = f(x)$. Thus, f is diffeotopic to the linear isomorphism Df_0.

3. If $f, g : X \to Y$ are ambient isotopic embeddings then it follows that the complement of their images are diffeomorphic. This can be used to produce embedded arcs in $\mathbb{R}^2$, which are not ambient isotopic. For example, take $f(t) = (t, 0)$ and $g(t) = (\tan \pi t/2, 0)$ on $(-1, 1)$. Directly write down an isotopy between them. The complement of $\mathbb{R}^2 \setminus (0,1) \times \{0\}$ is connected, whereas, $\mathbb{R}^2 \setminus \mathbb{R} \times 0$ is disconnected. So, f and g cannot be ambient isotopic.

As an immediate application of the notion of isotopy, let us derive the "uniqueness" of the collar neighborhoods, which itself is going to be very useful later. We begin with:

Lemma 6.3.3 Let M be a closed manifold and $\phi : M \times I \to M \times I$ be an embedding such that $\phi(x, 0) = x, \quad x \in M$. Then there exists $0 < \delta < 1$ such that $\phi|_{M \times [0,\delta]}$ is isotopic to $Id|_{M \times [0,\delta]}$.

Proof: Write $\phi(x,t) = (\phi_1(x,t), \phi_2(x,t))$. Then $\phi_1(x, 0) = x$ and $\phi_2(x, 0) = 0$. Moreover, we check that the derivative of ϕ at any point $(x, 0) \in M \times 0$ is of the form

$$\begin{pmatrix} Id & \sigma \\ 0 & \alpha \end{pmatrix}$$

where $\alpha > 0$. Consider the homotopy $H : M \times I \times I \to M \times I$ given by

$$(x, t, s) \mapsto (\phi_1(x, st), (1 - s)t + s\phi_2(x, st)).$$

Clearly,

$$H(x, t, 0) = (\phi_1(x, 0), t) = (x, t); \text{ and } H(x, t, 1) = (\phi_1(x, t), \phi_2(x, t)) = \phi(x, t).$$

We claim that there exists $0 < \delta < 1$ such that for each $s \in I$, $(x, t) \mapsto H(x, t, s)$ is an embedding on $M \times [0, \delta]$. The arguments involved are exactly similar to the one you have seen in the Stability Theorem 3.6.1.

Fix $s \in I$ and put $H_s(x, t) = H(x, t, s)$. At any point $(x, 0, s)$ the derivative of H_s is of the form

$$\begin{pmatrix} Id & s\sigma \\ 0 & 1 - s + \alpha \end{pmatrix}$$

and hence is invertible. By inverse function theorem, H_s is a local diffeomorphism at all points $(x, 0, s)$. Since $M \times \{0\} \times I$ is compact, it follows that there $0 < \delta < 1$ such that H_s is a local diffeomorphism on $M \times [0, \delta]$. Therefore, it suffices to show that for some $\delta > 0$, H_s is injective on $M \times [0, \delta]$ for all s.

If this were not true, then for each n there exists $x_{1,n}, x_{2,n} \in M, s_n \in I$ and $0 < t_{1,n}, t_{2,n} < 1/n$, such that $(x_{1,n}, t_{1,n}) \neq (x_{2,n}, t_{2,n})$ and $H_{s_n}(x_{1,n}, t_{1,n}) = H_{s_n}(x_{2,n}, t_{2,n})$. By passing to subsequences we may assume that $x_{i,n} \to x_i \in M$, and $s_n \to s \in I$. It then follows that $(x_1, 0) = H_s(x_1, 0) = H(x_1, 0, s) = H(x_2, 0, s) = (x_2, 0)$ and hence $x_1 = x_2$. But for large n, $(x_{i,n}, t_{i,n})$ are in a neighborhood of $(x_1, 0) = (x_2, 0)$, which violates the injectivity of H_s on this neighborhood. ♠

Theorem 6.3.4 *Let M be a compact component of ∂N. Then any two collar neighborhoods of M in N are isotopic.*

Proof: Let $f : M \times I \to N$ be an embedding such that $f(x, 0) = x, x \in M$ (i.e., f is a collar of M in N). Fix $0 < \delta < 1$. The map $F_\delta : M \times I \times I \to N$ given by $F(x, t, s) = f(x, t - st + st\delta)$ defines an isotopy of f with another collar of M, viz., $f_\delta(x, t) = f(x, t\delta)$. Given any neighborhood U of M, we can choose δ such that $f_\delta(M \times I) \subset V$. Therefore, in order to prove that any two collars f, g of M are isotopic to each other, we can replace f by f_δ and assume that the image of f is contained in the image of g. Now the problem is reduced to the situation of the above lemma, wherein, $\phi = g^{-1} \circ f : M \times I \to M \times I$. ♠

We shall now embark upon making good use of the existence of global solutions. To begin with the following lemma is easy to prove.

Lemma 6.3.4 Let $H : M \times \mathbb{R} \to N \times \mathbb{R}$ be such that $H(x, t) = (h(x, t), t)$. Then H is an embedding iff h is an isotopy.

Remark 6.3.4 Embeddings of the above form are called *level-preserving embeddings.* Often we call H as the *track* of h. Thus, isotopies of maps from M to N are in 1-1 correspondence with level-preserving embeddings of $M \times \mathbb{R}$ in $N \times \mathbb{R}$.

Definition 6.3.6 For any manifold X, recall that $T_{(x,t)}(X \times \mathbb{R}) = T_x X \oplus \mathbb{R}$. So, given a smooth vector field σ on $X \times \mathbb{R}$, we can decompose σ as a direct sum

$$\sigma := (\sigma_1, \sigma_2), \tag{6.13}$$

where $\sigma_2 : X \times \mathbb{R} \to \mathbb{R}$ is a smooth map and $\sigma_1(x, t) \in T_x(X)$. We call σ_1 as the space component of σ and σ_2 as the t-component (time component) of σ. We shall denote by ∂t the vector field whose space component σ_1 is 0 and whose time component σ_2 is the constant map $(x, t) \mapsto 1$.

Now let us take a close look at a level-preserving embedding. Fixing a point $x \in X$, we get the curves $t \mapsto H(x, t) = (h(x, t), t)$. The tangent field to this curve has its t-component equal to ∂t. The space component defines a vector field along the curve, which we can view as time dependent vector field on X.

We now want to reverse this process. We have a ready-made tool for this, viz., vector fields that admit global solutions (see Lemma 6.3.2).

The following theorem is independently due to R. Thom [Th] (compact case), R. Palais [Pa] and J. Cerf [Ce]. We begin with a lemma.

Lemma 6.3.5 Let σ be a vector field on $N \times \mathbb{R}$ whose time component is ∂t. Suppose $H : N \times \mathbb{R} \times \mathbb{R} \to N \times \mathbb{R}$ is the global solution for IVP(σ) as in (6.11). Then the mapping $G(x, t) = H(x, 0, t)$ is the track of an isotopy of the identity map of N.

Proof: We know that $H(x,s,0) = (x,s)$ for all $(x,s) \in N \times \mathbb{R}$ and $DH(\partial t) = \sigma$, since H is a solution of IVP(σ). Therefore, $G(x,0) = H(x,0,0) = (x,0)$.

Let $\pi_1 : N \times \mathbb{R} \to N$ and $\pi_2 : N \times \mathbb{R} \to \mathbb{R}$ be projection maps and let $H_i = \pi_i \circ H; G_i = \pi_i \circ G, i = 1,2$. Thus, $H = (H_1, H_2)$. Then $DH_2(\partial t)$ is nothing but the t-component of σ that is equal to ∂t. This just means that for each fixed (x,s), the curve $t \mapsto H_2(x,s,t)$ is of the form $t + \alpha(x,s)$. Consequently,

$$G_2(x,t) = \pi_2 \circ G(x,t) = t + \alpha(x,0).$$

Since $G(x,0) = (x,0)$, it follows that $\alpha(x,0) = 0$. This proves that G is level-preserving. To show that G is an embedding, we consider the map $L : N \times \mathbb{R} \to (N \times \mathbb{R}) \times \mathbb{R}$ defined by

$$L(x,t) = (H(x,t,-t),t).$$

Note that $(x,s,t) \mapsto (H(x,s,t),t)$ is a level-preserving diffeomorphism of $N \times \mathbb{R} \times \mathbb{R}$ with itself. Taking the composite with the embedding $(x,t) \mapsto ((x,t,-t),t)$, we get the map L. Therefore, L is an embedding. Therefore, it suffices to show that $L \circ G$ is an embedding. (See Exercise 3.7.17.)

Recall that H is a 1-parameter group of diffeomorphisms. Therefore,

$$H(H(p,s,r),t) = H(p,s,r+t).$$

Put $y = H_1(x,0,t)$. Therefore,

$$\begin{aligned} L(G(x,t)) &= L(H_1(x,0,t),t) = L(y,t) \\ &= (H(y,t,-t),t) = (H(H_1(x,0,t),t,-t),t) \\ &= (H(G_1(x,t),t,-t),t) = (H(G(x,t),-t),t) \\ &= (H(H(x,0,t),-t),t) \\ &= (H(x,0,t-t),t) = (H(x,0,0),t) = ((x,0),t). \end{aligned}$$

This proves that G is an embedding. Since we have already seen that G is level-preserving, it follows that G is the track of an isotopy of the identity map. ♠

Theorem 6.3.5 (Isotopy Extension Theorem) *Let $f : M \to N$ be an embedding of a manifold M in a manifold N with $\partial N = \emptyset$. Let $K \subset M$ be a compact subset and $F : M \times \mathbb{R} \to N \times \mathbb{R}$ be the track of an isotopy of f. Then there is $G : N \times \mathbb{R} \to N \times \mathbb{R}$ which is the track of an isotopy such that $G(f(x),t) = F(x,t)$ for all $x \in K$ and $t \in [0,1]$. Moreover, G is the identity map outside a compact subset of $N \times \mathbb{R}$.*

Proof: Consider the vector field $\sigma' = DF \circ \partial t$. [This is nothing but the tangent field to the curves $t \mapsto F(x,t)$.] This is defined on $F(M \times \mathbb{R})$ which is a smooth submanifold of $N \times \mathbb{R}$. Since $\hat{K} := F(K \times [0,1])$ is compact, we can extend $\sigma'|_{\hat{K}}$ to a vector fieldσ on all of $N \times \mathbb{R}$ in such a way that σ vanishes outside an open set V containing $\hat{K}$ and such that $\bar{V}$ is compact. We define a new vector field τ on $N \times \mathbb{R}$ by

$$\tau = (\sigma_1, \partial t),$$

i.e, the space-component of τ is the same as that of σ whereas the time-component has been changed to 1. Observe that $\sigma'|_{\hat{K}}$ has its time component ∂t and hence the extended σ also has this property. Therefore, it follows that

$$\tau|_{\hat{K}} = \sigma'|_{\hat{K}} = \sigma|_{\hat{K}}.$$

We have only to verify that τ admits global solutions. For this, we examine the behaviour of the solution curves of τ and find $\epsilon > 0$ such that for all $x \in N$ the solution curves are defined for $|t| < \epsilon$.

Let W be an open set such that $\bar{V} \subset W$ and $\bar{W}$ is compact. Then there exists $\epsilon > 0$ such that all solution curves originating in W are defined for $|t| < \epsilon$.

On the other hand, consider a solution curve originating at a point $(x_0, t_0) \notin \bar{V}$. Since σ_1 vanishes outside V, the curve is of the form $t \mapsto (x_0, t_0 + t)$. These curves may never hit $\bar{V}$, in which case, the solution is defined for all $t \in \mathbb{R}$. On the other hand, if it hits $\bar{V}$, say for some $t = t_1 \neq 0$, that is, now we have $(x_0, t_0 + t_1) \in \bar{V} \subset W, |t_1| > 0$. The solution curve can now be extended further for a time $|t| < \epsilon$ and hence is defined at least in the interval $|t| < \epsilon + |t_1|$.

Thus in either case, the solution curves are defined for $|t| < \epsilon$.

Let now $H : N \times \mathbb{R} \times \mathbb{R} \to N \times \mathbb{R}$ be the global solution given by σ and $G(x,t) = H(x, 0, t)$. For some $x_0 \in K$ let C_0 be the solution curve of σ which at $t = 0$ passes through $(f(x_0), 0)$. Then $C_0 : t \mapsto F(x_0, t)$. Since σ and τ agree on $\hat{K} = F(K \times [0,1])$, it follows that, C_0 is the solution curve for τ as well in $0 \leq t \leq 1$. Hence, $F(x_0, t) = H(f(x_0), 0, t) = G(f(x_0), t)$ for $0 \leq t \leq 1$. ♠

Remark 6.3.5 It is also important to note that if $V \subset N$ is any open set with compact closure and $\hat{K} = F(K \times I) \subset V$, then the isotopy G of N that we get is the identity map outside V. The hypothesis $\partial N = \emptyset$ itself is not crucially used; but if ∂N is nonempty we need to assume that the image of F avoids ∂N.

Remark 6.3.6 This theorem has several natural as well as surprising applications. One of the easy consequences is that any two collars of a compact boundary component of a manifold are ambient isotopic (see Theorem 6.3.4). We have many more to come.

Corollary 6.3.1 Let $Z \subset M$ be a compact submanifold. Let $f_0, f_1 : Z \to N \setminus \partial N$ be any two isotopic embeddings in $N \setminus \partial N$. If f_0 extends to an embedding of M into N then so does f_1.

Proof: Let $F : Z \times \mathbb{R} \to N$ be an isotopy of f_0 and f_1. Let $G : f_0(Z) \times \mathbb{R} \to N \times \mathbb{R}$ be defined by $G(y,t) = (F(f_0^{-1}(y), t), t)$. We can then apply Theorem 6.3.5 with $K = f_0(Z)$ to obtain the track of an isotopy $H : N \times \mathbb{R} \to N \times \mathbb{R}$ such that $H(y,t) = G(y,t)$ for all $y \in f_0(Z)$. Now consider $L : M \times \mathbb{R} \to N$ given by

$$L(x) = \pi \circ H(f_0(x), 1)$$

where $\pi : N \times \mathbb{R} \to N$ is the first coordinate projection. Clearly, $L(x) = \pi(G(f_0(x), 1)) = \pi(F(x,1), 1) = f_1(x),\ x \in Z$. ♠

Corollary 6.3.2 Let N be a connected manifold, and a, b be any two points in $N \setminus \partial N$. Then there exists a diffeomorphism $\phi : N \to N$ isotopic to identity such that $\phi(a) = b$.

Proof: A smooth path from a to b inside $N \setminus \partial N$ defines an isotopy of the singleton $\{a\} \subset N$. Now apply Theorem 6.3.5. ♠

Corollary 6.3.3 Let N be a connected manifold of dimension ≥ 2 and let k be any positive integer. For any two k-subsets $\{a_1, \ldots, a_k\}$ and $\{b_1, \ldots, b_k\}$ of $N \setminus \partial N$, there exists a diffeomorphism $\phi : N \to N$ isotopic to identity such that $\phi(a_i) = b_i, 1 \leq i \leq k$.

Proof: Induct on k, the case $k = 1$ being covered by the above corollary. Assume that the result holds for $k - 1$ where $k \geq 2$. Let $\psi : N \to N$ be a diffeomorphism isotopic to identity such that $\psi(a_i) = b_i,\ 1 \leq i \leq k-1$. If $\psi(a_k) = b_k$ there is nothing more to be done; simply take $\phi = \psi$. Otherwise, we can join $\psi(a_k)$ to b_k by a path $\gamma : \mathbb{R} \to N$ not passing through any of $b_1, \ldots, b_{k-1}$. Let $S = \{b_1, \ldots, b_{k-1}, \psi(a_k)\}$. Consider the isotopy of S, viz., $F : S \times \mathbb{R} \to N$ given by

$$F(s,t) = \begin{cases} s, & s \neq \psi(a_k), \\ \gamma(t), & s = \psi(a_k). \end{cases}$$

Now apply Theorem 6.3.5 with $M = K = S$ to obtain an isotopy $\tau : N \to N$ such that $\tau(b_i) = b_i, i \leq k-1$ and $\tau(\psi(a_k)) = b_k$. Now take $\phi = \tau \circ \psi$. ♠

Theorem 6.3.6 Orientation Reversing Isotopy: *Let N be a connected nonorientable manifold, $p \in N$ be any point. Then there exists a diffeomorphism $\phi : N \to N$ isotopic to Id_N such that $\phi(p) = p$ and $D\phi_p : T_pN \to T_pN$ is orientation reversing.*

Proof: Let γ be a loop at p that is orientation reversing. (See Exercise 4.1.1.) Then γ can be thought of as an isotopy of the embedding $\{p\} \hookrightarrow N$. Let $\Phi : N \times \mathbb{R} \to N \times \mathbb{R}$ be an isotopy extending this as in Theorem 6.3.5. Let us verify that $\phi(x) = \Phi(x, 1)$ is such that $D\phi_p : T_pN \to T_pN$ is orientation reversing. For this, we may choose any parameterization $\tau : \mathbb{R}^n \to N$ of a neighborhood of $p \in N$ such that $\tau(0) = p$ and fix an orientation on $T_p(N)$ say, $D\tau_0(E_n)$. Put $\phi_t(q) = \Phi(q,t)$, so that $\phi_0 = Id_N$ and $\phi_1 = \phi$. Also note that $\Phi(p,t) = (\gamma(t), t)$. Therefore, it follows that $\{\phi_t(\tau(\mathbb{R}^n))\}$ forms an open cover for the loop $\gamma[0,1]$. We can now extract a finite subcover $\{U_j\}$ out of this, where $U_j = \phi_{t_j}(\tau(\mathbb{R}^n))$ with $0 = t_0 < t_1 \cdots < t_k = 1$ and such that there are points $t_j < s_j < t_{j+1}$ so that $\gamma[s_j, s_{j+1}]) \subset U_j$. Orient U_0 by taking $D\tau_x(E_n)$ on $T_{\tau(x)}N$. Having oriented U_j, orient U_{j+1} so that the two orientations on $T_{\gamma(s_j)}N$ coincide. By continuity, and from the fact that $\phi_0 = Id$, it follows that for each j, and $s_j \leq t \leq s_{j+1}$, $D\phi_t$ preserves this orientation. Now γ is an orientation reversing path means that the orientation on $T_{\gamma(1)}N = T_pN$ coming from U_k is different from the one with which we started. Therefore, it follows that $D\phi_p$ reverses the orientation. ♠

Theorem 6.3.7 Disc Theorem: *Let $f_0, f_1 : \mathbb{D}^n \to N \setminus \partial N$ be any two embeddings, where N is a connected n-manifold. If N is orientable assume that either both f_0 and f_1 preserve orientation or both reverse it. Then they are ambient isotopic.*

Proof: Observe that by isotopy extension Theorem 6.3.5, since $\mathbb{D}^n$ is compact, it is enough to prove that f_i are isotopic to each other.

First consider the case, when N is orientable and f_i are both orientation preserving. By an ambient isotopy along a path joining $f_1(0)$ to $f_2(0)$ we can first make $f_1(0) = f_2(0) = p$. Let now U be a coordinate patch for N around p. Choose $\delta > 0$ such that $f_i(\mathbb{D}^n_\delta) \subset U$. Let $\Theta(x,t) = (1 - t + t\delta)x$. Then $\Theta : \mathbb{D}^n \times [0,1] \to \mathbb{D}^n$ is an isotopy such that $\Theta(\mathbb{D}^n \times 1) \subset U$. Therefore, $f_i \circ \Theta$ gives an isotopy of f_i with an embedding g_i such that $g_i(\mathbb{D}^n) \subset U$. Thus, we may as well assume that $f_i(\mathbb{D}^n) \subset U$. This is equivalent to assume that $N = \mathbb{R}^n$. The conclusion follows from the Example 6.3.1, 1 and 2.

The case when both f_i are orientation reversing is converted into the first case, by changing the orientation on the codomain N.

Now suppose N is nonorientable. We can follow the above arguments till we get to the case when $f_i(D^n) \subset U$, where U is a coordinate patch. But now it may happen that for any fixed orientation on U one of f_i is orientation preserving and the other revering! In this situation, we go back to the manifold N find a loop at p which is orientation reversing and perform an ambient isotopy of $f_2 \circ f^{-1}$ along this loop to an embedding h. Now f_2 and $\hat{f_2} = h \circ f_1$ are isotopic and $f_1, \hat{f_2}$ are both orientation preserving or orientation reversing. We are then back in the first case. ♠

Exercise 6.3

1. Justify the choice of the function ϵ in the proof of lemma 6.3.1.

2. Show that every orientation preserving diffeomorphism of $\mathbb{S}^1$ is isotopic to the identity map.

3. Show that the embedding $f : (0, 1) \to \mathbb{R}^2$ given by $f(t) = e^{2\pi \imath t}$ is not ambient isotopic to the inclusion map $t \mapsto (t, 0)$.

4. For each $n \geq 1$ construct at least one diffeomorphism $\mathbb{S}^n \to \mathbb{S}^n$ which reverses orientation.

5. Obtain an embedding of the Möbius band inside the solid torus in $\mathbb{R}^3$ which itself is obtained by rotating the disc

$$\{(x, y), 0) \in \mathbb{R}^3 \ : \ (x - 2)^2 + y^2 \leq 1\}$$

around the y-axis. (See Example 3.1.1.7.) Use this description to explicitly write down a diffeomorphism $\phi : \mathbb{M} \to \mathbb{M}$ which is isotopic to identity.

6. Let $f : \mathbb{R} \to \mathbb{R}^3$ be a proper embedding representing the trefoil knot as shown in the figure below:

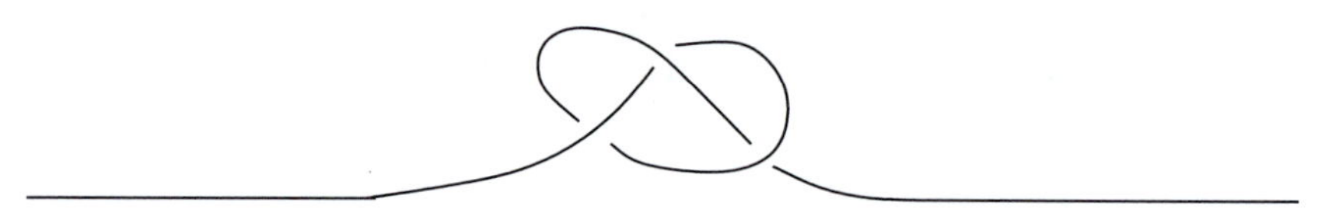

Figure 31 The trefoil knot.

Show that f is not ambient isotopic to the inclusion map $t \mapsto (t, 0, 0)$. Are the two embeddings isotopic?

7. Given $0 < \epsilon < 1/2$, construct a smooth map $\mu : I \to I$ such that
(i) $\mu(t) = t, \ 0 \leq t \leq \epsilon/2$ or $3/4 \leq t \leq 1$.
(ii) $\mu'(t) \geq 0$ for all t.
(iii) $\mu(1/2) = \epsilon$.

8. Define $h : \mathbb{D}^n \times I \to \mathbb{D}^n$ by

$$h(\mathbf{x}, t) = (t\mu(\|\mathbf{x}\|) + (1 - t)\|\mathbf{x}\|)\frac{\mathbf{x}}{\|\mathbf{x}\|}$$

where μ is as in the above exercise. Show that h defines an isotopy of $\mathbb{D}^n$ which is identity in a neighborhood of the boundary and maps $\mathbb{D}^n_{1/2}$ onto $\mathbb{D}^n_\epsilon$.

9. Let $f : \mathbb{D}^n \to \mathbb{R}^n$ be any embedding such that $f(0) = 0$. Let V be a neighborhood of $f(\mathbb{D}^n) \cup Df_0(\mathbb{D}^n)$ such that $\bar{V}$ is compact. Show that f is isotopic to a linear embedding by a diffeotopy of $\mathbb{R}^n$ which is the identity map outside V.

10. Why do we need dim $N \geq 2$ in Corollary 6.3.3?

6.4 Patching-up Diffeomorphisms

Let M be a closed m-manifold and $\phi : M \times I \to M \times I$ be any smooth map. Suppose for some $0 \le t \le 1$, $\phi(M \times \{t\}) \subset M \times \{s\}$ for some s. Then the derivative $D\phi$ at a point $(x,t) \in M \times I$ is of the form

$$\begin{pmatrix} A & v \\ 0 & \alpha \end{pmatrix}$$

where A is an invertible $m \times m$ matrix and α is a nonzero real number. In particular, if $\phi_1 : M \times [a,b] \to M \times [\alpha,\beta]$ and $\phi_2 : M \times [b,c] \to M \times [\beta,\gamma]$ are diffeomorphisms such that $\phi_1|_{M\times\{b\}} = \phi_2|_{M\times\{b\}}$, we can patch-up the two diffeomorphisms into a homeomorphism $\psi : M \times [a,c] \to M \times [\alpha,\gamma]$ but it may fail to be a $\mathcal{C}^1$-map in general. This is so because at points (x,b), the t-derivatives of ϕ_1 and ϕ_2 may not match. On the other hand the method we have followed to compose homotopies would produce a 1-to-1 map that is also $\mathcal{C}^1$ but fails to give a diffeomorphism since the t-derivatives at points (x,b) vanish. In this section we shall study a partial solution to this problem. It is of some practical importance for us in the diffeomorphic classification of surfaces.

In the following lemma, we have to use the fact that the solution generated by a smooth family of vector fields is smooth. (See Theorem 6.3.1.)

Lemma 6.4.1 Smoothing Lemma: Let M be a compact manifold and σ_0, σ_1 be any two vector fields on M which are tangential to the boundary at boundary points. There is a diffeomorphism $\Theta : M \times I \to M \times I$ such that
(a) $\Theta|_{M\times\{0,1\}} = Id$ and
(b) at points $(x,0)$ and $(x,1)$, $D\Theta$ is, respectively, of the form

$$\begin{bmatrix} Id & \sigma_0 \\ 0 & 1 \end{bmatrix}; \quad \begin{bmatrix} Id & \sigma_1 \\ 0 & 1 \end{bmatrix}$$

[Here Id denotes the identity map of the tangent space T_xM for an appropriate $x \in M$.]

Proof: Note that, it is enough to prove the result for the case $\sigma_1 = 0$. For then, we can put two such diffeomorphisms together to define a diffeomorphism $\Gamma : M \times [0,2] \to M \times [0,2]$ so that $D(\Gamma)$ at points $(x,0)$ and $(x,2)$ (instead of $(x,1)$) satisfies condition (b). It is then just a matter of reparameterizing $M \times [0,2]$ by $M \times [0,1]$. So, we shall prove the result for the case $\sigma_0 = \sigma$ and $\sigma_1 = 0$.

For any vector field τ on M admitting global solution, let us introduce the notation Φ^τ for the 1-parameter group of diffeomorphisms $M \times \mathbb{R} \to M$ generated by τ. We consider the family of vector fields $s\sigma$ and the associated 1-parameter group of diffeomorphisms $\Phi^{s\sigma}$, for $0 \le s \le 1$. (Note that $\Phi^0_t = Id$ for all t and $\Phi^{s\sigma}_0 = Id$ for all s.) Define $\Psi : M \times I \times I \to M \times I \times I$ by

$$\Psi(x,s,t) = (\Phi^{s\sigma}(x,t), s, t).$$

Check that Ψ defines a diffeomorphism that is doubly level-preserving.

Let $\eta : I \to I$ be a strictly monotonically decreasing smooth map with $\eta(0) = 1$, $\eta(1) = 0$ and $\eta'(0) = \eta'(1) = 0$ (see 1.78). Put $G(x,r) = (x, \eta(r), r)$ and let $\pi(x,s,t) = (x,t)$ be the projection. Put $\Theta = \pi \circ \Psi \circ G : M \times I \to M \times I$, i.e., $\Theta(x,r) = (\Phi^{\eta(r)\sigma}(x,r), r)$. It is easily seen that Θ is a diffeomorphism satisfying (a). We have to check (b). By the chain rule, it follows that the matrix form of $D\theta$ is the product:

$$\begin{bmatrix} 1 & 0 & 0 \\ 0 & 0 & 1 \end{bmatrix} \begin{bmatrix} \frac{\partial \Phi^{s\sigma}}{\partial x} & \frac{\partial \Phi^{s\sigma}}{\partial s} & \frac{\partial \Phi^{s\sigma}}{\partial r} \\ 0 & 1 & 0 \\ 0 & 0 & 1 \end{bmatrix} \begin{bmatrix} 1 & 0 \\ 0 & \eta'(r) \\ 0 & 1 \end{bmatrix}$$

which is equal to

$$\begin{bmatrix} \frac{\partial \Phi^{s\sigma}}{\partial x} & \frac{\partial \Phi^{s\sigma}}{\partial s}\eta'(r) + \frac{\partial \Phi^{s\sigma}}{\partial r} \\ 0 & 1 \end{bmatrix}.$$

By the definition, $\Phi^{s\sigma}(x,0) = x, \forall x$ and hence $\frac{\partial \Phi^{s\sigma}}{\partial x}(x,0) = Id$. On the other hand, since $s = \eta(r)$ and $\eta(1) = 0$ we have $\Phi^{s\sigma}(x,1) = \Phi^0(x,1) = x$. Therefore, $\frac{\partial \Phi^{s\sigma}}{\partial x}(x,1) = Id$. This takes care of the $(1,1)^{\text{th}}$ entry of the matrices in (b). Now use the fact that $\eta'(r) = 0$ for $r = 0, 1$ to see that the $(1,2)^{\text{th}}$ entry reduces to $\frac{\partial \Phi^{s\sigma}}{\partial r}(x,r) = \eta(r)\sigma(\Phi^{s\sigma}(x,r))$ for $r = 0, 1$. Again, for $r = 0$, since $s = \eta(0) = 1$, we have $\frac{\partial \Phi^{s\sigma}}{\partial r}(x,0) = \sigma(\Phi^{\sigma}(x,0)) = \sigma(x)$. Finally, for $r = 1$, since $s = \eta(1) = 0$ we have $\frac{\partial \Phi^{s\sigma}}{\partial r}(x,1) = 0\sigma(x,1) = 0$. This completes the proof of the lemma. ♠

Remark 6.4.1
(i) Given some smooth functions $\alpha, \beta : M \to (0,\infty)$, using Exercise 1.7.3, the conclusion in (b) of the above lemma can be strengthened to the following:
(b') at points $(x,0)$ and $(x,1)$, $D\Theta$ is, respectively, of the form

$$\begin{bmatrix} Id & \sigma_0 \\ 0 & \alpha \end{bmatrix}; \quad \begin{bmatrix} Id & \sigma_1 \\ 0 & \beta \end{bmatrix}$$

(ii) There is nothing special about the interval $I = [0,1]$, in the above lemma; we can replace $[0,1]$ by $[a,b]$ for any $a < b$.
(iii) Often in application, the vector fields σ, τ may be given on a closed subset of M. We can simply extend these on the whole of M smoothly and work with the extended vector fields.

Using the existence of the collar neighborhoods, we can now derive several interesting results from this lemma. The following corollary is immediate.

Corollary 6.4.1 Given a vector field $\lambda : \mathbb{S}^{n-1} \to T(\mathbb{D}^n)$ that is strictly pointing inwards, there exists a diffeomorphism $\phi : \mathbb{D}^n \to \mathbb{D}^n$, that is identity on the boundary and such that the radial component of $D\phi$ at any point on the boundary is equal to σ.

Proof: If σ and β are the spherical and radial components of λ, we can apply the above lemma, along with Remark 6.4.1 (a), to get a diffeomorphism $\Theta : S^{n-1} \times I \to S^{n-1} \times I$ so that $D(\theta)$ satisfies (b') with $\alpha \equiv 1$. Treating (x,r) as polar coordinates, this Θ defines a diffeomorphism $\phi : \mathbb{D}^n \to \mathbb{D}^n$ as required. ♠

Theorem 6.4.1 *Let $\phi : \mathbb{S}^{n-1} \to \mathbb{S}^{n-1}$ be a diffeomorphism which is diffeotopic to $Id_{\mathbb{S}^{n-1}}$. Then ϕ can be extended to a diffeomorphism of $\mathbb{D}^n$.*

Proof: If ϕ is diffeotopic to Id we can find a level-preserving diffeomorphism $\eta : \mathbb{S}^{n-1} \times [1/2, 1] \to \mathbb{S}^{n-1} \times [1/2, 1]$ such that $\eta(v, 1) = (\phi(v), 1)$ and $\eta(v, 1/2) = (v, 1/2)$. Via the polar coordinates this give a diffeomorphism ψ of the annulus $\{\mathbf{x} \in \mathbb{D}^n \ : \ 1/2 \leq \|\mathbf{x}\| \leq 1\}$ such that the radial component σ of $D\psi$ at any point $(x, 1/2) \in \mathbb{D}^n$ is pointing inward. Thus, by the above corollary, we can extend this diffeomorphism to a diffeomorphism of the disc to itself. Note that ψ restricts to ϕ on the boundary. ♠

Theorem 6.4.2 Patching-up diffeomorphisms: *Let M be a compact manifold, $\phi : M \times I \to M \times I$ be a continuous function such that $\phi_1 := \phi|_{M\times[0,1/2]}$ and $\phi_2 := \phi|_{M\times[1/2,1]}$ are self-diffeomorphisms of $M \times [0, 1/2]$ and $M \times [1/2, 1]$, respectively. Given $0 < \delta < 1/2$, there exists a diffeomorphism $\Phi : M \times I \to M \times I$ such that $\Phi|_{M\times[0,\delta]} = \phi|_{M\times[0,\delta]}$ and $\Phi|_{M\times[1-\delta,1]} = \phi|_{M\times[1-\delta,1]}$.*

Proof: Clearly, ϕ is a homeomorphism $M \times [0,1] \to M \times [0,1]$ and takes $M \times \{1/2\}$ into itself. Let us write $f(x) = \phi(x, 1/2)$. Then it follows that $f : M \to M$ is a diffeomorphism. The missing thing for ϕ to be a diffeomorphism on the whole of $M \times [0,1]$ is that ϕ may not be smooth at points in $M \times \{1/2\}$. Indeed, the only missing thing is that at points $(x, 1/2)$ the t-derivatives of ϕ from left and right may not be the same.

By composing with a variable reparameterization (see Exercise 1.7.3) we shall assume that the time component of the time derivative of ϕ at points $(x, 1/2)$ are both equal to 1 from either side of the interval. Thus, at each point $(x, 1/2)$, the left-hand and the right-hand t-derivatives of ϕ exist and are of the form $\begin{bmatrix} \sigma_1 \\ 1 \end{bmatrix}, \begin{bmatrix} \sigma_2 \\ 1 \end{bmatrix}$, respectively, where σ_j are some smooth vector fields on M. Note that since ϕ restricts to self-diffeomorphisms of $\partial M \times [0, 1/2]$ and $\partial M \times [1/2, 1]$, it follows that σ_j are tangential to the boundary. We shall now use the above lemma to make $\sigma_1 = \sigma_2$.

For $j = 1, 2$, let $\Theta_j = \theta(\sigma_j)$ be diffeomorphisms:

$$\Theta_1 : M \times [1/2, 1] \to M \times [1/2, 1]; \quad \Theta_2 : M \times [0, 1/2] \to M \times [0, 1/2]$$

such that $\Theta_1(x, 1/2) = x = \Theta_1(x, 1)$; $\Theta_2(x, 0) = x = \Theta_2(x, 1/2)$ and $D\Theta_1|_{(x,1/2)} = \begin{bmatrix} Id & \sigma_1(x) \\ 0 & 1 \end{bmatrix}$; $D\Theta_2|_{(x,1/2)} = \begin{bmatrix} Id & \sigma_2(x) \\ 0 & 1 \end{bmatrix}$;

$$D\Theta_1|_{(x,1)} = \begin{bmatrix} Id & 0 \\ 0 & 1 \end{bmatrix} = D\Theta_2|_{(x,0)}.$$

Let $T : M \times [0,1] \to M \times [1,2]$ be the translation $(x, t) \mapsto (x, t+1)$. Define $\Lambda : M \times [0,2] \to M \times [0,2]$ by:

$$\Lambda(x,t) = \begin{cases} \phi(x,t), & 0 \le t \le 1/2; \\ \Theta_1(f(x), t), & 1/2 \le t \le 1; \\ T \circ \Theta_2(f(x), t-1), & 1 \le t \le 3/2; \\ T \circ \phi(x, t-1), & 3/2 \le t \le 2. \end{cases}$$

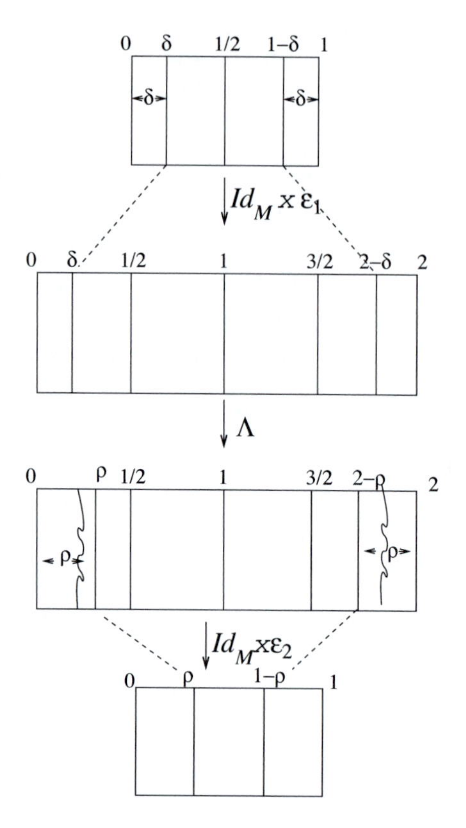

Figure 32 The t-derivatives of Φ are being adjusted.

Check that Λ is a diffeomorphism. Let $\epsilon_1 : [0,1] \to [0,2]$ be a diffeomorphism such that

$$\epsilon_1(t) = t,\ 0 \le t \le \delta; \quad \& \quad \epsilon_1(t) = t+1,\ 1-\delta \le t \le 1.$$

By the compactness of M, it follows that there is a $0 < \rho < 1/2$ such that $\phi(M \times [0,\delta]) \subset M \times [0,\rho]$ and $\phi(M \times [1-\delta, 1] \subset M \times [1-\rho, 1]$. Let $\epsilon_2 : [0,2] \to [0,1]$ be a diffeomorphism such that

$$\epsilon_2(t) = t,\ 0 \le t \le \rho \quad \& \quad \epsilon_2(t) = t-1,\ \ 2-\rho \le t \le 2.$$

Put $\Phi = (Id_M \times \epsilon_2) \circ \Lambda \circ (Id_M \times \epsilon_1)$ and verify that Φ is as required. ♠

Remark 6.4.2 Consider the converse of Theorem 6.4.1, viz., if $\Phi : \mathbb{D}^n \to \mathbb{D}^n$ is a diffeomorphism then the restriction map $\eta = \Phi|_{\partial\mathbb{D}^n}$ is isotopic to the identity diffeomorphism or to the reflection. This is "almost true" but extremely difficult to prove. What one can achieve without much difficulty is the following. If η is given as above, and is orientation preserving, then there exists a diffeomorphism $\Psi : \mathbb{S}^{n-1} \times [0,1] \to \mathbb{S}^{n-1} \times [0,1]$ which is the identity map on $\mathbb{S}^{n-1} \times 0$ and is equal to η on $\mathbb{S}^{n-1} \times 1$. However, this does not mean that η is isotopic to Id; the diffeomorphism Φ may not be level-preserving.

More generally, one can consider any closed manifold M and a diffeomorphism $\Phi : M \times [0,1] = M \times [0,1]$ such that $\Phi|_{M\times 0} = Id$. Such a diffeomorphism is called a *pseudo isotopy* of $\Phi_{M\times 1}$. If $\mathcal{P}$ denotes the space of all pseudo isotopies of diffeomorphisms of M one is interested in the number of path-connected components of $\mathcal{P}$. Cerf [C1] has proved that if M is simply connected and $\dim M \ge 5$ then $\mathcal{P}$ is path connected.

From this, Cerf [C1] deduced that the group A_n of all orientation preserving diffeomorphisms of $\mathbb{D}^n$ is path connected for $n \geq 6$. For $n = 1$, this is an easy exercise. For $n = 2$, this was a result of Smale [Sm]. For $n = 3$ and 4 this is again proved by Cerf [C2]. For $n = 5$, the problem is still open.

The following result which is a variant of Theorem 6.4.2 comes handy later in the classification of surfaces.

Theorem 6.4.3 *Let X, Y be n-dimensional smooth manifolds, $X_1, X_2 \subset X, Y_1, Y_2 \subset Y$ be such that*

$$X = X_1 \cup X_2, Y = Y_1 \cup Y_2, X_1 \cap X_2 = \partial X_1 \cap \partial X_2; \quad Y_1 \cap Y_2 = \partial Y_1 \cap \partial Y_2$$

where $X_i \subset X, Y_i \subset Y$ are closed subspaces. Let $f : X \to Y$ be a homeomorphism whose restrictions to X_1 and X_2 define diffeomorphisms $f_1 : X_1 \to Y_1$ and $f_2 : X_2 \to Y_2$. Then f can be altered in a neighborhood of $X_1 \cap X_2$ to a diffeomorphism $\tilde{f} : X \to Y$.

Proof: Put $X_1 \cap X_2 = M, Y_1 \cap Y_2 = N$. By restricting the discussion to collar neighborhoods of M, N we may assume that

$$X_1 = M \times [0, 1/2], X_2 = M \times [1/2, 1], X = M \times [0, 1];$$
$$Y_1 = N \times [0, 1/2], Y_2 = N \times [1/2, 1], Y = N \times [0, 1].$$

Note that f restricts to a diffeomorphism $g : M \to N$. Consider $h : N \times [0, 1] \to M \times [0, 1]$ given by $h(x, t) = (g^{-1}(x), t)$ and put $\phi = h \circ f : M \times [0, 1] \to M \times [0, 1]$. Apply the Theorem 6.4.2 to obtain a diffeomorphism $\Phi : M \times I \to M \times I$ as specified. Now $\tilde{f} = g \circ \Phi$ gives the required alteration of f. ♠

Example 6.4.1 Gluing Manifolds along Boundary Components: Let M be a closed manifold. Consider two embeddings $\phi_i : M \times I \to N_i$ where each $\phi_i|_{M \times \{0\}}$ is an embedding onto some boundary components M_i of N_i and $\phi_i(M \times [0, 1])$ is a collar neighborhood of M_i in N_i, $i = 1, 2$. Consider the quotient space

$$W = \frac{N_1 \setminus M_1 \coprod N_2 \setminus M_2}{\sim}$$

where $\phi_1(x, t) \sim \phi_2(x, 1 - t) \ \forall \ x \in M, 0 < t < 1$. Clearly, W is a smooth manifold that contains copies of both N_1 and N_2. Indeed, if we put $N_i' = N_i \setminus \phi_i(M \times [0, 1/2)$ then we see that $W = q(N_1') \cup q(N_2')$, with $q(N_1') \cap q(N_2') \cong M$. Also each N_i' is diffeomorphic to N_i and the quotient map q is also a diffeomorphism restricted to each N_i'.

Of course, the diffeomorphism class of W heavily depends on the embeddings $\phi_i, i = 1, 2$. Since any two collar neighborhoods are ambient isotopic (see Remark 6.3.6), ultimately it follows that the diffeomorphism class of W depends on the isotopy class of diffeomorphisms of M.

Example 6.4.2 Double of a Manifold Let X be a compact manifold with non empty boundary and $\phi : \partial X \times [0, 1] \to X$ be a collar neighborhood. Take $N_1 = X$ and N_2 to be another copy of X with the opposite orientation if X is orientable. Take $\phi_1 = \phi_2 = \phi$ and perform the gluing operation. The resulting manifold W in the above construction is called the double of the manifold X and is denoted by $2X$. Observe that if X is orientable then so is $2X$. In particular, the double of the disc $\mathbb{D}^n$ is $\mathbb{S}^n$, and the double of the cylinder is the torus, etc.

Example 6.4.3 Capping Off: Let X be an n-manifold with a boundary component C diffeomorphic to $\mathbb{S}^{n-1}$. By *capping off X at C* we mean the manifold obtained by gluing a disc $\mathbb{D}^n$ to X along C. More specifically, in Example 6.4.1, take $N_1 = X, N_2 = \mathbb{D}^n, M = \mathbb{S}^{n-1}$ and $\phi_1 : S^{n-1} \times [0,1] \to N_1$ be any collar neighborhood of C and $\phi_2(x,t) = (1-t/2)x$. We shall denote the resulting manifold W by $\hat{X}_C = W$.

When we cap-off all the boundary components of X that are diffeomorphic to $\mathbb{S}^{n-1}$, we get a manifold, that we simply denote by $\hat{X}$. This construction happens to be very important in the study of manifolds of dimension 2, 3, since any orientation-preserving diffeomorphism of $\mathbb{S}^1$ or S^2 is isotopic to the identity map. (This is an exercise for $n = 1$ and a nontrivial result due to Smale [Sm] for $n = 2$.) Therefore, the result of capping off is independent of the diffeomorphisms chosen and therefore is a well defined manifold. Notice that we can recover the original manifold X from $\hat{X}$ by removing as many discs as we have put. (Use Disc Theorem.)

Exercise 6.4

1. Directly using Corollary 6.3.1 deduce that any diffeomorphism $f : \mathbb{S}^{n-1} \to \mathbb{S}^{n-1}$ that is isotopic to Id extends to a diffeomorphism of $\mathbb{D}^n$.

2. Let M be any connected manifold of dimension $n \geq 2$. Let $D_i, D'_i, i = 1, 2, \ldots, k$ be embedded discs in int M such that $D_i \cap D_j = \emptyset = D'_i \cap D'_j, i \neq j$. Then there exists diffeomorphism $f : M \to M$ diffeotopic to Id such that $f(D_i) = D'_i, i = 1, 2, \ldots, k$.

3. Let W be a connected manifold with $S_1, \ldots, S_k$ as boundary components, all of them diffeomorphic to $\mathbb{S}^{n-1}$. Given any permutation σ of n letters, there exists a diffeomorphism $\lambda : W \to W$ such that $\lambda(S_i) = S_{\sigma(i)}$.

4. Let W be a smooth surface W_0, W_1 be such that $W = W_0 \cup W_1$. Suppose there are diffeomorphisms $\phi_i : I \times I \to W_i$ such that

   $$\phi_0(0 \times I) = \phi_1(1 \times I); \phi_0(1 \times I) = \phi_1(0 \times I); W_1 \cap W_2 = \phi_0(0 \times I) \cup \phi_0(1 \times I).$$

   Assume further that $\phi_0^{-1} \circ \phi_1|_{0 \times I}$ and $\phi_0^{-1} \circ \phi_1|_{1 \times I}$ are both orientation-preserving or both orientation-reversing. Then there exists a diffeomorphism $\Phi : \mathbb{S}^1 \times I \to W$, such that $\Phi(e^{\imath\pi\theta}, s) = \phi_0(\theta, s), 1/3 \leq \theta \leq 2/3$, and $\Phi(e^{-\imath\pi\theta}, s) = \phi_1(\theta, s), 1/3 \leq \theta \leq 2/3$.

5. In the above exercise assume that one of $\phi_0^{-1} \circ \phi_1|_{0 \times I}$ and $\phi_0^{-1} \circ \phi_1|_{1 \times I}$ is orientation-preserving and the other is orientation-reversing. Then show that there is a diffeomorphism $\Phi : \mathbb{M} \to W$ with similar properties as in the exercise, where $\mathbb{M}$ denotes the Möbius band.

6. State and prove similar results that describe a torus and a Klein bottle as a union of two cylinders.

6.5 Miscellaneous Exercises for Chapter 6

1. Let M be a closed submanifold of N. Show that any two tubular neighborhoods of M in N are ambient isotopic.

2. Let $A, B \subset M$ be closed submanifolds of a closed manifold M. Suppose $\dim A + \dim B < \dim M$. Then show that there exists an ambient isotopy ϕ of M such that $\phi(A) \cap B = \emptyset$.

3. **Some Consequences of the Tubular Neighborhood Theorem.**
(i) Let $X \subset \mathbb{R}^N$ be a smooth compact submanifold. Then X is a smooth retract of an open neighborhood (in fact of its ϵ-neighborhood, for all sufficiently small $\epsilon > 0$).
(ii) Let X be a smooth compact submanifold of a smooth manifold Y. Then X is a smooth retract of an open neighborhood of X in Y. (This is clear from (i) if Y is a submanifold of some $\mathbb{R}^N$.) Hence, any smooth map $X \to S$ (S an arbitrary smooth manifold) can be extended smoothly to a neighborhood of X in Y.
(iii) **Local triviality of a proper submersion:** Let $f : X \to Y$ be a smooth map, and $p \in S$ a regular value of f. Suppose $F := f^{-1}(0)$ is compact. Then show that there is a neighborhood U of p in Y and a diffeomorphism $\phi : f^{-1}(U) \to F \times U$ such that $\pi_U \circ \phi = f$. (Hint: Can take $\phi = (r, f)$, where r is a retraction onto F of a neighborhood of F in X, if U is sufficiently small.)
(iv) Conclude that if $f : X \to Y$ is a proper map, which is a submersion of connected manifolds, then there exists an open covering $\{U_\alpha\}$ of Y and diffeomorphisms $\phi_\alpha : f^{-1}(U_\alpha) \to U_\alpha \times F$ such that $f = \pi_1 \circ \phi_\alpha$, where $F = f^{-1}(p)$ where $p \in Y$ is fixed.

Chapter 7

Intersection Theory

The differentiable viewpoint of some topological concepts as expounded by Milnor in [M1] is fully realized in this chapter. We begin with some technical results about the transversal maps in Section 7.1. We then introduce the oriented intersection number in Section 7.2. In Section 7.3, the degree of maps between manifolds of the same dimension is introduced as a special case of intersection number. We then discuss, in Section 7.4, the mod 2 intersection number, which, while taking care of nonoriented cases, adds its own flavour to the recipe. In Section 7.5, the concept of winding number is introduced as a special case of the mapping degree. We then give applications of these results to the proofs of the Jordan-Brouwer Separation Theorem, the Borsuk-Ulam Theorem (Section 7.6), and the Hopf Degree Theorem (Section 7.7). We take up the study of the Lefschetz fixed point theory in Section 7.8. The local Lefschetz numbers are introduced, once again through the mapping degree. We then relate the Lefschetz number to the index of a vector field.

Of course, one can think of several applications at this stage. However, we shall give only two applications in the last section. We relate the integral of a pullback to the mapping degree of a smooth map between oriented manifolds of the same dimension. The second application is a proof of a simple version of the Gauss-Bonnet theorem.

7.1 Transverse Homotopy Theorem

Theorem 7.1.1 Transversality Theorem: *Let $Z \subset Y$ and S be manifolds without boundary. Let X be a manifold with or without boundary and $F : X \times S \to Y$ be a smooth map such that F and ∂F are both transversal to Z. Then for almost every $s \in S$, the mappings F_s and ∂F_s are transversal to Z. (Here $F_s(x) = F(x, s),\ x \in X, s \in S$.)*

Proof: Clearly, it suffices to prove that the set of points s for which F_s is **not** transversal is of measure zero. For then we can apply this result to the special case of $\partial F : \partial X \times S \to Y$ and conclude that the union of the two sets of points $s \in S$ for which F_s or ∂F_s is not transversal is of measure zero.

Our strategy is to employ Sard's theorem and hence, we shall convert the requirement of transversality into regularity of some other map. Recall that the transversality condition has to be verified only at points (x, s) so that $F(x, s) \in Z$. This means that we must consider the set $W = F^{-1}(Z)$. This is a proper submanifold of $X \times S$, since both F and ∂F are transversal to Z. Now the target manifold for the map of which we are going to look for regular values is obviously S. Thus, we may consider the map $\pi : W \to S$, which is the restriction of the second projection. We shall claim that s is a regular value of π implies $F_s \pitchfork Z$.

So, fix $s \in S$, a regular value of π. This implies that $\alpha := D\pi_{(x,s)} : T_{(x,s)}W \to T_s(S)$ is surjective for all $(x, s) \in W$. We want to show that $F_s \pitchfork Z$. So, let $(x, s) \in W$ and $F(x, s) = z$. Put

$$\lambda := DF_{(x,s)} : T_{(x,s)}(X \times S) \to T_zY.$$

Since $F \pitchfork Z$, we have,

$$Im(\lambda) + T_zZ = T_zY.$$

So, given $\mathbf{v} \in T_zY$, there exists $\mathbf{v}_1 \in T_zZ, \mathbf{u}_1 \in T_xX, \mathbf{u}_2 \in T_sS$ such that

$$\lambda(\mathbf{u}_1, \mathbf{u}_2) + \mathbf{v}_1 = \mathbf{v}. \tag{7.1}$$

From the surjectivity of α and the fact that it is the restriction of the second projection, it follows that there exists $(\mathbf{w}_1, \mathbf{u}_2) \in T_{(x,s)}W$ such that $\alpha(\mathbf{w}_1, \mathbf{u}_2) = \mathbf{u}_2$. Therefore, by adding and subtracting $\lambda(\mathbf{w}_1, \mathbf{u}_2)$ in (7.1), we have

$$\begin{aligned} \mathbf{v} &= \mathbf{v}_1 + \lambda(\mathbf{w}_1, \mathbf{u}_2) + \lambda((\mathbf{u}_1, \mathbf{u}_2) - (\mathbf{w}_1, \mathbf{u}_2)) \\ &= \mathbf{v}_1 + \lambda(\mathbf{w}_1, \mathbf{u}_2) + \lambda(\mathbf{u}_1 - \mathbf{w}_1, 0) \\ &= \mathbf{v}_2 + \lambda(\mathbf{u}_1 - \mathbf{w}_1, 0) \end{aligned}$$

where $\mathbf{v}_2 = \mathbf{v}_1 + \lambda(\mathbf{w}_1, \mathbf{u}_2) \in T_zZ$ since $\lambda(T_{(x,s)}(W)) \subset T_z(Z)$. This completes the proof that $F_s \pitchfork Z$. ♠

Remark 7.1.1 The steps in the above claim are completely reversible. Therefore, we can say that F_s is transversal to Z iff s is a regular value of $\pi : W \to S$.

Theorem 7.1.2 *Let $f : X \to Y$ be a smooth map, where Y is a proper submanifold of $\mathbb{R}^N$. Let B^N denote the open unit ball in $\mathbb{R}^N$. Then there exists a smooth map $F : X \times B^N \to Y$ such that $F(x, 0) = f(x)$, $x \in X$ and for each fixed $x \in X$, the mapping $s \mapsto F(x, s)$ is a submersion. In particular, F and ∂F are submersions.*

Proof: For any fixed real number $r > 0$, consider the map $G_r : X \times B^N \to \mathbb{R}^N$ defined by $(x, s) \mapsto f(x) + rs$, where $r \neq 0$ any fixed real number. Check that $G_r(x, 0) = f(x)$ and for each fixed x, $s \mapsto G(x, s)$ is a submersion of B^N into $\mathbb{R}^N$. Let $N_\epsilon(Y)$ be a tubular neighborhood of Y in $\mathbb{R}^N$ and let $\beta : N_\epsilon(Y) \to Y$ be the submersion as considered in Remark 6.1.2. Now observe that $G_\epsilon(X) \subset N_\epsilon(Y)$. Now, put $F = \beta \circ G_\epsilon$. ♠

Remark 7.1.2 Note that we need not be strict about the assumption that Y is a proper submanifold of $\mathbb{R}^N$. All that we need is that Y should have tubular neighborhoods in $\mathbb{R}^N$ as described in Remark 6.1.2.(iii). Thus, the result is applicable in a somewhat more general situation than the one stated in the above theorem.

Theorem 7.1.3 Transverse Homotopy Theorem: *Let $Z \subset Y$ be boundaryless manifolds. Given any smooth map $f : X \to Y$, on a compact manifold X, there exists a smooth map $g : X \to Y$ such that $g \pitchfork Z$, $\partial g \pitchfork Z$ and g is homotopic to f.*

Proof: We can always think of Y as a proper submanifold of some $\mathbb{R}^N$. Then by the above theorem, there exist $F : X \times B^N \to Y$ such that $F, \partial F$ are submersions and hence F_s and ∂F_s are both transversal to Z for almost all s. For any such value of $s \in B^N$, let $[0, s]$ denote the line segment joining 0 and s. Then F restricted to $X \times [0, s]$ defines a homotopy of f with $g = F_s$ as required. ♠

In applications, we need to strengthen this result to a relative version, viz., while taking the homotopy of f, we do not want to "move" f on parts where it is already transversal.

Theorem 7.1.4 Extended Transverse Homotopy Theorem: *Let Z be a closed submanifold of Y with $\partial Y = \partial Z = \emptyset$. Let X be any compact manifold and $W \subset X$ be closed. Suppose $f : X \to Y$ is a smooth map such that for all $x \in W$ both f and ∂f are transversal to Z. Then f is homotopic to a smooth map g relative to W, such that $g \pitchfork Z, \partial g \pitchfork Z$ and $f = g$ in a neighborhood of W.*

Proof: Observe that "being transverse" to a closed submanifold is an open condition. Hence f and ∂f are both transverse to Z in a neighborhood U of W. Let V be an open subset such that $W \subset V \subset \overline{V} \subset U$. By the smooth version of Urysohn's lemma (Corollary 1.7.1), we can find a smooth map $\gamma : X \to [0,1]$ such that $\gamma|_{\overline{V}} \equiv 0$ and $\gamma|_{X\setminus V} \equiv 1$. Let F be as in the proof of Theorem 7.1.2. Define $G : X \times B^N \to Y$ by

$$G(x, \mathbf{v}) = F(x, (\gamma(x))^2\mathbf{v}).$$

It is enough to show that $G \pitchfork Z$ and $\partial G \pitchfork Z$. For then, we can appeal to the transversality Theorem 7.1.1, to conclude that for almost all $\mathbf{v} \in B^N$, $G_{\mathbf{v}}$ and $\partial G_{\mathbf{v}}$ are transversal to Z. Fix one such $\mathbf{v}$, and take $g = G_{\mathbf{v}}$. Clearly, the map $H(x,t) = G(x, t\mathbf{v})$ defines a homotopy of f with g. Moreover, for $x \in W$, we have $H(x,t) = G(x, t\mathbf{v}) = F(x, t\gamma(x)^2\mathbf{v}) = F(x, 0) = f(x)$. Therefore, f is homotopic to g relative to W. It remains to show that $G \pitchfork Z$ and $\partial G \pitchfork Z$.

Let now $x \in X$. If $\gamma(x) \neq 0$, then $\mathbf{u} \mapsto \gamma(x)^2\mathbf{u}$ is a diffeomorphism of B^N onto an open ball in $\mathbb{R}^N$. Since $\mathbf{u} \mapsto F(x, \mathbf{u})$ is a submersion from B^N to Y, it follows that $\mathbf{u} \mapsto F(x, \gamma(x)^2\mathbf{u})$ is a submersion. In particular, G is transversal to Z at points $(x, \mathbf{u})$, where $\gamma(x) \neq 0$.

Next, consider a point x such that $\gamma(x) = 0$. We have to actually compute the derivative $DG_{(x,\mathbf{v})}$, now. For this we think of G as a composite of the two maps:

$$(x, \mathbf{u}) \mapsto (x, \mu(x, \mathbf{u}); \quad (x, \mathbf{v}) \mapsto F(x, \mathbf{v}),$$

where $\mu : X \times B^N \to B^N$ is given by $\mu(x, \mathbf{u}) = \gamma(x)^2\mathbf{u}$. Also,

$$D\mu_{(x,\mathbf{u})} = (2\gamma(x)D(\gamma)_x\mathbf{u}, \gamma(x)^2 Id) = (0, 0).$$

By the chain rule, it follows that

$$\frac{\partial G}{\partial x}|_{(x,\mathbf{u})} = Df_x,$$

as $F|_{X\times\{0\}} = f$. Since $f \pitchfork Z$, it follows that G is transverse to Z at these points also. The proof of transversality of ∂F to Z is similar. ♠

Corollary 7.1.1 Suppose $f : X \to Y$ is a smooth map such that $\partial f \pitchfork Z$. Then there exists a map $g : X \to Y$ homotopic to f, relative to the boundary such that $g \pitchfork Z$ and $g = f$ on ∂X.

Exercise 7.1 Supply details of Remark 7.1.1.

7.2 Oriented Intersection Number

We begin with the following situation: $X, Z \subset Y$ are smooth manifolds, X is compact and all these manifolds are boundaryless. Also, $dim\, X + dim\, Z = dim\, Y$. Such a data is called a data *appropriate for intersection theory.*

In this section, we also assume that all the manifolds are oriented. Later we shall consider the nonoriented case.

Definition 7.2.1 Let W be a compact 0-dimensional oriented manifold. Recall that each point of W receives a sign $\pm$. The number of $+$ signs minus the number of $-$ signs in W is called the *orientation number of* W.

Example 7.2.1 The orientation number on $\partial[a,b]$ is always zero irrespective of what orientation we take on $[a,b]$. More generally, if X is a compact oriented 1-dimensional manifold, then the orientation number on ∂X is zero. This follows from the classification of 1-dimensional manifolds that we have seen in the previous chapter: that every connected compact 1-dimensional manifold is diffeomorphic to either $\mathbb{S}^1$ or to the closed interval $[0,1]$. This result plays a key role in the theory of intersection numbers.

Definition 7.2.2 Assume that $f : X \to Y$ is smooth and $f \pitchfork Z$. Then it follows that $W = f^{-1}(Z)$ is a compact 0-dimensional oriented submanifold of X. The orientation number of W is called the *intersection number of* f *with* Z. We denote it by $I(f,Z)$.

Remark 7.2.1

1. Consider a situation in which f is an embedding. Through this embedding, we can identify X with a submanifold of Y. Then $W = X \cap Z$ and it consists of finitely many points. At each of these points, the transversality condition $X \pitchfork Z$ says that the tangent spaces of these manifolds are supplementary. Being of complementary dimensions, they are also, complementary. Now, if the orientation classes satisfy

$$[T_w(X)][T_w Z] = [T_w Y] \tag{7.2}$$

   the point w is assigned the number $+1$ at w; otherwise, -1 is assigned. These numbers are called *orientation numbers* and $I(X,Z)$ is nothing but the sum of these numbers.

2. What is the relation between $I(X,Z)$ and $I(Z,X)$?. To see this we need to examine what happens at each of the points $w \in W = X \cap Z$. Observe that we now have to consider the validity of the equation

$$[T_w Z][T_w(X)] = [T_w Y] \tag{7.3}$$

   which involves a permutation of the left-hand side of (7.2). The signature of this permutation equals $(-1)^{(\dim X)(\dim Z)}$. This then precisely gives the relation:

$$I(X,Z) = (-1)^{(\dim X)(\dim Z)} I(Z,X) \tag{7.4}$$

3. Suppose X is the disjoint union of two manifolds $X = X_1 \cup X_2$ and $f|_{X_i} = f_i$. Then, it follows easily that $I(f,Z) = I(f_1,Z) + I(f_2,Z)$.

4. If we change the orientation on only one of the manifolds X, Y, or Z, then $I(f,Z)$ changes its sign.

Theorem 7.2.1 *Let X be the boundary of an oriented manifold M and f be the restriction of a smooth map $F : M \to Y$. Suppose $f \pitchfork Z$. Then $I(f,Z) = 0$.*

Proof: By the extended transverse homotopy Theorem 7.1.4, there exists a homotopy of F to a map $G : M \to Y$ such that $G|_X = f$ and $G \pitchfork Z$. Then $W = f^{-1}(Z)$ is the boundary of an oriented compact 1-dimensional manifold $G^{-1}(Z)$ and hence as seen in Example 7.2.1 the orientation number of W is zero. ♠

The above theorem is the first step toward some meaningful way of computing the orientation numbers. For example, as a corollary we have,

Theorem 7.2.2 *If f and g are homotopic, and transversal to Z, then $I(f,Z) = I(g,Z)$.*

Proof: Recall that given an oriented manifold X, $I \times X$ is given the product orientation, wherein I is given the standard orientation. The two boundary components $0 \times X$ and $1 \times X$ then have the induced orientations. Under the diffeomorphisms $x \mapsto (0,x)$ and $x \mapsto (1,x)$, we have,

$$[\partial I \times X] = [1 \times X] - [0 \times X].$$

Put $\phi = \partial(F) : \partial(I \times X) \to Y$ where F is a homotopy from f to g. By Theorem 7.2.1, we have, $I(\phi, Z) = 0$. But $\phi|_{0 \times X} = f$ and $\phi|_{1 \times X} = g$. Hence, $I(g,Z) - I(f,Z) = 0$. ♠

Remark 7.2.2 The advantage of the above theorem is quite apparent: We can extend the definition of $I(f,Z)$ not only to all smooth maps $f : X \to Y$ without the assumption that $f \pitchfork Z$, but to any continuous map $f : X \to Y$ as well. For, by the Smooth Approximation Theorem and the Transverse Homotopy Theorem, we can choose g to be any map that is homotopic to f and $g \pitchfork Z$ and define $I(f,Z) = I(g,Z)$. If h is another such map then since $g \simeq h$, it follows that $I(g,Z) = I(h,Z)$. Therefore, the definition is unambiguous. Also it is clear that the above theorems and all other properties of the intersection number are now valid for all smooth maps.

Definition 7.2.3 Consider the special case when $X = Z$. We put $f = \iota : X \hookrightarrow Y$ to be the inclusion map and define

$$I(X;Y) := I(\iota, X).$$

This is called the *self-intersection* number of X inside Y. We shall return to the study of this important number a little later.

Exercise 7.2

1. Let M be a connected, closed, oriented manifold and take the product orientation on $M \times M$. Compute the self-intersection number $I(M \times \{p\}; M \times M)$ for any point $p \in M$. Also, compute $I(M \times \{p\}; \{p\} \times M; M \times M)$.

2. Compute the self-intersection number of the diagonal in $\mathbb{C}P^1 \times \mathbb{C}P^1$.

7.3 Degree of a Map

In this section, we continue to assume that X, Y, Z, etc., form an appropriate data for the intersection theory.

An important special case of the intersection number is when the manifold $Z = \{z\}$ is a singleton. Notice that this means that $\dim X = \dim Y$. We shall further assume that X and Y are boundaryless and compact and Y is connected. Under these conditions, we make the following definition:

Consider then a smooth map $f : X \to Y$ and a regular value $z \in Y$ of f. This is equivalent to say that $f \pitchfork \{z\}$. We can now consider the number $I(f, \{z\})$ and write the simpler notation $I(f,z)$ for it. Let us make a few observations:
(a) We know that if f is replaced by a map g homotopic to f and transversal to $\{z\}$, then this number remains the same: $I(g, \{z\}) = I(g, \{z\})$.
(b) Now given any $z \in Y$, we also know that f can be homotoped in such a way that it becomes transversal to $\{z\}$. Combining these two observations, we can define $I(f,z)$ for all $z \in Y$.

(c) Next, observe that if $z \in Y$ is a regular value of f by Stack-Record Theorem 3.4.5, it follows that there is a neighborhood V of $z \in Y$ such that $f^{-1}(V) = \coprod_i V_i$ is the disjoint union of open sets such that $f : V_i \to V$ is a diffeomorphism, for each $i = 1, 2, \ldots, k$. We can further assume that V is connected. As z varies over V, its inverse image will have exactly one point x_i in V_i and for each fixed i, the orientation number assigned to x_i is the same throughout V_i, since V is connected. Thus, it follows that $I(f, \{z\})$ has the same value for all $z \in V$.
(d) It follows that $I(f, z)$ is locally a constant function on Y. Since we have assumed Y is connected, it is actually a constant function. We can now make a definition:

Definition 7.3.1 Let X, Y be closed, oriented manifolds of same dimension n and let Y be connected. Then for any smooth map $f : X \to Y$ we define $\deg f$ to be equal to the value of the constant function $I(f, y)$ for $y \in Y$.

Remark 7.3.1

1. It follows easily that if f is not surjective, then its degree is zero. For then, we can choose $y \in Y \setminus f(X)$, which is clearly a regular value for f in the definition of $\deg f$.

2. If $X = \partial M$, where M is a compact, oriented manifold and f can be extended to $F : M \to Y$, then $\deg f = 0$. This is an immediate consequence of Theorem 7.2.1.

Example 7.3.1 Degree of the Antipodal Map: If X is a closed, connected, orientable manifold and $f : X \to X$ is a diffeomorphism then clearly, $\deg f = \pm 1$ according as f is orientation preserving or reversing. As a special case, let us compute the degree of the antipodal map $\eta : \mathbb{S}^n \to \mathbb{S}^n$ given by $\mathbf{x} \mapsto -\mathbf{x}$. For $n = 1$, it is easy to see that the antipodal map preserves orientation. For $n \geq 2$, we need to work it out carefully.

To see whether η preserves orientation or not, it is enough to do this at a single point $p \in \mathbb{S}^n$ and compute the derivative of η at this point and check whether $d(\eta)_p$ preserves orientation or not. So, let us take $p = (1, 0, \ldots, 0)$, say. The tangent space at p to $\mathbb{S}^n$ is the vector space spanned by $\{e_2, \ldots, e_{n+1}\}$. Also, the induced orientation from B^{n+1} on $\mathbb{S}^n$ at this point gives the ordered basis $[e_2, \ldots, e_{n+1}]$. Now $T_{-p}\mathbb{S}^n = T_p(\mathbb{S}^n)$. However, the induced orientation at $-p$ is the opposite of that at p. Moreover, $d\eta_p : T_p\mathbb{S}^n \to T_{-p}\mathbb{S}^n$ is also the antipodal map $v \mapsto -v$. Therefore, the ordered basis $[e_2, \ldots, e_{n+1}]$ is mapped onto $[-e_2, \ldots, -e_{n+1}]$, which is equal to $(-1)^n[e_2, \ldots, e_{n+1}]$. It follows that

$$\deg \eta = (-1)^{n+1}. \tag{7.5}$$

As a simple application of the computation of the degree of the antipodal map, let us prove the following:

Theorem 7.3.1 *On $\mathbb{S}^{2n}$, every vector field has to vanish at some point.*

Proof: We shall assume that there is a nowhere vanishing vector field ϕ on $\mathbb{S}^{2n}$ and arrive at a contradiction.

Dividing by the norm, we get a vector field $\hat{\phi}$ such that $\hat{\phi}(x)$ is of unit length for all x. Such a field is called a *unit vector field.* Using $\hat{\phi}$, we define a homotopy $H : \mathbb{S}^{2n} \times I \to \mathbb{S}^{2n}$ between the identity map and the antipodal map by

$$H(x, t) = (\cos \pi t)x + (\sin \pi t)\hat{\phi}(x).$$

[Observe that since x and $\hat{\phi}(x)$ are unit vectors orthogonal to each other, for each fixed

x, $H(x,t)$ defines a rotation in the plane spanned by $\{x, \hat{\phi}(x)\}$, through an angle πt in the direction indicated by $\hat{\phi}(x)$.]

We have proved that the degree of η is -1 on $\mathbb{S}^{2n}$. Under the assumption that $\mathbb{S}^{2n}$ has a nowhere vanishing vector field, we have obtained that η is homotopic to the identity map. Homotopy preserves the degree whereas the degree of the identity map is always $+1$! This contradiction proves the theorem. ♠

Remark 7.3.2

1. In particular, this implies that none of the even dimensional sphere is parallelizable.
2. In the case of 2-dimension, this theorem has the following amusing interpretation. Assume that you have a ball full of hairs sticking out of its surface. You will never be able to comb it "smoothly" without leaving any parting points. With this interpretation the above theorem is sometimes called **the hairy-ball theorem.**
3. It is clear that whenever we can compute the degree of a certain map, we will have rich consequences. Let us illustrate this point with another example that is one sure case where we can compute the degree.

Lemma 7.3.1 Let $\mathbb{S}^1_r \subset \mathbb{C}$ denote the circle with center 0 and radius r. For any integer m, the mapping $\eta_m : \mathbb{S}^1_r \to \mathbb{S}^1$ given by $z \mapsto z^m/|z^m|$ has degree m.

Proof: Intuitively, this is clear. However, we can directly verify this by first principles. First, observe that the map $\alpha : \mathbb{S}^1 \to \mathbb{S}^1_r$ given by $z \to rz$ is an orientation-preserving diffeomorphism. Hence, by considering $\eta_m \circ \alpha$, it is enough to prove the statement for $r = 1$. Consider the case $m \neq 0$. Look at the derivative of the map $z \mapsto z^m$ at a point ζ_j, where ζ_j are $|m|^{th}$ roots of unity. The tangent line at ζ_j is oriented by $\imath\zeta_j$. The derivative $d(\eta_m)_{\zeta_j}(\imath\zeta_j) = m\zeta_j^{m-1}\imath\zeta_j = m\imath$ which is the oriented tangent line at 1 or the opposite of it depending on m is positive or negative. In either case, since there are exactly $|m|$ number of points, there degree is equal to m. Of course, if $m = 0$ then the map is a constant and the degree is zero too. ♠

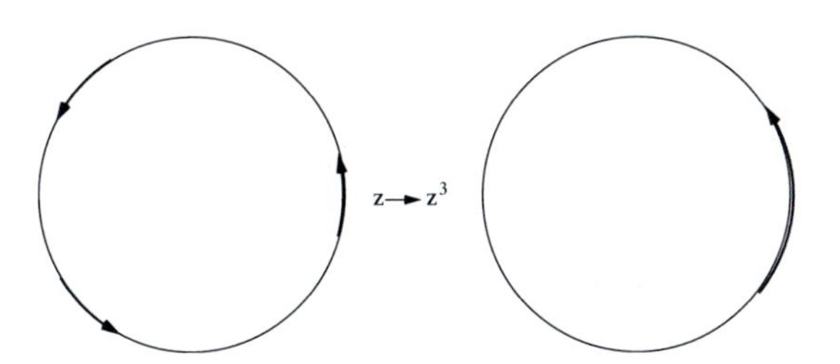

Figure 33 The degree of a map.

Lemma 7.3.2 Let $p(z)$ be any polynomial of degree $n \geq 1$ over complex numbers and let $P(z) = \dfrac{p(z)}{|p(z)|}$, wherever $p(z) \neq 0$. Then for sufficiently large $r > 0$, the mapping $P : \mathbb{S}^1_r \to \mathbb{S}^1$ has degree n.

Proof: By dividing by the coefficient of the leading term, we may as well assume that $p(z) = z^n + q(z)$ where $q(z)$ is a polynomial of degree $< n$. We know that for all large values of r, $p(z) \neq 0$ on $\mathbb{S}^1_r$. We shall show that P is homotopic to the map $\eta_n : \mathbb{S}^1_r \to \mathbb{S}^1$ defined by $z \mapsto z^n/|z|^n = z^n/r^n$. Since we have seen that $\deg \eta_n = n$, this will complete the proof.

So consider the obvious homotopy

$$H(z,t) = (1-t)p(z) + tz^n.$$

We first claim that $H(z,t) \neq 0$ for any $t \in [0,1]$ if $|z|$ is very large. For this, we observe that

$$\left|\frac{H(z,t)}{z^n}\right| = \left|1 + \frac{(1-t)q(z)}{z^n}\right| \geq 1 - \left|\frac{q(z)}{z^n}\right|.$$

As $z \to \infty$, $q(z)/z^n \to 0$. Therefore, for large $r > 0$, $H(z,t)$ does not vanish and so,

$$G(z,t) = \frac{H(z,t)}{|H(z,t)|}$$

is a well-defined homotopy $G : \mathbb{S}^1_r \times I \to \mathbb{S}^1$ as required. ♠

Theorem 7.3.2 Fundamental Theorem of Algebra (FTA): *Let p be a nonconstant polynomial with complex coefficients. Then p has at least one complex root.*

Proof: Consider the function $P : \mathbb{S}^1_r \to \mathbb{S}^1$, defined as above, for some large r. We have shown that $\deg P = m$. On the other hand, if p has no zeros at all, then P makes sense on the whole of the ball $B_r(0)$. From Theorem 7.2.1 $\deg P = 0$, which is a contradiction. ♠

We can strengthen the above arguments to arrive at a result similar to that of the argument principle in complex analysis.

Lemma 7.3.3 Let p be a polynomial and $z = z_0$ be a zero of order l. Then on the circle $\partial B_r(z_0)$ the mapping $p/|p|$ has degree $= l$ for sufficiently small r.

Proof: Choose $r > 0$ so that on the closed ball $B_r(z_0)$, p has no other zeros. Write $p = (z - z_0)^l q(z)$. Then on $B_r(z_0)$, q does not vanish. Hence, we can consider the homotopy

$$H(z,t) = \frac{(z-z_0)^l q(tz_0 + (1-t)z)}{|(z-z_0)^l q(tz_0 + (1-t)z)|}$$

for $z \in \partial B_r(z_0), t \in [0,1]$. Observe that

$$H(z,0) = \frac{p(z)}{|p(z)|}; \quad H(z,1) = c\frac{(z-z_0)^l}{|z-z_0|^l}.$$

Here $c = \dfrac{q(z_0)}{|q(z_0)|}$ is a complex number of unit length. Multiplication by such a number is a rotation and hence does not affect the degree. Thus, the map $z \mapsto H(z,1)$ is clearly of degree l. ♠

Theorem 7.3.3 Argument Principle: *Let R be a bounded region in $\mathbb{C}$ bounded by a smooth curve ∂R. Assume that p is a polynomial which has no zeros on ∂R. Then the number of zeros of p inside R counted with multiplicity is equal to the degree of $p/|p| : \partial R \to \mathbb{S}^1$.*

Proof: An elementary observation in algebra tells us that there are only finitely many zeros of a polynomial. Let z_i be the zeros of p inside R. We choose small enough discs D_i in R around each of the zeros of p so that they are mutually disjoint and restricted to the boundary of these discs, $P = p/|p|$ is of degree l_i where l_i is the multiplicity of the zero x_i. Now $p/|p|$ makes sense on the 2-dim manifold $M = R \setminus \cup_i D_i$ and hence the map restricted to ∂M should be of degree 0. Observe that $\partial M = \partial R \cup_i \partial D_i$ and the orientations induced on ∂D_i from M is the opposite of the standard one. Therefore, $0 = \deg P|_{\partial M} = \deg P|_{\partial R} - \sum_i \deg P|_{\partial D_i} = \deg P|_{\partial R} - \sum_i l_i$. ♠

Remark 7.3.3 Notice that in proving the above results, we have used only a little knowledge of complex numbers. Indeed, such a result is provable for any holomorphic map also, on similar lines. But for that, we need to use a little more knowledge of Complex Analysis, such as that the zeros of a holomorphic function are isolated etc.. So, we better leave this to Complex Analysis itself. For another proof (due to Euler) of the Fundamental Theorem of Algebra using the notion of degree, see Exercise 2 below.

We shall end this section with a recipe to determine the degree of a map $f : \mathbb{S}^1 \to \mathbb{S}^1$.

Example 7.3.2 Let $f : \mathbb{S}^1 \to \mathbb{S}^1$ be any smooth map. Consider the exponential map $exp : \mathbb{R} \to \mathbb{S}^1$ given by $t \mapsto e^{2\pi\imath t}$. We shall claim that there exists a smooth map $g : \mathbb{R} \to \mathbb{R}$ such that $f \circ exp = exp \circ g$. Let us examine what kind of map g should be. Since for a fixed t and for any integer k all points $t + k$ are mapped to the same point under exp, it follows that $g(t + k)$ should be points such that the difference between any two of them is an integer. Therefore, if we set $\alpha(t, k) = g(t+k) - g(t)$, then $\alpha(t, k)$ is a continuous function of $\mathbb{R} \times \mathbb{Z}$ taking integer values. Hence, it must be a constant for each $k \in \mathbb{Z}$. Let us denote this constant by $\alpha(k)$, i.e., $g(t + k) - g(t) = \alpha(k)$, for all $t \in \mathbb{R}$ and for all integers k. But then we have

$$\begin{aligned} g(t) + \alpha(k + l) &= g(t + (k + l)) \\ &= g(t + k + l) \\ &= g(t + k) + \alpha(l) \\ &= g(t) + \alpha(k) + \alpha(l) \end{aligned}$$

which shows that $\alpha(k+l) = \alpha(k)+\alpha(l)$. Therefore, denoting $\alpha(1) = d$, we have, $\alpha(k) = dk$. Thus, we have,

$$g(t + k) = g(t) + dk, \;\; k \in \mathbb{Z}, \;\; t \in \mathbb{R}. \tag{7.6}$$

We shall first of all show that d is indeed the degree of f. For this we shall prove that f is actually homotopic to the map $\eta : z \mapsto z^d$. Once again, we construct this homotopy in the space $\mathbb{R}$ and then pass down to $\mathbb{S}^1$. Clearly the power map η on $\mathbb{S}^1$ corresponds to the multiplication map $\mu : t \mapsto td$ on $\mathbb{R}$. So, consider $H : \mathbb{R} \times I \to \mathbb{R}$ given by $(t, s) \mapsto sg(t) + (1 - s)td$. This is a smooth homotopy between μ and g. It defines a homotopy $G : \mathbb{S}^1 \times I \to \mathbb{S}^1$ between η and f. For,

$$\begin{aligned} H(t + k, s) &= sg(t + k) + (1 - s)(t + k)d \\ &= sg(t) + skd + (1 - s)td + (1 - s)kd \\ &= H(t, s) + kd, \quad k \in \mathbb{Z}. \end{aligned}$$

Since we know that η is of degree d we are done.

Thus, it remains to prove the existence of the function g as claimed. Indeed, the properties of g require us to define it only in the interval $[0, 1]$. For, once the function is defined here, its values elsewhere are determined by the property (7.6).

This we shall do by proving a little more general result.

Lemma 7.3.4 Let $\gamma : [a, b] \to \mathbb{S}^1$ be any map, $t_0 \in \mathbb{R}$ be such that $e^{2\pi\imath t_0} = \gamma(a)$. Then there exists a unique map $\tilde{\gamma} : [a, b] \to \mathbb{R}$ such that $exp \circ \tilde{\gamma} = \gamma$ and $\tilde{\gamma}(a) = t_0$. Moreover, if γ is smooth then $\tilde{\gamma}$ is also smooth.

Proof: Since the uniqueness is not of our immediate concern, we leave its proof to you as an exercise. Let A be the set of all $c \in [a, b]$ such that there exists a map $\tilde{\gamma} : [a, c] \to \mathbb{R}$ as described above. Clearly, $a \in A$ and hence A is nonempty. Now, we shall show that A is both open and closed and hence $A = [a, b]$, which will complete the proof. Let c be a closure point of A. Our aim is to show that c is in the interior of A. Let U be any neighborhood of $f(c)$ in $\mathbb{S}^1$, which is not the whole of $\mathbb{S}^1$. For definiteness, we choose $U = \mathbb{S}^1 \setminus \{-f(c)\}$. Since f is continuous, there exists $\epsilon > 0$, such that if $I_\epsilon = (c-\epsilon, c+\epsilon) \cap [a,b]$, then $f(I_\epsilon) \subset U$. Since $c \in \bar{A}$, there exists $s \in A \cap I_\epsilon$. If $s > c$, this implies $c \in \text{int}\,(A)$. So assume $s \leq c$ and let $\tilde{\gamma} : [a,s] \to \mathbb{R}$ be as required. Now we know that $exp^{-1}(U)$ is a disjoint union of open intervals such that on each of these intervals J_r, $exp : J_r \to U$ is a diffeomorphism with its inverse ln_r. The point $\tilde{\gamma}(s)$ belongs to precisely one such interval J_r say. It follows that for points $t \in I_\epsilon$, we have $\tilde{\gamma}(t) = ln_r \circ f(t)$. Therefore, if we define $\tilde{\gamma} : [a,s] \cup I_\epsilon \to \mathbb{R}$ by the formula

$$\tilde{\gamma}(t) = \begin{cases} \tilde{\gamma}(t), & a \leq t < s, \\ ln_r \circ f(t), & t \in I_\epsilon, \end{cases}$$

then we obtain a continuous $\tilde{\gamma} : [a,s] \cup I_\epsilon \to \mathbb{R}$ as required. This proved that $[a,s] \cup I_\epsilon \subset A$. Since c is in the interior of I_ϵ this proves that $c \in \text{int}\,A$.

If γ is smooth, since locally $\tilde{\gamma}$ can be expressed as $\ln \circ \gamma$ for some branch of the logarithm, it follows that $\tilde{\gamma}$ is also smooth. ♠

Remark 7.3.4 Using Lemma 7.3.4, given $t_0 \in \mathbb{R}$ such that $\exp t_0 = f(1)$, we can now get a map $g : [0,1] \to \mathbb{R}$ such that $g(0) = t_0$. It follows =that $g(1) = t_0 + d$ for some integer d. We now extend g to the whole of $\mathbb{R}$ by the formula (7.6) as a continuous function.

For future reference, we shall state whatever we have proved just now as a theorem.

Theorem 7.3.4 *Given any map $f : \mathbb{S}^1 \to \mathbb{S}^1$ and a point $x_0 \in \mathbb{R}$ such that $exp(x_0) := e^{2\pi\imath x_0} = f(1)$, there exists a unique map $g : \mathbb{R} \to \mathbb{R}$ such that $g(0) = x_0$ and $exp \circ g = f \circ exp$.*

As an immediate consequence, we derive an interesting result which is the first case of the well-known Borsuk-Ulam Theorem.

Theorem 7.3.5 *Let $f : \mathbb{S}^1 \to \mathbb{S}^1$ be a smooth map that is odd, i.e., such that $f(-x) = -f(x), x \in \mathbb{S}^1$. Then $\deg f$ is odd.*

Proof: We have a map $g : \mathbb{R} \to \mathbb{R}$ such that $exp \circ g = f \circ exp$. Since f is an odd map, it follows that $g(t+1/2) = g(t) + \lambda(t)$, where $=e^{2\pi\lambda(t)\imath} = -1$. As in Example 7.3.4, $\lambda(t)$ can be shown to be locally constant and hence does not depend upon t. This means $\lambda(t) = 1/2 + m$ for some integer m. But then, if $d = \deg f$ we have,

$$\begin{aligned} g(t) + d &= g(t+1) \\ &= g(t + 1/2 + 1/2) \\ &= g(t + 1/2) + 1/2 + m \\ &= g(t) + 1 + 2m \end{aligned}$$

This proves that $d = 1 + 2m$ is odd. ♠

What we have seen so far, is the tip of a great iceberg called the *theory of covering spaces,* which you will learn in a first course in Algebraic Topology.

Exercise 7.3

1. Compute the degree of $f : \mathbb{S}^n \to \mathbb{S}^n$ given by

$$(x_1, \ldots, x_{n+1}) \mapsto (-x_1, \ldots, -x_k, x_{k+1}, \ldots, x_{n+1}).$$

2. **Euler's Proof of Fundamental Theorem of Algebra:**[1] Consider the affine space V_m of all monic polynomials

$$f(x) = x^n - a_1 x^{n-1} + \cdots + (-1)^n a_n$$

[1]The idea behind this proof of FTA is due to Euler.

with real coefficients and identify it with $\mathbb{R}^n$ via

$$f \mapsto (a_1, \ldots, a_n).$$

The Cauchy product of two polynomials

$$(f, g) \mapsto fg$$

then defines a map

$$\phi_{m,n} : V_m \times V_m \to V_{m+n}.$$

(a) Show that $\phi_{m,n}$ is a smooth, proper map.
(b) Show that the Jacobian of $\phi_{m,n}$, i.e., the determinant of $D(\phi_{m,n})$ at any point (f, g) is equal to the resultant $R(f, g)$ of the two polynomials f, g with respect to the variable x. (See [Lang] p. 200.)
(c) Show that (f, g) is a regular point of $\phi_{m,n}$ iff f and g have no common roots in $\mathbb{C}$.
(d) Put $f = \prod_{i=1}^{k}(x^2 + a_i^2), g = \prod_{j=1}^{l}(x^2 + b_j^2)$, for some $a_i, b_j \in \mathbb{R}$. If $a_1^2, \ldots, a_k^2, b_1^2 \ldots, b_l^2$ are distinct real numbers, then show that $v = \phi(f, g)$ is a regular value and $\#(\phi^{-1}(v)) = \binom{k+l}{k}$ and the Jacobian of $\phi_{m,n}$ at all these points is positive.
(e) Conclude $\phi_{2,2l}$ is a surjective map.[2]
(f) For $f(x) = x^n + c_1 x_{n-1} + \cdots + c_n \in \mathbb{C}[x]$, define $\bar{f}(x) = x^n + \bar{c}_1 x^{n-1} + \cdots + \bar{c}_n$. Observe that $h(x) = f(x)\bar{f}(x) \in \mathbb{R}[x]$ and f has a root iff h has a root.
(g) Deduce the Fundamental Theorem of Algebra.

7.4 Nonoriented Case

We shall now consider the case when the orientation information on one or the other manifolds appearing in the data for intersection theory is missing. This could simply mean that we have not decided on any orientation on one of them, or that such a data is not available to us or that indeed one of these manifolds may not be orientable. It also includes the case when all the orientation data were available (as before) but we prefer not to use them for awhile. Nevertheless, as soon as the dimensionality condition is met, viz., $\dim X + \dim Z = \dim Y$ and $f \pitchfork Z$, we know that $W = f^{-1}(Z)$ is a (compact) 0-dimensional manifold and we can simply count the number of points in it. We define

$$I_2(f, Z) = \#(W) \bmod (2)$$

and call it the *mod 2 intersection number*. It follows that all the earlier theorems, available for the oriented intersection number are available for mod 2 intersection number as well, with trivial modifications. Moreover, if $I(f, Z)$ were defined then $I(f, Z) = I_2(f, Z)$ mod(2). It is also obvious, on general considerations, that the mod 2 intersection theory is weaker than the oriented intersection theory, when the latter is available. For instance, using only the mod 2 intersection theory, the arguments used in proving FTA 7.3.2 above would only lead us to conclude that every odd degree polynomial has a root, instead of the full assertion of FTA. However, there are situations in which, actual computation of the intersection number could be impossible or very cumbersome, whereas, the mod 2 intersection is easily computable. Certainly, the mod 2 intersection is stronger in one sense, viz., it has wider range of applicability. Let us now illustrate this point.

[2] This step uses the fact that a positive degree map has to be necessarily surjective, and that the degree is independent of the regular value chosen to compute it, which in turn, uses Sard's theorem. During Euler's time, perhaps one could not have expected such justification to be provided.

Theorem 7.4.1 *Let Y be a manifold of dimension $4m+2, m \geq 0$. Let Z be a closed submanifold of Y of dimension $2m+1$. If mod 2 self-intersection $I_2(Z;Y)$ of Z in Y is nonzero then Z or Y is nonorientable.*

Proof: Assuming that both Z, Y are orientable, we fix some orientations on them and consider the number $I(Z;Y) \in \mathbb{Z}$. Recall that this is nothing but $I(X,Z;Y)$ where $X = Z$. In (7.4), we have seen that $I(X,Z) = (-1)^{(\dim X)(\dim Z)} I(Z,X)$. Putting $X = Z$ and using the fact that $\dim Z$ is odd, this means that $I(Z;Y) = -I(Z;Y)$. Hence, $I(Z;Y) = 0$. On the other hand, $I(Z,Y) = I_2(Z;Y) \bmod(2)$, which contradicts the hypothesis. ♠

As an application, we have a neat proof of the following "obvious fact":

Theorem 7.4.2 *The Möbius band is nonorientable.*

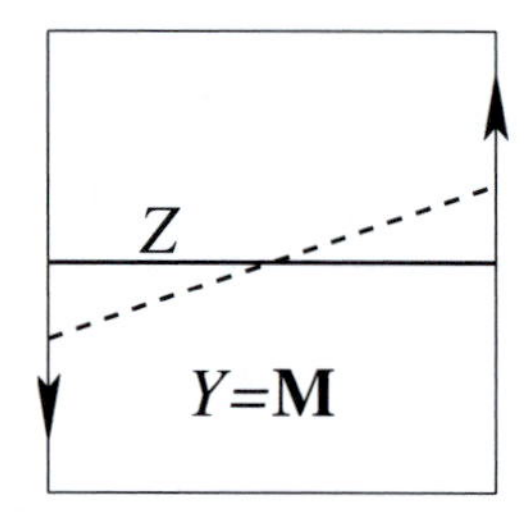

Figure 34 The central circle has nontrivial self-intersection.

Proof: Take Y to be the Möbius band and Z to be the central circle in the above theorem. By pushing Z a little bit all over, we can obtain a homotopy of Z to another circle $X \subset Y$ such that $X \cap Z$ is a single point and the intersection is transversal. (Ex. Prove this rigorously.) Therefore, $I_2(Z;Y) = 1$. Since $Z = \mathbb{S}^1$ is orientable, from Theorem 7.4.1, it follows that Y is not orientable. ♠

Remark 7.4.1 For more application of mod 2 intersection, see the next section.

Exercise 7.4 Compute the (mod 2) self-intersection number of the diagonal in $\mathbb{P}^2 \times \mathbb{P}^2$.

7.5 Winding Number and Separation Theorem

We shall now study two more important geometric concepts using the intersection theory.

The classical Jordan curve theorem says that every simple closed curve in $\mathbb{R}^2$ separates $\mathbb{R}^2$ into two parts. Restricting ourselves to the case of smooth curves, this can be stated as: *every embedding of $\mathbb{S}^1$ separates $\mathbb{R}^2$ into two components.* This somewhat "self-evident" result, as we all know is hard to prove. On the other hand, intuitively this should also be true in dimension three, i.e., any embedding of $\mathbb{S}^2$ should separate $\mathbb{R}^3$ into two components. Indeed, such a result is true in all dimensions. In order to prove this and more, we shall consider the situation in a general setup. However, for motivation and geometric understanding of what is going on, you may keep the picture of a simple closed curve in $\mathbb{R}^2$ in your mind.

Let X be a compact, orientable, boundaryless manifold of dimension $n-1$ and let $f : X \to \mathbb{R}^n$ be a smooth map. Let us take a point $z \in \mathbb{R}^n$ which does not belong to the image of f. We are interested in studying how f "wraps" around z. For this, we consider the behavior of the mapping $\tau_f : X \to \mathbb{S}^{n-1}$ given by

$$x \mapsto \frac{f(x) - z}{\|f(x) - z\|}.$$

Hold on then. This is the setup that is ideal for the study of the degree of a map. In what follows, we assume that X is oriented and talk about the degree d of τ_f and its properties. In the nonorientable case, corresponding statements with appropriate modifications will make sense and will be true.

The experience that we have with the mappings $\mathbb{S}^1 \to \mathbb{S}^1$ suggests the following definition:

Definition 7.5.1 We define the *winding number* $W(f, z)$ *of* f *around* z to be the degree of τ_f.

One of the simplest properties of the winding number is that it is locally a constant as a function of z. Hence, it is constant on each component of $\mathbb{R}^n \setminus f(X)$. This remark will become quite handy soon.

We begin with a result similar to the argument principle which is also proved in a similar way. So let us call it the *generalized argument principle.*

Theorem 7.5.1 Generalized Argument Principle: *Let* $X = \partial M$*, where* M *is a compact oriented manifold and let* $F : M \to \mathbb{R}^n$ *be a smooth map such that* $\partial F = f$*. Let* $z \in \mathbb{R}^n \setminus f(X)$ *be a regular value of* F*. Then* $W(f, z)$ *is equal to the sum of the intersection numbers on* $F^{-1}(z)$*. In particular,* $W_2(f, z) \equiv \#(F^{-1}(z))$ *(mod 2).*

Proof: The proof is similar to the one that we gave for the argument principle (Theorem 7.3.3). Without loss of generality we may choose the origin of $\mathbb{R}^n$ to be at z itself. Choose small neighborhood B_i around each of the points $z_i \in F^{-1}(z)$ in X so that
(i) $F : B_i \to U$ is a diffeomorphism onto a neighborhood of z.
(ii) Each B_i is diffeomorphic to a n-disc.
(iii) The closures of B_i are mutually disjoint.
If ϵ_i is the number assigned to z_i, we must show that $W(f, z) = \sum_i \epsilon_i$. Recall that $\epsilon_i = \pm 1$ according as $F|_{B_i}$ is orientation preserving or reversing. But this is the same as whether $F|_{\partial B_i}$ is orientation preserving or reversing. Put $g_i = F|_{B_i}$ and let $h_i = g_i/|g_i|$. Then $\deg h_i = \deg g_i = \deg(F|_{\partial B_i}) = \epsilon_i$. On the other hand, consider $A = M \setminus \cup_i B_i$. Then $F(A)$ does not contain 0 and hence, $H = F/|F| : A \to \mathbb{S}^{n-1}$ is defined. Therefore, $0 = \deg H = \deg f - \sum_i \deg h_i = \deg f - \sum \epsilon_i$. ♠

Remark 7.5.1 We would like to bring to your notice once again that, in the statement of the above theorem, we have assumed that M is orientable. However, if this is not the case, the result is still valid provided you replace $W(f, z)$ by $W_2(f, z)$, etc. This remark is applicable to almost all situations that we discuss in future.

We shall now give a method that works well in computing the winding number. First, prepare yourself with two lemmas, the proof of the first one being an exercise for you.

Lemma 7.5.1 Let $\Theta : \mathbb{R}^n \setminus \{0\} \to \mathbb{R}^n$ be defined by $x \mapsto x/\|x\|$. At any point $a \in \mathbb{R}^n \setminus \{0\}$, show that $d(\Theta)_a = \frac{1}{\|a\|}\pi_a$, where $\pi_a : \mathbb{R}^n \to \mathbb{R}^n$ is the projection of $\mathbb{R}^n$ on the orthogonal complement of a. In particular $Ker\, d(\Theta)_a = L(\{a\})$, the linear span of a.

Lemma 7.5.2 Let $v \in \mathbb{S}^{n-1}$ and let $L_v = \{xv \ : \ r > 0\}$ be the ray along the unit vector v, $\tau_f(x) = f(x)/\|f(x)\|$. Then for any smooth map $f : X \to \mathbb{R}^n \setminus \{0\}$, v is a regular value of τ_f iff $f \pitchfork L_v$. Further, we have, $W(f, 0) = I(L_v, f)$.

Proof: Clearly, $x \in \tau_f^{-1}(v) \iff f(x) \in L_v$. Now, v is a regular value of τ_f iff $d(\tau_f)_x : T_xX \to T_v(\mathbb{S}^{n-1})$ is an isomorphism. Also, $ker\,(d(\theta)_{f(x)}) = L(\{v\})$, the subspace spanned by v, which is also equal to $T_{f(x)}(L_v)$. Since the dimension of T_xX is $n-1$, it follows that $d(\tau_f)_x : T_xX \to T_v(\mathbb{S}^{n-1})$ is an isomorphism iff $d(\tau_f)_x$ is injective and its image is complementary to $L(\{v\})$. This is again equivalent to say that $df_x(T_xX) + T_{f(x)}(L_v) = T_{f(x)}(\mathbb{R}^n) = \mathbb{R}^n$, which, in turn, is equivalent to say that f is transversal to L_v.

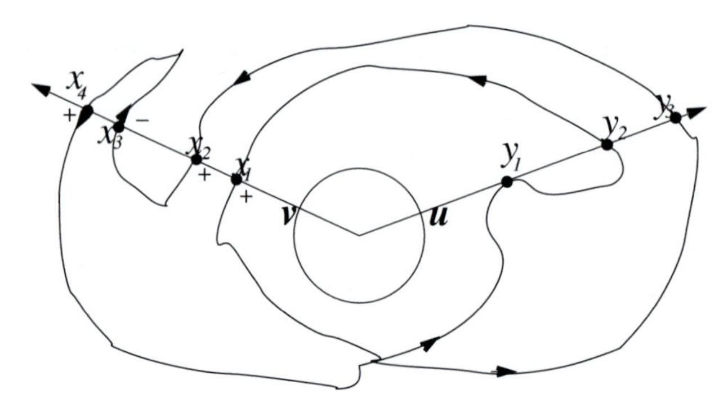

Figure 35 Computing winding number.

Now, recall that the orientation taken on L_v is in the direction of v. Let $[a]$ denote the orientation class of T_xX. In computing the number $I(L_v, f)$, the orientation number ϵ_x attached to x is $+1$ iff $[v]df_x([a]) = E_n$ is the standard orientation on $\mathbb{R}^n$. On the other hand, if $[b]$ denotes the orientation class of $T_v(\mathbb{S}^{n-1})$, then we also have $[v][b] = E_n$. Therefore, it follows that $\epsilon_x = +1$ iff $d(\tau_f)$ maps $[a]$ onto $[b]$. This is the same as saying that τ_f preserves orientation at x iff $\epsilon_x = +1$. (In Figure 35, the vector $\mathbf{u}$ is not a regular value of τ_f because of the point y_1. The vector $\mathbf{v}$ is a regular value with $x_i \in \tau_f^{-1}(\mathbf{v})$. The orientation numbers add up to $+2$.)

Upon taking the sum over all $x \in \tau_f^{-1}(v)$ the last assertion of the lemma follows. ♠

Theorem 7.5.2 Jordan-Brouwer Separation Theorem: *Let X be a connected, compact, boundaryless submanifold of codimension 1 in $\mathbb{R}^n$. Then $\mathbb{R}^n \setminus X$ consists of two connected components, A, B of which one is bounded. Moreover, X is the common boundary of A and B.*

Proof: The proof will be divided into several easy steps.

Step 1: We first prove the local version of the above theorem: *At every point $x \in X$, there exist arbitrary small neighborhood U of x in $\mathbb{R}^n$ such that $U \setminus X$ has precisely two components.* To prove this, first consider the simplest situation, when X is replaced by $\mathbb{R}^{n-1} \times 0$. In this case, all that we have to do is to take U to be a ball around x. Then, clearly $U \setminus X$ has two components, viz., $U \cap \mathbf{H}^n_{\pm}$. Now, in the general case, let $x \in X$ be any point, U_x be a neighborhood of $x \in \mathbb{R}^n$ and let $\phi_1, \ldots, \phi_n$ be coordinate functions such that

$$U \cap X = \{z \in U \ : \ \phi_n(z) = 0\}; \quad \phi_i(x) = 0, \ i = 1, \ldots, n.$$

If $\phi = (\phi_1, \ldots, \phi_n)$ defines a diffeomorphism of U onto the unit ball B^n say, then it follows easily that $U \setminus X$ has precisely two components C_1, C_2 say, which are diffeomorphic to the upper-half and the lower-half balls.

Step 2 : *Let $z \in \mathbb{R}^n \setminus X$. Let $x \in X$ be any point and U_x be a neighborhood of x in $\mathbb{R}^n$. Then there exists a path $\gamma \subset \mathbb{R}^n \setminus X$, which starts at z and ends in a point of U_x.* To prove this, we have to use the connectivity of X. If Y is the set of all points in X which have the above property, it is easily seen that Y is both open and closed. So, it remains to see why Y is nonempty. Take any regular value $\mathbf{v} \in \mathbb{S}^{n-1}$ of the function

$$x \mapsto \frac{z - x}{\|z - x\|}.$$

Then the ray $L = \{z + t\mathbf{v} \, : \, t > 0\}$ intersects X transversely. (See lemma 7.5.2.) Let x_0 be the point on this ray which belongs to X and nearest to z. We can then choose a point $y \in L \cap U_{x_0}$ such that the line segment $[z, y]$ does not intersect X at all. This means $x_0 \in Y$. Hence, Y is nonempty as required. This completes the proof of Step 2.

Step 3: $\mathbb{R}^n \setminus X$ *has at most two connected components.* Fix $x \in X$. If U_x is chosen as in Step 1, it follows from Step 2, that every element $z \in \mathbb{R}^n \setminus X$ can be joined to a point either in C_1 or in C_2 inside $\mathbb{R}^n \setminus X$. Hence, the assertion.

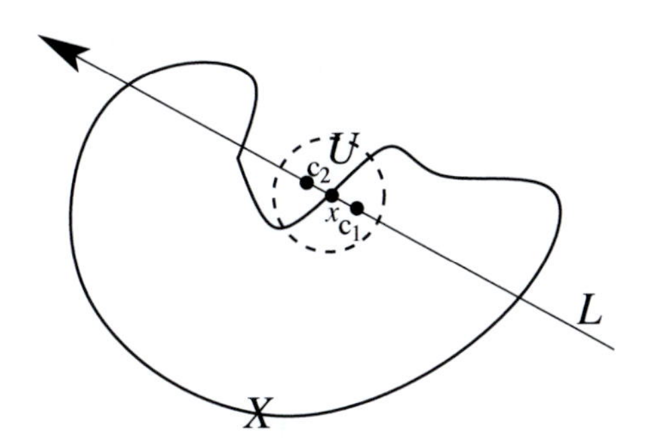

Figure 36 The complement of X has at most two components.

Step 4: *Let* $x \in X$ *and* U*, etc., be as in Step 1. Let* $f : X \to \mathbb{R}^n$ *be the inclusion map. Then the winding number* $W_2(f, z)$ *takes different values on* C_1 *and* C_2*. In particular,* $\mathbb{R}^n \setminus X$ *has at least two components.* To see this, draw a line L through x that is transversal to X. (L exists by Sard's theorem). Then on this line, pick up points c_1, c_2 on either side of x so that $[c_1, c_2] \subset U$ and $[c_1, c_2] \cap X = \{x\}$. It follows that c_i are in different $C_j's$. Hence, we may assume that $c_1 \in C_1$ and $c_2 \in C_2$. Now we can use the ray $[c_1, \infty)$ that includes the point c_2 to compute the winding number $W_2(f, c_1)$. Likewise, we can use the sub-ray of this $[c_2, \infty)$ (which excludes c_1) to compute $W_2(f, c_2)$. It follows that $W_2(f, c_1) = W_2(f, c_2) + 1$. Therefore, c_1, c_2 must be in different components of $\mathbb{R}^n \setminus X$. This completes the proof of the assertion in Step 4.

Combining this with Step 3, we have completed the proof that $\mathbb{R}^n \setminus X$ has precisely two components.

Observe that X is a closed subset of $\mathbb{R}^n$. Therefore both A and B are open sets in $\mathbb{R}^n$ such that $\overline{A} = A \cup X$ and $\overline{B} = B \cup X$. Combining this with Step 1, we conclude that A and B are manifolds with their common boundary as X. Of course at least one of them must be unbounded say, A. On the other hand, there cannot be more than one unbounded component, since X is compact. So, B is bounded. ♠

Remark 7.5.2

1. Consider the special case when X is diffeomorphic to $\mathbb{S}^{n-1}$. For $n = 2$, and for a topological embedding of X, this is the classical Jordan Curve Theorem. Indeed, a "not so easy" application of Riemann mapping theorem, tells you that the bounded component U of $\mathbb{R}^2 \setminus X$ is diffeomorphic to (biholomorphic to) the open unit disc. For $n = 3$, the famous Alexander's *horned sphere* [A] gives an example of a topological embedding of $\mathbb{S}^2$ in $\mathbb{S}^3$ such that the complements are not 3-cells. Thus, a certain "flatness" condition on the embedding is necessary, such a condition being always satisfied in case of smooth embeddings. And then these results are popularly know as Schoenfly-type theorems which need special techniques to handle. (See [Bn], [M].)

2. What happens when we take a submanifold of codimension > 1? To understand this, consider a simple situation first. If U is a connected open subset of $\mathbb{C}$, then you know that $U\setminus F$ is still connected where F is a finite set. Likewise if we take away a line from $\mathbb{R}^3$ it is easily seen that the resulting space is still connected. Indeed, this holds in much generality. *Let Y be a connected manifold, and $X \subset Y$ be submanifold of codimension > 1. Then $Y \setminus X$ is connected.* We can use the transverse homotopy theorem to prove this as follows. Let $a, b \in Y\setminus X$ be any. Since Y is connected, there exists a smooth path $\gamma : [0,1] \to Y$ connecting a, b. By the extended transverse homotopy Theorem 7.1.4, we may assume that γ is transversal to X. But then $W = \gamma^{-1}(X)$ is a submanifold of $[0,1]$, which is of codimension > 1. This means $W = \emptyset$, which is the same as saying that $Im\,\gamma_1 \cap X = \emptyset$.

Exercise 7.5

1. Prove Lemma 7.5.1

2. For $n \geq 2$ and for any connected manifold X, the set of all homotopy classes of maps $\mathbb{S}^n \to X$ is denoted by $\pi_n(X)$. It can be shown that these form an abelian group. For any continuous map $f : X \to Y$, taking composition induces a group homomorphism $f_{\#} : \pi_n(X) \to \pi_n(Y)$.
(a) Let N be a closed connected submanifold of codimension $k \geq 2$, in a manifold M. Show that the inclusion induced map $\eta_{\#} : \pi_i(M \setminus N) \to \pi_i(M)$ is an isomorphism for $i \leq k-2$ and is surjective for $i = k-1$.
(b) In the above situation suppose further that $\pi_k(M) = (0)$. Prove $\pi_{k-1}(M\setminus N) \neq (0)$.

7.6 Borsuk-Ulam Theorem

In this section, we shall apply the notion of winding number to prove one of the most popular results in Topology.

Recall that a map $f : \mathbb{S}^n \to \mathbb{R}^m$ is a called *odd* if it satisfies the following symmetry condition:

$$f(-x) = -f(x), \quad \forall\, x \in \mathbb{S}^n.$$

A typical example of such a map is the antipodal map itself, which takes each x to $-x$. Standard examples of odd maps are $\sin x$ and polynomial maps with only odd degree terms. In particular $z \mapsto z^d$ is an odd map if the degree d is odd. The Borsuk-Ulam theorem can be loosely stated as a partial converse to this, viz.,

every odd map is of odd degree.

Lemma 7.6.1 Let $k \geq$ be an integer. The following statements are equivalent to each other.
(k1) The image of any odd map $f : \mathbb{S}^k \to \mathbb{R}^{k+1} \setminus \{0\}$ has to intersect every line in $\mathbb{R}^{k+1}$ through the origin.
(k2) The image of any odd map $g : \mathbb{S}^k \to \mathbb{R}^k$ contains $(0,\ldots,0)$.
(k3) There is no odd map $g : \mathbb{S}^k \to \mathbb{S}^{k-1}$.
(k4) Given any map $h : \mathbb{S}^k \to \mathbb{R}^k$, there exists $v \in \mathbb{S}^k$ such that $h(v) = h(-v)$.

Proof:
(k1) $\Longrightarrow$ (k2) Define $f(x) = (g(x), 0)$. Then f is odd and hence there exists $v \in \mathbb{S}^k$ such that $f(v)$ belongs to the x_{k+1}-axis, i.e., $f(v) = (0, \ldots, 0, t) = (g(v), 0)$. This implies $g(v) = (0, \ldots, 0)$.
(k2) $\Longrightarrow$ (k3) An odd map $g : \mathbb{S}^k \to \mathbb{S}^{k-1}$ can be thought of as an odd map $f : \mathbb{S}^k \to \mathbb{R}^k \setminus \{0\}$ contradicting (k2).
(k3) $\Longrightarrow$ (k4) Define $\phi(x) = h(x) - h(-x)$. Then ϕ is an odd map. If its image contains no zeros then $g = h/||h||$ makes sense and is an odd map. This contradicts (k3).
(k4) $\Longrightarrow$ (k1) Given any line L through the origin, consider the projection $\pi_L : \mathbb{R}^{k+1} \to \mathbb{R}^k$ parallel to L. Then π_L being linear, it follow that $h = \pi_L \circ f$ is an odd map. By (k4), there exists a point $v \in \mathbb{S}^k$, such that $h(x) = h(-x) = -h(x)$. Hence, $h(x) = 0$. This implies that $f(x) \in Ker\, \pi_L = L$. ♠

Theorem 7.6.1 Borsuk-Ulam Theorem: *Consider the following statements, in which $k \geq 1$ is a fixed integer:*
(ka) *Given an odd map $f : \mathbb{S}^k \to \mathbb{R}^{k+1} \setminus \{0\}$, we have $W_2(f, 0) = 1$.*
(kb) *Given an odd map $\phi : \mathbb{S}^k \to \mathbb{S}^k$, we have, $\deg_2 \phi = 1$.*
These two statements are equivalent to one another. Each of them implies all the four statements in the previous lemma. Moreover, statements (ka) and (kb) are true, for all $k \geq 1$.

Proof: We shall first prove that statements (ka) and (kb) are equivalent, prove (ka) $\Longrightarrow$ (k1) and then prove the statement (ka).
(ka) $\Longrightarrow$ (kb) Consider ϕ as an odd map into $\mathbb{R}^{k+1} \setminus \{0\}$. Then by (ka), $1 = W_2(\phi, 0)$ and by definition, this is equal to $\deg_2 (\phi/||\phi||) = \deg_2 \phi$.
(kb) $\Longrightarrow$ (ka) Put $\phi = f/||f||$.
(ka) $\Longrightarrow$ (k1) If L is a line that does not intersect image of f, then using this we can compute the mod 2 intersection. So, if L^+ is a ray on L from the origin, $W_2(f, 0) = I_2(f, L^+) = 0$.

We shall now turn our attention to the proof of (ka), which is achieved by induction on k. Observe that once we have an induction hypothesis that (ka) is true say, then corresponding statements (k1)–(k4) as well as (kb) are available for us in proving ((k+1)a). Also, we can choose to prove either ((k+1)a) or (k+1)b at our will.

Consider the statement (1b). This is precisely what we have proved in Theorem 7.3.5. Thus, we have (1a) and (1b) are true.

Now assume that ((n–1)a) is true for some $n \geq 2$. Let $f : \mathbb{S}^n \to \mathbb{S}^n$ be an odd map. We have to prove that $\deg f$ is odd. Treat $\mathbb{S}^{n-1}$ as a submanifold of $\mathbb{S}^n$ via the equator given by $x_{n+1} = 0$ and consider the map $g = f|_{\mathbb{S}^{n-1}}$. Let v be a regular value for both f and g. By symmetry, it follows that $-v$ is also a regular value of f and g. It also follows that $\pm v$ do not belong to the image of g since the domain of g is of dimension one lower than that of its codomain. Therefore, if L is the line spanned by v, then L does not intersect $Im\, g$. Moreover, it also follows that $f \pitchfork L$. Therefore, by definition, $\deg_2 f = W_2(f, 0) = \#(f^{-1}(v))$, $(mod\, 2)$.

Now, under the antipodal mapping, there is a one-to-one correspondence between the sets $f^{-1}(v)$ and $f^{-1}(-v)$, which follows from the symmetry property of f. Therefore,

$$\#(f^{-1}(v)) = \frac{1}{2}\#(f^{-1}\{v, -v\}) = \frac{1}{2}\#(f^{-1}(L)).$$

Also observe that $f^{-1}(L) \cap \mathbb{S}^{n-1} = g^{-1}(L) = \emptyset$. Once again, by the symmetry, there is a one-to-one correspondence of the sets $f^{-1}(L) \cap \mathbb{S}^n_+$ and $f^{-1}(L) \cap \mathbb{S}^n_-$ where $\mathbb{S}^n_\pm$ denote the upper and lower hemispheres. Therefore,

$$\#f^{-1}(v) = \#(f^{-1}(L) \cap \mathbb{S}^n_+).$$

Now, consider the map $h = \pi_L \circ f : \mathbb{S}^n_+ \longrightarrow \mathbb{R}^n$. It follows that h is an odd map. Moreover, $0 \notin h(\mathbb{S}^{n-1})$, since L does not intersect $g(\mathbb{S}^{n-1})$. Hence, $W_2(h|_{\mathbb{S}^{n-1}}, 0) = 1$, by induction hypothesis ((n-1)a). On the other hand, 0 is a regular value of h, since $f \pitchfork L$. Therefore, from Theorem 7.5.1, we have, $W_2(h|_{\mathbb{S}^{n-1}}, 0) \equiv \#h^{-1}(0) \equiv \#(f^{-1}(L) \cap \mathbb{S}^n_+) = 1 \pmod 2$. ♠

Remark 7.6.1

1. The 2-dimensional version of (k4) in Lemma 7.6.1 has the following meteorological interpretation. At any given time, there exists a pair of antipodal points on the globe at which temperature and pressure are equal.

2. **Ham-Sandwich problem:** In dimension 3, the Borsuk-Ulam theorem gives an affirmative answer to the following question: Consider three sandwiches of arbitrary size and shape, situated in the space arbitrarily. Can one cut each of the three sandwiches into two equal parts, by a single stroke of the knife? The mathematical model of this question will read as follows: Suppose A_1, A_2, A_3 are bounded regions (open and connected) in $\mathbb{R}^3$. Does there exist a plane P that separates each A_i into two portions of equal measure? (It is not asserted that the two halves of A_i are connected.) One converts this problem into the setup of the Borsuk-Ulam theorem, the sketch of which is as follows. (We leave it to you to fill up all the necessary arguments.)

   Fix a vector $v \in \mathbb{S}^2$. To each $r \in R$, let P_r be the plane perpendicular to v and passing through rv. For each $A = A_i$, let $f_A(r)$ be the difference of the measures of portion of A lying on either side of P_r. First, show that $f(r)$ is continuous. Use the intermediate value theorem to show that there is $r_v \in \mathbb{R}$ such that $f(r_v) = 0$. Prove that r_v is unique. This is where connectivity of A has to be used.

   We get a map $g_A : \mathbb{S}^2 \longrightarrow \mathbb{R}$ defined by $v \mapsto r_v$. Show that g_A is continuous, by using an appropriate property of integration. Verify that g_A is odd. Now take $g := (g_{A_1}, g_{A_2}, g_{A_3})$ and complete the argument.

3. It is also easily seen that the same arguments as above can be used to solve the higher dimensional analogue of the Ham-Sandwich Problem.

Exercise 7.6 Fill in all the details in the proof of the Ham-Sandwich Theorem outlined in the above remark.

7.7 Hopf Degree Theorem

Lemma 7.7.1 Let $f : \mathbb{S}^n \longrightarrow \mathbb{S}^n, n \geq 1$ be any smooth map of degree 0. Then f is null homotopic.

Proof: We prove it by induction. For $n = 1$, consider a map $g : \mathbb{R} \longrightarrow \mathbb{R}$ such that $g \circ exp = exp \circ f$ as given by Example 7.3.2. Since $\deg f = 0$, it follows that ($\alpha(1) = 0$ and hence) g has the property $g(t + n) = g(t)$ for all t and for all integer n. So, we consider $g_s(t) = sg(t)$, $0 \leq s \leq 1, t \in \mathbb{R}$. Then each g_s has the property $g_s(t+n) = sg(t+n) = sg(t) = g_s(t)$ and hence defines a unique smooth homotopy $f_t : \mathbb{S}^1 \longrightarrow \mathbb{S}^1$ such that $g_t \circ exp = f_t \circ exp$. Of course this is a homotopy of the constant map 1 with f.

We now assume the result for n and go on to prove it for $n + 1$. Observe that the induction hypothesis is equivalent to say that any smooth map $\mathbb{S}^n \longrightarrow \mathbb{R}^{n+1} \setminus \{0\}$ having winding number 0 around the origin is null homotopic in $\mathbb{R}^{n+1} \setminus \{0\}$.

We know that the punctured $\mathbb{S}^{n+1}$ is diffeomorphic to $\mathbb{R}^{n+1}$, which is contractible.

Therefore, it suffices to prove that f is homotopic to a (continuous) map $\alpha : \mathbb{S}^{n+1} \to \mathbb{S}^{n+1} \setminus \{v_1\}$ for some point $v_1 \in \mathbb{S}^{n+1}$.

Let $f : \mathbb{S}^{n+1} \to \mathbb{S}^{n+1}$ be a smooth map of degree 0. Choose two regular values $v_1, v_2 \in \mathbb{S}^{n+1}$ for f. By an isotopy, push all the points in $f^{-1}(v_1)$ into $\mathbb{S}^{n+1}_+$, the upper hemisphere, and $f^{-1}\{v_2\}$ inside $\mathbb{S}^{n+1}_-$, the lower hemisphere. (See Corollary 6.3.3.) Next, by an isotopy in the codomain (or otherwise), we may assume that $v_1 = N, v_2 = S$, the north and the south pole, respectively. Let $\rho : \mathbb{S}^{n+1} \setminus \{N\} \to \mathbb{R}^{n+1}$ be the stereographic projection and $h = \rho \circ f : \mathbb{S}^{n+1}_- \to \mathbb{R}^{n+1}$.

Now, for each $a_i \in f^{-1}(\{S\}) = h^{-1}(\{0\})$, choose disjoint neighborhoods V_i of a_i in $\mathbb{S}^{n+1}_-$ such that each is diffeomorphic to a disc neighborhood V of 0 under h (Stack-Record Theorem 3.4.5). Let $W = \mathbb{S}^{n+1}_- \setminus \cup_{i\geq 0} V_i$. Then clearly $\partial W = \mathbb{S}^n \cup_{i=1} \partial V_i$. Consider the map $h : W \to \mathbb{R}^{n+1} \setminus \{0\}$. It follows that the winding number $W(h|_{\mathbb{S}^n}, 0)$ is equal to the sum of all the winding numbers $W(f|_{\partial V_i}, 0)$. On the other hand, this sum is equal to the degree of f up to sign, and hence is equal to zero. By induction hypothesis, $h|_{\mathbb{S}^n}$ is homotopic to a constant map in $\mathbb{R}^{n+1} \setminus \{0\}$. From this, we get a smooth map

$$g : \mathbb{S}^{n+1}_- \to \mathbb{R}^{n+1} \setminus \{0\}$$

which extends the map $h|_{\mathbb{S}^n}$. Now, $\rho^{-1} \circ g$ will patch-up with $f|_{\mathbb{S}^{n+1}_+}$ to define a continuous map $\hat{f} : \mathbb{S}^{n+1} \to \mathbb{S}^{n+1}$. It easily follows that the south-pole S is not in the image of $\hat{f}$. Finally define $\widehat{F} : \mathbb{S}^{n+1} \times I \to \mathbb{S}^{n+1}$ by

$$\widehat{F}(x,t) = \begin{cases} f(x), & x \in \mathbb{S}^{n+1}_+, \\ \rho^{-1}(tg(x) + (1-t)h(x)), & x \in \mathbb{S}^{n+1}_-. \end{cases}$$

Then F is a homotopy of f to a map $\hat{f}$ as required. ♠

Theorem 7.7.1 Hopf Degree Theorem: *Let X be a compact oriented boundaryless manifold of dimension n and let $f, g : X \to \mathbb{S}^n$ be any two maps. Then $f \simeq g$ iff $\deg f = \deg g$.*

Proof: We have only to prove the "if" part here.

Observe that it is enough to get a map $G : X \times I \to \mathbb{R}^{n+1} \setminus \{0\}$ such that $G_0 = f, G_1 = g$. So, first define $H : X \times I \to \mathbb{R}^{n+1}$ by $H(x,t) = (1-t)f(x) + tg(x)$. By extended transverse homotopy Theorem 7.1.4, we may assume that 0 is a regular value for H. By an isotopy of $X \times I$ which is identity near $\partial(X \times I)$ we can push the finite set $H^{-1}(0)$ inside a subset A of $X \times (0,1)$, which is diffeomorphic to a disc. (See Corollary 6.3.3.) Now consider $W = X \times I \setminus \operatorname{int} A$ and let $B = \partial A$. Then

$$\partial W = (X \times 0) \cup (X \times 1) \cup B.$$

Since $H(W) \subset \mathbb{R}^{n+1} \setminus \{0\}$, it follows that

$$0 = W(g,0) - W(f,0) - W(H|_B, 0) = \deg g - \deg f - W(H|_B, 0).$$

Hence, $W(H|_B, 0) = 0$ and by the lemma above, it follows that $H|_B$ can be extended to a map $\hat{H} : A \to \mathbb{R}^{n+1} \setminus \{0\}$. Together with the rest of H on W, this defines the map G that we wanted. ♠

Remark 7.7.1 Thus, for a compact connected, oriented boundaryless manifold X of dimension $n \geq 1$, the homotopy class of a map $f : X \to \mathbb{S}^n$ is determined by an integer, viz., the degree of the map. A natural question that arises is: for each integer d, does there exist a map $f : X \to \mathbb{S}^n$ with its degree equal to d? The answer is "YES".

Of course, for $d = 0$, we simply take f to be any constant map. So let $d \neq 0$. Now consider the special case where $X = \mathbb{S}^n$. For $n = 1$, we know that the map $\tau_d : z \mapsto z^d$ is of degree d on $\mathbb{S}^1$. By successively spinning τ_d (equivalently, taking supsension), we obtain degree d maps on higher dimensional spheres. The catch is that the "spinned" maps are not smooth at the north and south poles. One can change them "slightly" by a homotopy to get smooth maps. However, this technique will **not** be available on other manifolds.

We can try another geometric idea. Pick disjoint copies of balls in the manifold, as many as the degree of the map that you want to construct; pinch the rest of the manifold to a single point to obtain a bouquet of spheres; map each of these spheres to a single sphere by orientation-preserving diffeomorphisms. It is clear that such a map is continuous and is of degree d. The only problem seems to be the smoothness of the map at the boundary points of the discs. This can be resolved in several ways, e.g., one can appeal to smooth approximation Theorem 6.1.3. Here is another.

Theorem 7.7.2 *Let X be an oriented, connected, closed n-dimensional manifold. For every integer d, there exists a smooth map $f : X \to \mathbb{S}^n$ such that* $\deg f = d$.

Proof: We first observe that if $\sigma : \mathbb{R}^n \to \mathbb{S}^n \setminus \{N\}$ is the inverse of the stereographic projection, then $\lim\limits_{x\to\infty} d\sigma_x = 0$. For

$$\sigma(x_1, \ldots, x_n) =: (\sigma_1, \ldots, \sigma_{n+1}) = \left(\frac{2x_1}{1+\|x\|^2}, \ldots, \frac{2x_n}{1+\|x\|^2}, \frac{1-\|x\|^2}{1+\|x\|^2}\right).$$

Now take partial derivatives of each σ_i and see that the limit is zero as $x \to \infty$. (Indeed, all the higher derivatives also have limit zero.)

Let now $\phi_i : U_i \to \mathbb{R}^n$ be orientation-preserving diffeomorphisms where U_i are disjoint open subsets of X. We define $f : X \to \mathbb{S}^n$ by

$$f(x) = \begin{cases} \sigma \circ \phi_i(x), & x \in U_i,\ i = 1, \ldots, d; \\ N, & x \in X \setminus \cup_i U_i. \end{cases}$$

From the lemma above, it follows that as x approaches a boundary point of U_i from inside U_i, df_x tends to 0. It follows that f is smooth. Now observe that σ is orientation preserving or reversing according as n is odd or even. Therefore, it follows that $\deg f = (-1)^{n-1}d$. When n is even, we also know that the antipodal map is of degree -1. So, in this case we compose the above f with the antipodal map to get a map of degree d. ♠

Remark 7.7.2 The Hopf degree theorem is available when X is nonorientable also. The basic result that we need is that in a connected nonorientable manifold, there exist embedded loops ω which reverse orientation. (See Exercise 4.1.1).

Theorem 7.7.3 *Let X be a connected nonorientable manifold of dimension $n \geq 2$. Then two maps $f, g : X \to \mathbb{S}^n$ are homotopic iff* $\deg_2 f = \deg_2 g$.

Proof: Again, we need to prove the case "if" only. As in the orientable case, it suffices to prove that if $\deg_2 f = 0$ then f is null homotopic. Choose a regular value $v \in \mathbb{S}^n$ for f and let $F = f^{-1}(v) = \{a_1 \ldots, a_{2k}\}$. We shall show that f is homotopic to a map f_1 with v as a regular value but $f_1^{-1}(v) = \{a_1, \ldots, a_{2k-2}\}$. Repeating this argument, it would follow that f is homotopic to a map that is not surjective. As before, this will imply that f is null homotopic.

We shall prove that there is an embedded closed disc $B^n \subset X$ such that $a_{2k-1}, a_{2k} \in \operatorname{int} B^n$ and $W(f|_{\partial B^n}, v) = 0$. It then follows as in Lemma 7.7.1 that $f|_{B^n}$ can be homotoped so as not to contain the point v in its image. This homotopy can then be extended to a homotopy of f all over X to a map f_1 as required.

By an ambient isotopy of X, first push a_{2k} and a_{2k-1} inside a co-ordinate neighborhood U and the rest of the $a_i's$ inside a disjoint coordinate neighborhood. Now fix an orientation for U and obtain the intersection numbers $\epsilon(a_{2k-1})$ and $\epsilon(a_{2k})$. If these two are different then their sum is 0. Well and good—you can take $B = \bar{U}$.

If not, choose an embedded loop $\gamma : I \to X$ at a_{2k} such that γ is orientation reversing and does not pass through any of the points $a_i, i \neq 2k$. Extend this to an ambient isotopy of X that is a constant outside a tubular neighborhood of γ, which does not contain any of the points $a_i, i \neq 2k$. The new map f is obviously homotopic to the old one and has the property that the intersection number at a_{2k} with respect to the orientation that we have fixed on U is the opposite of $\epsilon(a_{2k-1})$. Therefore, we are back in the above case. ♠

Exercise 7.7 For smooth maps $f : M \to N$ and $g : N \to P$, where M, N, P are oriented closed n-manifolds, show that $\deg(g \circ f) = (\deg g)(\deg f)$.

7.8 Lefschetz Theory

Given a self-map $f : X \to X$ of a compact smooth manifold X, consider the problem of studying the set of fixed points of f. Of course, many questions can be raised in this regard. The very first question is: *Is the set nonempty?* Second, if there are finitely many such points, *how many of them are there?* Moreover, we also like to know the local behavior of the function f itself in a neighborhood of a fixed point. Lefschetz's theory comes up with a fairly good answer to many of these questions.

First of all, we make a completely natural observation, viz., $f(x) = x$ for some $x \in X$ iff $(x, f(x)) \in \Delta_X$ the diagonal in $X \times X$. Thus, the fixed points of f are in one-to-one correspondence with the points in the intersection of the graph Γ_f of f with the diagonal Δ in $X \times X$. So, the intersection theory must have a role to play in the fixed point theory. This simple observation is the key to the entire discussion that follows.

Definition 7.8.1 Let X be a compact, oriented manifold and $f : X \to X$ be a smooth map. Then the number $I(\Delta_X, \Gamma_f; X \times X)$ is denote by $L(f)$ and is called the *Lefschetz number of f*. In particular, we define the (Poincaré) index $I(X)$ of X by the formula

$$I(X) := L(ID_X) = I(\Delta_X; X \times X) = I(\Delta_X, \Delta_X; X \times X). \tag{7.7}$$

Remark 7.8.1 Since intersection number is a homotopy invariant, it follows that $L(f)$ is also an invariant of the homotopy class of f. Also, it is invariant under conjugation, viz, if $\phi : X \to X$ is a diffeomorphism, then $L(f) = L(\phi^{-1} \circ f \circ \phi)$.

Remark 7.8.2 If the fixed point set of f is empty, then clearly $L(f) = 0$. Thus, at least we know when the fixed point set is nonempty, viz., if $L(f) \neq 0$.

Is the converse true? How does the number $L(f)$ measure the number of fixed points? These questions are subtle, for a number of reasons. First of all, we know that the set theoretic intersection is not a homotopy invariant. Next, certain signs are attached to the points and the summation of these signed numbers is taken as the intersection number. Since cancellations are possible while taking the summation, it is not at all clear why $L(f)$ need not vanish, even when f may have fixed points. This is indeed true, yet, the scenario is not so bad, as we shall soon see.

Let us first examine the simplest case when x is a fixed point and Γ_f is transversal to Δ at (x, x). This means that

$$T_{(x,x)}(\Gamma_f) + T_{(x,x)}(\Delta) = T_{(x,x)}(X \times X).$$

This is the same as saying

$$Im(Id, df_x) + \Delta_{T_x X} = T_x(X) \times T_x(X).$$

Here Id denotes the identity map of $T_x X$. Observe that (Id, df_x) is injective and hence the image of (Id, df_x) is of dimension n. Also, the diagonal in $T_x X$ is of dimension n. It follows that the above transversality condition is equivalent to say that $Im(Id, df_x) \cap \Delta_{T_x(X)} = \{0\}$. This is the same as saying that no eigenvalue of the linear map df_x is equal to 1.

Definition 7.8.2 A fixed point x of f is said to be of *Lefschetz type* if df_x has all its eigenvalues different from 1. Also, we call f itself a map of *Lefschetz type* if all of its fixed points are of the Lefschetz type.

Remark 7.8.3 From the above analysis, it is immediate that fixed points of maps of the Lefschetz type are isolated. Since the condition is open, we would anticipate that fixed points are almost always of the Lefschetz type. In any case, the importance of Lefschetz type maps is enhanced by the following result.

Lemma 7.8.1 Every map is homotopic to a map of the Lefschetz type.

Proof: Recall that given $f : X \to X$, there exists a map $F : X \times B^N \to X$ such that for each fixed x, the map $v \mapsto F(x, v)$ itself is a submersion (Theorem 7.1.2). Consider the map $G : X \times B^N \to X \times X$ given by $G(x, v) = (x, F(x, v))$. Then it is fairly obvious that G is also a submersion (see Exercise 3.6). Hence, by transversality Theorem 7.1.1, for almost all $v \in B^N$, the map $x \mapsto G(x, v) = (x, F(x, v))$ is transversal to Δ. But this map is the graph of the map $x \mapsto F(x, v)$, which is clearly homotopic to f. ♠

For a Lefschetz type map, let us denote the intersection number at a fixed point x_i by $\eta(x_i)$. Then by definition, $L(f) = \sum \eta(x_i)$. The following lemma tells us how to identify this. Since the transversality condition is the same as saying that 1 is not an eigenvalue of df_{x_i}, this means that $df_{x_i} - Id$ is an isomorphism of $T_{x_i} X$.

Lemma 7.8.2 Let x be a Lefschetz type fixed point of f. Then the intersection number $\eta(x)$ is ± 1 according as $df_x - Id$ preserves or reverses orientation at x.

Proof: Let $A = df_x : T_x X \to T_x X$. Let $[a] = [a_1, \ldots, a_n]$ denote the orientation class for $T_x X$. Then for $T_{(x,x)}(X \times X)$, $T_{(x,x)}(\Delta)$, and $T_{(x,x)}(\Gamma_f)$, we have the orientation classes given by

$$\begin{array}{lcl} [(a,0)][(0,a)] & := & [(a_1, 0), \ldots, (a_n, 0), (0, a_1), \ldots, (0, a_n)], \\ [(a,a)] & := & [(a_1, a_1), \ldots, (a_n, a_n)], \\ [(a, A(a)] & = & [(a_1, Aa_1), \ldots, (a_n, Aa_n)], \end{array}$$

respectively. Let ϵ denote ± 1 according as $A - Id$ preserves orientation or reverses orientation, i.e.,

$$(A - Id)[a] = [(A - Id)(a_1), \ldots, (A - Id)a_n] = \epsilon[a_1, \ldots, a_n] = \epsilon[a].$$

Then

$$\begin{array}{lcl} [(a,a)][(a, Aa)] & = & [(a,a)][(0, Aa - a)] = [(a,a)][(0, (A - Id)(a))] \\ & = & \epsilon[(a,a), (0,a)] = \epsilon[(a,0), (0,a)]. \end{array}$$

Therefore, the intersection number at x is also equal to ϵ (see (7.2)). ♠

Remark 7.8.4 In the above discussion, notice that we do not need X to be an orientable manifold, the entire discussion is local in nature: To obtain the numbers $\eta(x_i)$ we need to fix some orientation at the point x_i and it does not matter which orientation is taken, the value of $\eta(x_i)$ will be the same. Therefore,

Corollary 7.8.1 The number $L(f) = I(\Delta_X, \Gamma_f; X \times X)$ is well defined for all closed manifolds X and for all smooth maps $f : X \to X$ irrespective of whether X is orientable or not, and is an invariant of the homotopy class of f. (In particular, it does not even depend upon the smooth structure of X either.)

Remark 7.8.5 In particular, notice that the self-intersection number

$$I(X) = I(\Delta_X, \Delta_X; X \times X) \tag{7.8}$$

is defined as an integer even in the case when X is not orientable.

Example 7.8.1 Let us examine what it means to say that the intersection number at a point (x, x) is ± 1 in the 2-dimensional case. By passing to a coordinate neighborhood and adding a suitable constant to f, we may assume that $x = 0$ and f is defined in a neighborhood of $0 \in \mathbb{R}^2$ and $f(0) = 0$. Then $df_0 = A : \mathbb{R}^2 \to \mathbb{R}^2$ is a linear map that approximates f and hence $f(t, s) = A(t, s) + \epsilon(t, s)$, where $\epsilon(t, s) \to (0, 0)$ as $(t, s) \to (0, 0)$. Therefore, an approximate picture of f can be obtained by studying A.

Let us assume that A has two independent eigenvectors, and λ_1, λ_2 are two real eigenvalues of A. By a coordinate transformation, we may assume that these eigenvectors are in the direction of x-axis and y-axis. Now consider the case, when both λ_i are > 1. This implies that each nonzero vector is mapped to a vector of length larger than itself. Thus, f is an expanding map (near 0) and we call 0 a "source". Exactly the opposite happens when both λ_i are < 1 and we call 0 a "sink". In either case, it is clear that $\det(A - Id)$ is positive and hence $\eta(x) = 1$.

Now consider the case when $\lambda_1 > 1$ and $\lambda_2 < 1$. Then f is expanding along the x-axis and contracting along the y-axis. In this case, we call 0 a *saddle point*. Also $\det(A - Id) < 0$ and hence $\eta(x) = -1$.

Of course, it may happen that A has no real eigenvalues in which case, the picture may be more complicated. We shall come back to it later.

Definition 7.8.3 Let x be a Lefschetz type isolated fixed point of a smooth map $f : X \to X$. We say x is a *source* (resp. a *sink*) if $df_x : T_xX \to T_xX$ has all eigenvalues > 1 (resp. < 1). If some eigenvalues > 1 and some other < 1, then we say x is a saddle point.

Remark 7.8.6 Observe that the local Lefschetz number $L_x(f)$ at a source is always equal to 1. At a sink it is equal to $(-1)^n$ and at a saddle point it is equal to $(-1)^k$ where k is the number of eigenvalues of df_x, which are less than 1.

As seen above we know that most maps are of the Lefschetz type. However, in practice, we do come across maps that are not. This phenomenon is similar to the fact that most polynomials have distinct roots but sometimes we have to deal with those that have multiple roots.) We shall now use the homotopy invariance property of $L(f)$ to define the local Lefschetz numbers at any isolated fixed point that is not necessarily of the Lefschetz type. The first step is:

Lemma 7.8.3 Let U be an open subset of $\mathbb{R}^n$, $f : U \to \mathbb{R}^n$ be a smooth map with a

Lefschetz type fixed point at $z \in U$. Then for a suitable $\epsilon > 0$, $f(x) \neq x$ for any $x \in \partial B_\epsilon(z)$. Moreover, the Gauss map associated to f, viz., $F : \partial B_\epsilon(z) \to \mathbb{S}^{n-1}$ given by

$$F(x) = \frac{f(x) - x}{\|f(x) - x\|}$$

has its degree equal to $\eta(z)$.

Proof: By linear approximation, we first reduce the problem to the case when f itself is linear. First try it yourself and only then read the following details.

Put $A = df_z$. Then by Taylor's' Theorem, $f(x) = f(z) + A(x - z) + \psi(x)$, where

$$\left\| \frac{\psi(x)}{x - z} \right\| \to 0 \quad \text{as } x \to z.$$

Put $c = 1/\|(A - I)^{-1}\|$. Then for all $\mathbf{v} \in \mathbb{R}^n$, we have $\|(A - I)(\mathbf{v})\| \geq c\|\mathbf{v}\|$. Now choose $\epsilon > 0$ such that $\|\psi(x)\| \leq c\|x - z\|/2$ for all x such that $\|x - z\| \leq \epsilon$. Then the first part follows. Now define $g_t(x) = (A - I)(x - z)) + t\psi(x)$. Then g_t never vanishes in $B_\epsilon(z)$ and we have $g_0(x) = (A - I)(x - z)$ and $g_1(x) = f(x) - x$. After dividing out by the norm, this yields a homotopy of F with the map

$$x \mapsto \frac{(A - I)(x - z)}{\|(A - I)(x - z)\|}$$

on $\partial B_\epsilon(z)$. Therefore, the two maps have the same degree. It remains to show that the degree of the map

$$\mathbf{v} \mapsto \frac{(A - I)(\mathbf{v})}{\|(A - I)(\mathbf{v})\|}$$

is equal to ± 1 according as $(A - I)$ preserves orientation or reverses it. From lemma 7.8.2 we are through. ♠

Theorem 7.8.1 Splitting Principle: *Let $f : X \to X$ have an isolated fixed point at x_0 inside a neighborhood U of x. Then there exists a homotopy H of f that is constant outside a compact subset K of U to a map g such that all fixed points of g inside U are of the Lefschetz type.*

Proof: It suffices to prove the result for the case when U is an open subset of $\mathbb{R}^n$ and 0 is the only fixed point of f in U. Choose a bump function $\rho : U \to [0, 1]$ such that $\rho(B_r(0)) = \{1\}$ and $\rho(U \setminus B_{2r}(0)) = \{0\}$, with $B_{2r}(0) \subset U$. For any fixed $v \in \mathbb{R}^n$, consider the mapping

$$H(x, t) = f(x) + t\rho(x)v =: h_t(x).$$

We shall show that for all $\epsilon > 0$ sufficiently small, there exist plenty of v with $\|v\| < \epsilon$ such that H defined as above gives the required homotopy.

First, we want to ensure that h_t does not have any fixed points outside $B_r(0)$. In the compact set $r \leq ||x|| \leq 2r$, since f has no fixed points, we can find $\delta > 0$, such that $||f(x) - x)|| > \delta$. Now take v such that $||v|| < \delta/2$. Then

$$||h_t(x) - x|| \geq ||f(x) - x|| - t\rho(x)||v|| > \delta/2.$$

On the other hand, for $||x|| \geq 2r$, we have, $\rho(x) = 0$ and hence $H(x, t) - x = f(x) - x \neq 0$. Now by Sard's theorem, there exists $v \in \mathbb{R}^n$ such that $0 < ||v|| < \delta/2$ which is a regular value for $f - Id$. Now suppose x is a fixed point of $g = h_1$. Then $x \in B_r(0)$ and hence $g(y) = f(y) + v$ in $B_r(0)$. Therefore, $dg_x = df_x$. Hence, x is of the Lefschetz type iff $df_x - Id$

is an isomorphism. Now, since we have assumed that x is a fixed point of g, we have $x = g(x) = f(x) + v$, which implies $(f - Id)(x) = v$. Since v is a regular value, it follows that $d(f - Id)_x = df_x - Id$ is an isomorphism. Hence, we have proved that f is homotopic to a map g which has all its fixed points inside $B_r(0)$ and they are all of the Lefschetz type. Moreover, $g = f$ outside $B_{2r}(0)$. ♠

Definition 7.8.4 Given a map $f : U \to \mathbb{R}^n$ with an isolated fixed point z, we consider the map $F(x) = \frac{f(x)-x}{\|f(x)-x\|}$ on $\partial B_r(z)$ for sufficiently small $r > 0$ and put $L_z(f) = \deg F$. It is easily verified that if we choose any other sphere around z to define $L_z(f)$, then we get the same number. As remarked earlier, if $\phi : V \to U$ is a diffeomorphism then $L_z(\phi^{-1} f \phi) = L_{\phi(z)}(f)$. This enables us to define the local Lefschetz number of self-maps of any manifold.

Thus, suppose $f : X \to X$ has an isolated fixed point $z \in X$. Choose a parameterization $\phi : \mathbb{R}^n \to U$ with $\phi(0) = z$ around z. Define the *local Leschetz number of f at z* by

$$L_z(f) = L_0(\phi^{-1} \circ f \circ \phi). \tag{7.9}$$

From the splitting theorem, using similar arguments as in the argument principle, we obtain:

Lemma 7.8.4 Let $f : X \to X$ be any smooth map where X is a compact manifold. Suppose f has finitely many fixed points $\{x_1, \ldots, x_k\}$. Then $L(f) = \sum_i L_{x_i}(f)$.

Proof: Choose disjoint neighborhoods U_i of x_i each diffeomorphic to a ball and apply the splitting principle in each of the these neighborhoods simultaneously, to obtain a Lefschetz map g that is homotopic to f and agrees with f outside $X \setminus \cup_i U_i$. So, $L(f) = L(g)$. By definition, since g is a map of the Lefschetz type, the global Lefschetz number is the sum of the local ones, i.e., we have $L(g) = \sum_{ij} L_{x_{ij}}(g)$. Now by argument principle, $L(g|_{U_i})$ is the sum total of the degrees at $x'_{ij}s$, i.e., $L(g|_{U_i}) = \sum_j L_{x_{ij}}(g)$. On the other hand, on the boundary of U_i, $f = g$ and hence $L(g|_{U_i})$ is equal to the degree of $\frac{f(x)-x}{\|f(x)-x\|}$. Since x_i is the only fixed point of f inside U_i, this degree is equal to $L_{x_i}(f)$. ♠

Remark 7.8.7 Thus, the above lemma gives a method to determine the Lefschetz number, even without actually homotoping the given map to a Lefschetz map. We shall now use this in a very special case, viz., to compute the Lefschetz number of the identity map.

Let us begin with an example.

Example 7.8.2 Consider the following mapping $f : \mathbb{S}^n \to \mathbb{S}^n$, which fixes the north pole and the south pole, given by

$$f(x) = \frac{x + S/2}{||x + S/2||}$$

where S denotes the south pole. Check that the north pole is a source and the south pole is a sink. Of course there are no other fixed points of f. Therefore, $L(f) = 1 + (-1)^n$. On the other hand, f is homotopic to identity (why?) and hence

$$L(f) = L(Id) = I(\mathbb{S}^2) = I(\Delta_{\mathbb{S}^2}; \mathbb{S}^2 \times \mathbb{S}^2). \tag{7.10}$$

Thus, $I(\mathbb{S}^2) = 2$. Using similar ideas, it is possible to compute $I(S)$ for any 2-dimensional closed manifold S. This we shall do in the next chapter.

Remark 7.8.8

1. The discussion that we had on the Lefschetz number is going to be of some use. For this, we should choose the map homotopic to Id so that it has only isolated fixed points. Already, we have a method to generate homotopies of the identity map, viz., by choosing a vector field σ and generating a 1-parameter group of diffeomorphisms, $\{h_t\}$ from it.

2. It is fairly easy to see that for sufficiently small $t > 0$, each h_t will have finitely many fixed points, provided the vector field σ that we choose has finitely many zeros. In such a situation, we have, $I(X) = L(h_t)$.

3. A natural question that arises is: Can we read off this number directly from the vector field? The answer is *yes indeed* and this is the major part of the celebrated Poincaré-Hopf Index Theorem (see Theorem 7.8.2 below). We begin with:

Definition 7.8.5 Let σ be a smooth vector field on $\mathbb{R}^n$ such that $\sigma(x) = (x, 0)$ iff $x = 0$. Recall that the tangent space to $\mathbb{R}^n$ is a product $T(\mathbb{R}^n) = \mathbb{R}^n \times \mathbb{R}^n$. Therefore, a vector field σ on $\mathbb{R}^n$ corresponds simply to a map $\hat{\sigma} : \mathbb{R}^n \to \mathbb{R}^n$, by the formula:

$$\sigma(x) = (x, \hat{\sigma}(x)). \tag{7.11}$$

In this case $\hat{\sigma}$ restricts to a smooth map $\hat{\sigma} : \mathbb{R}^n \setminus 0 \to \mathbb{R}^n \setminus 0$. Consider the winding number $W(\hat{\sigma}, 0)$ of $\hat{\sigma}|_{\mathbb{S}^{n-1}}$ around 0 and call this the *local index of* σ *at* x and denote it by $\mathrm{ind}_x \sigma$.

Remark 7.8.9 Of course, we could take the restriction of $\hat{\sigma}$ to any sphere around 0 and they would all have same winding number around 0.

Lemma 7.8.5 Let $f, g : \mathbb{R}^n \setminus \{0\} \to \mathbb{R}^n \setminus \{0\}$ be any two smooth maps. Then $W(g \circ f, 0) = W(g, 0)W(f, 0)$.

Proof: Recall that $W(f, 0)$ is defined to be the degree of the map $\hat{f} : \mathbb{S}^{n-1} \to \mathbb{S}^{n-1}$ given by $\hat{f}(x) = \frac{f(x)}{\|f(x)\|}$. Now consider the homotopy

$$H(x,t) = g\left(t\frac{f(x)}{\|f(x)\|} + (1-t)f(x)\right)$$

between $g \circ f$ and $g \circ \hat{f}$ which yields $W(g \circ f, 0) = W(g \circ \hat{f}, 0)$. Now $\widehat{g \circ \hat{f}} = \hat{g} \circ \hat{f}$ and hence $W(g \circ \hat{f}, 0) = \deg(\widehat{g \circ \hat{f}}) = \deg(\hat{g} \circ \hat{f})$. We now appeal to Exercise 7.7 to conclude the lemma. ♠

Lemma 7.8.6 The local index, $\mathrm{ind}_0\, \sigma$ is invariant under diffeomorphisms:

Proof: If $\phi : \mathbb{R}^n \to \mathbb{R}^n$ is a diffeomorphism such that $\phi(0) = 0$. We have to show that $\mathrm{ind}\, \sigma = \mathrm{ind}\, \phi^*(\sigma)$. First of all note that

$$\phi^*(\sigma)(x) = (x, (d(\phi)_x)^{-1}(\hat{\sigma}(\phi(x))) =: (x, \tau(x)), \text{ say.}$$

Therefore, we must show that $W(\tau, 0) = W(\hat{\sigma}, 0)$. Now consider the homotopy $H(x,t) = (d\phi_{tx})^{-1}(\hat{\sigma}(\phi(x)))$, which yields that $W(\tau, 0) = W(A^{-1} \circ \hat{\sigma} \circ \phi, 0)$, where $A = d\phi_0$ is a linear isomorphism. Now, we know that A preserves orientation iff the diffeomorphism ϕ preserves orientation. Therefore, $W(A^{-1}, 0) = W(\phi, 0) = \pm 1$. We can now appeal to Lemma 7.8.5 to conclude that

$$W(\tau, 0) = W(A^{-1}, 0)W(\hat{\sigma}, 0)W(\phi, 0) = W(\hat{\sigma}, 0).$$

This completes the proof of the lemma. ♠

Definition 7.8.6 Index of a Vector Field: Let σ be a smooth vector field on a manifold X with finitely many zeros $\{z_1, \ldots, z_k\}$. By the above lemma, it follows that the "local index" $\text{ind}_{z_i}(\sigma)$ of σ at each of the point z_i is well-defined, being the index of the pullback vector field under a parameterization around z_i. We define the (global) *index of* σ by the formula:

$$\text{Ind}\,(\sigma) = \sum_{z_i} \text{ind}_{z_i}(\sigma). \tag{7.12}$$

Remark 7.8.10 The well-definedness of the local index is a very important property. In particular, it does not depend on the local orientations. The fallout is that the sum of all these local indices is well-defined even for nonorientable manifolds also. This is something that is not so obvious at all since all this depended on our theory of oriented intersection numbers, or for that matter, on the notion of degree.

Example 7.8.3 There is a close relation between the discussion we had on Lefschetz type maps and vector fields. Suppose σ is a vector field on a 2-disc with an isolated zero. We can draw arrows to indicate this vector field and look at the various possible patterns as in the figure below.

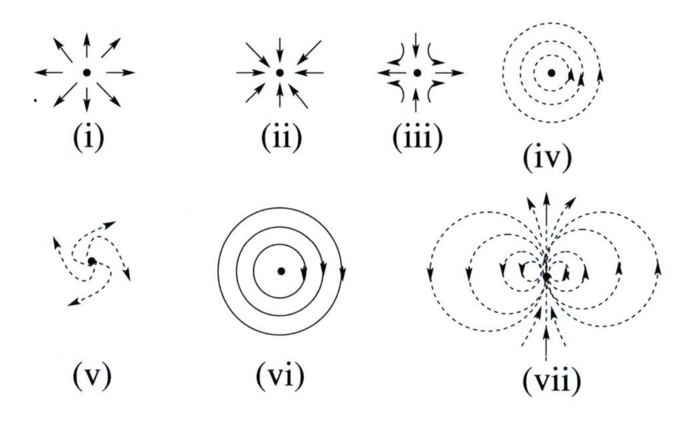

Figure 37 Types of 2-dimensional fluid flows.

The first one is called a *source,* the second one a *sink,* terms we have used to describe certain types of Lefschetz fixed points. We have more here. The third one is clearly a *saddle point.* The fourth one may be called a *vortex,* and the fifth one is a *spiral.* The sixth one is again a vortex but with an opposite orientation to that of the fourth one. Finally, you are welcome to choose a name for the last one (*butterfly?*) Compute the index in each case. Draw a small circle C around each zero of the field, draw a unit circle S on the side and watch how $\sigma(x)/\|\sigma(x)\|$ moves on S as your point x moves on C. The indices are respectively, $1, 1, -1, 1, -1, 1, 2$. ARE YOU SURPRISED TO SEE THAT the first and second and similarly, the third and the fifth have the same index?

Theorem 7.8.2 *Let σ be a smooth vector field with finitely many zeros on a closed manifold X. Then the sum of all the indices of σ at these points is equal to the self-intersection number $I(X)$, i.e.,*

$$Ind(\sigma) = I(X).$$

Proof: Let h_t be the 1-parameter group of diffeomorphisms generated by σ. Since $h_0 = Id_X$, it follows that $I(X) = L(h_t)$ for all t. Now for sufficiently small $t > 0$, the fixed points of h_t are precisely, the zeros of σ. Therefore, we can compute $L(h_t)$ as the sum of finitely many local Lefschetz numbers. Thus, it is enough to prove that $L_p(h_t) = \text{ind}_p(\sigma)$ for each of these fixed points. This local problem can be transferred to a neighborhood of 0 in the Euclidean space. Then by Taylor's expansion, we have

$$h_t(x) = h_0(x) + t\sigma(x) + t^2\sigma_1(x)$$

where $\sigma_1(x)$ is some smooth function. Therefore,

$$\frac{h_t(x) - x}{\|h_t(x) - x\|} = \frac{h_t(x) - h_0(x)}{\|h_t(x) - h_0(x)\|} = \frac{t\sigma(x) + t^2\sigma_1(x)}{\|t\sigma(x) + t^2\sigma_1(x)\|} = \frac{\sigma(x) + t\sigma_1(x)}{\|\sigma(x) + t\sigma_1(x)\|}$$

The above equation is valid for sufficiently small $t > 0$. The winding number of the map on the LHS restricted to a small sphere is the local Lefschetz number $L_p(h_t)$ of the map h_t. On the RHS, we have a homotopic family of maps defined even for $t = 0$. Therefore, they all have the same degree. Putting $t = 0$, we get the map $\hat{\sigma} : x \mapsto \sigma(x)/\|\sigma(x)\|$, the winding number of which is $\text{ind}_p(\sigma)$. ♠

Remark 7.8.11 As an immediate fallout of this theorem, we now have a better definition of $I(X)$, that is valid for all closed manifolds, viz., take any smooth vector field with finitely many zeros and take the sum total of local indices. For compact manifolds with boundary, the choice of vector field should be such that at the boundary points it is strictly outward. All this will be futile of course, if we do not have such vector fields. We shall see more about them in a later chapter.

Example 7.8.4 On the projective plane $\mathbb{P}^2$, we shall construct a vector field σ that has only one zero. Also, this zero happens to be a vortex. Consequently, we obtain: $I(\mathbb{P}^2) = 1$. The vector field σ is obtained by constructing a vector field τ on $\mathbb{S}^2$ which is invariant under antipodal action. We use spherical coordinates

$$(\psi, \theta) \mapsto (\cos\psi\cos\theta, \cos\psi\sin\theta, \sin\psi), -\pi/2 \le \psi \le \pi/2, \quad 0 \le \theta \le 2\pi$$

for $\mathbb{S}^2$, and take

$$\tau(\psi, \theta) = (\pi^2/4 - \psi^2)(-\sin\theta, \cos\theta, 0).$$

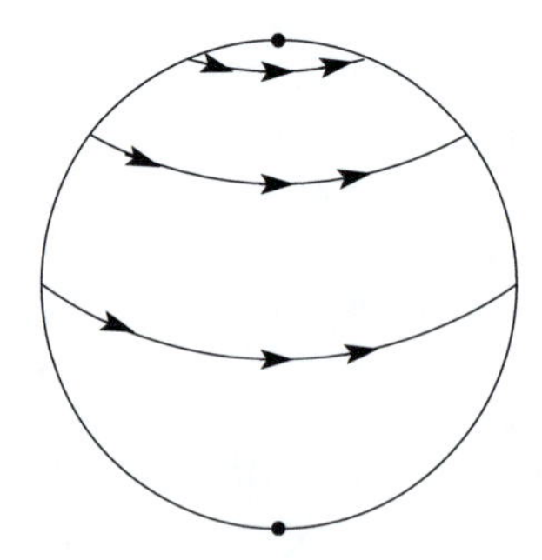

Figure 38 An invariant vector field on $\mathbb{S}^2$.

The two zeros of τ are precisely at the south and north pole and the index at these points is $+1$, they being vortices (see example 7.8.3). Passing down to $\mathbb{P}^2$, this will produce a vector field σ with index 1.

Exercise 7.8

1. Let σ be a vector field on $\mathbb{D}^n$ with finitely many zeros all of which are in the interior. Suppose the sum of the indices at these zeros is equal to zero. Then there exists a vector field on $\mathbb{D}^n$ that agrees with σ on a neighborhood of $\partial\mathbb{D}^n$ and which does not vanish anywhere.

2. Let X be a connected manifold and $P \subset X$ be a finite subset. Then there exists an open set U in X diffeomorphic to an open disc that contains P.

3. Let X be a closed manifold. Then $I(X) = 0$ iff there exists a nowhere vanishing smooth vector field on X.

4. Construct a nowhere vanishing vector field on the Möbius band.

7.9 Some Applications

We shall now bring our knowledge of degree of a map together with the integration theory that we have developed so far. The very first result may be viewed as a further generalization of the change of variable formula. We then give an application of this to the Gauss-Bonnet theorem, a classical result in differential geometry.

Of course, the Gauss-Bonnet formula for the curvature form can be viewed as another way to compute the Euler characteristic itself.

Theorem 7.9.1 Degree Formula for Integration: *Let X, Y be any two n-dimensional smooth compact oriented connected manifolds without boundary and $f : X \to Y$ be any smooth map. Then for any n-form ω on Y, we have*

$$\int_X f^*(\omega) = (deg\, f) \int_Y \omega. \tag{7.13}$$

As a first step toward the proof of this theorem we have:

Lemma 7.9.1 Let $y \in Y$ be a regular value for f. Then there exists a neighborhood U of y in Y such that for all n-forms ω with support contained in U, formula (7.13) holds.

Proof: By Stack-Record Theorem 3.4.5, we have a neighborhood U of y such that $f^{-1}(U) = \cup_{i=1}^k V_i$, with each $f : V_i \to U$ being a diffeomorphism. Then by the change of variable formula 4.10, for any ω with support inside U, we have,

$$\int_{V_i} f^*(\omega) = \eta(x) \int_Y \omega$$

where $\eta(x) = \pm 1$ according as $f : V_i \to U$ is orientation preserving or not. Therefore,

$$\int_X f^*\omega = \sum_i \int_{V_i} f^*\omega = \sum_{f(x)=y} \eta(x) \int_Y \omega = (deg\, f) \int_Y \omega.$$

This completes the proof of the lemma. ♠

We now complete the proof of the theorem as follows. Fix a regular value $y \in Y$ for f and a neighborhood U as above. Recall that, given any point $z \in Y$ we can find a diffeotopy h_t of Y such that $h_1(y) = z$ (see Disc Theorem 6.3.7). Using the compactness of Y, we can find finitely many diffeomorphisms $g_i : Y \to Y$ such that $Y = \cup g_i(U)$. Using the partition of unity, we can write any form ω as a finite sum of forms each of which has support inside one of $g_i(U)$. By the additive property of the integrals, it is now enough to prove the theorem, for a form ω with its support inside one of $g_i(U)$. Since $g_i \simeq Id_Y$, it follows that $f \simeq g_i \circ f$. By Exercise 4.3, we have,

$$\int_X f^*\omega = \int_X (g_i \circ f)^*\omega = \int_X f^* g_i^*(\omega) = (deg\, f) \int_Y g_i^*(\omega) = (deg\, f) \int_Y \omega.$$

The proof of the theorem is complete. ♠

Definition 7.9.1 Let $X \subset \mathbb{R}^{n+1}$ be a compact connected smooth hyper-surface. Consider the map $x \mapsto \nu(x)$, where $\nu(x)$ denotes outward unit normal to X at x. (This makes sense, because of the Jordan-Brouwer separation theorem.) This map is called the *Gauss map* of X and let us denote it by $g_X = g$. Its Jacobian, i.e, the determinant of the derivative is called the *curvature form* (which is an n-form) and is denoted by κ. Indeed, let $\mathcal{V}_n$ denote the volume form on the unit sphere. Then $\kappa = g^*\mathcal{V}_n$.

Example 7.9.1 If X is a sphere of radius r, then $g(x) = x/r$ for each x and hence κ is equal to $1/r^n$ times the volume form.

Lemma 7.9.2 The degree of the Gauss map is equal to half of $\chi(X)$ for an even dimensional hypersurface X in $\mathbb{R}^{n+1}$.

Proof: Choose $a \in \mathbb{S}^n$ so that both $\pm a$ are regular values of the Gauss map $g : X \to \mathbb{S}^n$. (This is possible, by choosing a regular value for the composite map $\pi \circ g : X \to \mathbb{P}^n$ where $\pi : \mathbb{S}^n \to \mathbb{P}^n$ is the quotient map onto the real projective space.) Consider the vector field Θ_a on $\mathbb{S}^n$ given by

$$\Theta_a(v) = a - \langle a, v\rangle v. \tag{7.14}$$

The zeros of this vector field are precisely $\pm a$. One is a sink and the other is a source. Since n is even, the index at both points is equal to 1. (This is the only place where evenness of n is used and it is crucial.)

Now consider the pullback vector field $\sigma = g^*(\Theta_a)$ on X. Since $\Theta_a(v) = 0$ iff $v = \pm a$, $\sigma(x) = 0$ iff $g(x) = \pm a$. In particular, the zero set Z of σ is finite and we know that

$$\chi(X) = \sum_{z\in Z} ind_z(\sigma) = \sum_{g(z)=a} ind_z(\sigma) + \sum_{g(z)=-a} ind_z(\sigma).$$

Also, observe that in a neighborhood of each $z \in Z$, the Gauss map g is a diffeomorphism, z being a regular point. Moreover, since $g(x) \perp T_xX$, it follows that $T_x(X) = T_{g(x)}\mathbb{S}^n$. One can easily check that $ind_z(\sigma) = \pm ind_{g(z)}(\Theta_a)$ according as dg_x preserves the orientation or not. Therefore,

$$\sum_{g(z)=a} ind_z(\sigma) = \deg g = \sum_{g(z)=-a} ind_z(\sigma).$$

The conclusion of the lemma follows. ♠

Remark 7.9.1 Note that in the odd dimensional case, this formula is not valid since $\chi(X) = 0$ whereas κ could be positive everywhere. That is the reason why the following theorem is also available for even dimensional manifolds only. However, if you are interested in computing the total curvature somehow, computing the degree of the Gauss map somehow is a solution.

Theorem 7.9.2 Gauss-Bonnet: *If X is a compact even dimensional hypersurface in $\mathbb{R}^{n+1}$ then*

$$\int_X \kappa = \frac{1}{2}\gamma_n \chi(X)$$

where γ_n is the volume of the unit n-sphere.

Proof:

$$\int_X \kappa = \int_X g^\star \mathcal{V}_n = (deg\, g)\int_{\mathbb{S}^n} \mathcal{V}_n = (deg\, g)\gamma_n = \frac{1}{2}\gamma_n\chi(X).$$

This completes the proof. ♠

Exercise 7.9

1. Let X, Y be connected, oriented, closed n-manifolds, and $f : X \to Y$ be a smooth map. We know that $H^n_{DR}(X) \approx \mathbb{R} \approx H^n_{DR}(Y)$. Therefore, there is a real number λ such that $f^* : H^n_{DR}(Y) \to H^n_{DR}(X)$ is given by $f^*(r) = \lambda r$. Show that $\lambda = \deg f$.

2. Let X be a nonorientable closed n-manifold. Show that every n-form on X is exact by using the fact that every n-form ω on $\tilde{X}$ such that $\int_{\tilde{X}} \omega = 0$ is exact. (Compare Exercise 4.5.4.)

7.10 Miscellaneous Exercises for Chapter 7

1. Let X, Y be submanifolds of $\mathbb{R}^n$. Show that for almost all $\mathbf{v} \in \mathbb{R}^n$, the translates $X + \mathbf{v}$ intersect Y transversely.

2. If X, Z are submanifolds of Y such that dim X+dim $Z <$ dim Y then show that X can be isotoped away from Z by an isotopy that is constant outside a given neighborhood of Z.

3. Let X be a submanifold of $\mathbb{R}^n \setminus \{0\}$. For any $1 \le k \le n$, let $G_{k,n}$ denote the space of all k-dimensional linear subspaces of $\mathbb{R}^n$. (See Exercise 10 in 5.8. Show that almost all elements of $G_{k,n}$ intersect X transversely.

4. Let $n \ge 2$ and $f : \mathbb{R}^n \to \mathbb{R}^n$ be any smooth map. Show that given $\epsilon > 0$ and a compact subset K of $\mathbb{R}^n$, there exists a smooth map $g : \mathbb{R}^n \to \mathbb{R}^n$ such that dg is never zero and $|f - g| < \epsilon$ on K. The same statement is false for $n = 1$.

5. A manifold X is called *s-parallelizable* if $TX \oplus \Theta_k$ is a trivial bundle, for some k, where $\Theta_k = X \times \mathbb{R}^k$ denotes the trivial bundle of rank k.
(i) Show that product of a finite number (more than one) of spheres is s-parallelizable and is actually parallelizable if one of the sphere is odd dimensional.
(ii) Show that every codimension 1 submanifold of $\mathbb{R}^n$ is s-parallelizable.
(iii) Show that if X is a product of spheres then it can be embedded as a codimension 1 submanifold of a Euclidean space.

6. Show that TX is orientable for any manifold X.

7. Let $M \subset N$ be a codimension 1 submanifold.
(i) Suppose $N = \mathbb{R}^{n+1}$, $n \ge 1$ and $\partial M = \emptyset$. Then M bounds a unique compact submanifold of $\mathbb{R}^{n+1}$.
(ii) Suppose $\partial M \ne \emptyset$ and $\partial N = \emptyset$. If M, N are connected then so is $N \setminus M$.
(iii) Suppose $\partial M = M \cap \partial N$. Then the normal bundle of M in N is trivial iff there exist arbitrary small neighborhoods of M in N that are separated by M.

8. Show that any map $f : \mathbb{S}^n \to \mathbb{S}^n$ such that $f(x) = f(-x)$ has an even degree.

9. Let X, Y be orientable closed connected manifolds of the same dimension and let $f : M \to N$ be of degree k. Then $f_\#(\pi_1(X))$ is a subgroup of $\pi_1(Y)$ of index that divides $|k|$. In particular, if $k = 1$, then $f_\#$ is surjective on the fundamental groups.

10. Let M^n be a closed submanifold of $\mathbb{R}^{n+1}$. For each $x \in \mathbb{R}^{n+1} \setminus M$, define $\tau_x : M \to \mathbb{S}^n$ by
$$\tau_x(y) = \frac{y - x}{\|y - x\|}.$$
Show that
(i) $x_i, i = 0, 1$ are in the same component of $\mathbb{R}^{n+1} \setminus M$ iff $\tau_{x_0} \simeq \tau_{x_1}$.
(ii) x is in the unbounded component of $\mathbb{R}^{n+1} \setminus M$ iff τ_x is homotopic to a constant map.
(iii) Suppose M is connected. Then x belongs to the bounded component of $\mathbb{R}^{n+1} \setminus M$ iff $\deg \tau_x = \pm 1$.

11. Let M, N be closed, oriented submanifolds of dimension m, n respectively of $\mathbb{R}^q$ where, $m + n + 1 = q$. The linking number $Lk(M, N)$ is defined to be the degree of the map $\tau : M \times N \to \mathbb{S}^{m+n}$, given by
$$(x, y) \mapsto \frac{x - y}{\|x - y\|}.$$
Show that
(i) $Lk(M, N) = (-1)^{mn} Lk(N, M)$
(ii) If M can be deformed to a point in $\mathbb{R}^q \setminus N$ or if it bounds an oriented compact submanifold in $\mathbb{R}^q \setminus N$ then $Lk(M, N) = 0$.
(iii) If $M = \{x, y\}$ then N separates x and y iff $Lk(M, N) = \pm 1$.

12. Prove the following generalization of the fundamental theorem of algebra: Let U be an open set in $\mathbb{R}^n$ and $f : U \to \mathbb{R}^n$ be a proper smooth map such that outside a compact set $Det(Df_x) > 0$ always or < 0 always. Then f is surjective.

Chapter 8

Geometry of Manifolds

Recall that for a real valued smooth function of a real variable, the local extrema were identified by conditions on the values of the derivative of the function. In particular, the first derivative itself necessarily vanished at these points. Further, in order to determine the nature of the function at an extremum point, viz., whether it has a local maximum or a local minimum at that point, it was necessary to study the second derivative of the function. There is no reason why a similar theory should not exist in the case of functions of several variables.

Indeed, such is precisely the case, viz., the "nonsingularity" of the second derivative at a point is a sufficient condition to determine the local behavior of the function at a critical point. The key result is called the Morse[1] lemma. While dealing with the second derivative, the Morse lemma plays a role similar to the one played by the inverse function theorem while dealing with the first derivative. The importance of this result is enhanced by the fact that it gives a lot of information on the topology of the manifold itself. Smale[2] developed this idea into a powerful tool and used it to solve the Poincaré conjecture for dimension ≥ 5 [Sm1]. In this chapter, we shall introduce the easier part of Morse theory and use it to present a classification of compact smooth manifolds.

8.1 Morse Functions

In order to see the effectiveness of the Morse theory, one needs to know that there are "enough" smooth functions which have all their critical points nondegenerate. Such functions are called *Morse functions.* In this section, we shall see that Morse[3]-Sard's theorem can be effectively used to prove that there are plenty of Morse functions on any given manifold.

Definition 8.1.1 Let U be an open subset of $\mathbb{R}^n$ and $f : U \to \mathbb{R}$ be a smooth map and $x \in U$ be a critical point of f. (Recall that this simply means that $Df_x = 0$.) We say x is a

[1]Marston Morse (1892-1977) was born in Waterville, Maine, USA. He did his Ph. D. under the guidance of G. D. Birkhoff. The so called Morse theory came out of his paper 'Relations between the critical points of a real functions of n variables in 1925. He has contributed more than 180 research papers and eight books on a whole range of topics such as minimal surfaces, topological methods in the theory of functions of a complex variable, differential topology and dynamics.

[2]Stephen Smale (1930–) is a US mathematician, who has out-standing contributions in differential topology, dynamical systems, mathematical economics and theoretical computer science. He received Fields medal in 1966 for solving Poincaré conjecture in dimensions greater than 4.

[3]This is not Marston Morse but A. P. Morse (1811-1984) whose major contributions are in the area of probability and measure theory and mathematical foundation.

nondegenerate critical point if the *Hessian matrix*

$$H_f(x) = \left(\frac{\partial^2 f}{\partial x_i \partial x_j}(x)\right) \tag{8.1}$$

is nonsingular.

Lemma 8.1.1 Nondegenerate critical points are isolated.

Proof: Consider the map $g = \nabla f : U \to \mathbb{R}^n$, i.e.,

$$g(x) = \left(\frac{\partial f}{\partial x_1}, \ldots, \frac{\partial f}{\partial x_n}\right)_x.$$

Then the critical points of f are nothing but the zeros of g. We look at the derivative of g at a point x, which is nothing but the linear map $\mathbb{R}^n \to \mathbb{R}^n$ given by the Hessian matrix (8.1). Assuming that this is nonsingular at $x = x_0$, we can apply the Inverse Function Theorem to conclude that g is a local diffeomorphism at x. Therefore, $g(x) = 0$ has a unique solution in a neighborhood of x_0. This means that there is no other critical points of f in a neighborhood of x_0. ♠

Remark 8.1.1 Observe that under diffeomorphisms, the nondegeneracy of critical points is preserved. (See the exercise at the end of this section.) This observation enables us to define nondegenerate critical points of a smooth function f defined on any manifold X, viz., $x \in X$ is a nondegenerate critical point of f if for some parameterization ϕ of X around x, $\phi^{-1}(x)$ is a nondegenerate critical point of $f \circ \phi : \mathbb{R}^n \to \mathbb{R}$.

Definition 8.1.2 A smooth function on a manifold is called a *Morse function* if all its critical points are nondegenerate.

Let us first of all prove a result that will guarantee that there are plenty of Morse functions. We now need the observation that we made in Exercise 3.3.(6). Since we are going to use it now, let us give a proof of this here.

Lemma 8.1.2 Let X be an n-dimensional smooth submanifold of $\mathbb{R}^N$. Given any point $x \in X$, there exist n coordinate projections $x_{i_1}, \ldots, x_{i_n}$, which are such that restricted to a neighborhood of $x \in X$, they form a coordinate system for X.

Proof: This follows directly from the corresponding linear algebra result: If L is any n-dimensional vector subspace of $\mathbb{R}^N$, then some n of the coordinate functions (treated as linear functionals) restricted to L are linearly independent. Now, choose $L = T_x X$ and observe that the derivative of x_i restricted to X is nothing but x_i restricted to $T_x X$. Choose $x_{i_1} \ldots, x_{i_n}$ so that on L they are independent and define $\phi_k = x_{i_k}, 1 \leq k \leq n$. Then $\phi : X \to \mathbb{R}^n$ is such that $D(\phi)_x$ is an isomorphism. Now, by appealing to the inverse function theorem, get the conclusion of the lemma. ♠

Theorem 8.1.1 *Let $X \hookrightarrow \mathbb{R}^N$ be a smooth manifold and $f : X \to \mathbb{R}$ be any smooth function. Then for almost all vectors $u \in \mathbb{R}^N$, the mapping f_u defined by:*

$$f_u(x) = f(x) + \langle x, u \rangle$$

is a Morse function on X.

Proof: We are required to prove that the set of all vectors $u \in \mathbb{R}^N$ for which f_u is not a Morse function is of the Lebesgue measure zero in $\mathbb{R}^N$. Suppose, we cover X by a countable family of open subsets $\{U_i\}$ such that the statement is true for each U_i in place of X. Since the countable union of measure zero sets is of measure zero, the result follows for X itself.

Let us first consider the case when $N = n$ and X is an open set in $\mathbb{R}^n$. Let $g : \mathbb{R}^n \to \mathbb{R}^n$ be the derivative of f, i.e.,

$$g(x) = \left(\frac{\partial f}{\partial x_1}, \ldots, \frac{\partial f}{\partial x_n}\right)(x).$$

It follows that

$$D(f_u)(x) = g(x) + u.$$

Therefore, x is a critical point of f_u iff $g(x) = -u$. Since $H_f(x) = Dg(x) = H_{f_u}(x)$, we want $Dg(x)$ to be nonsingular, for all $x \in g^{-1}(-u)$. Since the dimensions of the domain and the codomain of g are the same, this is the same as saying that $-u$ is a regular value for the function g. Morse-Sard's theorem says that this is the case for almost all vectors u. This completes the proof in this special case.

Now consider the general case. By the previous lemma, given $x \in X$, we can choose a neighborhood U such that some n coordinate projections restrict to a coordinate system on U. Without loss of generality, for convenience of writing down the proof, we assume that these are the first n coordinate projections and let $\psi : V \to U$ be the inverse parameterization, where V is some open subset of $\mathbb{R}^n$. Then clearly, $\psi(y) = (y, \gamma(y))$ for some smooth function $\gamma : V \to \mathbb{R}^{N-n}$.

For each $w \in \mathbb{R}^{N-n}$, consider the mapping

$$F(w)(x) = f(x) + \langle x, (0, w)\rangle$$

and let $h := F(w) \circ \psi : V \to \mathbb{R}$. Then from the earlier case, for almost all $v \in \mathbb{R}^n$, we have, h_v is a Morse function.

Now we observe that for any $u = (v, w) \in \mathbb{R}^n \times \mathbb{R}^{N-n}$ and $x = \psi(y), y \in V$,

$$\begin{aligned}
f_u(\psi(y)) &= f(\psi(y)) + \langle \psi(y), u\rangle \\
&= f(\psi(y)) + \langle (y, \gamma(y)), (v, w)\rangle \\
&= f(\psi(y)) + \langle y, v\rangle + \langle \gamma(y), w\rangle \\
&= f(\psi(y)) + \langle \psi(y), (0, w)\rangle + \langle y, v\rangle \\
&= h(y) + \langle y, v\rangle = h_v(y).
\end{aligned}$$

Let S be the set of all $u \in \mathbb{R}^N$ such that f_u is not a Morse function on U. Then the above argument together with the first case of the theorem we have proved, shows that for every $w \in \mathbb{R}^{N-n}$, the intersection $S \cap (\mathbb{R}^n \times \{w\})$ is a subset of measure zero in $\mathbb{R}^n \times \{w\}$. Since this is true for all $w \in \mathbb{R}^{N-n}$, it follows from Fubini's theorem that S itself is of measure zero in $\mathbb{R}^N$, as required. ♠

Remark 8.1.2 Recall that every smooth manifold X can be embedded in a Euclidean space. Therefore, you get plenty of Morse functions on any manifold. Indeed, taking $f \equiv 0$, we get plenty of linear functions on the ambient Euclidean space, which when restricted to X give Morse functions. Also, if X is compact, given any $\epsilon > 0$, we can choose $u \in \mathbb{R}^N$ in such a way that $|\langle x, u\rangle| < \epsilon$ for all $x \in X$ and thereby, get a Morse function f_u on X such that $|f(x) - f_u(x)| < \epsilon$ for all $x \in X$.

When X is noncompact, it is desirable to have Morse functions $f : X \to \mathbb{R}$, which are proper, i.e., $f^{-1}(K)$ is compact for all compact subsets $K \subset \mathbb{R}$. If we stick to linear functions, this is more or less not possible. However, if we allow "quadratic" functions, then the answer is in the affirmative.

Theorem 8.1.2 *Let $X \subset \mathbb{R}^N$ be any smooth closed manifold. Then for almost all points $z \in \mathbb{R}^N$, the square of the distance function from y restricted to X is a proper Morse function.*

Proof: Given $z \in \mathbb{R}^N$ put $L_z(x) = \sum_i (z_i - x_i)^2$. Since $L_z^{-1}[0,r]$ is nothing but the closed ball of radius r with center z, it follow that $X \cap L_z^{-1}[0,r]$ is compact. From this, the properness of $L_z|_X$ follows. Note that $x \in X$ is a critical point of L_z iff $z - x$ is perpendicular to the tangent space T_xX, iff $v = z - x \in N_xX$. Therefore, we consider the space of normals $N(X) \subset \mathbb{R}^N \times \mathbb{R}^N$ and the function $\alpha : (x, v) \mapsto x + v$ on it. The image of α consists of those points z for which the critical set of $L_z : X \to \mathbb{R}$ is nonempty. Put $x + v = z$. We claim that the Hessian of L_z at $x \in X$ is nonsingular iff $D\alpha_{(x,v)}$ is nonsingular. Appealing to Morse-Sard theorem, it follows that almost all points $z \in \mathbb{R}^N$ are regular values of α and for these values of z, L_z is a Morse function.

As in the Theorem 6.1.1, let $\Theta : \mathbb{R}^n \times \mathbb{R}^{N-n} \to N(X)$ be a parameterization covering $(x, v) \in N(X)$, say $\Theta(\mathbf{t}, \mathbf{s}) = (\phi(\mathbf{t}), \mathbf{s} \cdot \mathbf{u}(\mathbf{t}))$, where $\phi : \mathbb{R}^n \to U$ is a parameterization for X near x, and $\mathbf{u}(\mathbf{t})$ is an orthonormal frame for $N_{\phi(\mathbf{t})}(X)$. Here $\mathbf{t} = (t_1, \ldots, t_n), \mathbf{s} = s_1, \ldots, s_{N-n})$; and the vectors $\{\mathbf{u}_1(\mathbf{t}), \ldots, \mathbf{u}_{N-n}(\mathbf{t})\}$ form an orthonormal frame orthogonal to $T_{\phi(t)}X$. Therefore, $\{\mathbf{u}_j\}$ are orthogonal to the tangent vectors $\frac{\partial \phi_i}{\partial t_j}$. Now $\alpha \circ \Theta(\mathbf{t}, \mathbf{s}) = \phi(\mathbf{t}) + \mathbf{s} \cdot \mathbf{u}(\mathbf{t})$ with its N partial derivatives

$$\frac{\partial \phi}{\partial t_i} + \sum_{k=1}^{N-n} s_k \frac{\partial \mathbf{u}_k}{\partial t_i}, 1 \le i \le n, \text{ and } \mathbf{u}_1, \mathbf{u}_2, \cdots \mathbf{u}_{N-n}.$$

So, $D\alpha_{(x,v)}$ is nonsingular, iff these N vectors are independent at the corresponding $(\mathbf{t}, \mathbf{s})$. Taking the dot product with the N independent vectors $\frac{\partial \phi}{\partial t_j}, 1 \le j \le n$, and $\mathbf{u}_1, \mathbf{u}_2, \cdots \mathbf{u}_{N-n}$, we get a matrix

$$\begin{pmatrix} \left(\frac{\partial \phi}{\partial t_i} \cdot \frac{\partial \phi}{\partial t_j} + \sum_k s_k \frac{\partial \mathbf{u}_k}{\partial t_i} \cdot \frac{\partial \phi}{\partial t_j}\right) & \left(\sum_k s_k \frac{\partial \mathbf{u}_k}{\partial t_i} \cdot \mathbf{u}_j\right) \\ 0 & Id_{N-n} \end{pmatrix} \tag{8.2}$$

Clearly (8.2) is nonsingular, iff its first block is nonsingular. Since $\frac{\partial \phi}{\partial t_j} \cdot \mathbf{u}_k = 0$, it follows that

$$0 = \frac{\partial}{\partial t_i}\left(\frac{\partial \phi}{\partial t_j} \cdot \mathbf{u}_k\right) = \frac{\partial^2 \phi}{\partial t_i \partial t_j} \cdot \mathbf{u}_k + \frac{\partial \phi}{\partial t_j} \cdot \frac{\partial \mathbf{u}_k}{\partial t_i}.$$

Therefore, the first block of (8.2) is equal to the matrix

$$\left(\frac{\partial \phi}{\partial t_i} \cdot \frac{\partial \phi}{\partial t_j} - \sum_k s_k \frac{\partial^2 \phi}{\partial t_i \partial t_j} \cdot \mathbf{u}_k\right) = D(\phi) \cdot D(\phi) - \sum_k s_k \mathbf{u}_k \cdot D^2(\phi).$$

On the other hand, to compute the Hessian of L_z, we can use the parameterization $\phi : \mathbb{R}^n \to U$ and consider the map $\beta = L_z \circ \phi : \mathbf{t} \mapsto \|\phi(\mathbf{t}) - z\|^2$. The first derivative $D\beta = 2(\phi(\mathbf{t}) - z) \cdot D(\phi)$ and the second derivative is $2[D(\phi) \cdot D(\phi) + (\phi(\mathbf{t}) - z) \cdot D^2(\phi)]$. Since $\phi(\mathbf{t}) - z = x - z = v = \sum_k s_k \mathbf{u}_k$ we are through. ♠

Remark 8.1.3 There is a deep relation between the Morse functions L_z and the geometry of the submanifold X. (See [M2].)

Exercise 8.1 Let $\phi : U \to V$ be a diffeomorphism of open subsets of $\mathbb{R}^n$ and $f : V \to \mathbb{R}$ be a smooth function. Show that $\mathbf{x} \in U$ is a nondegenerate critical point of $g \circ \phi$ iff $\phi(\mathbf{x})$ is a nondegenerate critical point of f. [Hint: See (1.46).]

8.2 Morse Lemma

One of the landmarks in the theory of Differential Topology is the characterization of a function in a neighborhood of a nondegenerate critical point. This result is due to Marston Morse. Before we proceed, let us consider a simple example.

Example 8.2.1 Let $\mathbb{S}^2$ denote the unit sphere in $\mathbb{R}^3$ and let $\pi : \mathbb{S}^2 \to \mathbb{R}$ be the projection to the last coordinate, $\pi(x,y,z) = z$. Since π is a linear map $d\pi_p = \pi$ for all p and hence vanishes identically on the plane $z = 0$. Since the only points at which the plane $z = 0$ is tangent to $\mathbb{S}^2$ are $(0,0,\pm1)$, it follows that $(0,0,\pm1)$ are the only critical points of π. Let now p be one of these two points, say $p = N = (0,0,1)$ the north pole. In order to determine the second derivative of π at p, we must express $d\pi$ in terms of a coordinate system for $\mathbb{S}^2$ around p. Since the restrictions of (x,y) coordinates themselves define a coordinate system for $\mathbb{S}^2$ around p, we may write

$$h(x,y) = \pi(x,y,z) = \sqrt{(1 - x^2 - y^2)}$$

valid in a neighborhood of p. Therefore,

$$dh_{(x,y)} = -(x(1 - x^2 - y^2)^{-1/2}, y(1 - x^2 - y^2)^{-1/2}).$$

Therefore, the Hessian at p is given by $\begin{bmatrix} -1 & 0 \\ 0 & -1 \end{bmatrix}$.

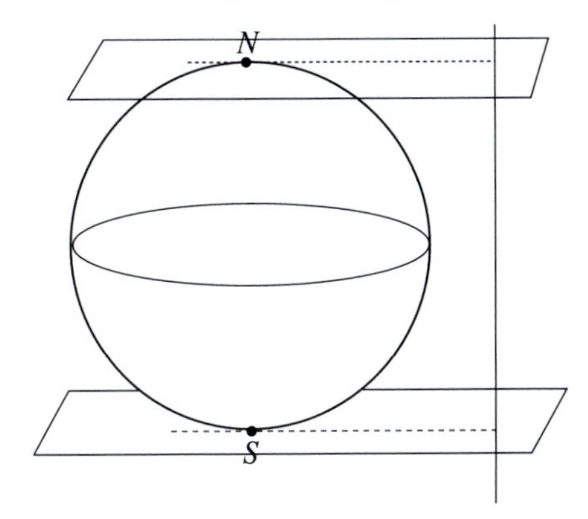

Figure 39 The z-coordinate as a Morse function.

A similar computation, (which differs only in the sign) is valid at $-p$ as well. This shows that both the critical points $(0,0,\pm1)$ are nondegenerate.

Notice how the topology of the space $h^{-1}(-\infty, z]$ changes: for $z < -1$, this set is empty; for $z = -1$, it is a singleton space; for $-1 < z < 1$, it is diffeomorphic to $\mathbb{D}^2$ and for $z \geq 1$, it is the whole space $\mathbb{S}^2$. In sections that follow, we shall see a precise formulation of this phenomenon.

We begin with the celebrated result of Marston Morse.

Theorem 8.2.1 Morse Lemma: *Let $f : X \to \mathbb{R}$ be a smooth map and $p \in X$ be a nondegenerate critical point of f. Then there exists a chart (U, ϕ) for X near p such that $\phi(p) = 0$ and*

$$f \circ \phi^{-1}(x) = f(p) - \sum_{i=1}^{k} x_i^2 + \sum_{i=k+1}^{n} x_i^2 \tag{8.3}$$

for all $x \in \phi(U)$ and for some k.

We shall first prove a weaker version of the above theorem, which was an exercise to you earlier (see Exercise 1.2.10).

Lemma 8.2.1 Mini-Morse Lemma: Let $f : X \to \mathbb{R}$ be a smooth function from a convex open subset X of $\mathbb{R}^n$, $0 \in X$, $f(0) = 0$ and 0 be a critical point of f. Then there exist smooth maps $\tau_{i,j} : X \to \mathbb{R}$, $1 \leq i, j \leq n$, such that $\tau_{i,j}(x) = \tau_{j,i}(x)$, and

$$f(x) = \sum_{1 \leq i,j \leq n} \tau_{ij}(x) x_i x_j \tag{8.4}$$

for all $x \in X$.

Proof: We are going to apply (1.30), first to f and then to its partial derivatives $\dfrac{\partial f}{\partial x_i}$. Since $f(0) = 0$ and $\dfrac{\partial f}{\partial x_i}(0) = 0$, it follows that

$$f(x) = \sum_{j=1}^{n} \left(\int_0^1 \frac{\partial f}{\partial x_j}(tx) dt \right) x_j = \sum_{i,j=1}^{n} \left(\int_0^1 \int_0^1 \frac{\partial^2 f}{\partial x_j \partial x_i}(stx) ds\, dt \right) x_i x_j.$$

Now, take $\tau_{ij}(x) = \displaystyle\int_0^1 \int_0^1 \frac{\partial^2 f}{\partial x_j \partial x_i}(stx) ds\, dt$, to get (8.4). ♠

Remark 8.2.1 Note that one could directly obtain the above lemma via Taylor's expansion (exercise). Also, we have $H_f(0) = ((\frac{1}{2!}\tau_{ij}(0)))$. Now by (1.46), if $\phi : V \to U$ is a diffeomorphism, $\phi(\mathbf{y}_0) = \mathbf{x}_0$, and $f' : U \to \mathbb{R}$ is a smooth map with a nondegenerate critical point at $\mathbf{x}_0$, then we have

$$H_{f \circ \phi}(\mathbf{y}_0) = D(\phi)^t_{\mathbf{y}_0} H_f(\mathbf{x}_0) D(\phi)_{\mathbf{y}_0}.$$

Recall that any symmetric matrix can be diagonalized. Since H_f is symmetric, we can hope to make $H_{f\circ\phi}$ into a diagonal matrix by appropriately choosing ϕ. For this we need to perform the diagonalization in a parameterized form. This is the gist of the following proof.

Proof of Morse Lemma: Replacing $f \circ \phi^{-1}$ by f, for a suitable coordinate chart ϕ at p, it is enough to prove the lemma under the assumption that X is a convex open subset of $\mathbb{R}^n, p = 0, f(0) = 0$ and $Df_0 = 0$. By Lemma 8.2.1, there exist smooth functions τ_{ij} such that $\tau_{ij} = \tau_{ji}$ and

$$f(\mathbf{x}) = \sum_{1 \leq i,j \leq n} \tau_{ij}(\mathbf{x}) x_i x_j. \tag{8.5}$$

Put $\tau(\mathbf{x}) = ((\tau_{ij}(\mathbf{x})))$. Then $\tau(0) = \dfrac{1}{2} H(f)(0)$ and hence $\tau(0)$ is nonsingular, by hypothesis. Therefore, at least one of the $\tau_{nj}(0) \neq 0$, say $\tau_{nk}(0) \neq 0$. Make a linear change of coordinates,

$$y_i = x_i, i \neq k; y_k = tx_n + x_k.$$

It is easy to check that

$$f(y_1, \ldots, y_n) = \sum_{ij} \sigma_{ij}(\mathbf{y}) y_i y_j$$

where

$$\sigma_{nn}(\mathbf{y}) = \tau_{nn}(\mathbf{x}) + 2t\tau_{nk}(\mathbf{x}) + t^2 \tau_{kk}(\mathbf{x}).$$

Therefore, for a suitable choice of $t \in \mathbb{R}$ we can assume that $\sigma_{nn}(0) \neq 0$. All this amounts to saying that after a linear change of coordinates we may as well assume that $\tau_{nn}(0) \neq 0$ in (8.5).

We now follow the simple method of completing the square.

$$\begin{aligned} f(\mathbf{x}) &= \tau_{nn}[x_n^2 + 2\sum_{j<n} \frac{\tau_{nj}}{\tau_{nn}} x_1 x_j] + \sum_{i,j<n} \tau_{ij} x_i x_j \\ &= \tau_{nn}[x_n + \textstyle\sum_{j<n} \frac{\tau_{nj}}{\tau_{nn}} x_j]^2 + \sum_{i,j<n} \lambda_{ij} x_i x_j \end{aligned}$$

by adding and subtracting suitable terms. Since $\tau_{nn}(0) \neq 0$, we can take the squareroot of the modulus of this function in a suitable neighborhood of 0 :

$$g(\mathbf{x}) = \sqrt{|\tau_{nn}(\mathbf{x})|}.$$

We now make the substitution

$$y_i = x_i, i < n;\ y_n = g(\mathbf{x}) \left(x_n + \frac{\sum_{j<n} \tau_{nj}(\mathbf{x}) x_j}{\tau_{nn}(\mathbf{x})} \right).$$

Why does this substitution define a change of coordinates and where is it valid? We have to merely compute the Jacobian matrix $J(\mathbf{y}; \mathbf{x})$ and check that it is invertible at $\mathbf{x} = 0$. By the inverse function theorem, it then follows that in a smaller neighborhood, the substitution is invertible, i.e., there exists smooth functions $\psi_i(\mathbf{y})$ such that $x_i = \psi_i(\mathbf{y}), i = 1, 2, \ldots, n$. If $\psi = (\psi_1, \ldots, \psi_n)$ then it follows that

$$f \circ \psi(y_1, y_2, \ldots, y_n) = \pm y_n^2 + \sum_{i,j<n} \rho_{ij}(\mathbf{y}) y_i y_j,$$

for some smooth functions $\rho_{ij}(\mathbf{y})$, say. Once again, since the Hessian of f is nonsingular at 0, it follows that the $(n-1) \times (n-1)$ matrix $((\rho_{ij}(0)))$ is nonsingular. All this amounts to saying that by a suitable change of coordinates in a suitable neighborhood of 0 we may assume that

$$f(\mathbf{x}) = \sum_{i,j<n} \tau_{ij}(\mathbf{x}) x_i x_j \pm x_n^2.$$

By repeated application of this process it follows that by a suitable change of coordinates in a suitable neighborhood of 0 we may assume that

$$f(\mathbf{x}) = \pm \sum_{i=1}^{n} x_i^2.$$

We now make a permutation of coordinates to get the form (8.3). ♠

Following such a fundamental result, we make a few definitions:

Definition 8.2.1 At a nondegenerate critical point p of a smooth function f, the Hessian $H_f(p)$ is nonsingular and the number k that occurs in (8.3) is nothing but the number of negative eigenvalues of $H_f(p)$. This number is called *the index* of f at p.

Remark 8.2.2

1. Now, if the index is zero, then it is clear that $f(p)$ is a local minimum and if the index is equal to n then $f(p)$ is a local maximum. For any other value of the index, $f(p)$ fails to be either a minimum or a maximum. Such points are collectively called *saddle points*.

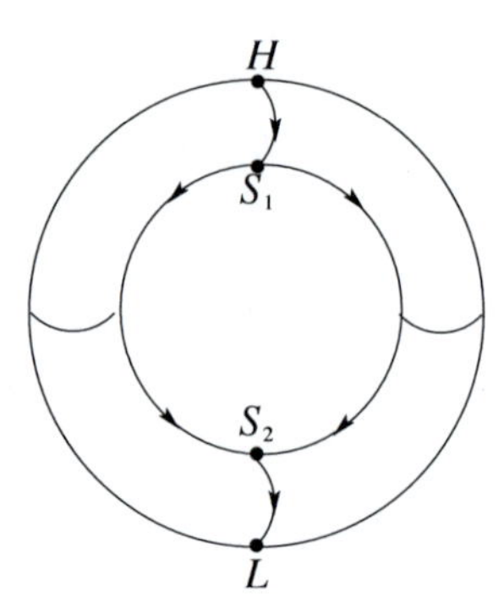

Figure 40 A maximum, two saddles, and a minimum.

In the above picture we have the torus on which we are considering the height function $f(x, y, z) = z$. H, L indicate the points at which the maximum and the minimum occur. S_1 and S_2 are saddle points. At S_1, the function f decreases in the direction of the x-axis and increases in the direction of the y-axis, whereas, the reverse holds at S_2. Clearly, the indices at the critical points H, S_1, S_2, L are respectively $2, 1, 1, 0$.

2. Consider the vector field grad f where f is a Morse function. By definition, the critical points of f are precisely the zeros of this vector field. Moreover, the local index of the vector field is precisely equal to $(-1)^k$ where k is the index of the critical point as will be seen below. If ν_k denotes the number of critical points of f of index k, let us denote the alternate sum

$$e(f) := \sum_k (-1)^k \nu_k. \tag{8.6}$$

Lemma 8.2.2 The number $e(f)$ is equal to the index of the vector grad f and hence is equal to the self-intersection number of the manifold X.

Proof: Indeed if p is a critical point of f of index k we shall show that $ind_p(\text{grad}\, f) = (-1)^k$, from which the lemma would follow.

By Morse Lemma, we can choose coordinates at p for X so that

$$f(\mathbf{x}) = f(p) - \sum^k x_i^2 + \sum_{k+1} x_i^2.$$

Therefore, in this neighborhood of p, grad f has the form

$$\mathbf{x} \mapsto (-2x_1, \ldots, -2x_k, 2x_{k+1}, \ldots, 2x_n).$$

Therefore, the local index of grad f around this point is $(-1)^k$ as required. ♠

We shall soon see that the Morse functions give much more information on the manifold.

Exercise 8.2

1. Let X be the subspace of $\mathbb{R}^3$ consisting of points

$$((2 + \cos\theta)\cos\psi, (2 + \cos\theta)\sin\psi, \sin\theta), \quad 0 \leq \theta, \psi \leq 2\pi.$$

Observe that this is the surface of revolution of the circle

$$(x-2)^2 + z^2 = 1$$

about the z-axis.
(i) Draw a decent picture of the surface.
(ii) Show that the projection on the x-axis is a Morse function on X. Determine the critical points and whether they are local maxima, minima or saddle points.
(iii) Are the projections to the y-axis and the z-axis Morse functions?
(iv) Determine all orthogonal projections $f : \mathbb{R}^3 \to \mathbb{R}$, which are Morse functions restricted to X.
(v) Show that all Morse functions which occur in (iv) have the property that different critical points occur at different levels.
(vi) Write down an algebraic equation for this surface.
(vii) Observe that portion of the surface lying in $x \leq 0$ forms a copy of the cylinder. Use this to obtain a Morse function on the cylinder with exactly two critical points, one of index 0 and the other of index 1.
(viii) Use a similar parameterization to obtain a Morse function on the Möbius band with exactly the same property as in (vii).

2. If $f : X \to \mathbb{R}$, $g : Y \to \mathbb{R}$ are Morse functions, show that $h(x,y) = f(x) + g(y)$ is a Morse function on $X \times Y$.

3. Show that $\sin x, \cos x : \mathbb{R} \to \mathbb{R}$ are Morse functions and because of periodicity, they define Morse functions on $\mathbb{S}^1$.

4. Use the above two exercises to obtain a Morse function on $\mathbb{S}^1 \times \mathbb{S}^1$ with $(\nu_0, \nu_1, \nu_2) = (1, 2, 1)$ and with precisely three critical values, $-2, 0, 2$.

5. If X is a compact manifold without boundary, show that every smooth function $f : X \to \mathbb{R}$ has at least two critical points.

6. Give an example of an isolated critical point that need not be nondegenerate.

7. Show that $f(x,y) = y^2 + x^4 - 2x^2$ is a Morse function, with three critical points of index $0, 0, 1$ at the points $(-1,0), (0,0), (1,0)$, respectively. Find another Morse function with similar critical indices and which sends the three critical points to three distinct values.

8.3 Operations on Manifolds

The basic idea here is to build new manifolds out of old ones. The central theme is Lemma 5.3.1 that was employed in Chapter 5 for defining the tangent bundle of an abstract manifold and also to classify 1-dimensional manifolds. Here we shall see this lemma being used again and again.

We shall first introduce the notion of connected sum, then the noting of gluing manifolds along common boundary components, and finally the notion of "attaching handles". In the next section, we shall revert back to the study of Morse functions.

Definition 8.3.1 Connected Sum: Recall the notation: for $r > 0$,

$$\mathbb{D}^n_r := \{x \in \mathbb{R}^n : \|x\| \leq r\}; \quad \mathbb{D}^n = \mathbb{D}^n_1.$$

Let M_1 and M_2 be any two connected manifolds of dimension n and let $f_i : \mathbb{D}^n \to \operatorname{int} M_i$ be embeddings, $i = 1, 2$. Consider the quotient space of the disjoint union

$$M_1 \setminus \{f_1(0)\} \coprod M_2 \setminus \{f_2(0)\} \tag{8.7}$$

by the relation

$$f_1(x) \simeq f_2 \circ \eta(x), \quad 0 < \|x\| < 1, \tag{8.8}$$

where η is the inversion map

$$\eta(x) = \frac{\sqrt{(1 - \|x\|^2)}x}{\|x\|}. \tag{8.9}$$

Temporarily, we shall denote this quotient space by $[f_1, f_2]$.

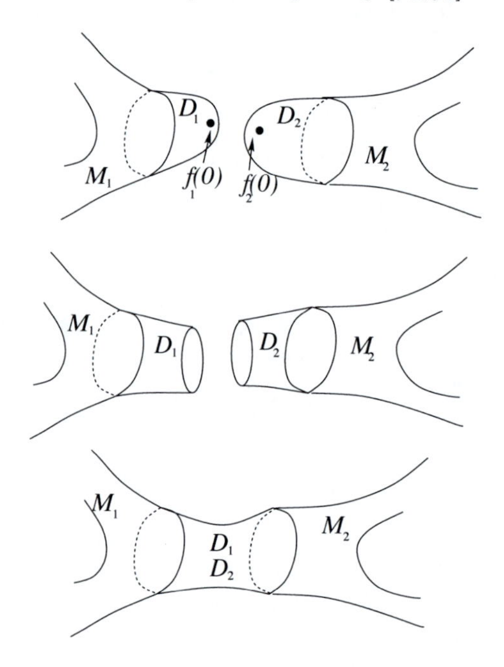

Figure 41 The connected sum.

Lemma 8.3.1 The diffeomorphism class of $[f_1, f_2]$ depends on the isotopy classes of the embeddings f_i, i.e., if f_i is isotopic to g_i in M_i, $i = 1, 2$ then $[f_1, f_2]$ is diffeomorphic to $[g_1, g_2]$.

Proof: We shall prove that both of them are diffeomorphic to $[f_1, g_2]$. An isotopy of f_2 to g_2 yields an isotopy $H : K \times I \to M_2$ of the inclusion map to $g_2 \circ f_2^{-1}$, where $K = f_2(\mathbb{D}^n)$. By the isotopy extension theorem, we get $\hat{H} : M_2 \times I \to M_2$, an isotopy of Id_{M_2} which extends H. Now in Lemma 5.3.1, put

$$M = M_1 \setminus \{f_1(0)\}) \coprod (M_2 \setminus \{f_2(0)\})$$

$$\alpha(x) = \begin{cases} x, & x \in M_1; \\ \hat{H}(x, 1), & x \in M_2. \end{cases}$$

Then for $x \in U =: f_1(\mathbb{D}^n \setminus \{0\})$, we have, $\alpha \circ \phi(x) = \psi \circ \alpha(x)$, where $\phi = f_2 \circ \eta \circ f_1^{-1}$, and $\psi = g_2 \circ \eta \circ f_1^{-1}$. Therefore, by (iii) of Lemma 5.3.1, $[f_1, f_2]$ is diffeomorphic to $[f_1, g_2]$. Similarly, we can prove that $[f_1, g_2]$ is diffeomorphic to $[f_2, g_2]$. ♠

Remark 8.3.1

1. Appealing to the Disc Theorem 6.3.7, the hypothesis in the above lemma can be assured always if M_i are connected nonorientable, and whenever M_i is orientable, by assuming that $g_i \circ f_i^{-1}$ is orientation preserving. Whenever, both M_i are oriented, we shall actually assume that one of the two f_i is orientation preserving and the other is orientation reversing, which is the same as saying $f_2 \circ f_1^{-1}$ is orientation reversing. In this case, it is clear that $f_2 \circ \eta \circ f_1^{-1}$ is orientation preserving and hence $[f_1, f_2]$ will inherit an orientation compatible with those of M_i. With these conventions, we can now say that the quotient space $[f_1, f_2]$ is well-defined for connected manifolds M_1, M_2. We denote this space by $M_1\#M_2$. The notion easily gets extended to the case when M_i are not connected, since only one component each of M_i will be involved in the entire operation.

2. Observe that the quotient map q given by (8.7) and (8.8) restricted to the smaller set
$$M_1 \setminus f_1(\mathbb{D}^n_{\epsilon_1}) \coprod M_2 \setminus f_2(\mathbb{D}^n_{\epsilon_2}),$$
where $0 < \epsilon_1, \epsilon_2 < 1/2$, is also surjective and hence defines the same manifold below.

3. The simpler way of identifying "$f_1(x) \sim f_2(x)$" will not produce a Hausdorff space. This is the reason why we have to involve the inversion map η. Since η is orientation reversing (in all dimensions), it is necessary to choose one of the embeddings f_1 or f_2 to be orientation reversing. Beyond this, there is nothing very sacrosanct about the inversion map η either. Observe that η fails to be a diffeomorphism at 0 and at the boundary. Therefore, it is better to restrict this to a suitable open annular region $\{\mathbf{x} : 0 < s < \|x\| < r < 1\}$. This is where the previous remark comes handy. Then in the construction of $M_1\#M_2$, we can replace η also by any diffeomorphism isotopic to η. We leave the details of this to you as an exercise.

Definition 8.3.2 The space $M_1\#M_2$ is called the *connected sum* of M_1 and M_2.

Remark 8.3.2
(i) Put $B_i^n = f_i(\mathbb{D}^n_{1/2})$. We may then think of $M_1\#M_2$ as obtained by gluing $M_1 \setminus \text{int}\, B_1^n$ with $M_2 \setminus \text{int}\, B_2^n$ along the boundaries ∂B_1^n and ∂B_2^n via a diffeomorphism $f : \mathbb{S}^{n-1} \to \mathbb{S}^{n-1}$. This description of the connected sum is much more intuitive and helpful. It is perfectly suited in case we are dealing with topological manifolds (or the so called "piecewise linear manifolds" which we have not introduced here). However, in the category of smooth manifolds, first of all, this approach does not display the smooth structure on the connected sum. One has to prove this separately and indeed, there is no way we get a canonical smooth structure. All that one can hope to prove is that there do exist smooth structures on the quotient that extends the smooth structure on open parts $M_i \setminus B_i^n$. Second, the diffeomorphism type of the new manifold depends on the isotopy class of f and not all diffeomorphisms of a sphere are isotopic to Id. Hence, even the diffeomorphism structure on the glued manifold is not well-defined. However, in dimensions less than four, we do not have this problem, since there are just two isotopy classes of diffeomorphisms of $\mathbb{S}^n$ for $n \leq 3$.
(ii) The operation of taking connected sum is associative on the collection of diffeomorphism classes of connected n-dimensional manifolds. i.e., $(M_1\#M_2)\#M_3$ is diffeomorphic to $M_1\#(M_2\#M_3)$ for any three n-manifolds. (Apply the Disc Theorem appropriately.)

Clearly this operation is commutative.
(iii) The standard sphere $\mathbb{S}^n$ plays the role of identity for this operation: $M\#\mathbb{S}^n = M$.
(iv) When we allow nonconnected manifolds, it is necessary to mention the components along which the operation is being performed.
(v) We remark again that the connected sum is orientable iff both the original manifolds are so. Also, if you are not particular about retaining the original orientations on the manifolds, then it is not necessary to assume that $f_1 \circ f_2^{-1}$ is orientation reversing, for the simple reason that this can always be arranged by choosing appropriate orientations on M_i. This remark is applicable whenever we are gluing along single connected components taken from each connected manifold.
(vi) For some interesting results on connected sums, see [Sh1].

Example 8.3.1 Recall that we have constructed the Klein bottle as the quotient $K = M_\psi$ of $M = (-1/2, 1) \times \mathbb{S}^1$ where $\psi : (-1/2, 0) \times \mathbb{S}^1 \to (1/2, 1) \times \mathbb{S}^1$ is the map $(t, v) \mapsto (1+t, \bar{v})$. Check that the image of the two lines $(-1/2, 1) \times \{-\imath, \imath\}$ in the Klein bottle K is an embedded loop that separates K into two copies of the Möbius band. If you recall that capping off the Möbius band yields $\mathbb{P}^2$, you have just proved that $K = \mathbb{P}^2\#\mathbb{P}^2$.

We shall now describe an operation that is also a special case of gluing.

Definition 8.3.3 Attaching Handles: Fix integers n, k such that $0 \le k \ne n$. Write $\mathbb{R}^n = \mathbb{R}^k \times \mathbb{R}^{n-k}$ and elements $\mathbf{v} \in \mathbb{R}^n$ in the form $(\mathbf{x}, \mathbf{y})$. Put $S^{k-1} = \mathbb{S}^{k-1} \times 0$ and $S^{n-k-1} = \{0\} \times \mathbb{S}^{n-k-1}$. Put $T = \mathbb{D}^n \setminus 0 \times \mathbb{D}^{n-k}$. Along with the projection map $(\mathbf{x}, \mathbf{y}) \mapsto \mathbf{x}/\|\mathbf{x}\|$, we shall refer to T as a tubular neighborhood of S^{k-1} in $\mathbb{D}^n$. There is an obvious diffeomorphism $\lambda : \mathbb{D}^n \setminus S^{k-1} \to B^k \times \mathbb{D}^{n-k}$ given by

$$(\mathbf{x}, \mathbf{y}) \mapsto (\mathbf{x}, (1 - \|\mathbf{x}\|^2)^{-1/2}\mathbf{y}). \tag{8.10}$$

Consider the inversion

$$\alpha : \mathbb{D}^n \setminus (S^{k-1} \cup 0 \times \mathbb{D}^{n-k}) \to \mathbb{D}^n \setminus (S^{k-1} \cup 0 \times \mathbb{D}^{n-k}) \tag{8.11}$$

given by $\alpha = \lambda^{-1} \circ (\eta, Id) \circ \lambda$, where $\eta : B^k \setminus \{0\} \to B^k \setminus \{0\}$ is the inversion map as given in (8.9). Indeed, we have,

$$\alpha(\mathbf{x}, \mathbf{y}) = \left(\frac{\sqrt{1 - \|\mathbf{x}\|^2}}{\|\mathbf{x}\|}\mathbf{x}, \frac{\|\mathbf{x}\|}{\sqrt{1 - \|\mathbf{x}\|^2}}\mathbf{y} \right). \tag{8.12}$$

Let now M be a n-dimensional manifold with boundary and $f : T \to M$ be an embedding such that $f(\partial T) \subset \partial M$. Consider the quotient space of the disjoint union

$$\left[(M \setminus f(S^{k-1}) \coprod (\mathbb{D}^n \setminus S^{k-1}) \right] / \sim$$

where for each $x \in T \setminus S^{k-1}$, we have the identification $x \sim f\alpha(x)$. Clearly this is a n-manifold with boundary. We shall call this space *"M with a k-handle attached along the attaching-sphere* $f(S^{k-1})$" and denote it by $M \cup H^k$. Note that the quotient map is an embedding restricted to $M \setminus f(S^{k-1})$ as well as $\mathbb{D}^n \setminus S^{k-1}$, which we shall use for identifying these manifolds as submanifolds of $M \cup H^k$. With this understanding, $\mathbb{D}^n \setminus S^{k-1}$ is called the *k-handle,* the $(n-k)$-disc $0 \times \mathbb{D}^{n-k}$ is called the *belt disc* and its boundary, the *belt sphere,* which we shall denote by Σ^{n-k-1}.

Remark 8.3.3
(a) As in the case of the connected sum, it is clear that the diffeomorphism class of the result of attaching a handle depends on the isotopy class of the attaching map. This being a tubular neighborhood of the attaching sphere, it further depends on the isotopy class of the attaching sphere itself except perhaps in the case of 1-handles, when the attaching sphere $\mathbb{S}^0$ consists of two points. In this special case, the two points with "the same orientation" and with opposite orientation' has to be distinguished. Simply speaking, now the orientations on the tubular neighborhood of these points enter into the picture.
(b) One of the drawbacks of the above definition of attaching handles is that it does not readily exhibit the original manifold M as a subspace of $M \cup H^k$. The other is, of course, it is less intuitive. The first reason indeed produces a serious handicap in the computation of homology and homotopy. In order to tackle this, it is necessary to see the relation of this construction with the more intuitive definition of attaching a handle, which we shall call *combinatorial attaching*.

Definition 8.3.4 Combinatorial Attachment: Let $0 \leq k \leq n$. Let M be a (topological) n-manifold with boundary and $h : \mathbb{S}^{k-1} \times \mathbb{D}^{n-k} \to \partial M$ be an embedding. The quotient space of the disjoint union $M \coprod \mathbb{D}^k \times \mathbb{D}^{n-k}$ by the identification $x \sim h(x)$ for all $x \in \mathbb{S}^{k-1} \times \mathbb{D}^{n-k}$ is called *the space obtained by combinatorial attachment of a k-handle to M*. We shall denote it by $M \cup_h \mathbb{D}^k \times \mathbb{D}^{n-k}$.

Remark 8.3.4 It is clear that $M \cup_h \mathbb{D}^k \times \mathbb{D}^{n-k}$ is a topological manifold that contains M and $\mathbb{D}^k \times \mathbb{D}^{n-k}$ as closed subspaces such that their intersection is $h(\mathbb{S}^{k-1} \times \mathbb{D}^{n-k})$. The biggest disadvantage of this construction is when we start with a smooth manifold M, the construction does not provide any way to put a smooth structure on the space so constructed. The following proposition gives a clear relation between combinatorial attachment and attaching (smooth) handles so that we can go from one to the other as and when necessary. This result can be viewed as a "smoothing" of the combinatorial attachment.

Proposition 8.3.1 Let $M \cup H^k$ be obtained by attaching a k-handle to M via the attaching map f. Then there exists a diffeomorphism $h : \mathbb{S}^{k-1} \times \mathbb{D}^{n-k} \to \partial M$ and a homeomorphism $\Theta : M \cup H^k \to M \cup_h \mathbb{D}^k \times \mathbb{D}^{n-k}$, which is the identity map on the boundary of $M \cup H^k$ and on the belt disc.

Proof: The first step is similar to Remark 8.3.2.(i), viz., under the quotient map q we take the image of some smaller subsets restricted to which q is surjective. So, put $A = \{(\mathbf{x}, \mathbf{y}) \in \mathbb{D}^n \ : \ \|\mathbf{x}\|^2 \leq 1/2\}$ and $B = \hat{T}(1/2) = \{(\mathbf{x}, \mathbf{y}) \in \mathbb{D}^n \ : \ \|\mathbf{x}\|^2 \geq 1/2\}$. Clearly α interchanges $A \setminus S^{k-1}$ and $B \setminus S^{k-1}$ and is the identity map on $A \cap B$. Therefore, $M \cup H^k$ is equal to the identification space $M \setminus f(B) \cup_g A$, where $g = f|_{A \cap B}$.

The second step is to "rewrite" $M \setminus f(B) \cup_g A$ as $M \cup_h \mathbb{D}^n$. Consider $\gamma_1 : A \to \mathbb{D}^n, \gamma_2 : A \to \mathbb{D}^k \times \mathbb{D}^{n-k}$ given by

$$\gamma_1(\mathbf{x}, \mathbf{y}) = \begin{cases} (\mathbf{x}, \mathbf{y}), & \text{if} \quad \|y\|^2 \geq 1/2, \\ (\sqrt{2(1 - \|\mathbf{y}\|^2)}\mathbf{x}, \mathbf{y}), & \text{if} \quad \|y\|^2 \leq 1/2; \end{cases}$$

$$\gamma_2(\mathbf{x}, \mathbf{y}) = \begin{cases} \sqrt{2}(\mathbf{x}, \mathbf{y}), & \text{if} \quad \|y\|^2 \leq 1/2, \\ (1 - \|\mathbf{x}\|^2)^{-1/2}(\mathbf{x}, \mathbf{y}), & \text{if} \quad \|y\|^2 \geq 1/2. \end{cases}$$

Check that γ_1 is a homeomorphism whereas γ_2 is a diffeomorphism and both are equal to the identity map on the belt-disc $0 \times \mathbb{D}^{n-k}$. Now consider $\omega : M \setminus f(B) \to M$ given by

$$\omega(q) = \begin{cases} q, & \text{if} \quad q \in M \setminus f(A), \\ f \circ \gamma_1 \circ f^{-1}(q), & \text{if} \quad q \in f(T \setminus B). \end{cases}$$

Once again, the above said properties of γ imply that ω is a homeomorphism. Now ω and γ_2 together define a homeomorphism

$$\Theta : M \cup H^k = M \setminus f(B) \cup_g A \to M \cup_h \mathbb{D}^k \times \mathbb{D}^{n-k}$$

where $h : \mathbb{S}^{k-1} \times \mathbb{D}^{n-k} \to M$ is given by $h = \gamma_1 \circ g \circ \gamma_2^{-1}$. Check that Θ is as required. ♠

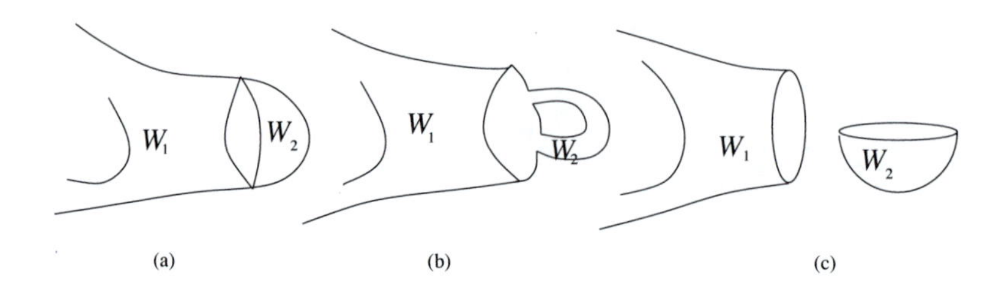

Figure 42 The three cases of attaching handles in dimension 2.

Remark 8.3.5
(i) The figures (a),(b),(c) above show attaching a k-handle, for $k = 2, 1, 0$ respectively, in 2-dimension.
(ii) In the above definition, the role of the integer k, $0 \leq k \leq n$ is important. The two extreme cases are the easiest to understand. For $k = 0$, since the sphere $\mathbb{S}^{-1}$ is empty, this just means that $W_1 \cap W_2 = \emptyset$. Equivalently, M is the disjoint union of W_1 and $\mathbb{D}^n$. We next take up the case $k = n$ which turns out to be nothing but "capping off" which we have discussed earlier in Chapter 6.

Theorem 8.3.1 *Let M be a smooth n-manifold. Suppose $M = W_1 \cup W_2$ with $W_1 \cap W_2 = \partial W_1 = \partial W_2$ and both W_1 and W_2 are diffeomorphic to the closed unit disc $\mathbb{D}^n$.*
(i) Then M is homeomorphic to $\mathbb{S}^n$.
(ii) M is diffeomorphic to $\mathbb{S}^n$ iff there are diffeomorphisms $f_j : \mathbb{D}^n \to W_j$ such that $f_2^{-1} \circ f_1 : \mathbb{S}^{n-1} \to \mathbb{S}^{n-1}$ extends to a diffeomorphism $\mathbb{D}^n \to \mathbb{D}^n$.

Proof: (i) Let $f_j : \mathbb{D}^n \to W_j$ be homeomorphisms. Consider the restrictions to the boundary and the composite homeomorphism $f_2^{-1} \circ f_1 : \mathbb{S}^{n-1} \to \mathbb{S}^{n-1}$. Extend this to a homeomorphism $\phi : \mathbb{D}^n \to \mathbb{D}^n$ (via the cone construction, for instance). Let $\eta_\pm : \mathbb{D}^n \to \mathbb{S}^n$ be embeddings given by

$$\mathbf{x} \mapsto (\mathbf{x}, \pm\sqrt{1 - \|\mathbf{x}\|^2}).$$

Define $\psi : M \to \mathbb{S}^n$ by the formula:

$$\psi(x) = \begin{cases} \eta_+ \circ f_1^{-1}(x), & x \in W_1, \\ \eta_- \circ \phi^{-1} \circ f_2^{-1}(x), & x \in W_2. \end{cases}$$

To see that ψ is a homeomorphism, the only thing we need to check is that its two definitions on the boundary ∂W_j agree.
(ii) Now assume that f_j are diffeomorphisms and that the diffeomorphism $f_2^{-1} \circ f_1$ extends to a diffeomorphism $\lambda : \mathbb{D}^n \to \mathbb{D}^n$. Let $\tau : \mathbb{D}^n \to \mathbb{D}^n$ be a diffeomorphism, which is the identity map on the boundary and the radial component of its derivative at any point on the boundary is equal to the radial component of $D(f^{-1}) \circ D(f_2) \circ D(\lambda^{-1})$. Take $\phi = \tau \circ \lambda$ and follow the steps in (i). The extra work that is needed here is to check the smoothness of ψ

along the equator. This is precisely the role played by τ here, so that the radial components of the derivatives of f_1^{-1} and $\phi^{-1} \circ f_2^{-1}$ coincide.

Conversely, suppose we have a diffeomorphism $\psi : M \to \mathbb{S}^n$. Consider the embedding $\psi \circ f_1 : \mathbb{D}^n \to \mathbb{S}^n$. By an ambient isotopy of the sphere $\mathbb{S}^n$, we may (change ψ and) assume that the image of this embedding is the upper hemisphere and it is the identity map on the boundary. This means that $\psi(W_2)$ is the lower hemisphere and $\psi(\partial W_1) = \psi(\partial W_2)$ is equal to the equator in $\mathbb{S}^n$. Therefore, we can take diffeomorphism $\phi = f_2^{-1} \circ \psi^{-1} \circ \eta_- : \mathbb{D}^n \to \mathbb{D}^n$ and see that it extends $f_2^{-1} \circ f_1$. ♠

Corollary 8.3.1 With M as in the above theorem if $n = 2$, then M is diffeomorphic to $\mathbb{S}^2$.

Proof: It is enough to prove that every diffeomorphism $\mathbb{S}^1 \to \mathbb{S}^1$ extends to a diffeomorphism $\mathbb{D}^2 \to \mathbb{D}^2$. [This depends on the result given in Exercise 6.3.2. Since this step is crucial, we offer a solution here.] In view of Theorem 6.4.1, we need to prove that every orientation preserving diffeomorphism $\phi : \mathbb{S}^1 \to \mathbb{S}^1$ is isotopic to the identity map. Without loss of generality, (by composing with a rotation), we will assume that $\phi(1) = 1$. Let $f : \mathbb{R} \to \mathbb{R}$ be a map such that $\exp \circ f = \phi \circ \exp$ and $f(0) = 0$ as in Theorem 7.3.4. Then it follows that $f(1) = 1$ and $f : [0, 1] \to [0, 1]$ is an orientation preserving diffeomorphism. In particular, f is strictly increasing. Therefore, for each fixed t, $x \mapsto tf(x) + (1 - t)x$ is a diffeomorphism of $[0, 1]$. Check that the diffeotopy

$$(x, t) \mapsto tf(x) + (1 - t)x$$

actually factors down to give a diffeotopy of Id with ϕ on $\mathbb{S}^1$. ♠

Remark 8.3.6 The above theorem deals with a very special case of attaching an n-handle to an n-manifold. The arguments, however, go through in the general case as well with a weaker conclusion, viz., the homeomorphism type of a manifold obtained by capping off at a boundary component which is homeomorphic to a sphere is unique. Indeed, we can go one step further. If W is connected, among such boundary components of W, it does not matter at which component we are attaching the n-handle—the homeomorphism type of the new manifold is the same. This is an easy consequence of Exercise 6.4.3. For ready reference, below we state the corresponding result for the case $n = 2$.

Theorem 8.3.2 *Let M be a connected (compact) smooth 2-manifold obtained by attaching a 2-handle to a connected surface W_1. Then the diffeomorphism class of M is uniquely determined by the diffeomorphism class of W_1 irrespective of which boundary component of W is being used in the attaching process.*

We shall consider one more special case when $n = 2$ and $k = 1$.

Theorem 8.3.3 *Attaching a 1-handle to a 2-disc produces either the cylinder or the Möbius band.*

Proof: The attaching sphere consists of two points on $\partial\mathbb{D}^2$, which, we may assume, are $\{(\pm 1, 0)\}$, up to isotopy. The tubular neighborhoods of these points can be chosen to be $\mathbb{D}^2 \setminus (0 \times [-1, 1])$. Again, up to isotopy, the attaching diffeomorphism has two choices : it may preserve (or reverse) orientation on both the components or it may preserve orientation on one component and reverse it on the other. In the first type, by changing the orientation on the handle, we can assume that the orientation is preserved on both components. Accordingly, we get two possibly distinct diffeomorphism classes of surfaces. One class will be orientable and other nonorientable and hence the two **are** diffeomorphically distinct.

We need to determine these two manifolds. Here, Exercises 8.2.1.(vii)-(viii) come to our aid. We have constructed Morse functions on the cylinder as well as the Möbius band with exactly two critical points one of index 0 and the other of index 1. Therefore, both are obtained by the process of attaching a 1-handle to a 2-disc. Therefore, of the two surfaces obtained, the orientable one must be the cylinder and the nonorientable one must be the Möbius band. ♠

We shall now take up a little bit of the general theory of attaching handles.

Definition 8.3.5 By *a (finite) handle presentation for an n-manifold M*, we mean a sequence $M_0 \subset M_1 \subset \cdots \subset M_p = M$ of submanifolds, where M_0 is diffeomorphic to $\mathbb{D}^n$ and each M_i is obtained by attaching a k_i-handle to $M_{i-1}, 0 \leq k \leq n, 1 \leq i \leq p$. The presentation is said to be monotonic if $i < j$ implies $k_i \leq k_j$.

Remark 8.3.7 The following theorem immediately tells us that any handle presentation can be converted to a monotonic one.

Theorem 8.3.4 *Let W be a manifold obtained from M by attaching a k-handle first and then attaching a l-handle. If $k \geq l$, then W is diffeomorphic to a manifold obtained from M by attaching a l-handle first and then a k-handle.*

Proof: On the boundary $\partial(M \cup H^k)$ look at the attaching sphere S^{l-1} of the l-handle and the belt-sphere Σ^{n-k-1}. Since $l - 1 + n - k - 1 < n - 1$, it follows that (see Exercise 6.5.2) there is an ambient isotopy of $\partial(M \cup H^k)$ that moves S^{l-1} away from Σ^{n-k-1}. Of course, this isotopy can be extended to the whole of $M \cup H^k$. But any compact subset that is disjoint from Σ^{n-k-1} can further be isotoped into $\partial M \setminus S^{k-1}$. We can now perform the attaching of the two handles in any order. ♠

Corollary 8.3.2 If M admits a finite handle presentation then it admits a monotone one.

Remark 8.3.8 Caution is needed in the correct interpretation of this result when $k = l = 1$ and $n = 2$. No doubt the two handles can be attached in whichever order you like. However, the relative location of the four points on the boundary of the 2-manifold comes into play now since you cannot shuffle them any which way you like by an isotopy.

Remark 8.3.9 We shall prove in the next section that every compact smooth manifold has a handle presentation and hence a monotonic one. We shall now see the homotopy theoretic aspect of attaching a k-handle. This was the original point of view of M. Morse. This part may demand a little bit more familiarity with elementary homotopy theory on your part.

Definition 8.3.6 Let X be a topological space and Y be a subspace of X. By *a strong deformation retraction* $r : X \to Y$ we mean a continuous map r such that there is a homotopy $H : X \times I \to X$ such that $H(x, 0) = x, H(x, 1) = r(x)$ and $H(y, t) = y$ for all $y \in Y$ and $0 \leq t \leq 1$. If such a strong deformation retraction r exists then Y is called a *strong deformation retract of X*.

Lemma 8.3.2 There is a strong deformation retraction $R : [0, 1] \times \mathbb{D}^l \to \{1\} \times \mathbb{D}^l \cup [0, 1] \times \{0\}$ with the additional property $R(0, y) = (0, 0)$ for all $y \in \mathbb{D}^l$.

Proof: See the picture and try to write down an expression for R, before reading further.

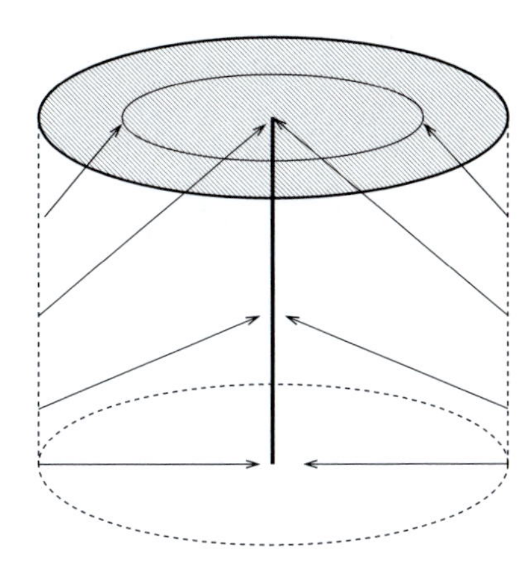

Figure 43 A typical deformation.

For $\theta \in \mathbb{S}^{l-1}$, $0 \leq r \leq 1$, $0 \leq s \leq 1$ take

$$R(r, \theta, s) = \begin{cases} \left(\frac{2s}{2-r}, \theta, 0\right), & r + 2s \leq 2, \\ \left(\frac{r+2(s-1)}{s}, \theta, 1\right), & r + s \geq 2. \end{cases}$$

Verify that R has the required properties. ♠

By "spinning" the above result, we get:

Lemma 8.3.3 $\mathbb{S}^{k-1} \times \mathbb{D}^l \cup \mathbb{D}^k \times \{0\}$ is a strong deformation retract of $\mathbb{D}^k \times \mathbb{D}^l$.

Proof: Express elements of $\mathbb{D}^k$ in polar coordinates (θ, t) where $0 \leq t \leq 1$ and $\theta \in \mathbb{S}^{k-1}$. Let $R : [0,1] \times \mathbb{D}^l \to \{1\} \times \mathbb{D}^l \cup [0,1] \times \{0\}$ be a strong deformation retraction as provided by the above lemma. Put $\hat{R}((\theta, t), y) = (\theta, R(t, y))$. Check that $\hat{R}$ is the required strong deformation retract. ♠

Corollary 8.3.3 If W' is obtained from W by attaching a k-handle, then $W \cup \mathbb{D}^k \times \{0\}$ is a strong deformation retract of W'.

Proof: This is because, wherever the identification takes place, that part is not disturbed by the strong deformation retraction and so the two operations can be performed interchangeably. ♠

Definition 8.3.7 By an *n-cell,* or a *cell of dimension n*, we mean a topological space homeomorphic to the closed unit disc $\mathbb{D}^n$ in $\mathbb{R}^n$. Attaching an n-cell to a topological space means that we have a continuous map $\alpha : \partial\mathbb{D}^n \to X$ and we are taking the quotient space of the disjoint union $X \coprod \mathbb{D}^n$ by the relation $x \sim \alpha(x)$ for all $x \in \partial\mathbb{D}^n$. Note that a 0-cell is a singleton space and attaching a 0-cell means just taking the disjoint union with a singleton space. We say a topological space X is a *finite CW-complex* if it can be obtained by attaching successively finitely many cells of nondecreasing dimensions, beginning with a 0-cell.

Corollary 8.3.4 Every monotone handle presentation of a manifold M defines a homotopy equivalence of M with a finite CW complex X with one cell of dimension k for each k-handle in the presentation.

Proof: This follows from the following observations and a simple induction:
(i) If $f : X \to Y$ is a homotopy equivalence, $\lambda : \mathbb{S}^{k-1} \to X$ is a continuous map, then the space obtained by attaching a k-cell to X via λ is homotopy equivalent to the space obtained by attaching a k-cell to Y via $f \circ \lambda$. Indeed, the homotopy f extends to a map $\hat{f} : X \cup_\lambda \mathbb{D}^k \to Y \cup_{f\circ\lambda} \mathbb{D}^k$ by sending the interior of the first k-cell in a one-to-one fashion into the interior of the second k-cell. A similar comment applies to a homotopy inverse of f and defines the homotopy inverse of $\hat{f}$.
(ii) If M' is obtained from M by attaching a k-handle, then M' has the homotopy type of attaching a k-cell to M. This is a direct consequence of Corollary 8.3.3. ♠

Definition 8.3.8 Let X be a finite CW-complex, with the number of k-cells equal to n_k. Then the number

$$\chi(X) := \sum_k (-1)^k n_k$$

is called the Euler characteristic of the space X. Let M be a manifold with a handle presentation $\mathcal{P}$. Let h_k denote the number of k-handles in $\mathcal{P}$. We define the *Euler characteristic of the presentation* $\mathcal{P}$ to be the alternate sum:

$$\chi(\mathcal{P}) := \sum_k (-1)^k h_k.$$

Remark 8.3.10 It follows immediately from corollary 8.3.4 that $\chi(X) = \chi(\mathcal{P})$, where X is the CW-complex associated with the handle presentation $\mathcal{P}$ of a given smooth manifold M We shall soon relate this number with $I(M)$ via the Morse theory and that is the celebrated Poincaré-Hopf index theorem.

Remark 8.3.11 Closely associated with the operation of attaching handles is another notion called "surgery", or "spherical modification". Roughly speaking this is what happens to the boundary $M = \partial W$ of a manifold W to which you are attaching a handle. In particular, given a closed manifold M, we can consider $M \times [0,1]$ and perform the operation of attaching a handle along the boundary part $M \times \{1\}$ (i.e., without involving $M \times \{0\}$). If W is the resulting manifold, we then consider $\partial W \setminus M \times \{0\}$ as the result of performing a surgery on M. We shall not go into the details here.

Of particular interest to us here is the case of 1-surgery on a surface F, which is the result of attaching a 1-handle to $F \times [0,1]$. For this reason, this operation also justifiably goes under the name "attaching a 1-handle", though technically incorrect. This limited study will be taken up in the next section, where it will help us in completely classifying the result of attaching 1-handles to surfaces.

Exercise 8.3 Show that $\mathbb{R}^n \# \mathbb{R}^n$ is diffeomorphic to $\mathbb{S}^{n-1} \times \mathbb{R}$.

8.4 Further Geometry of Morse Functions

In the example 8.2.1, we had hinted at a certain change in the topology of the level sets $h^{-1}(-\infty, r]$ where h was the height function on the sphere. We shall now consolidate this idea.

Definition 8.4.1 For simplicity of the exposition, we take M to be a manifold without boundary. Let $f : M \to \mathbb{R}$ be a Morse function. For any $r \in \mathbb{R}$, put $M_r = f^{-1}(-\infty, r]$ and for $r < s$, put $M_{r,s} = f^{-1}[r,s]$. If r and s happen to be regular values, then it follows that M_r and $M_{r,s}$ are submanifolds of M with boundary $f^{-1}(r)$ and $f^{-1}(\{r,s\})$ respectively.

Theorem 8.4.1 Regular Interval Theorem *If f has no critical values in $[r,s]$, and $M_{r,s}$ is compact, then $M_{r,s}$ is diffeomorphic to $f^{-1}(r)\times[r,s]$. Moreover, M_r is diffeomorphic to M_s.*

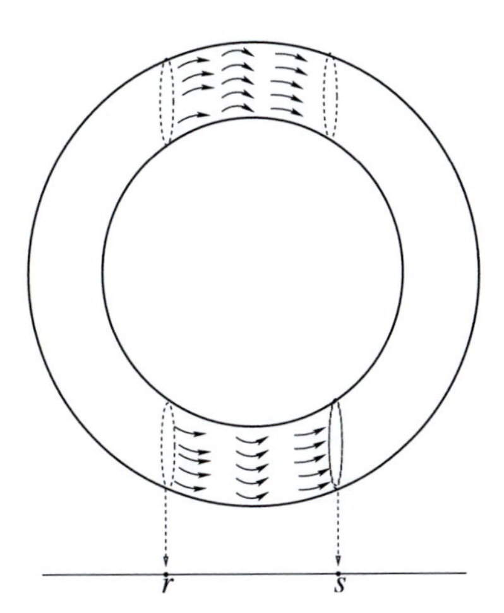

Figure 44 No change occurs in a regular interval.

Proof: We observe that $f(M_{r,s}) = [r,s]$. (why?) Now, consider the vector field $\sigma = \operatorname{grad} f/\|\operatorname{grad} f\|^2$ on $M_{r,s}$. We take a smooth function $c : M \to \mathbb{R}$ which is identically 1 on $M_{r,s}$ and vanishes outside a compact neighborhood of $M_{r,s}$. We then consider the vector field $\rho = c\sigma$. Let $\phi : M\times\mathbb{R} \to M$ be the 1-parameter group of diffeomorphism generated by ρ. Then we have the identity

$$\frac{\partial\phi}{\partial t}(x,t) = \rho(\phi(x,t)).$$

For any fixed $x \in M$, consider the map $\tau : t \mapsto f(\phi(x,t))$. Then by the chain rule, we have,

$$\frac{d\tau}{dt}(t) = \langle \operatorname{grad} f(\phi(x,t)), \rho(\phi(x,t))\rangle$$

This is equal to 1 for all values of t such that $\phi(x,t) \in M_{r,s}$, for, then $\rho = \sigma = \operatorname{grad} f/\|\operatorname{grad} f\|^2$. Therefore, it follows that $\tau(t+s) = \tau(t)+s$ in this range. In particular, it follows that $\phi(-,t)$ carries M_r to M_{r+t} diffeomorphically for all $0 \le t \le s-r$. Clearly, then the map $\psi : f^{-1}(r)\times[r,s] \to M_{r,s}$ defined by

$$(x,t) \mapsto \phi(x,t-r)$$

is a diffeomorphism. Also, the map $\phi_{s-r} : M \to M$ takes M_r to M_s diffeomorphically. ♠

Remark 8.4.1

1. Observe that the diffeomorphism ψ satisfies the property $f\circ\psi(x,t) = t+r$.

2. The compactness assumption on $M_{r,s}$ is necessary. Usually, this is ensured by assuming that f is a proper map. By simply deleting a point from $M_{r,s}$ we can break the conclusion of the theorem whereas, in the hypothesis, only the compactness of $M_{r,s}$ is not satisfied.

As an immediate corollary let us prove:

Theorem 8.4.2 *Let M be a closed n-manifold with a smooth function $f : M \to \mathbb{R}$ having exactly two critical points. Suppose the critical points are nondegenerate. Then M is homeomorphic to $\mathbb{S}^n$.*

Proof: Since M is compact, f must attain its minimum, say, at p with $f(p) = a$. Likewise, f attains its maximum also, say, at $q \neq p$ with $f(q) = b$. Since M is boundaryless, these points must be critical points of f. It also follows that $f^{-1}(a) = \{p\}$ and $f^{-1}(b) = \{q\}$. Now by the Morse lemma, it follows that for sufficiently small $\epsilon > 0$, $f^{-1}[a, a+\epsilon]$ and $f^{-1}[b-\epsilon, b]$ are closed n-cells. In particular, $f^{-1}(a+\epsilon)$ is diffeomorphic to $\mathbb{S}^{n-1}$. From Theorem 8.4.1, it follows that $f^{-1}[a+\epsilon, b-\epsilon]$ is diffeomorphic to $\mathbb{S}^{n-1} \times I$. This means that M is actually obtained by gluing two discs to $\mathbb{S}^{n-1} \times I$. Hence, we can obtain a homeomorphism between $\mathbb{S}^n$ and M. ♠

Remark 8.4.2

1. In general, we cannot say that M is diffeomorphic to $\mathbb{S}^n$ since we do not have any control over the diffeomorphisms that we have used in gluing the discs. Indeed, there are examples of closed connected 7-dimensional manifolds, which have a Morse function with precisely two critical points but not diffeomorphic to $\mathbb{S}^7$ (see [M3]).

2. However, we know that any diffeomorphism of a circle is isotopic to either the identity map or to the antipodal map. Using this, it would follow that, if $n = 2$ then M is actually diffeomorphic to $\mathbb{S}^2$.

3. The hypothesis "nondegenerate" may be dropped from the above theorem but the proof will be much harder since we do not have the ready made Morse Lemma here. Therefore, careful analysis of the fact that there are only two critical points has to be made. Indeed, on $\mathbb{S}^n$ itself, there do exist Morse functions with exactly two critical points neither of which is a nondegenerate one (exercise).

4. Since we would like to concentrate on one critical point at a time, it helps to have a Morse function that maps different critical points to different values. That is the content of the next lemma.

Lemma 8.4.1 There exists Morse function $f : M \to \mathbb{R}$ which maps different critical points to different values.

Proof: Add suitable "bump functions" to modify a Morse-function near a critical point. We leave the details to the reader as an exercise. ♠

Theorem 8.4.3 *Let M be a compact manifold without boundary and $f : M \to \mathbb{R}$ be a Morse function such that $p \in M$ is the only critical point inside $f^{-1}(c)$, where $c = f(p)$. Suppose the index of p is k. Then there exists $\epsilon > 0$ such that $M_{c+\epsilon}$ is obtained by attaching a k-handle to M_ϵ.*

Proof: Replacing f by $f - c$ we shall assume $c = 0$.

The idea is to find a new Morse function $F : M \to \mathbb{R}$, which is equal to f outside a small neighborhood U of p and $F < f$ inside a smaller neighborhood V of p so that the region $F^{-1}(-\infty, -\epsilon]$ will consist of $M_{-\epsilon}$ together with a small portion H inside V and $F^{-1}(-\infty, \epsilon] = M_\epsilon$. Moreover, F will have no critical values in $[-\epsilon, \epsilon]$, so that, from Theorem 8.4.1, we have $F^{-1}(-\infty, -\epsilon]$ is diffeomorphic to $F^{-1}(-\infty, \epsilon]$. It then remains to describe the portion H and how it is glued to $M_{-\epsilon}$ and to prove that $F^{-1}(-\infty, -\epsilon]$ is obtained by attaching the k-handle to $M_{-\epsilon}$.

Using the Morse Lemma, we get a diffeomorphism $\phi : \mathbb{R}^k \times \mathbb{R}^{n-k} \to U \subset M$ onto an open subset of M such that $\phi(0) = p$ and

$$f(\phi(\mathbf{x}, \mathbf{y})) = -\sum_{i=1}^{k} x_i^2 + \sum_{i=1}^{n-k} y_i^2 = -\|\mathbf{x}\|^2 + \|\mathbf{y}\|^2$$

for all points $(\mathbf{x}, \mathbf{y})$ inside a disc of radius 1, say. Choose $1/3 > \epsilon > 0$ so that $[-\epsilon, \epsilon]$ does not have any critical values of f other than 0.

Let $\mu : \mathbb{R} \to \mathbb{R}$ be a smooth function so that
(i) $\mu(0) > \epsilon$;
(ii) $\mu(t) = 0, \quad \forall\, t \geq 2\epsilon$;
(iii) $-1 < \mu'(t) \leq 0$, for all t.
(See Exercise 1.7.8.)

For $(\mathbf{x}, \mathbf{y}) \in \mathbb{R}^l \times \mathbb{R}^{n-k}$ put $q = \phi(\mathbf{x}, \mathbf{y})$. Define $F : M \to \mathbb{R}$ by

$$F(q) = \begin{cases} f(q), & q \in M \setminus \phi(\mathbb{D}^n), \\ f(q) - \mu(\|\mathbf{x}\|^2 + 2\|\mathbf{y}\|^2), & q = \phi(\mathbf{x}, \mathbf{y}) \in U. \end{cases}$$

Observe that the second formula yields $F = f$ outside $\phi(\mathbb{D}^n_{2\epsilon})$. Therefore, it is well defined and smooth, all over M.

Note that $F^{-1}(-\infty, -\epsilon] \setminus M_{-\epsilon}$ is contained in $\phi(\mathbb{D}^n)$. Therefore, by replacing $F^{-1}(-\infty, -\epsilon]$ and $M_{-\epsilon}$ by their inverse image under $\phi : \mathbb{D}^n \to U$ and replacing f, F by $f \circ \phi, F \circ \phi$, etc., we can also replace M by $\mathbb{D}^n$.

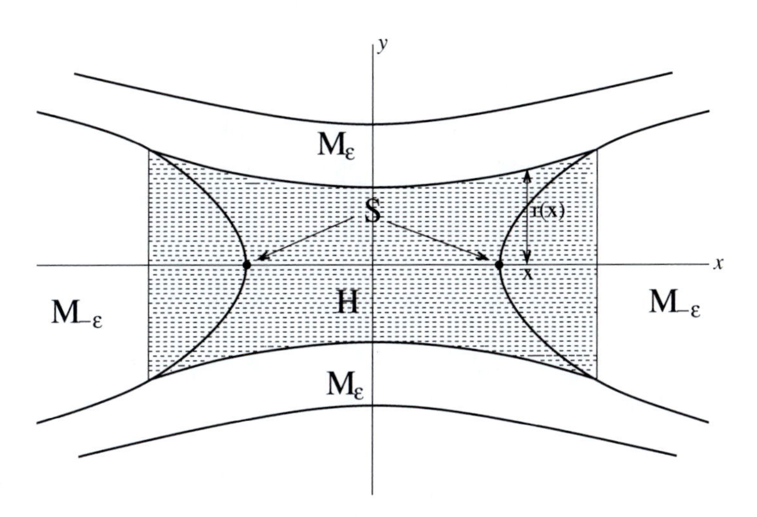

Figure 45 The k-handle H.

Thus, we now have

$$M_{-\epsilon} = \{(\mathbf{x}, \mathbf{y}) \in \mathbb{D}^n \ : \ -\|\mathbf{x}\|^2 + \|\mathbf{y}\|^2 \leq -\epsilon\}$$

and

$$F^{-1}(-\infty, -\epsilon] = \{\{(\mathbf{x}, \mathbf{y}) \in \mathbb{D}^n \ : \ -\|\mathbf{x}\|^2 + \|\mathbf{y}\|^2 + \mu(\|\mathbf{x}\|^2 + 2\|\mathbf{y}\|^2) \leq -\epsilon\},$$

etc.

Claim I: $F^{-1}(-\infty, \epsilon] = M_\epsilon = f^{-1}(-\infty, \epsilon]$.
Since $F \le f$, it follows that $f^{-1}(-\infty, \epsilon] \subset F^{-1}(-\infty, \epsilon]$. Now, suppose $F(q) \le \epsilon$. If $q = (\mathbf{x}, \mathbf{y})$ and $\|\mathbf{x}\|^2 + 2\|\mathbf{y}\|^2 > 2\epsilon$, then $F(q) = f(q)$ and we have nothing to do. Otherwise,

$$f(q) = -\|\mathbf{x}\|^2 + \|\mathbf{y}\|^2 \le \frac{\|\mathbf{x}\|^2 + 2\|\mathbf{y}\|^2}{2} \le \epsilon.$$

Therefore, $F^{-1}(-\infty, \ \epsilon] = f^{-1}(-\infty, \ \epsilon]$.

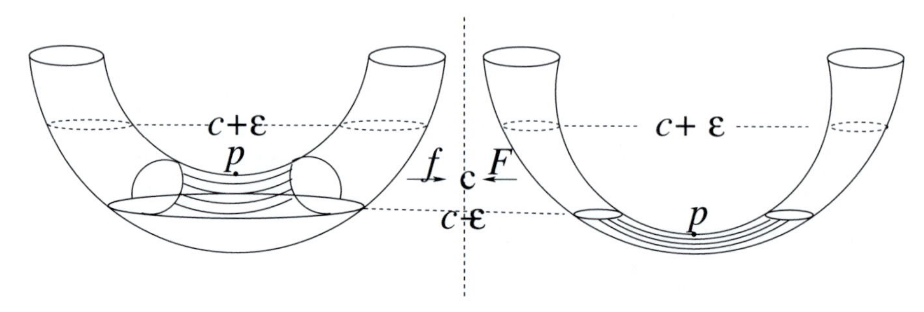

Figure 46 Modification of the function near a critical point.

Claim II: The critical points of F are the same as that of f.
We have to compare the two functions:

$$-\|\mathbf{x}\|^2 + \|\mathbf{y}\|^2; \quad -\|\mathbf{x}\|^2 + \|\mathbf{y}\|^2 - \mu(\|\mathbf{x}\|^2 + 2\|\mathbf{y}\|^2).$$

The derivatives of these are respectively,

$$2(-\mathbf{x}, \mathbf{y});$$

$$2(-(1+\mu')\mathbf{x}, (1-2\mu')\mathbf{y})$$

each of which vanishes iff $(\mathbf{x}, \mathbf{y}) = (0, 0)$, since $-1 < \mu' \le 0$. Hence the claim.
Claim III: $F^{-1}(-\infty, -\epsilon]$ is diffeomorphic to M_ϵ.
Since F and f have same set of critical points and q is the only critical point of f inside $f^{-1}[-\epsilon, \ \epsilon] \supset F^{-1}[-\epsilon, \ \epsilon]$, and $F(q) = f(q) - \mu(0) < -\epsilon$, it follows that F has no critical points in $F^{-1}[-\epsilon, \ \epsilon]$. Therefore, $F^{-1}(-\infty, -\epsilon]$ is diffeomorphic to $F^{-1}(-\infty, \ \epsilon] = M_\epsilon$.
Claim IV: For each $\mathbf{x} \in \mathbb{R}^k$, the intersection $F^{-1}(-\infty, -\epsilon] \cap (\{\mathbf{x}\} \times \mathbb{R}^{n-k}) = \mathbf{x} \times \mathbb{D}^{n-k}_{r(\mathbf{x})}$ for some $r(\mathbf{x}) > 0$. The function $\mathbf{x} \mapsto r(\mathbf{x})$ is smooth and if $\|\mathbf{x}\|^2 > 2\epsilon$ then $r(\mathbf{x}) = (\|\mathbf{x}\|^2 - \epsilon)^{1/2}$. In particular, $F^{-1}(-\infty, -\epsilon] = M_{-\epsilon} \cup H$ where,

$$H = \{(\mathbf{x}, \mathbf{y}) \in \mathbb{R}^n \ : \ \|\mathbf{x}\|^2 < 2\epsilon, \|\mathbf{y}\|^2 \le r(\mathbf{x})^2\}.$$

For a fixed $\mathbf{x}$, we are looking at points $(\mathbf{x}, \mathbf{y})$, which satisfy

$$-\|\mathbf{x}\|^2 + \|\mathbf{y}\|^2 - \mu(\|\mathbf{x}\|^2 + 2\|\mathbf{y}\|^2) \le -\epsilon.$$

Putting $t = \|\mathbf{x}\|^2 + 2\|\mathbf{y}\|^2$, this is the same as

$$\mu(t) - \frac{t}{2} + \frac{3\|\mathbf{x}\|^2}{2} \ge \epsilon. \tag{8.13}$$

Since $\mu(t) - \frac{t}{2}$ is a strictly monotonically decreasing function, there is a unique t_0 such that equality occurs in (8.13) and (8.13) holds iff $t \le t_0$ iff $\|\mathbf{x}\|^2 + 2\|\mathbf{y}\|^2 \le t_0$ iff $\|\mathbf{y}\|^2 \le (t_0 - \|\mathbf{x}\|^2)/2$.

Since $\mu(t) > \epsilon$ for $t \leq 0$, it follows that $t_0 > 0$. Also since $\mu'(t) > -1$, it follows that $\mu(t) + t > \mu(0) = \epsilon$ for all $t \geq 0$. Therefore,

$$\epsilon = \mu(t_0) - \frac{t_0}{2} + \frac{3\|\mathbf{x}\|^2}{2} = \mu(t_0) + t_0 - \frac{3}{2}(t_0 - \|\mathbf{x}\|^2) > \epsilon - \frac{3}{2}(t_0 - \|\mathbf{x}\|^2)$$

which proves that $t_0 > \|\mathbf{x}\|^2$.

Putting $r(\mathbf{x}) = \sqrt{(t_0 - \|\mathbf{x}\|^2)/2}$, the first part of this claim is done.

Now consider the function $\mu(t) - \frac{t}{2} + \frac{3\|\mathbf{x}\|^2}{2} - \epsilon$, and apply the implicit function theorem. It follows that the function $\mathbf{x} \mapsto t_0$ is smooth. Therefore, $\mathbf{x} \mapsto r(\mathbf{x})$ is smooth.

[At this stage, it is clear that $F^{-1}(-\infty, -\epsilon]$ is obtained from $M_{-\epsilon}$ by a combinatorial attachment of a k-handle. The following details are necessary to see that the attachment is indeed smoothly performed.]

Claim V: The space $F^{-1}(-\infty, -\epsilon]$ is obtained from $M_{-\epsilon}$ by attaching a k-handle along the $(k-1)$-sphere $S := \mathbb{S}^{k-1}_{\sqrt{\epsilon}} \times 0$.

The k-handle is attached to $M_{-\epsilon}$ along the $(k-1)$-sphere

$$S = \{(\mathbf{x}, 0) \ : \ \|\mathbf{x}\|^2 = \epsilon\}.$$

The attaching map is the diffeomorphism $h : T \to T'$:

$$h(\mathbf{x}, \mathbf{y}) = \sqrt{2\epsilon}\left(\frac{(3/2 - \|\mathbf{x}\|^2)^{1/2}}{\|\mathbf{x}\|}\mathbf{x}, \mathbf{y}\right)$$

where $T' = \{(\mathbf{x}, \mathbf{y}) \in M_{-\epsilon} \ : \ \|\mathbf{x}\|^2 < 3\epsilon\}$ is a tubular neighborhood of S in $M_{-\epsilon}$. Therefore, $M_{-\epsilon} \cup H^k$ is the quotient of the disjoint union

$$\left[(M_{-\epsilon} \setminus S) \coprod (\mathbb{D}^n \setminus \mathbb{S}^{k-1})\right] / \sim$$

where

$$(\mathbf{x}, \mathbf{y}) \sim h\alpha(\mathbf{x}, \mathbf{y}) = \sqrt{2\epsilon}\left(\mathbf{x}, \left(\frac{\|\mathbf{x}\|^2 - 1/2}{1 - \|\mathbf{x}\|^2}\right)^{1/2} \mathbf{y}\right).$$

[See (8.12) for the definition of α.]

In order to define a diffeomorphism $g : M_{-\epsilon} \cup H^k$ with $F^{-1}(-\infty, -\epsilon]$, we consider two smooth maps $\sigma : M_{-\epsilon} \setminus S \to F^{-1}(-\infty, -\epsilon]$ and $\tau : \mathbb{D}^n \setminus \mathbb{S}^{k-1} \to F^{-1}(-\infty, -\epsilon]$ given by

$$\sigma(\mathbf{x}, \mathbf{y}) = \left(\mathbf{x}, \frac{r(\mathbf{x})}{\|\mathbf{x}\|}\mathbf{y}\right); \quad \tau(\mathbf{x}, \mathbf{y}) = \left(\sqrt{2\epsilon}\mathbf{x}, \frac{r(\sqrt{2\epsilon}\mathbf{x})}{\sqrt{1 - \|\mathbf{x}\|^2}}\mathbf{y}\right).$$

It is readily checked that both σ and τ are embeddings. Since $\sigma \circ h \circ \alpha = \tau$, it follows that these two patch up to define a smooth embedding g of $M_{-\epsilon} \cup H^k$. It remains to show that $g(M_{-\epsilon} \cup H^k) = F^{-1}(-\infty, -\epsilon]$.

For each fixed $\mathbf{x}$ such that $\|\mathbf{x}\| < 1$, let $L_{\mathbf{x}} = \{(\mathbf{x}, \mathbf{y}) \in \mathbb{D}^n\}$. Then $L_{\mathbf{x}}$ is a $(n-k)$-disc of radius $(1 - \|\mathbf{x}\|^2)^{1/2}$. Under τ, this is mapped onto a $(n-k)$-disc in $L_{\sqrt{2\epsilon}\mathbf{x}}$ of radius $r(\sqrt{2\epsilon}\mathbf{x})$. Now by Claim IV, it follows that $\tau(L_x) = F^{-1}(-\infty, -\epsilon] \cap L_{\sqrt{2\epsilon}\mathbf{x}}$. Therefore,

$$\tau(\mathbb{D}^n \setminus \mathbb{S}^{k-1}) = F^{-1}(-\infty, -\epsilon] \cap \{(\mathbf{x}, \mathbf{y}) \in \mathbb{D}^n \ : \ \|\mathbf{x}\|^2 < 2\epsilon\}. \tag{8.14}$$

Now suppose that $\|\mathbf{x}\|^2 > \epsilon$ and put $S_{\mathbf{x}} = M_{-\epsilon} \cap L_{\mathbf{x}}$. Then $S_{\mathbf{x}}$ is a $(n-k)$-disc of radius $(\|\mathbf{x}\|^2 - \epsilon)^{1/2}$. Its image under σ is a disc of radius $r(\mathbf{x})$. As seen in Claim IV, it follows that $\sigma(S_{\mathbf{x}}) = L_{\mathbf{x}} \cap F^{-1}(-\infty, -\epsilon]$. Since $M_{-\epsilon} \setminus S \subset \{(\mathbf{x}, \mathbf{y}) \ : \ \|\mathbf{x}\|^2 > \epsilon\}$, it follows that

$$\sigma(M_{-\epsilon} \setminus S) = \{(\mathbf{x}, \mathbf{y}) \in F^{-1}(-\infty, -\epsilon] \ : \ \|\mathbf{x}\|^2 > \epsilon\}. \tag{8.15}$$

Together with (8.14), this implies

$$g(M_{-\epsilon} \cup H^k) = F^{-1}(-\infty, -\epsilon].$$

That completes the proof of the theorem. ♠

Theorem 8.4.4 Handle Decomposition: *Every closed manifold M has a handle presentation.*

Proof: Begin with a Morse function $f : M \to [a, b]$, which sends distinct critical points to distinct values $a = c_0 < c_1 \cdots < c_{k-1} = b$. Put $M_i = f^{-1}[a, c_i + \epsilon]$, where ϵ is half the minimum of all $c_i - c_{i-1}$. If $p \in M$ is such that $f(p) = a$ then p must be a local minimum and hence the index at p is 0. Therefore, by the Morse Lemma M_0 is a disc. The rest of the claim follows from the previous lemma. ♠

Remark 8.4.3 Combined with Corollary 8.3.4, we conclude that every compact manifold has a monotone handle presentation and hence is the homotopy type of a finite CW-complex.

Note that starting with a Morse function f on a closed manifold M, we get a handle presentation $\mathcal{P}$ for M, which, in turn, yields a homotopy equivalence of M with a CW-complex X. We also note that for each $0 \leq k \leq n$, the number ν_k of critical points of index k of f is equal to the number h_k of k-handles in $\mathcal{P}$, which, in turn, is equal to the number of k-cells in X. Therefore, we get

$$e(f) = \chi(\mathcal{P}) = \chi(X). \tag{8.16}$$

On the other hand, we also get the vector field $\sigma = \operatorname{grad} f$ for which we have seen $I(M) = e(f)$ (see (8.6)). Combining these two we have:

$$I(M) = e(f) = \chi(X). \tag{8.17}$$

Thus we have the following celebrated result:

Theorem 8.4.5 Poincaré-Hopf: *The Euler characteristic of a closed manifold M is equal to the index of any vector field on M with finitely many zeros.*

Remark 8.4.4
(i) In particular, it follows that the combinatorial quantity $\chi(X)$ is a diffeomorphism invariant. Indeed, it is an easy consequence of homology theory that $\chi(X)$ is a homology (and hence a homotopy) invariant of the underlying topological space of any finite CW-complex.
(ii) Classically, Euler-characteristic is defined for any finite simplicial complex as the alternate sum of number of *face numbers*. You will learn this in a first course in Algebraic Topology.
(iii) It may be noted that the underlying topological space X (geometric realization) of a finite simplicial complex K can be treated as a finite CW complex in a natural way. Then our definition of $\chi(X)$ for this CW-complex coincides with the classical definition $\chi(K)$. Thus, when X denotes the CW complex obtained via a Morse function on a manifold M, we can define $\chi(X)$ to be the Euler characteristic of the manifold M itself and denote it by $\chi(M)$.

We shall end this section with a result that is useful for computing the Euler characteristic and that will be used in the classification of surfaces in the next section.

Theorem 8.4.6 *Let M_1, M_2 be any two n-dimensional manifolds. Then*

$$\chi(M_1 \# M_2) = \chi(M_1) + \chi(M_2) + a(n) \tag{8.18}$$

where $a(n) = 0$ if n is odd and $= -2$ if n is even.

Proof: Choose Morse functions $f_i : M_i \to \mathbb{R}, i = 1, 2$ such that $p_1 \in M_1$ is the only maximum for f_1 and $p_2 \in M_2$ is the only minimum for f_2. Choose coordinate neighborhoods (U_i, ϕ_i), around p_i, $i = 1, 2$ such that

$$f_1(\phi_1^{-1}(x)) = f(p_1) - \sum_i x_i^2; \;\; f_2(\phi_2^{-1}(x)) = f_2(p_2) + \sum_i x^2, \;\; 0 \leq \|x\| \leq 1.$$

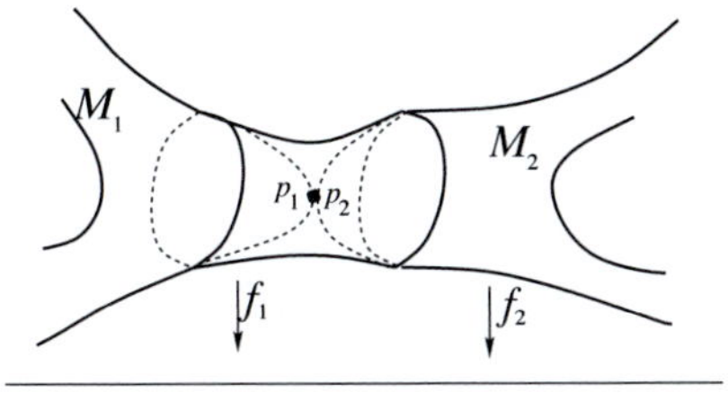

Figure 47 Connected sum of two Morse functions.

Now take $c = f_1(p_1) - 1 + f_2(p_2)$ and $g_2 : M_2 \to \mathbb{R}$ to be $g_2(x) = f_2(x) + c$. Verify that $f_1 \coprod g_2 : M_1 \setminus \{p_1\} \coprod M_2 \setminus \{p_2\} \to \mathbb{R}$ factors down to define a smooth function h on $M_1 \# M_2$ (see Definition 8.3.1). The critical points of h are precisely those of f_1 and f_2 minus the set $\{p_1, p_2\}$. Indeed, in a neighborhood of any of these critical points, h coincides with either f_1 or with $f_2 + c$. The index at p_1 for f_1 is 1 whereas the index at p_2 for f_2 is $(-1)^n$. This proves

$$e(h) = e(f_1) + e(f_2) + a(n).$$

Now use (8.17) to derive formula (8.18). ♠

Remark 8.4.5

1. In summary, we have related half-a-dozen different versions of "Euler characteristic": (i) as the alternate sum $\sum_k (-1)^k \nu_k$ of the number of k-cells of a CW-structure (or the number of k-faces of a simplicial structure),
   (ii) as $\sum_k (-1)^k h_k$ of the numbers h_k of k-handles of a handle presentation,
   (iii) as self-intersection number of a manifold,
   (iv) as the index of a vector field with finitely many zeros,
   (v) as the Lefschetz number of a self-map homotopic to identity, and
   (vi) as the alternate sum of number of critical points of a Morse function.
   We have also discussed a simple version of Gauss-Bonnet Theorem, which relates the Euler characteristic of a hypersurface in $\mathbb{R}^{2n+1}$ to its curvature. These ideas have motivated a large number of deep results which go under the generic name index theorems, such as Atiyah-Singer Index Theorem. What we have seen so far is just the tip of an ice-berg.

2. For a compact manifold M with nonempty boundary, one can find a Morse function $f : M \to [a, b]$ such that $f^{-1}(a) = \partial M$. From this, information similar to the above can be extracted.

Example 8.4.1 Consider the unit sphere

$$\mathbb{S}^{2n+1} = \{(z_0, \ldots, z_n) \in \mathbb{C}^{n+1} \; : \; \sum_j |z_j|^2 = 1\}.$$

The group of unit complex numbers $\mathbb{S}^1$ acts on $\mathbb{S}^{2n+1}$ as scalars and the quotient space is called the *complex projective space* $\mathbb{C}P^n$. Writing an element of $\mathbb{C}P^n$ in the form

$$z = [z_0 : z_0 : \cdots : z_n],$$

we have $\sum |z_j|^2 = 1$, and $[z_0 : z_0 : \cdots : z_n] = [wz_0 : wz_0 : \cdots : wz_n]$, for all $w \in \mathbb{S}^1$. Let $\lambda_1 < \lambda_2 < \cdots < \lambda_n$ be any real numbers and consider $f(z) = \sum_j \lambda_j |z_j|^2$. Let us verify that f is indeed a nice Morse function on $\mathbb{C}P^n$.

Put $U_j = \{z \in \mathbb{C}P^n \; : \; z_j \neq 0\}$. Let B^{2n} denote the open unit disc in $\mathbb{C}^n$ and consider the map $\phi_0 : B^{2n} \to \mathbb{C}P^n$ given by

$$(z_1, \ldots, z_n) \mapsto \left[\left(1 - \sum_j |z_j|^2 \right)^{1/2} : z_1 : \cdots : z_n \right].$$

This map defines a diffeomorphism of B^{2n} onto U_0 and so defines a chart around the point $[1 : 0 : \cdots : 0]$. Likewise, we can consider diffeomorphisms of B^{2n} onto U_j for other j's and get an atlas for $\mathbb{C}P^n$.

Clearly $f \circ \phi_0(z) = \lambda_0 + \sum_j (\lambda_j - \lambda_0)|z_j|^2$. Hence, $[1 : 0 : \cdots : 0]$ is the only critical point of f in U_0; it is nondegenerate and is of index 0. Likewise, we see that in U_j, f takes the form

$$c_j + \sum_{i \neq j} (\lambda_j - \lambda_i)|z_i|^2.$$

Hence, $[0 : 0 : \cdots : 1 : \cdots : 0]$ is the only critical point of f in U_j, is nondegenerate and is of index $2j$. Observe that the value of the function at this point is λ_j. Thus, we obtain

$$\mathbb{C}P^n \simeq e^0 \cup e^2 \cup \cdots \cup e^{2n}$$

as a finite CW-complex, the cells being attached in the increasing order of their dimension.

Exercise 8.4

1. Prove that M_r is a strong deformation retract of M_s.

2. Read a proof of Poincaré duality for smooth manifolds based on Theorem 8.4.4 from [B] or from [K].

8.5 Classification of Compact Surfaces

In any classification problem, we must first of all have a sufficiently large list of the type of objects that we want to classify. The second step is to get rid of redundant elements from this list. The final task is to prove that our list is "complete".

Naturally, the boundary surfaces of solid objects that we are familiar with are the first set of examples of 2-dimensional manifolds that we come across– the surface of a football, of a bicycle tube, of a teacup, of a Swiss cheese, and so on. Surprisingly, as we shall see soon, these are all the surfaces that could be there, had we made just one restriction, viz., orientability. That is, all smooth orientable, (closed) surfaces occur as the boundary of some smooth solid objects in $\mathbb{R}^3$.

Beginning with a connected solid and assuming that it is somewhat symmetric in shape, let us choose our coordinates so that the solid is situated symmetrically about the XY-plane. We then see that the boundary of this solid intersects the XY-plane in a number of Jordan curves, arranged in a certain pattern. What is this pattern? We may further assume that these curves are all circles. It is important to notice that there is one large circle and all other circles are interior to it. The smaller circles themselves are external to each other. We can easily write down equations for these circles. Let $p_i = 0, \; i = 1, 2, \ldots, g$ be the equations for smaller circles and $p_0 = 0$ be the equation for the large one. Then the curve C defined by $p_0 \cdots p_g = 0$ is the union of all these circles.

Now observe that the closed and bounded region A defined by these circles is a disc with g holes in it. We can then think of the portion of the surface above the XY-plane being the graph $z = f(x, y)$ of a suitable smooth function f, which takes positive values in the interior of A and which vanishes on $\partial A = C$. By symmetry, we can say that the lower portion is given by the graph of $z = -f(x, y)$. These two can be combined into a single equation $z^2 = f(x, y)^2$.

Where do we look for such a function f? Luckily, we need not search for it much long: we can simply take the function f so that $f^2 = -p_0 p_1 \cdots p_g$.

More specifically, we may take

$$(x^2 + (y - g - 1)^2 - (g + 1)^2) \prod_{k=1}^{g} (x^2 + (y - k)^2 - 1/9) + z^2 = 0 \tag{8.19}$$

to represent our surface. (Use the Pre-image Theorem to see that this actually represents a smooth surface.)

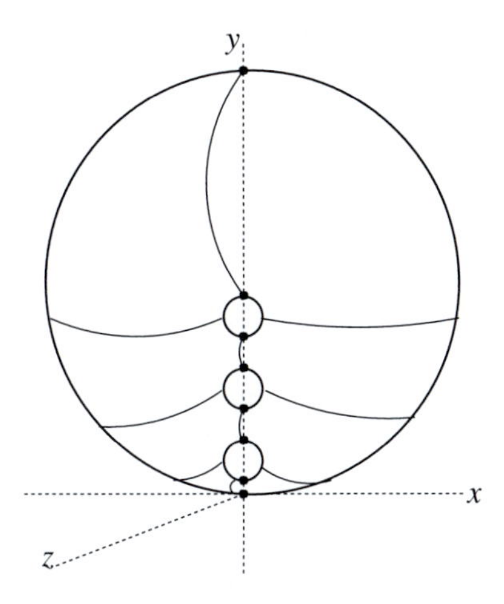

Figure 48 A surface of genus 3.

We can think of these surfaces as obtained by taking two copies of a disc with some holes and identifying each boundary circle of the first copy with one boundary of the second copy in a one-to-one fashion. In order to get an orientable surface, we need to be a bit careful in choosing the identifying homeomorphisms. What happens if we are careless? Well, the result would still be a surface, but probably not orientable. So, this is then a good way of producing nonorientable surfaces. Indeed, it turns out that all connected, nonorientable closed surfaces can also be obtained in this way. It seems that we have found our list. What to do next?

An interesting fact about surfaces given by (8.19) is that the projection to the y-axis is a Morse function on it with critical points at $(0, 0, 0), (0, 2(g+1), 0)$, and $(0, k \pm 1/3, 0), k = 1, 2, \ldots, g$. The indices at the first and the second one are $0, 2$ respectively and all others are of index 1. It turns out that this information is enough to distinguish homologically, and hence diffeomorphically, one surface from the other.

However, in our classification scheme, we shall take a route through an entirely different geometric description of the surfaces.

Starting with a teacup, suppose we just break its handle. The boundary of the resulting object is then seen to be diffeomorphic to the surface of a ball. Reversing this procedure, we would like to think of the surface of the teacup as being obtained from the surface of the ball by "attaching a handle". We can then repeat this process to obtain other surfaces. This single idea turns out to be sufficient to classify all compact orientable surfaces. Of course, we have to take care of a few technical details, for which, we have done enough preparation in the previous chapters. The key role is played by the fact that attaching handles in dimension 2 can more or less be completely described by another operation, viz., taking "connected sum".

In what follows, we are speaking about diffeomorphism classes of manifolds, even though we may simply refer to them as manifolds.

A Family $\mathcal{M}$ of Model Surfaces

We generate a family $\mathcal{M}$ of compact surfaces by the following simple procedure:
(i) We put the sphere $\mathbb{S}^2$, the torus $\mathbb{T} := \mathbb{S}^1 \times \mathbb{S}^1$ and the projective plane $\mathbb{P} := \mathbb{P}^2$ in $\mathcal{M}$.
(ii) Given any two members X_1 and X_2 in $\mathcal{M}$, we take $X_1 \# X_2$ to be also a member of $\mathcal{M}$. It should be noted that for this operation, we do not insist that X_1 and X_2 are distinct.
(iii) Given a member X of $\mathcal{M}$, we allow the surface obtained by deleting the interior of finitely many disjoint discs in X to be a member of $\mathcal{M}$. Also, the reverse operation of "capping off" any number of boundary components is allowed.
(iv) Disjoint union of finitely many copies of finitely many members of $\mathcal{M}$ is again a member of $\mathcal{M}$.

Since any compact surface is the disjoint union of its connected components, it is enough to consider only those that are connected. Also, observe that the two operations (ii) and (iii) commute with each other. For, discs that we want to remove can be, first of all, isotoped away from the scene of action of taking the connected sums and vice versa. Therefore, the operation of taking connected sum needs to be performed only on pairs of members of $\mathcal{M}$ that are closed and connected manifolds. Let us denote the orientable surface obtained by taking the connected sum of g copies of $\mathbb{T}$ by $\mathbb{T}_g$ ($g \geq 0$, with the convention: $\mathbb{T}_0 = \mathbb{S}^2$) and the (nonorientable surface obtained by taking the connected sum of k copies of $\mathbb{P}^2$ by $\mathbb{P}_k, k \geq 1$. Next, let us denote the surface obtained by making h holes in $\mathbb{T}_g$ (respectively, in $\mathbb{P}_k$) by $\mathbb{T}_{g,h}$ (resp. by $\mathbb{P}_{k,h}$.)

The classification theorem for compact surfaces states:

Theorem 8.5.1 *Every compact connected surface is diffeomorphic to precisely one member of the following list:*
(a) $\mathbb{T}_{g,h}$ $g, h \geq 0$;
(b) $\mathbb{P}_{k,h}$, $k \geq 1, h \geq 0$.

The proof of this will be broken up into proving the following three propositions:

Proposition 8.5.1 Every compact surface is diffeomorphic to a member of $\mathcal{M}$.

Proposition 8.5.2 Every connected member of $\mathcal{M}$ is diffeomorphic to either a $\mathbb{T}_{g,h}$, $g, h \geq 0$, or to $\mathbb{P}_{k,h}, k \geq 1, h \geq 0$.

Proposition 8.5.3 Distinct members of the family

$$\{\mathbb{T}_{g,h} \ : \ g, h \geq 0\} \cup \{\mathbb{P}_{k,h}, \ : \ k \geq 1, h \geq 0\}$$

are nondiffeomorphic.

We shall take up the proof of Proposition 8.5.3 first. Here the Euler characteristic plays a crucial role.

Lemma 8.5.1 $\chi(\mathbb{T}_g) = 2 - 2g$; $\chi(\mathbb{P}_k) = 2 - k$.

Proof: We already know that $\chi(\mathbb{T}_0) = \chi(\mathbb{S}^2) = 2$ (see Example 7.8.2). On the torus $\mathbb{T}_1 = \mathbb{S}^1 \times \mathbb{S}^1$, we have plenty of nowhere vanishing vector fields, e.g., $(x, y) \mapsto (\imath x, \imath y)$. Therefore, $\chi(\mathbb{T}_1) = 0$. Now appeal to (8.18) (with $\alpha(n) = -2$ to get,

$$\chi(\mathbb{T}_g) = \chi(\mathbb{T}_{g-1}) + \chi(\mathbb{T}_1) - 2.$$

Therefore, by induction

$$\chi(\mathbb{T}_g) = 2 - 2(g-1) - 2 = 2 - 2g.$$

Likewise, starting with the fact that $\chi(\mathbb{P}) = 1$ (see Example 7.8.4), it follows that $\chi(\mathbb{P}_k) = 2 - k$.

Proof of Proposition 8.5.3: Since orientability is a diffeomorphism invariant, it follows that no member of $\{\mathbb{T}_{g,h}\ g, h \geq 0\}$ is diffeomorphic to any member of $\{\mathbb{P}_{k,h}, k \geq 1, h \geq 0\}$. Since the number of connected components of the boundary is a diffeomorphism invariant, it follows that different values of h will mean different diffeomorphism types in each of these sets. Now if $\mathbb{T}_{g,h}$ is diffeomorphic to $\mathbb{T}_{g',h}$ by capping of all the boundary components on either side we conclude that $\mathbb{T}_g$ is diffeomorphic to $\mathbb{T}_{g'}$. Since the Euler characteristic is a diffeomorphism invariant, we have

$$2 - 2g = \chi(\mathbb{T}_g) = \chi(\mathbb{T}_{g'}) = 2 - 2g'$$

and hence $g = g'$. Similarly, if $\mathbb{P}_k$ is diffeomorphic to $\mathbb{P}_{k'}$ then $2-k = \chi(\mathbb{P}_k) = \chi(\mathbb{P}_{k'}) = 2-k'$ and hence $k = k'$. ♠

We now move toward the proof of Proposition 8.5.1.

Lemma 8.5.2 Let W' be obtained by attaching a 0-handle or a 2-handle to a member W of $\mathcal{M}$. Then $W' \in \mathcal{M}$.

Proof: Suppose we are attaching a 0-handle. This is the same as taking the disjoint union with a 2-disc. By (i) and (iii) it follows that a 2-disc is in $\mathcal{M}$. Therefore, by (iv) it follows $W' = W \coprod \mathbb{D}^2 \in \mathcal{M}$, in this case.

Similarly, attaching a 2-handle is the same as capping off a boundary component and so, $W' \in \mathcal{M}$. ♠

We now consider the case of attaching a 1-handle. The following lemma is self-evident in view of the isotopy theorems of Section 6.3.

Lemma 8.5.3 Let W be the result of attaching a 1-handle to a surface M.
(i) If the attaching sphere belongs to two different components of M or
(ii) M is nonorientable.
Then the diffeomorphism type of W is uniquely determined by M and the isotopy class of the attaching sphere.
Otherwise there are precisely two isotopy classes of attaching maps which determine two distinct possibilities for the diffeomorphism class for W. In the first case, the attaching diffeomorphisms are both orientation preserving (or reversing); this case will be called *orientation preserving attaching maps.* In the second case, one attaching diffeomorphism is orientation preserving and the other is orientation reversing; this case will be called the case of *orientation reversing attaching maps.*

Proposition 8.5.4 Let W be a manifold obtained by attaching a 1-handle to a disjoint union of two copies of the 2-disc in such a way that the two points of the attaching 0-sphere belong to the two different components. Then W is a disc.

Proof: With the specific condition on the attaching sphere mentioned, we know that the resulting surface is unique. So, it is enough to show that a 2-disc can be obtained in this fashion.

This is the same as finding a Morse function $f : \mathbb{D}^2 \to \mathbb{R}$ with three critical points of indices $0, 0, 1$ respectively. For then we know that the domain of such a function can be obtained by attaching a 1-handle to the disjoint union of two 2-discs. Of course then there are two possibilities for the attaching sphere. The first one is as described above. The second possibility actually gives a surface which is disconnected and hence ruled out.

Finally, we just appeal to the Exercise 8.2.7, where we have constructed such a Morse function. ♠

Example 8.5.1 We consider the parameterization of the torus $\mathbb{T}$:

$$(\theta, \psi) \mapsto ((2 + \cos\theta)\cos\psi, (2 + \cos\theta)\sin\psi, \sin\theta),$$

where $-\pi/2 \le \theta \le 3\pi/2$ and $-\pi/2 \le \psi \le 3\pi/2$. The coordinate projection onto the x-axis defines a Morse function f with precisely four critical points $(-3, 0, 0), (-1, 0, 0), (1, 0, 0)$, and $(3, 0, 0)$ of indices $0, 1, 1$, and 2, respectively. Let us denote T_r the subspace of T consisting of points with x-coordinate $\le r$. It follows that $T_{-2} \subset T_0 \subset T_2 \subset T_3 = \mathbb{T}$ is a monotone handle presentation of $\mathbb{T}$.

Going back to the parameterizing space $[-\pi/2, 3\pi/2] \times [-\pi/2, 3\pi/2]$, we can consider the function $f(\theta, \psi) = (2 + \cos\theta)\cos\psi$. This has four critical points precisely at

$$(\theta, \psi) = (0, \pi), (\pi, \pi), (\pi, 0), (0, 0).$$

The identification on the boundary $(\theta, -\pi/2) \sim (\theta, 3\pi/2), (-\pi/2, \psi) \sim (3\pi/2, -\psi + 2\pi)$ produces the Klein bottle $\mathbf{K}$. The function f respects this relation and hence defines a smooth map $\tilde{f} : \mathbf{K} \to \mathbb{R}$. Since the quotient map is a local diffeomorphism, it follows that $\tilde{f}$ is a Morse function with the critical behaviour similar to that of f.

We shall first discuss the case of the Klein bottle along with the Morse function $\tilde{f}$. Let us denote $\tilde{f}^{-1}(-\infty, r] := K_r$. It then follows that

$$K_{-2} \subset K_0 \subset K_2 \subset K_3 = \mathbf{K}$$

is a monotone handle presentation of $\mathbf{K}$. Clearly K_{-2} is a 2-disc. Observe that K_0 is the image $[-\pi/2, 3\pi/2] \times [\pi/2, 3\pi/2]$ and hence is the Möbius band. Since K_0 is obtained by attaching a 1-handle to the disc K_{-2}, there were only two possibilities for K_0 corresponding to the two cases whether the 1-handle is oriented or not. Since the orientable handle case would have produced an orientable manifold and since the Möbius band is nonorientable, we conclude that the result of attaching a nonorientable 1-handle to a 2-disc is precisely the Möbius band.

It now follows that the diffeomorphism type of K_2 is unique. Since attaching a 2-handle to it, or which is the same as capping off the boundary component, produces the Klein bottle K_3, K_2 is nothing but the surface obtained by removing a disc from $K_3 = \mathbf{K}$.

We now consider the torus $\mathbb{T}$ and the Morse function f. As before, let $T_r = f^{-1}(-\infty, r]$. We then have a monotone handle presentation

$$T_{-2} \subset T_0 \subset T_2 \subset T_3 = \mathbb{T}.$$

As usual, T_0 is a disc and T_2 is clearly a cylinder. It follows that attaching an orientable 1-handle to a 2-disc produces the cylinder.

Now suppose we are attaching a 1-handle to a cylinder. As seen before, there are two distinct cases to be considered: (i) orientable handle and (ii) nonorientable handle.

Since T_2 is an orientable surface obtained by attaching a 1-handle to the cylinder, it follows that this is the unique surface corresponding to case (i). What is the surface corresponding to (ii)?

By Theorem 8.3.4, we can first perform the two operations of attaching handles in the reverse order. We already know that attaching a nonoriented 1-handle to a 2-disc gives the Möbius band $\mathbb{M}$. Once a manifold is nonorientable, we also know that the orientability of the 1-handle does not matter any more. So the diffeomorphism type of the surface obtained by attaching another 1-handle to $\mathbb{M}$ is unique, which is nothing but K_2 as seen before.

Let us summarize the results that we have seen in the above discussion.

Proposition 8.5.5 Let M be a surface obtained by attaching a 1-handle to a cylinder or a Möbius band. Then the surface $\hat{M}$ obtained by capping off M is diffeomorphic to either (i) a 2-sphere, (ii) a projective space, (iii) a torus, or (iv) a Klein bottle. Equivalently, M is either (i) $\mathbb{T}_{0,1}$, (ii) $\mathbb{P}_{1,1}$, (iii) $\mathbb{T}_{1,1}$ or (iv) $\mathbb{P}_{2,1}$ respectively.

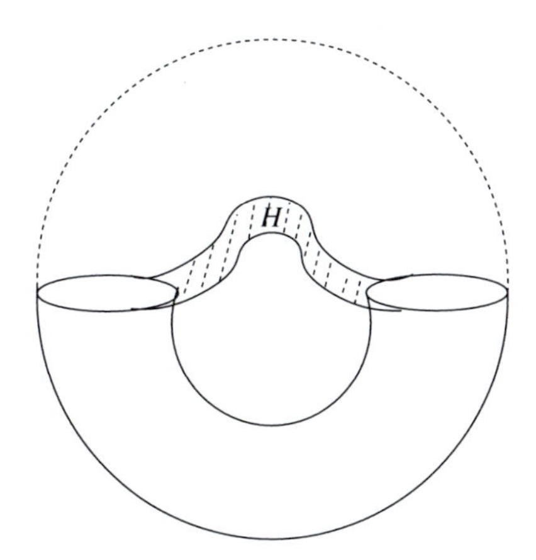

Figure 49 Attaching a 1-handle to a cylinder.

Proof: The first two cases correspond to the case when we start with a cylinder and attach the 1-handle along a 0-sphere contained in the same boundary component of the cylinder, which can be converted into attaching a 1-handle to a disc, by first capping off the other boundary component of the cylinder. The rest of the cases are as in the above example. ♠

We are now ready to discuss attaching 1-handle to a general surface.

Lemma 8.5.4 Let W be a surface with a connected boundary $\partial W = C$. Let M be a smooth surface obtained by attaching a 1-handle to W along C. Then $\hat{M}$ is diffeomorphic to either $\hat{W}$ or $\hat{W}\#\mathbb{P}^2$.

Proof: Choose a collar neighborhood $C \times [0,1]$ of C in W. Then $W' = W \setminus C \times [0,1)$ is diffeomorphic to W itself (ambient isotopy of collar neighborhoods). Now $W = \hat{W}'\#D$ where the 2-disc D is obtained from the collar $C \times [0,1]$ by capping of at the boundary component $C \times \{1\}$. Note that the operation of attaching a 1-handle is performed on the collar $C \times [0,1]$ along the boundary component $C \times \{0\}$ away from the scene of operation of connected sum. Therefore, we may interchange the order of these two operations and think of M as obtained by taking the connected sum of $\hat{W}'$ with the result Z of attaching a 1-handle to a disc. We know that Z is either a cylinder or a Möbius band.

To obtain $\hat{M}$ we have to further cap off Z. Clearly $\hat{Z}$ is either $\mathbb{S}^2$ or $\mathbb{P}^2$. The conclusion follows. ♠

Lemma 8.5.5 Let W be a connected surface with precisely two boundary components C_1, C_2. Let M be a smooth surface obtained by attaching 1-handle to W where the attaching 0-sphere consists of points one from each C_i. Then M is diffeomorphic to either $\hat{W}\#\mathbb{T}_{1,2}$ or $\hat{W}\#\mathbb{P}_{2,1}$.

Proof: Since W is connected, $\hat{W}$ is connected. By the Disc Theorem, the two discs D_1, D_2 in $\hat{W}$ which we have used for capping of C_1, C_2, may be assumed to be inside a larger disc D. This produces as in the previous lemma a diffeomorphism of W with $W'\#\mathbb{T}_{0,2}$, where $\hat{W}'$ itself is diffeomorphic to $\hat{W}$. Now the operation of attaching the handle is carried out on $\mathbb{T}_{0,2}$ part and we have already seen that this produces either $\mathbb{T}_{1,1}$ or $\mathbb{P}_{2,1}$. ♠

Lemma 8.5.6 Let W_1 and W_2 be two connected surfaces with connected boundary components C_1 and C_2 respectively. Let M be the result of attaching a 1-handle to $W_1 \cup W_2$ where the new handle intersects both C_1 and C_2. Then $\hat{M}$ is diffeomorphic to $\hat{W}_1\#\hat{W}_2$.

Proof: By taking collar neighborhoods of C_1, C_2, etc., as before we conclude that $\hat{M}$ is diffeomorphic to $\hat{W}_1\#\hat{W}_2\#\hat{W}_3$, where W_3 is the result of attaching a handle to a disjoint union of two discs D_1, D_2 wherein both the components $\partial D_1, \partial D_2$ are involved in the attaching process. We know that this produces a disc again. Therefore, $\hat{W}_3$ is $\mathbb{S}^2$ and the conclusion follows. ♠

Lemma 8.5.7 Let W be a member of $\mathcal{M}$ and W' be obtained by attaching a 1-handle to W. Then W' is also in $\mathcal{M}$.

Proof: There are three distinct possibilities.
(a) The attaching 0-sphere is contained in a single boundary component of W. This is the case described in Lemma 8.5.4 and we conclude that $W \in \mathcal{M}$.
(b) The attaching sphere intersects two boundary components that are both in the same component of W. This case is taken care by Lemma 8.5.5 and we are done.
(c) Finally, the attaching sphere intersects two boundary components, that belong to different components of W. This is described in Lemma 8.5.6 and we are through again. ♠

Proof of Proposition 8.5.1 : Given a compact surface M, to show that it is a member of $\mathcal{M}$, we can first cap-off all its boundary components and assume that it is a closed manifold. Then by Theorem 8.4.4, M has a handle decomposition,

$$M_0 \subset M_1 \subset \cdots \subset M_k = M.$$

Since M_0 is a disc, $M_0 \in \mathcal{M}$. Also, each M_i is obtained by attaching k-handle $(k = 0, 1, 2)$ to M_{i-1}. By Lemmas 8.5.2 and 8.5.7 each M_i is in $\mathcal{M}$ and we are done. ♠

Now toward the proof of Proposition 8.5.2, we first prove:

Theorem 8.5.2 *Let W be a connected nonorientable closed surface. Then $W\#\mathbb{T}$ is diffeomorphic to $W\#\mathbf{K}$, where $\mathbb{T}, \mathbf{K}$ denote the torus and the Klein bottle, respectively.*

Proof: In the Gluing Lemma 5.3.1, let $M = (-1/2, 1) \times \mathbb{S}^1, U = (-1/2, 0) \times \mathbb{S}^1, V = (1/2, 1) \times \mathbb{S}^1$. Take $\phi, \psi : U \to V$ given by

$$\phi(t, v) = (1 + t, v); \quad \psi(t, v) = (1 + t, \bar{v}).$$

Then we know that the quotient spaces M_ϕ, M_ψ are respectively the torus and the Klein bottle (see Example 8.5.1). Consider the connected sum $W\#M$. We may assume that the connected sum is being performed via a disc in M which is disjoint from $\bar{U}$ and $\bar{V}$. It then follows that $W\#\mathbb{T}$ (respectively, $W\#\mathbf{K}$) is diffeomorphic to the quotient space $(W\#M)_\phi$

(respectively $(W\#M)_\psi$). By (ii) of Lemma 5.3.1, if we find a diffeomorphism $\alpha : W\#M \to W\#M$ such that $\phi \circ \alpha = \alpha \circ \psi$ on U, then it follows that $(W\#M)_\phi$ is diffeomorphic to $(W\#M)_\psi$.

So, in order to produce a diffeomorphism α as required, we consider the manifold X obtained by filling up the two holes in $W\#M$ at the two boundary components $\{-1/2\} \times \mathbb{S}^1$ and $\{1\} \times \mathbb{S}^1$. Let the two discs be denoted by D_1, D_2 and let $U' = D_1 \cup U, V' = D_2 \cup V$. Then $\bar{U'}$ and $\bar{V'}$ are themselves disjoint embedded discs in X. Let $f, g : \mathbb{D}^2 \to X$ be embeddings such that $f(\mathbb{D}^2) = \bar{U'}$ and $g(\mathbb{D}^2) = \bar{V'}$. Note that the diffeomorphism $\psi \circ \phi^{-1} : V \to V$ extends to a diffeomorphism $\lambda : \bar{V'} \to \bar{V'}$. Put $f' = f$ and $g' = g \circ \lambda : B^2 \to V'$. Then by the Disc Theorem, we get a diffeomorphism $\alpha : X \to X$ such that $\alpha \circ f = f$ and $\alpha \circ g = g'$. It follows that α restricts to a diffeomorphism of $W\#M \to W\#M$ such that $\alpha|_U = Id_U$ and on U

$$
\begin{aligned}
\alpha \circ \phi &= (\alpha \circ g) \circ g^{-1} \circ \phi = (g' \circ g^{-1}) \circ \phi \\
&= (\psi \circ \phi^{-1}) \circ \phi = \psi = \psi \circ \alpha
\end{aligned}
$$

as required. This completes the proof. ♠

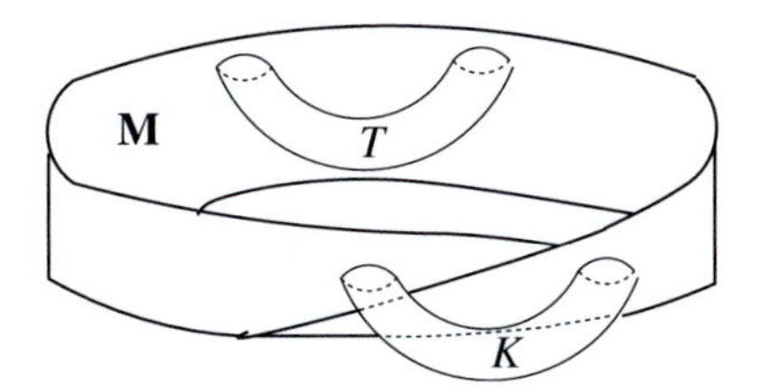

Figure 50 An oriented handle T can be moved onto K, an unoriented one.

Remark 8.5.1 In the figure above, we have taken $W = \mathbb{P}^2$ but have drawn only a Möbius band (which is all that one can draw on a piece of paper). One can either attach an oriented 1-handle as shown by T or an unoriented 1-handle as shown by K. The result is the same since the Möbius band has no sides. One can actually "move" the handle K onto the handle T.

Proof of Proposition 8.5.2: By capping off boundary components, it is enough to consider only closed manifolds in $\mathcal{M}$. As we have seen these objects are obtained by taking connected sums of copies of $\mathbb{T}$ and $\mathbb{P}$. If only copies of $\mathbb{T}$ are involved or only copies of $\mathbb{P}$ are involved, we are within this list. It remains to identify $\mathbb{T}_g\#\mathbb{P}_k$ the connected sum of g copies of $\mathbb{T}$ and k copies of $\mathbb{P}$ $g, k > 0$. But then by the above theorem along with the fact $\mathbf{K} = \mathbb{P}_2$, we get

$$\mathbb{P}_k\#\mathbb{T}_g = (\mathbb{P}_k\#\mathbb{T})\#\mathbb{T}_{g-1} = (\mathbb{P}_k\#\mathbf{K})\#\mathbb{T}_{g-1} = \mathbb{P}_{k+2}\#\mathbb{T}_{g-1} = \cdots = \mathbb{P}_{k+2g}.$$

This completes the proof of the proposition 8.5.2. ♠

Remark 8.5.2 To summarize, given any connected compact surface without boundary, by the Morse theory, it has a handle decomposition and hence is a connected sum of finitely many copies of the torus and the projective space. If it is orientable then there cannot be any projective space involved and hence it is diffeomorphic to $\mathbb{T}_g$ for some $g \geq 0$. Otherwise, it has to be diffeomorphic $\mathbb{P}_k$, for some $k \geq 1$. This completes the proof of the classification of smooth compact surfaces.

Remark 8.5.3

1. The unique number g in $\mathbb{T}_g$ is called the *genus* of the surface. The Euler characteristic of this surface is equal to $2 - 2g$ which determines g. In fact, using the technique of Mayer-Vietoris sequence, one can easily find all the homology groups of this surface. It turns out that the first homology group is a free abelian group of rank $2g$ and the 0th and the 2nd one are isomorphic to $\mathbb{Z}$. The number k in $\mathbb{P}_k$ is also significant. It is equal to $2 - \chi(\mathbb{P}_k)$. The first Betti number (the rank of the first homology group) of $\mathbb{P}_k$ is equal to $k - 1$ and the 2nd homology group vanishes.

2. It turns out that the homeomorphic classification of all compact surfaces coincides with the diffeomorphism classification. In some sense, topological classification should be easier since we have to deal with homeomorphisms rather than diffeomorphisms at each stage. However, while dealing with topological manifolds, we do not have the tools of differential topology either, such as Morse functions. So, after all, the problem seems to be not so easy. One could first try to prove that every topological surface carries a smooth structure. Though this happen to be a fact, (at least as a consequence of the classification) solving this problem independently seems to be not easy at all.

3. A well-known approach to topological classification of surfaces is as follows: One first proves that every compact 2-dimensional topological manifold (surface) is triangulable. (This classical result is attributed to Tibor Rado.) Using the triangulation one then proves that every connected surface is the quotient of a regular polygon with even number of sides, by certain edge identifications on the boundary. One then gets rid off all superfluous identifications and bring them to a number of "standard form" (what are called "canonical polygons") each of which happens to produce a topologically distinct (Euler characteristic again) surface. This approach also helps to actually write down generators and relations for the fundamental group of the surface.

4. Slightly on a different track, by "the smoothing theory" developed by several authors, it follows that every triangulated surface admits a "unique" smooth structure. Combined with the above cited result of T. Rado, it follows that every topological surface has a smooth structure. Our classification theorem then shows that the smooth structure itself is unique of the topological surface. Of course, there is also the combinatorial equivalence of any two triangulations on a given surface. None of these results are elementary. The reader is urged to go through the "Foreword" to this book once again at this stage.

Exercise 8.5

1. Show that the tubular neighborhood of any embedded circle in a surface is diffeomorphic to either $\mathbb{S}^1 \times (-1, 1)$ or an open Möbius band.

2. Show that every nonorientable surface contains a Möbius band.

Chapter 9

Lie Groups and Lie Algebras: The Basics

In this last chapter, we introduce the reader to some basic results in Lie groups. On the one hand, it serves the purpose of supplying a large number of interesting examples of smooth manifolds and on the other, it gives the reader an opportunity to see some basic tools of differential topology being employed fruitfully. In the first section, we quickly recall some results from matrix theory and in the second some results from topological groups. Mastering everything in these sections is not all that necessary to go ahead. In Section 9.3, we introduce the Lie groups and in Section 9.4, the Lie algebras. In section 9.5 and 9.6, the fundamental inter-relation between the Lie group and its Lie algebra is discussed. The next two sections exploit this relation to get information on the structure of Lie groups such as subgroups, normal subgroups, conjugation action and so on. The last two sections deal with the fundamental problem of existence of Lie subgroups and we take this opportunity to introduce yet another differential topological/geometrical notion–*foliation.* The treatment in this chapter is somewhat different from other chapters, in the sense that we assume more maturity and indulgence on the part of the reader. As a result, exercises are spread out all over the chapter, instead of at the end of each section.

9.1 Review of Some Matrix Theory

This section is a quick review of some linear algebra. We hope that you are familiar with most of the stuff here and will be able to figure out a lot of routine checking which we are going to leave as exercises for you.

The Quaternions

Consider the group of quaternions with eight elements: $Q = \{\pm 1, \pm\mathbf{i}, \pm\mathbf{j}, \pm\mathbf{k}\}$ with the following group law:

$$\mathbf{i}^2 = \mathbf{j}^2 = \mathbf{k}^2 = -1;\ \mathbf{ij} = \mathbf{k};\ \mathbf{jk} = \mathbf{i};\ \mathbf{ki} = \mathbf{j}.$$

The "associated group algebra $\mathbb{H}$ over $\mathbb{R}$" consisting of linear combinations

$$\mathbf{a} = a_0 + a_1\mathbf{i} + a_2\mathbf{j} + a_3\mathbf{k},\quad a_r \in \mathbb{R}$$

is a four dimensional vector space of over $\mathbb{R}$ with a multiplication that extends the group law of Q linearly.

We define the conjugation in $\mathbb{H}$ by

$$\bar{\mathbf{a}} = a_0 - a_1\mathbf{i} - a_2\mathbf{j} - a_3\mathbf{k}$$

Clearly $\mathbf{a}\bar{\mathbf{a}} = \sum_r a_r^2$ and hence we define

$$|\mathbf{a}| := \sqrt{\sum_r a_r^2}$$

It follows that $\mathbf{a} = 0$ iff $|\mathbf{a}| = 0$.

We say $\mathbf{a} \in \mathbb{H}$ is purely imaginary if its real part $a_0 = 0$.

The following two results are easy to prove.

Theorem 9.1.1 *$\mathbb{H}$ is a skew field and it contains the field of complex numbers.*

Theorem 9.1.2 *The set of elements of modulus 1 in $\mathbb{H}$ form a group denoted by $\mathbb{S}^3$.*

Remark 9.1.1 Note that $\mathbb{H}$ is a 2-dimensional vector space over $\mathbb{C}$ in two different ways: the left- and right-module structures do not coincide. On the other hand, there is no such problem with $\mathbb{H}$ as a 4-dimensional vector space over $\mathbb{R}$.

Exercise 9.1.1

1. Express $\mathbf{a} \in \mathbb{H}$ as $z_1 + z_2\mathbf{j}$, where $z_i \in \mathbb{C}$.
   (a) Verify that

$$\mathbf{ai} = \mathbf{i}(z_1 - z_2\mathbf{j}); \;\; \mathbf{aj} = \mathbf{j}(\bar{z_1} + \bar{z_2}\mathbf{j}); \;\; \mathbf{ak} = \mathbf{k}(\bar{z_1} - \bar{z_2}\mathbf{j}) \tag{9.1}$$

   (b) Use the cosine rule and the fact that $|z_1 + z_2\mathbf{j}|^2 = |z_1|^2 + |z_2|^2$ to show that

$$|\mathbf{ab}| = |\mathbf{a}|\,|\mathbf{b}|, \quad \mathbf{a}, \mathbf{b} \in \mathbb{H}.$$

2. View $\mathbb{S}^3$ as the space of unit quaternions and $\mathbb{S}^2$ as the space of unit quaternions that are purely imaginary.
   (a) Show that $q \in \mathbb{S}^2 \implies pqp^{-1} \in \mathbb{S}^2$ for all $p \in \mathbb{S}^3$.
   (b) Show that for each $p \in \mathbb{S}^3$, the map $q \mapsto pqp^{-1}$ is the restriction of a rotation $\mathbb{R}^3 \to \mathbb{R}^3$.
   (c) Denote by A_p, the 3×3 matrix representing the rotation $q \mapsto pqp^{-1}$. Show that $p \mapsto A_p$ defines a continuous homomorphism $\psi : \mathbb{S}^3 \to S0(3)$. What is the kernel of this homomorphism?
   (d) Show that $p \mapsto p\mathbf{i}p^{-1}$ defines a smooth map $h : \mathbb{S}^3 \to \mathbb{S}^2$.
   (e) Show that the map h given above is a submersion.
   (f) Show that $(p, q) \mapsto pqp^{-1}$ defines a transitive action of $\mathbb{S}^3$ on $\mathbb{S}^2$. Determine the isotropy subgroup I_q of any element $q \in \mathbb{S}^2$.
   (g) For any element $q \in \mathbb{S}^2$, and $p \in I_q$, show that q is an eigenvector corresponding to the eigenvalue 1 of A_p.
   (h) Show that the axis of rotation A_p is the line spanned by a vector q where $p \in I_q$.
   (i) If $p = r + p'$ where p' is purely imaginary, then show that $p \in I_q$ where $q = p'/|p'|$.

From now onward $\mathbb{K}$ will denote $\mathbb{R}, \mathbb{C}$, or $\mathbb{H}$. Each of them is endowed with the topology of the underlying Euclidean space.

We shall reserve the notation c to denote the vector space dimension of $\mathbb{K}$ over $\mathbb{R}$. Thus, $c = 1, 2, 4$ according as $\mathbb{K} = \mathbb{R}, \mathbb{C}, \mathbb{H}$.

The General Linear Group

We consider finite dimensional vector spaces over $\mathbb{K}$. When $\mathbb{K} = \mathbb{H}$, we have to specify whether V is a left-vector space or a right-vector space. We have to make a choice here between two equally good things and our choice is to take left vector spaces V over $\mathbb{K}$ always so that we don't have to mention it again and again. In particular, $\mathbb{H}^n$ will denote the left-vector space of row vectors of size n with entries in $\mathbb{H}$.

The set of $n \times n$ matrices with entries over $\mathbb{K}$ will be denoted by $M(n, \mathbb{K})$. The space of all $\mathbb{K}$-linear endomorphisms on a $\mathbb{K}$-vector space V will be denoted by $\mathrm{End}\,(V)$. The right multiplication by a matrix $A \in M(n, \mathbb{K})$ on $\mathbb{K}^n$ gives an element of $\mathrm{End}\,(\mathbb{K}^n)$:

$$R_A : \mathbf{v} \mapsto \mathbf{v}A.$$

The map $A \mapsto R_A$ itself is a $\mathbb{K}$-linear isomorphism, by which we identify $M(n, \mathbb{K})$ with $\mathrm{End}\,(\mathbb{K}^n)$.

Thus, topologically, $\mathrm{End}\,(\mathbb{K}^n)$ is homeomorphic to $\mathbb{R}^{cn}$.

Remark 9.1.2 The left multiplication $\mathbf{v} \mapsto A\mathbf{v}$ is $\mathbb{K}$-linear only if $\mathbb{K}$ is commutative. This is the reason why we would like to deal with right-multiplication instead of the left multiplication, even though the latter is more conventional in linear algebra textbooks. Thus, caution is needed while dealing with the skew field $\mathbb{H}$.

Definition 9.1.1 We define the *general linear group* $GL(V)$ of V to be the group of all $\mathbb{K}$-linear automorphisms of V. The group of all invertible elements of $M(n, \mathbb{K})$ is denoted by $GL(n, \mathbb{K})$.

Exercise 9.1.2 There is an obvious multiplication rule on $M(n, \mathbb{K})$, that makes it a $\mathbb{K}$-algebra. Check that $R_A \circ R_B = R_{BA}$. Deduce that $A \in GL(n, \mathbb{K})$ iff $R_A \in GL(n, \mathbb{K})$. In the cases $\mathbb{K} = \mathbb{R}, \mathbb{C}$, we have,

$$GL(n, \mathbb{K}) = \{A \in M(n, \mathbb{K}) \ : \ \det A \neq 0\}.$$

Recall that $\mathbb{C}^n$ can be thought of as a $2n$-dimensional vector space over $\mathbb{R}$. Let us fix such a structure on $\mathbb{C}^n$, e.g.,

$$\mathcal{C}_n : (x_1 + y_1\imath, \ldots, x_n + y_n\imath) \mapsto (x_1, y_1, \ldots, x_n, y_n).$$

(Some authors may prefer to use another structure, viz.,

$$(x_1 + y_1\imath, \ldots, x_n + y_n\imath) \mapsto (x_1, x_2, \ldots, x_n, y_1, y_2, \ldots, y_n).$$

There will be corresponding changes throughout the following discussion that are not difficult to figure out.)

Given $A \in M(n, \mathbb{C})$, the $\mathbb{C}$-linear map $R_A : \mathbb{C}^n \to \mathbb{C}^n$ is also a $\mathbb{R}$-linear map. Let $\mathcal{C}_n(A)$ denote the unique $2n \times 2n$ real matrix that represents R_A. Thus, we have a commutative diagram

$$\begin{array}{ccc} \mathbb{C}^n & \xrightarrow{\mathcal{C}_n} & \mathbb{R}^{2n} \\ {\scriptstyle R_A}\downarrow & & \downarrow{\scriptstyle R_{\mathcal{C}_n(A)}} \\ \mathbb{C}^n & \xrightarrow{\mathcal{C}_n} & \mathbb{R}^{2n} \end{array}$$

Example 9.1.1 For $n = 1$, recall how complex numbers were embedded in $M(2, \mathbb{R})$, viz., $a + b\imath \mapsto \begin{pmatrix} a & b \\ -b & a \end{pmatrix}$. Check that this coincides with $\mathcal{C}_1$. Take $J_2 = \begin{pmatrix} 0 & 1 \\ -1 & 0 \end{pmatrix}$. Prove that $\{A \in M(2, \mathbb{R}) \ : \ AJ_2 = J_2A\}$ is precisely the image of $\mathcal{C}_1$.

Exercise 9.1.3 Put $J_{2n} = diag(J_2, \ldots, J_2)$ the block matrix in which each 2×2 matrix along the main diagonal are equal to J_2 and the rest of the entries are zero. Show that

$$\mathcal{C}_n(M(n, \mathbb{C})) = \{A \in M(2n, \mathbb{R}) \ : \ AJ_{2n} = J_{2n}A\}. \tag{9.2}$$

In a similar manner, taking $\{1, \mathbf{j}\}$ as a $\mathbb{C}$-basis for $\mathbb{H}$, the right multiplication by an element $z + w\mathbf{j} \in \mathbb{H}$ on $\mathbb{H}$ gets identified with

$$\begin{pmatrix} z & w \\ -\bar{w} & \bar{z} \end{pmatrix} \in M(2; \mathbb{C}).$$

This in turn induces the identification of $\mathbb{H}^n$ with $\mathbb{C}^{2n}$ via:

$$\mathcal{Q}_n : (z_1 + w_1\mathbf{j}, \ldots, z_n + w_n\mathbf{j}) \mapsto (z_1, w_1, \ldots, z_n, w_n)$$

and the corresponding identification of $M(n, \mathbb{H})$:

$$\mathcal{Q}_n(M(n, \mathbb{H})) = \{A \in M(2n, \mathbb{C}) \ : \ AJ_{2n} = J_{2n}\bar{A}\}. \tag{9.3}$$

Theorem 9.1.3 *$\mathcal{C}_n : GL(n, \mathbb{C}) \to GL(2n, \mathbb{R})$ and $\mathcal{Q}_n : GL(n, \mathbb{H}) \to GL(2n, \mathbb{C})$ define monomorphisms of groups.*

Definition 9.1.2 We define the *determinant function on* $M(n, \mathbb{H})$ by the formula

$$\det A = \det(\mathcal{Q}_n(A)). \tag{9.4}$$

Exercise 9.1.4

1. Show that $A \in GL(n, \mathbb{H})$ iff $\det A \neq 0$.
2. Show that $\det A \in \mathbb{R}$ for all $A \in M(n, \mathbb{H})$.
3. Show that $GL(n, \mathbb{K})$ is an open subspace of $M(n, \mathbb{K})$.
4. Show that $GL(n, \mathbb{K})$ is path connected for $\mathbb{K} = \mathbb{C}, \mathbb{H}$ and has two connected components for $\mathbb{K} = \mathbb{R}$. (Hint: Use induction and the Gauss Elimination Method).
5. Prove that $\det A > 0$ for all $A \in GL(n, \mathbb{H})$.

The Orthogonal Group

We fix the standard inner product $\langle\, , \rangle_{\mathbb{K}}$ on $\mathbb{K}^n$ defined as follows:

$$\langle \mathbf{a}, \mathbf{b} \rangle_{\mathbb{K}} = \sum_r a_r \bar{b}_r$$

where $\mathbf{a} := (a_1, \ldots, a_n) \in \mathbb{K}^n$ etc.. We define the norm on $\mathbb{K}^n$ by

$$\|\mathbf{a}\|_{\mathbb{K}} = \sqrt{\langle \mathbf{a}, \mathbf{a} \rangle_{\mathbb{K}}}.$$

Note that for $\mathbf{a} \in \mathbb{C}^n$, $\|\mathbf{a}\|_{\mathbb{C}} = \|\mathcal{C}_n(\mathbf{a})\|_{\mathbb{R}}$. Similarly, for $\mathbf{b} \in \mathbb{H}^n$, $\|\mathbf{b}\|_{\mathbb{H}} = \|\mathcal{Q}_n(\mathbf{b})\|_{\mathbb{C}}$. Thus, the embeddings $\mathcal{C}_n$ and $\mathcal{Q}_n$ are norm preserving. For this reason, we may soon drop the subscript $\mathbb{K}$, from $\| - \|_{\mathbb{K}}$ unless we want to draw your attention to it.

Check that $\langle\, , \rangle_{\mathbb{K}}$ is sesqui-linear, conjugate symmetric, non degenerate and positive definite. (In case $\mathbb{K} = \mathbb{R}$, the conjugation is the identity map and hence it is bilinear and symmetric.) Orthogonality, orthonormal basis, etc., are taken with respect to this norm.

Theorem 9.1.4 *$\mathcal{C}_n$ preserves the inner product. In particular, $\{\mathbf{x}_1, \ldots, \mathbf{x}_n\} \subset \mathbb{C}^n$ is an orthonormal basis for $\mathbb{C}^n$ iff*

$$\{\mathcal{C}_n(\mathbf{x}_1), \mathcal{C}_n(\imath\mathbf{x}_1), \ldots, \mathcal{C}_n(\mathbf{x}_n), \mathcal{C}_n(\imath\mathbf{x}_n)\}$$

is an orthonormal basis for $\mathbb{R}^{2n}$.

Exercise 9.1.5

1. State and prove a result similar to Theorem 9.1.4, for $\mathcal{Q}_n$.

2. Prove Schwartz's inequality: $\|\langle \mathbf{x}, \mathbf{y}\rangle\| \le \|\mathbf{x}\|\,\|\mathbf{y}\|$.

For any $A \in M(n, \mathbb{K})$ we shall use the notation $A^* := \overline{A^T} = \bar{A}^T$. Observe that the identity

$$\langle \mathbf{x}A, \mathbf{y}\rangle = \langle \mathbf{x}, \mathbf{y}A^*\rangle, \quad \mathbf{x}, \mathbf{y} \in \mathbb{K}^n$$

defines A^*.

Theorem 9.1.5 *For $A \in M(n, \mathbb{K})$ the following conditions are equivalent:*
(i) $\langle \mathbf{x}A, \mathbf{y}A\rangle_{\mathbb{K}} = \langle \mathbf{x}, \mathbf{y}\rangle_{\mathbb{K}}$.
(ii) *R_A takes orthonormal sets to orthonormal sets.*
(iii) *The row-vectors of A form an orthonormal basis for $\mathbb{K}^n$.*
(iv) $AA^* = Id$.
(v) $A^*A = Id$.

Definition 9.1.3 $\mathcal{O}_n(\mathbb{K}) = \{A \in M_n(\mathbb{K}) \ : \ AA^* = I\}$ is called the *orthogonal group* of the standard inner product on $\mathbb{K}^n$. For $\mathbb{K} = \mathbb{R}, \mathbb{C}, \mathbb{H}$, it is also denoted by $O(n), U(n), Sp(n)$ respectively and called *the orthogonal group, the unitary group,* and *the symplectic group,* respectively.

Observe that $\mathcal{C}_n(A^*) = (\mathcal{C}_n(A))^*$ for all $A \in M(n, \mathbb{C})$. Similarly, $\mathcal{Q}_n(B^*) = (\mathcal{Q}_n(B))^*$ for all $B \in M(n, \mathbb{H})$. The following theorem is then an easy consequence.

Theorem 9.1.6 *For each $n \ge 1$, we have:*
(i) $\mathcal{C}_n(U(n)) = O(2n) \cap \mathcal{C}_n(GL(n, \mathbb{C}))$;
(ii) $\mathcal{Q}_n(Sp(n)) = U(2n) \cap \mathcal{Q}_n(GL(n, \mathbb{H}))$;
(iii) $\mathcal{C}_{2n} \circ \mathcal{Q}_n(Sp(n)) = O(4n) \cap \mathcal{C}_{2n} \circ \mathcal{Q}_n(GL(n, \mathbb{H}))$.

Corollary 9.1.1 $A \in \mathcal{O}_n(\mathbb{K})$ iff $\mathbb{R}_A$ is norm preserving.

Proof: For $\mathbb{K} = \mathbb{R}$, this is a consequence of the fact that the norm determines the inner product via the **polarization identity:**

$$\langle \mathbf{x}, \mathbf{y}\rangle = \frac{1}{2}(\|\mathbf{x} + \mathbf{y}\|^2 - \|\mathbf{x}\|^2 - \|\mathbf{y}\|^2).$$

For $\mathbb{K} = \mathbb{C}, \mathbb{H}$ we can now use the above theorem and the fact that $\mathcal{C}_n$ and $\mathcal{Q}_n$ are norm preserving. ♠

Exercise 9.1.6

1. Show that $|\det A| = 1$ for $A \in \mathcal{O}_n(\mathbb{K})$.

2. Show that $\mathcal{O}_n(\mathbb{K})$ is compact.

3. Show that the Gram-Schmidt process is valid in a finite dimensional vector space over $\mathbb{H}$.

4. Show that $Sp(n, \mathbb{C}) := \{A \in M(2n, \mathbb{C}) : A^t J_{2n} A = J_{2n}\}$ forms a subgroup of $GL(2n, \mathbb{C})$. This is called the *complex symplectic group of order n*.

5. Show that $Sp(n, \mathbb{C}) \cap U(2n) = Sp(n, \mathbb{C}) \cap \mathcal{Q}_n(M(n, \mathbb{H})) = \mathcal{Q}_n(Sp(n))$.

The Special Orthogonal Group

Definition 9.1.4 $SL(n; \mathbb{K}) = \{A \in M(n, \mathbb{K}) : \det A = 1\}$. This forms a subgroup of $GL(n, \mathbb{K})$. We define $SO(n) = SL(n, \mathbb{R}) \cap O(n)$; $SU(n) = SL(n, \mathbb{C}) \cap U(n)$. These are called the *special orthogonal* and the *special unitary group,* respectively.

Remark 9.1.3 We do not need to define the special symplectic groups. Why?

Example 9.1.2

1. $O(1) = \{\pm 1\} \approx \mathbb{Z}_2$ and $SO(1) = (1)$ is the trivial group. It is not difficult to see that

$$SO(2) = \left\{ \begin{pmatrix} \cos\theta & \sin\theta \\ -\sin\theta & \cos\theta \end{pmatrix} : \theta \in [0, 2\pi) \right\}$$

and

$$0(2) = \left\{ \begin{pmatrix} \cos\theta & \sin\theta \\ \pm\sin\theta & \mp\cos\theta \end{pmatrix} : \theta \in [0, 2\pi) \right\}$$

Thus, $SO(2)$ can be identified with the group of unit complex numbers under multiplication. Topologically this is just the circle. Similarly, $O(2)$ just looks like disjoint union of two circles. However, the emphasis is on the group structure rather than just the underlying topological space.

2. Again, it is easy to see that $U(1)$ is just the group of complex numbers of unit length and hence is isomorphic to $SO(2)$. Indeed, it is not hard to see that $\mathcal{C}_1$ defines this isomorphism. Clearly, $SU(1) = (1)$.

3. The group $Sp(1)$, is nothing but the group of unit quaternions that we have met. As a topological space, it is $\mathbb{S}^3$. Thus, we have now got three unit spheres, viz., $\mathbb{S}^0, \mathbb{S}^1, \mathbb{S}^3$ endowed with a group multiplication. A very deep result in topology says that spheres of other dimensions have no such group operations on them.

4. We also have $\mathcal{Q}_1 : Sp(1) \to SU(2)$ is an isomorphism. It is easily seen that $\mathcal{Q}_1(Sp(1)) \subset SU(2)$. The surjectivity of $\mathcal{Q}_1$ is the only thing that we need to prove. This we shall leave as an exercise.

Exercise 9.1.7

1. A function $f : \mathbb{R}^n \to \mathbb{R}^n$ is called a *rigid motion* (*isometry*) if

$$d(f(x), f(y)) = d(x, y) \ \forall \ x, y \in \mathbb{R}^n.$$

(Recall that the Euclidean distance is defined by $d(x, y) = \|x - y\|$.) Show that for $A \in O(n)$, $R_A : \mathbb{R}^n \to \mathbb{R}^n$ is an isometry.

2. Show that every isometry of $\mathbb{R}^n$ is continuous and injective. Can you also show that it is surjective and its inverse is an isometry?

3. Show that composite of two isometries is an isometry.

4. Show that $x \mapsto x + \mathbf{v}$ is an isometry for any fixed $\mathbf{v} \in \mathbb{R}^n$.

5. Let f be an isometry of $\mathbb{R}^n$. If $\mathbf{v}, \mathbf{u} \in \mathbb{R}^n$ are such that $f(\mathbf{v}) = \mathbf{v}$ and $f(\mathbf{u}) = \mathbf{u}$ then show that f fixes the entire line L passing through $\mathbf{v}$ and $\mathbf{u}$ and keeps invariant every hyperplane perpendicular to L.

6. Let $f : \mathbb{R}^n \to \mathbb{R}^n$ be an isometry such that $f(0) = 0$. If $f(e_i) = e_i, i = 1, 2, \ldots, n$, where e_i are the standard orthonormal basis for $\mathbb{R}^n$, then show that $f = Id$.

7. Given any isometry f of $\mathbb{R}^n$ show that there exists a vector $\mathbf{v} \in \mathbb{R}^n$ and $A \in O(n)$ such that $f(x) = xA + \mathbf{v}, x \in \mathbb{R}^n$. Thus, isometries of $\mathbb{R}^n$ are all "affine transformations". (In particular, this answers Exercise 2 completely.)

8. Show that the set of all $A \in GL(n+1, \mathbb{R})$, which keep the subspace $\mathbb{R}^n \times \{1\}$ invariant forms a subgroup isomorphic to the group of affine transformations of $\mathbb{R}^n$. Is it a closed subgroup?

9. Show that $GL(n, \mathbb{K})$ is isomorphic to a closed subgroup of $GL(n + 1, \mathbb{K})$.

10. Show that $A \in O(3)$ is an element of $SO(3)$ iff the rows of A form a right-handed orthonormal basis, i.e., $R_A(e_3) = R_A(e_1) \times R_A(e_2)$.

11. Show that every element of $A \in O(n)$ is the product $A = BC$ where $B \in SO(n)$ and R_C is either Id or the reflection in the hyperplane $x_n = 0$, i.e.,

$$R_C(x_1, \ldots, x_{n-1}, x_n) = (x_1, \ldots, x_{n-1}, -x_n).$$

12. Show that eigenvalues of any $A \in O(n)$ are of unit length.

13. Show that one of the eigenvalues of $A \in SO(3)$ is equal to 1.

14. Given $A \in SO(3)$, fix $\mathbf{v} \in \mathbb{S}^2$ such that $\mathbf{v}A = \mathbf{v}$. Let P be the plane perpendicular to $\mathbf{v}$.
    (a) Show that $R_A(P) = P$.
    (b) Choose any vector $\mathbf{u} \in P$ of norm 1. Let $\mathbf{w} = \mathbf{v} \times \mathbf{u}$. Show that every element of P can be written as $(\cos\theta)\mathbf{u} + (\sin\theta)\mathbf{w}$, for some θ.
    (c) Show that there exists ϑ such that

$$R_A(\cos\theta\mathbf{u} + \sin\theta\mathbf{w}) = \cos(\theta + \vartheta)\mathbf{u} + \sin(\theta + \vartheta)\mathbf{w}.$$

    Thus, every element of $SO(3)$ is a rotation about some axis.

The Exponential Map

We can endow $M(n, \mathbb{K})$ with various norms. The Euclidean norm is our first choice. If we view $M(n, \mathbb{K})$ as the space of linear maps, then the so-called L^2-norm becomes quite handy. There are other norms such as row-sum norm, column-sum norm, maximum norm, etc. For the discussion that follows, you can use any one of them. But let us concentrate on the Euclidean norm.

Lemma 9.1.1 For any $x, y \in M(n, \mathbb{K})$, we have $\|xy\| \leq \|x\|\|y\|$.

Proof: Straightforward. ♠

Definition 9.1.5 By a *formal power series in one variable T with coefficients in $\mathbb{K}$* we mean a formal sum $\sum_{r=0}^{\infty} a_r T^r$, with $a_r \in \mathbb{K}$. In an obvious way, the set of all formal power series in T forms a module over $\mathbb{K}$ denoted by $\mathbb{K}[[T]]$. We define the *Cauchy product* of two power series $p = \sum_r a_r T^r, q = \sum_r b_r T^r$ to be another power series $s = \sum_r c_r T^r$ where $c_r = \sum_{l=0}^{r} a_l b_{r-l}$. (Note that except for being noncommutative, $\mathbb{H}[[T]]$ has all other arithmetic properties of $\mathbb{C}[[T]]$ or $\mathbb{R}[[T]]$.)

Theorem 9.1.7 *Suppose for some $k > 0$, the series $\sum_r |a_r| k^r$ is convergent. Then for all $A \in M(n, \mathbb{K})$ with $\|A\| < k$, the series $\sum_r a_r A^r$ is convergent.*

Proof: The convergence of the series is the same as the convergence of each of n^2 series $\sum_r a_r A^r_{ij}$, formed by the entries. Since $|A^r_{ij}| \leq \|A^r\| \leq \|A\|^r < k^r$, we are done. ♠

Definition 9.1.6 Taking $p(T) = exp(T) = 1 + T + \frac{T^2}{2!} + \cdots$, which is absolutely convergent for all $T \in \mathbb{R}$, we define the *exponential of* $A \in M(n, \mathbb{K})$ to be the value of the convergent sum

$$\exp(A) = Id + A + \frac{A}{2!} + \cdots .$$

Lemma 9.1.2 For any invertible $B \in M(n, \mathbb{K})$ we have $B \exp(A) B^{-1} = \exp(BAB^{-1})$.

Proof: Check this first on the partial sums. ♠

Lemma 9.1.3 If $AB = BA$, then $\exp(A + B) = \exp(A) \exp(B)$.

Proof: In this special case, the Cauchy product becomes commutative and hence binomial expansion holds for $(A + B)^n$. The rest of the proof is similar to the case when the matrix is replaced by a complex number. ♠

Corollary 9.1.2 For all $A \in M(n, \mathbb{K})$, $\exp(A) \in GL(n, \mathbb{K})$.

Proof: We have $\exp(A) \exp(-A) = \exp(A - A) = \exp(0) = Id$. ♠

Theorem 9.1.8 *The function* $\exp : M(n, \mathbb{K}) \to GL(n, \mathbb{K})$ *is smooth (indeed, real analytic). The derivative at 0 is the identity transformation.*

Proof: The analyticity follows since the n^2-entries are all given by convergent power series. To compute the derivative at 0, we fix a "vector" $A \in M(n, \mathbb{K})$ and take the directional derivative of exp in the direction of A : viz.,

$$D(\exp)_0(A) = \lim_{t \to 0} \frac{\exp(tA) - \exp(0)}{t} = A.$$

Corollary 9.1.3 The function exp defines a diffeomorphism of a neighborhood of 0 in $M(n, \mathbb{K})$ with a neighborhood $Id \in GL(n, \mathbb{K})$.

Proof: Appeal to Inverse Function Theorem 1.4.2.

Polar Decomposition

Lemma 9.1.4 Given any $A \in M(n, \mathbb{C})$, there exists $U \in U(n)$ such that UAU^{-1} is a lower triangular matrix.

Proof: If λ is a root of the characteristic polynomial of A there exists a unit vector $\mathbf{v}_1 \in \mathbb{C}^n$ such that $\mathbf{v}_1 A = \lambda \mathbf{v}_1$. The Gram-Schmidt process allows us to complete this vector to an orthonormal basis $\{\mathbf{v}_1, \ldots, \mathbf{v}_n\}$. Take U to be the matrix with these as row vectors. Then $\mathbf{e}_1 UAU^{-1} = \mathbf{v}_1 AU^{-1} = \lambda \mathbf{e}_1$. Hence, UAU^{-1} is of the form

$$\begin{pmatrix} \lambda, & 0 \\ \star & B \end{pmatrix}.$$

Now a simple induction completes the proof. ♠

Definition 9.1.7 We say A is *normal* if $AA^* = A^*A$. A square matrix A is called *symmetric, skew-symmetric, Hermitian, skew-Hermitian,* respectively, if $A = A^T$ $A = -A^T$, $A = A^*$, $A = -A^*$.

Corollary 9.1.4 Spectral Theorem If $A \in M(n, \mathbb{C})$ is normal, then there exists $U \in U(n)$ such that UAU^{-1} is a diagonal matrix.

Proof: If A is normal then so is UAU^{-1} for any $U \in U(n)$. On the other hand a lower triangular normal matrix is a diagonal matrix. ♠

Remark 9.1.4 In particular, if A is hermitian or symmetric or skew symmetric etc., then it is diagonalizable. The entries on the diagonal are necessarily the characteristic roots of the original matrix. Moreover, if A is real symmetric matrix, then all its eigenvalues are real with real eigenvectors and hence U can be chosen to be inside $O(n)$.

Definition 9.1.8 A Hermitian matrix A defines a sesqui-linear (Hermitian) form on $\mathbb{C}^n$:

$$(\mathbf{u}, \mathbf{v}) \mapsto \mathbf{u}A\mathbf{v}^*,$$

i.e., linear in the first slot $\mathbf{u}$ and conjugate linear in the second slot $\mathbf{v}$. Recall that A^* satisfies the property

$$\langle \mathbf{u}A, \mathbf{v}\rangle_{\mathbb{C}} = \langle \mathbf{u}, \mathbf{v}A^*\rangle_{\mathbb{C}}.$$

Therefore, for a Hermitian matrix A, $\langle \mathbf{u}A, \mathbf{u}\rangle_{\mathbb{C}}$ is always a real number. In particular, all eigenvalues of a Hermitian matrix are real. We say A is *positive semidefinite (positive definite)* if $\langle \mathbf{u}A, \mathbf{u}\rangle_{\mathbb{C}} \geq 0$ for all $\mathbf{u}$ (respectively, > 0 for all nonzero $\mathbf{u}$.)

Lemma 9.1.5 A Hermitian matrix is positive semidefinite (positive definite) iff all its eigenvalues are nonnegative (respectivey, positive).

Lemma 9.1.6 If A is Hermitian so is, Exp(A).

Theorem 9.1.9 *The space of all $n \times n$ complex Hermitian matrices is a real vector space of dimension n^2. Exponential map defines a diffeomorphism of this space onto the space of all positive definite Hermitian matrices.*

Proof: The first part is obvious. Given a positive definite Hermitian matrix B let U be a unitary matrix such that

$$UBU^{-1} = diag\,(\lambda_1, \ldots, \lambda_n).$$

Then we know that $\lambda_j > 0$ and hence we can put $\mu_j := log\,\lambda_j$. Put

$$A := U^{-1} diag\,(\mu_1, \ldots, \mu_n) U.$$

Then Exp$(A) = B$. This shows Exp is surjective. For the proof of the injectivity of Exp on the space of hermitian matrices, see the exercise 9.1.8 below, which is completely elementary linear algebra modulo the spectral theorem. ♠

Theorem 9.1.10 Polar Decomposition: *Every element A of $GL(n,\mathbb{C})$ can be written in a unique way as a product $A = UH$ where U is unitary and H is positive definite Hermitian. The decomposition defines a diffeomorphism of $\varphi : GL(n,\mathbb{C}) \to U(n) \times \mathbb{R}^{n^2}$. Furthermore, if A is real, then U is orthogonal and H is symmetric and hence φ restricts to a diffeomorphism $\varphi : GL(n,\mathbb{R}) \to O(n) \times \mathbb{R}^{n(n-1)/2}$.*

Proof: Consider the matrix $B = A^*A$, which is Hermitian. Since A is invertible, so is A^*. Therefore, for a nonzero vector $\mathbf{v}$, $\langle \mathbf{v}A^*A, \mathbf{v}\rangle = \langle \mathbf{v}A^*, \mathbf{v}A^*\rangle > 0$, which shows that B is positive definite. Choose $C \in GL(n,\mathbb{C})$ such that $CBC^{-1} = diag(\lambda_1, \ldots, \lambda_n)$ and put $H = C^{-1}diag(\sqrt{\lambda_1}, \ldots, \sqrt{\lambda_n})C$. Then H is clearly a positive definite Hermitian matrix and $H^2 = B$. Put $U = AH^{-1}$. Then $A = UH$ and we can directly verify that U is unitary.

Finally, if $A = U_1H_1$ where U_1 is unitary and H_1 is positive definite Hermitian, then $X = U^{-1}U_1 = HH_1^{-1}$ is both unitary and positive definite Hermitian. Therefore, X has all its eigenvalues of unit length, as well as, positive. Thus, all its eigenvalues are 1. Since it is diagonalizable also, it follows that $X = Id_n$.

The construction of H from A is indeed a smooth process, though this is not clear from the way we have done this. But we can simply write

$$H = \mathrm{Exp}\,(\frac{1}{2}log\, A^*A)$$

where log is the (local) inverse map of Exp in Corollary 9.1.3. It follows that the assignment $\varphi : A \mapsto (U, H)$ is smooth. The inverse map is clearly smooth. ♠

Exercise 9.1.8

1. Let $D = diag\,(\lambda_1 I_{k_1}, \ldots, \lambda_r I_{k_r})$. If $AD = DA$ then show that A is of the form $diag\,(A_{k_1}, \ldots, A_{k_r})$ where A_{k_j} are $k_j \times k_j$ matrices.

2. Let $D =$ diag $(\lambda_1, \ldots, \lambda_n), D' =$ diag $(\lambda'_1, \ldots, \lambda'_n)$ where $\lambda_i \geq \lambda_{i+1}, \lambda'_i \geq \lambda'_{i+1}$ for all i. If U is an invertible matrix such that $U\mathrm{Exp}\,D_1U^{-1} = \mathrm{Exp}\,D_2$ then show that $D_1 = D_2$ and $UD_1U^{-1} = D_1$.

3. Let A, B be hermitian matrices such that $\mathrm{Exp}\,A = \mathrm{Exp}\,B$. Then show that $A = B$.

9.2 Topological Groups

The interaction of the group multiplication and the Euclidean topology in the discussion of matrix groups in the previous section motivates the study of so-called topological groups in an abstract setup. This section is just a brief introduction to topological groups.

Definition 9.2.1 Let G be a group together with a topology on it such that the function $G \times G \to G$ given by $(x, y) \mapsto xy^{-1}$ is continuous, where $G \times G$ is given the product topology. We then call G a *topological group.* A subgroup together with the subspace topology will be called a topological subgroup. (However, when the context is clear, we may simply mention this as a subgroup.) Homomorphisms $g : G \to H$ between topological groups are always assumed to be continuous.

We shall consider topological groups that are Hausdorff (T_2) only. (Extra caution should be taken when you are studying topological groups that arise in algebraic geometry with Zariski topology–these are non-Hausdorff.)

Example 9.2.1

1. Any group together with the discrete topology is a topological group.

2. The real numbers, the complex numbers etc., along with the standard addition and standard topology are topological groups. The nonzero real (complex) numbers and the positive real numbers form topological groups under multiplication. The complex (real) numbers of unit length form a closed subgroup of the respective multiplicative topological groups.

3. The skew field of quaternions is a topological group under addition with the standard Euclidean topology.

4. Any finite dimensional vector space over $\mathbb{K}$ is also a topological group. Indeed, there is an obvious notion of topological vector spaces, and these are easy examples of topological groups.

5. $GL(n, \mathbb{K})$ is a topological group for any of the three (skew) fields $\mathbb{K}$. The orthogonal groups are all subgroups of $GL(n, \mathbb{K})$.

For any two subsets A, B of a group G, we set up the notation

$$AB = \{ab \; : \; a \in A, b \in B\}; \quad A^{-1} = \{a^{-1} \; : \; a \in A\}.$$

The following easily proved fundamental result is at the heart of various special topological properties of a topological group that we are going to see. Most of them are in the exercises that follow.

Lemma 9.2.1 Let U be an open subset of topolgical group G.
(a) If $e \in U$, then there exist neighborhoods V, W of e such that $VV^{-1} \subset U, WW \subset U$.
(b) Given any $x \in G$, there a neighborhood A of e such that $xAx^{-1} \subset U$.
(c) UX and XU are open subsets for any arbitrary subset X of G.

Exercise 9.2.1 Let G denote a topological group.

1. For each $g \in G$, the left (and right) multiplication $L_g : G \to G$ ($R_g : G \to G$) is a homeomorphism.

2. A group homomorphism $f : G \to G'$ of topological groups is continuous iff it is continuous at $e \in G$.

3. The commutator subgroup, the center, etc., are all topological subgroups. If H is a topological subgroup then its centralizer, the normalizer, etc., are topological subgroups.

4. Every open subgroup is a closed subgroup also.

5. Closure of a subgroup is a subgroup.

6. If a subgroup H is not closed then $\bar{H} \setminus H$ itself is dense in $\bar{H}$.

7. If a topological group is T_0, then it is T_1. This is equivalent to the condition that the intersection of all neighborhoods of e is equal to $\{e\}$.

8. There are plenty of interesting topological groups which are T_1 but not T_2. They especially occur in the theory algebraic groups. (For example you can take the co-finite topology on the multiplicative group $\mathbb{S}^1$ of unit complex numbers.) However, we will be interested mostly in topological groups which are T_2 (Hausdorff).

9. The connected component of G containing the identity element e is a closed, normal topological subgroup of G. If G is locally connected, then it is also an open subgroup.

10. The product of any family of topological groups is a topological group, with the product topology.

11. Let V be a neighborhood of $e \in G$. If G is connected, then V is a set of generators for G.

12. Let G be a connected topological group and H be a normal discrete subgroup. Then H is contained in the center. (Hint: If N is a neighborhood of $x \in H$ such that $N \cap H = \{x\}$, choose a neighborhood V of e such that $VxV^{-1} \subset N$. Now use the previous exercise.)

Definition 9.2.2 Let H be a closed subgroup of G. Then the set of left (or right) cosets is given the quotient topology and is called a *homogeneous space* and is denoted by G/H (or by $H\backslash G$.)

Remark 9.2.1 Observe that the group G acts on the left of G/H transitively. (This is the reason for the name "homogeneous".) Therefore, local topological properties of the space G/H are the same at any of its points.

Exercise 9.2.2 Let H be a closed subgroup of a topological group G.

1. Show that the quotient map $q : G \to G/H$ is an open mapping.

2. If H is a closed subgroup of G then the homogeneous space G/H is a Hausdorff space.

3. If G is compact then so are H and G/H. The converse is also true. (Hint: First prove the following version of the tube lemma, using compactness of H : Given an open set $W \supset H$, there exists a neighborhood U of e in G such that $UH \subset W$.)

4. If G is connected then G/H is connected. Conversely, if both H and G/H are connected then G is connected.

5. The special orthogonal group $SO(n)$ acts on $\mathbb{S}^{n-1}$ transitively. The isotropy subgroup (stabilizer) of $(0, \ldots, 0, 1)$ is equal to $SO(n-1)$. Therefore, there is a bijective continuous map $f : SO(n)/S0(n-1) \to \mathbb{S}^{n-1}$. Since $SO(n)$ is compact, so is $SO(n)/SO(n-1)$. Since $\mathbb{S}^{n-1}$ is Hausdorff, f is a homeomorphism. Similarly, it follows that $U(n)/U(n-1) \approx \mathbb{S}^{2n-1}$ and $Sp(n)/Sp(n-1) \approx S^{4n-1}$. Write full details for the above claims.

6. Use the results in the above exercise to prove that for $n \geq 1$, the groups $GL(n, \mathbb{C}), U(n), SL(n, \mathbb{R}), SO(n, \mathbb{R}), Sp(n)$ are all connected. Show that $GL(n, \mathbb{R}), O(n), n \geq 1$ have precisely two connected components.

Example 9.2.2 Consider the action of $x \in Sp(1)$ on $\mathbb{H}$ by conjugation: $C_x : y \mapsto x^{-1}yx$. Clearly the action is linear. Let T be the linear span of $\{\mathbf{i}, \mathbf{j}, \mathbf{k}\}$ in $\mathbb{H}$. Using Exercise 9.1.1.2a, show that C_x maps $\mathbf{i}, \mathbf{j}, \mathbf{k}$ into T and hence $C_x(T) = T$. Moreover, since $|\mathbf{ab}| = |\mathbf{a}||\mathbf{b}|$ for $\mathbf{a}, \mathbf{b} \in \mathbb{H}$, it follows that C_x is norm preserving. Therefore, C_x defines an element $\Theta(x) \in O(3)$. Check that $\Theta : Sp(1) \to O(3)$ is a homomorphism of groups. Clearly, it is a smooth map and hence is a homomorphism of topological groups. Since $Sp(1) = \mathbb{S}^3$ is connected, the image of Θ is contained in the connected component of $Id \in O(3)$. Therefore, we have $\Theta : Sp(1) \to S0(3)$.

What is the kernel of this homomorphism? To answer this, observe that $x \in Ker\,\Theta$

iff x commutes with $\mathbf{i}$, and $\mathbf{j}$. (why?) That $x = (a_0, a_1, a_2, a_3)$ commutes with $\mathbf{i}$ implies $a_2 = a_3 = 0$. Similarly, $a_1 = 0$. Therefore, $x = \pm 1$. Since $\{-1, 1\}$ is the center of $Sp(1)$, it follows that the group $Sp(1)/\{-1, 1\}$ is isomorphic to a subgroup of $SO(3)$.

Since the action of the subgroup $\{-1, 1\}$ on $Sp(1) = \mathbb{S}^3$ is nothing but the antipodal, we know that the quotient map $q : Sp(1) \to Sp(1)/\{-1, 1\}$ is the double cover $\mathbb{S}^3 \to \mathbb{P}^3$. Thus, we have shown that $\mathbb{P}^3$ is a closed subgroup of $S0(3)$. Since both the groups are of the same dimension $= 3$, it follows that $\mathbb{P}^3$ is also an open subgroup of $SO(3)$. But $SO(3)$ is connected. Therefore, $\mathbb{P}^3 = SO(3)$.

Exercise 9.2.3

1. An element $\sigma \in S0(3)$ is a rotation about x_i-axis iff $\sigma(x_i) = x_i$.
2. Show that elements of $S0(3)$ can be written as $\sigma_1 \sigma_2 \sigma_1'$ where σ_i, σ_i' are rotations about the x_i-axis, $i = 1, 2$.
3. Show that $\Theta(a_0 + a_1\mathbf{i})$ is a rotation about the x_1-axis. Similarly, obtain an element $x \in Sp(1)$ such that $\Theta(x)$ is a rotation about x_2-axis.
4. Deduce that $\Theta : Sp(1) \to S0(3)$ is surjective. (This gives an alternative proof of the last part of Example 9.2.2.)
5. Show that $\mathbb{P}^3$ is diffeomorphic to the space $UT(\mathbb{S}^2)$ of unit tangents to $\mathbb{S}^2$. [Hint: Consider the mapping $x \to (x\mathbf{i}x^{-1}, x\mathbf{j}x^{-1})$ from $\mathbb{S}^3$ to $UT(\mathbb{S}^2) \subset \mathbb{R}^3 \times \mathbb{R}^3$.]
6. Use the above exercise to construct an explicit embedding of $\mathbb{P}^3$ in $\mathbb{R}^5$.

In the remaining part of this section, we assume that the reader is familiar with rudiments of fundamental group and covering space theory. (See any basic algebraic topology book such as [Spa] or [Hat]). She may choose to skip this part until she acquires such knowledge and come back to it later. Only some results in the sections to follow will depend on it.

Theorem 9.2.1 *Let G be a locally connected and connected topological group, H be a locally connected closed subgroup and H_0 be the connected component of H containing e. Then the quotient map $\bar{q} : G/H_0 \to G/H$ is a covering projection.*

Proof: Observe that H_0 is a normal subgroup. Since H is locally connected, H_0 is an open subgroup of H. Therefore, we can find an open neighborhood V of e in G such that $V^{-1}V \cap H \subset H_0$. Let $q : G \to G/H$ and $q_0 : G \to G/H_0$ be the respective quotient maps. Both are open maps and $q = \bar{q} \circ q_0$. Hence, $\bar{q}$ is a surjective continuous open mapping.

Given any $g \in G$, we claim that the open neighborhood $p(gV)$ of $p(g)$ is evenly covered by $\bar{q}$. Let us choose a complete set of representatives R of the left-cosets of H_0 in H so that $H = \coprod_{h \in R} hH_0$, a disjoint union. It then follows that

$$(\bar{q})^{-1}(p(gV)) = \bigcup_{h \in R} q_0(gVh),$$

where each member on the right-hand side is open in G/H_0. First, we claim that any two of them are disjoint.

Suppose $q_0(gVh_1) \cap q_0(gVh_2) \neq \emptyset, h_i \in R, i = 1, 2$. This means $gv_1h_1 = gv_2h_2h_0$ for some $v_i \in V, i = 1, 2$ and $h_0 \in H_0$. Therefore, $v_2^{-1}v_1 = h_2h_0h_1^{-1} \in V^{-1}V \cap H \subset H_0$. Since H_0 is normal in H this means that $h_2h_1^{-1} = v_2^{-1}v_1h_1h_0^{-1}h_1^{-1} \in H_0$. That means $h_2H_0 = h_1H_0$ and hence $h_2 = h_1$ since they are both in R.

It remain to prove that $\bar{q} : q_0(gVh) \to p(gV)$ is a homeomorphism for each $h \in R$. We already know that it is a surjective, continuous, open mapping. So, we check that it is injective. If for some $h \in R$, $\bar{q}(gv_1hH_0) = \bar{q}(gv_2hH_0)$, i.e,, $gv_1H = gv_2H$, then $v_2^{-1}v_1 = h_0 \in H_0$. But then $gv_1hH_0 = gv_2h_0hH_0 = gv_2hH_0$, since H_0 is normal in H. This proves the injectivity of $\bar{q}$ restricted to each $q_0(gVh)$. ♠

Corollary 9.2.1 If G/H is simply connected, then H is connected.

Corollary 9.2.2 For any discrete normal subgroup H of a connected locally connected topological group G, the quotient map $G \to G/H$ is a covering group homomorphism.

Lemma 9.2.2 Let $f : X \to Y, g : Y \to Z$ be maps of locally connected, connected topological spaces. Suppose $g \circ f$ is a covering projection. Then f is a covering projection iff g is.

Proof: Exercise.

Theorem 9.2.2 *Every locally simply connected topological group admits a simply connected covering group.*

Remark 9.2.2 We shall not prove this. For our purpose, it suffices to remark that, in general, every locally simply connected space admits a simply connected covering. In particular, every topological manifold has this property. Further, if it is a topological group then we know how to make the covering into a covering group.

Theorem 9.2.3 *Let $H \subset G$ be locally simply connected and connected topological groups. Suppose H is a closed subgroup and G/H is simply connected. Then $\pi_1(G)$ is isomorphic to a quotient of $\pi_1(H)$.*

Proof: Consider the following commutative diagram,

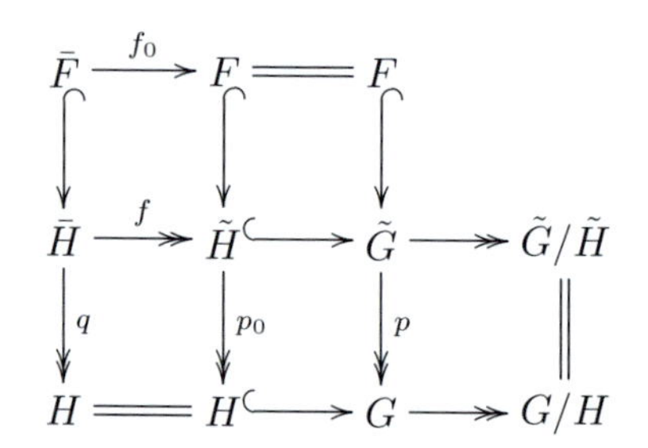

in which $p : \tilde{G} \to G$ is the (universal) simply connected covering, $\tilde{H} = p^{-1}(H) \subset \tilde{G}$, and $p_0 = p|_{\tilde{H}}$. It follows easily that $\tilde{G}/\tilde{H}$ is naturally homeomorphic to G/H. Let $q : \bar{H} \to H$ be the universal cover. By the lifting property, it follows that there is a unique map $f : \bar{H} \to \tilde{H}$ such that $p_0 \circ f = q$ and $f(\bar{e}) = \tilde{e}$, where $\bar{e}, \tilde{e}$, etc., denote the identity elements in the corresponding group. It follows that f is a group homomorphism. From Theorem 9.2.1, it follows that $\tilde{H}$ is connected. From Lemma 9.2.2, it follows that f is a covering projection. Therefore, f is surjective. Let $F, \bar{F}$ denote the kernel of p and q, respectively. Then $\pi_1(G)$ is isomorphic to F and $\pi_1(H)$ is isomorphic to $\bar{F}$. Also, it follows easily that f restricts to a surjective homomorphism $f_0 : \bar{F} \to F$. ♠

Exercise 9.2.4

1. Let X be a locally path connected and connected topological space, which admits a (simply connected) universal covering projection $p : \bar{X} \to X$. Let Y be a subspace of X and $\eta : Y \to X$ be the inclusion map. Then $\eta_\# : \pi_1 Y \to \pi_1 X$ is surjective iff $p^{-1}(Y)$ is path connected.

2. Use this to give an alternative proof of Theorem 9.2.3.

3. Let G be a topological group. Suppose $\hat{G}$ is a simply connected topological space and $p : \hat{G} \to G$ is a covering projection. Show that $\hat{G}$ can be made into a topological group in such a way that p is a homomorphism of topological groups. Show further that a simply connected covering of a topological group is unique up to isomorphisms.

4. Show that if G admits a simply connected covering then it is the quotient of a simply connected topological group by a central subgroup.

5. Let G, G' be topological groups. By a local homomorphism f from G to G' we mean a continuous mapping $f : V \to G'$ defined on a neighborhood V of e in G such that

$$x, y, xy \in V \Longrightarrow f(xy) = f(x)f(y).$$

(a) **Principle of Monodromy** Show that if G is simply connected and V is connected then there is a unique continuous homomorphism $\hat{f} : G \to G'$ such that $\hat{f}|_V = f$. (Hint: By Exercise 9.2.1.(11), V is a set of generators for G and hence if at all $\hat{f} : G \to G'$ is defined then we must have $f(g_1 \cdots g_k) = f(g_1) \cdots f(g_k)$ where $g_i \in V$. The problem is precisely, in verifying why $\hat{f}$ is well defined. This is where the simply connectedness has to be used. Assume that G is path connected and follow arguments similar to the one that you have used in analytic continuation.)
(b) If a connected topological group G' has a neighborhood of e' isomorphic to a neighborhood of e in a simply connected topological group G, then G' is the quotient of G by a normal discrete subgroup of G.

9.3 Lie Groups

Definition 9.3.1 By a *Lie group* we mean a topological group whose underlying topological space is a smooth manifold and whose group operations are smooth with respect to the smooth structure of this manifold. Mappings between two Lie groups will be always assumed to be smooth homomorphisms. A submanifold, which happens to be a topological subgroup, is automatically a Lie group on its own and is called a *Lie subgroup.*

The proto-type of Lie groups for us are matrix groups some of which you have seen in Section 9.1. We shall now deal with a "general method" which gives you many more examples.

Definition 9.3.2 We say $A \in M(n, \mathbb{K})$ is *nonexceptional* if $\det(I + A) \neq 0$. For such a matrix its *Cayley transformation* $A^{\#}$ is defined by the formula

$$A^{\#} := (I - A)(I + A)^{-1}. \tag{9.5}$$

Let us denote the set of all nonexceptional matrices in $M(n, \mathbb{K})$ by $\mathbb{K}_0^n$.

Lemma 9.3.1 $\mathbb{K}_0^n$ is an open set in $M(n, \mathbb{K})$ on which the map $A \mapsto A^{\#}$ is smooth and is an involution, i.e., $A \in \mathbb{K}_0^n \Longrightarrow A^{\#} \in \mathbb{K}_0^n$ and $(A^{\#})^{\#} = A$.

Theorem 9.3.1 *Let G be a matrix group in $M(n, \mathbb{K})$ and $G_0 = G \cap \mathbb{K}_0^n$. Then G is a Lie group if the image of G_0 under the Cayley map is an open subset of a real vector subspace of $M(n, \mathbb{K})$.*

Proof: Being a closed subgroup of $GL(n, \mathbb{K})$, G is a Hausdorff topological group. Therefore, the only thing that is needed for it to be a Lie group is that G is a smooth manifold. For this, it is enough to prove that $Id \in G$ has a neighborhood that is diffeomorphic to an open subset of a finite dimensional vector space over $\mathbb{R}$. This is precisely what is being provided by the hypothesis. ♠

Definition 9.3.3 Let $J \in GL(n, \mathbb{K})$ be fixed. Consider the set $O_J(n)$ of all matrices in $M(n, \mathbb{K})$ such that

$$A^t J A = J. \tag{9.6}$$

Such a matrix A is called a *J-orthogonal matrix*. Let $ss_J(n, \mathbb{K})$ denote the set of all B such that

$$B^t J = -JB. \tag{9.7}$$

Such a matrix B is called a J-skew-symmetric matrix.

Lemma 9.3.2 $A \in O_J(n, \mathbb{K}) \cap \mathbb{K}_0^n \iff A^{\#} \in ss_J(n; \mathbb{K}) \cap \mathbb{K}_0^n$.

Theorem 9.3.2 *$O_J(n; \mathbb{K})$ forms a closed Lie subgroup of $GL(n, \mathbb{K})$ and $ss_J(n, \mathbb{K})$ forms a real vector subspace of $M(n, \mathbb{K})$.*

Proof: That it is a topological closed subgroup is easy to see. Similarly, the fact that $ss_J(n, \mathbb{K})$ is a vector subspace of $M(n, \mathbb{K})$ is easily verified. The previous lemma together with Theorem 9.3.1 complete the proof. ♠

Taking $J = Id_n$, we get the orthogonal group $\mathcal{O}(n; \mathbb{K})$. In this case, since $ss_{Id}(n, \mathbb{K})$ is the space of skew-symmetric matrices, with real dimension equal $cn(n-1)/2$, it follows that $\mathcal{O}(n; \mathbb{K})$ is a Lie group of dimension $cn(n-1)/2$. For $\mathbb{K} = \mathbb{R}$, clearly $\mathcal{O}(n; \mathbb{R}) = O(n)$. But for $\mathbb{K} = \mathbb{C}$, or $\mathbb{H}$, this is not the same as $U(n)$ or $Sp(n)$.

Similarly, taking $J = J_{2n}$, we get various symplectic matrices as Lie groups.

Remark 9.3.1 It follows easily that composites of two Lie group homomorphisms is again a Lie group homomorphism. Also, the Cartesian product of finitely many Lie groups is again a Lie group. At this stage, it is not at all clear that if H is normal Lie subgroup of G, whether G/H is a Lie group. So, let us take up this problem now.

Theorem 9.3.3 *Let H be a closed Lie subgroup of a Lie group and $q : G \to G/H$ be the quotient map onto the space of left-cosets. Then G/H is a smooth manifold of dimension equal to the codimension of H in G such that the left action of G on G/H is smooth. Moreover, the quotient map is a smooth locally trivial fiber bundle. Further, if H is normal, then G/H is a Lie group and q is a surjective Lie-homomorphism with $Ker\, q = H$.*

Proof: The quotient space G/H is a Hausdorff space since H is closed [see Exercise 9.2.2.(2)]. Choose a local coordinate system $x_1, \ldots, x_n$, $(\dim G = n)$ around a neighborhood U of $e \in G$ for G such that $x_i(e) = 0,\ i = 1, \ldots, n$, and

$$H \cap U = \{g \in U \ : \ x_i(g) = 0, i = 1, \ldots, k\}$$

where $k = \operatorname{codim} H$. Put

$$T := \{g \in U \ : \ x_i(g) = 0, \text{ for } i = k+1, \ldots, n\}.$$

Consider $\mu : T \times H \to G$ given by $\mu(t, h) = th$. Then $\mu(t, e) = t, \mu(e, h) = h$. Therefore,

$$d\mu_e(\mathbf{u}, \mathbf{v}) = d\mu_e(\mathbf{u}, 0) + d\mu_e(0, \mathbf{v}) = \mathbf{u} + \mathbf{v}.$$

Hence, by the inverse function theorem, $\mu : T' \times H' \to U'$ is a diffeomorphism where T', H', U' are some open neighborhoods of e in T, H, G, respectively. Choose a neighborhood $S \subset T$ of e such that $S^{-1}S \subset U'$. Put $V = \mu(S \times H)$. We then claim that $q : S \to q(V)$ is a homeomorphism and $\mu : S \times H \to V$ is a diffeomorphism such that the following diagram

$$\begin{array}{ccc} S \times H & \xrightarrow{\mu} & V \\ {\scriptstyle \pi_1}\downarrow & & \downarrow{\scriptstyle q} \\ S & \xrightarrow{q} & q(V) \end{array}$$

is commutative, where, π_1 denotes the projection to the first factor.

Assuming this, we can then transfer the smooth structure of S to $q(V)$. Then for any point $g \in G$, the neighborhood $g(q(V)) = q(gV)$ gets the smooth structure from gS. If $g_1 S \cap g_2 S \neq \emptyset$, then the two transition functions on $S \cap g_1^{-1} g_2 S$ differ by the multiplication by $g_1^{-1} g_2$, which is a diffeomorphism. Hence, this would define a smooth structure on G/H as desired. The rest of the claims are all straightforward. Clearly, $q : S \to q(V)$ is a continuous bijection. Since $q : G \to G/H$ itself is an open mapping, it follows that $q : S \to q(V)$ is also open. The commutativity of the above diagram is obvious. It remains to see why $\mu : S \times H \to V$ is a diffeomorphism. That it is a surjective submersion is clear. Suppose $\mu(s_1, h_1) = \mu(s_2, h_2)$. This means $s_1 h_1 = s_2 h_2 \Longrightarrow s_1^{-1} s_2 = h_1 h_2^{-1} \in H \cap U' = H' \Longrightarrow s_2 = s_1 h$ for some $h \in H'$. Since $s_i \in T'$ and $\mu : T' \times H' \to U'$ is injective, it follows that $s_1 = s_2$ and $h_1 = h_2$. This shows that $\mu : S \times H \to V$ is injective. ♠

Corollary 9.3.1 Let H be a closed subgroup of a Lie group G. Then corresponding to the exact sequence $H \hookrightarrow G \to G/H$, there is long a exact sequence of homotopy groups and sets:

$$\begin{aligned} \cdots \to \pi_i(H) \to \pi_i(G) \to \pi_i(G/h) \to \quad & \pi_{i-1}(H) \to \\ \cdots \to \quad & \pi_1(G/H) \to \pi_0(H) \to \pi_0(G). \end{aligned} \tag{9.8}$$

Proof: This is an immediate consequence of the fact that the quotient map $G \to G/H$ is a locally trivial fiber bundle over a manifold G/H and hence is a fibration, i.e., it possesses the homotopy lifting property. (See [Spa].) ♠

Corollary 9.3.2 Let H be a closed subgroup of a connected Lie group G such that G/H is simply connected. Then H is connected.

Exercise 9.3.1

1. Let X, Y, Z be smooth manifolds. Suppose $\phi : X \times Z \to Y$ is a smooth map such that for each $z \in Z$ the map $\phi_z : X \to Y$ given by $\phi_z(x) = \phi(x, z)$ is a diffeomorphism. Then show that the map $\psi : Y \times Z \to X$ defined by $\psi(y, z) = \phi_z^{-1}(y)$ is a smooth map. (Hint: Consider the track of ϕ and apply the inverse function theorem.)

2. Let G be a smooth manifold together with a group operation such that the multiplication $G \times G \to G$ given by $(g, h) \mapsto gh$ is smooth. Then G is a Lie group. (Note that similar statement is false for arbitrary topological groups.)

3. Let $\mu : G \times G \to G$ the multiplication in a Lie Group G. Compute the derivative of μ at (e, e). Show that $D(\eta)_e = -Id$, where $\eta : G \to G$ is the inverse map $g \to g^{-1}$.

4. Let $f, g : X \to G$, where X is a manifold and G is a Lie group. Put $h(x) = f(x)g(x), x \in X$. Assume that $f(x_0) = g(x_0) = e$. Show that $d(h)_{x_0} = d(f)_{x_0} + d(g)_{x_0}$.

5. By an *action* α of a Lie group G on a smooth manifold X we mean a group homomorphism $\alpha : G \to \text{Diff}\,(X)$, such that the map $(g, x) \mapsto (\alpha(g))(x)$ is smooth. (For example, typically, G acts on itself in three different ways: left action, right action, and by conjugation.) For any $x \in X$ the sets

$$G^x := \{g \in G \;:\; \alpha(g)x = x\};\; \alpha(G)x := \{\alpha(g)x \;:\; g \in G\}$$

are called *stabilizer* of x and *orbit* of x, respectively.
(a) Show that for each $x \in X$, the map $\alpha_x : G \to X$ defined by $g \mapsto \alpha(g)x$ is a smooth map of constant rank k and hence deduce (from the rank theorem) that $G^x = \alpha_x^{-1}(x)$ is a submanifold of G of dimension $=$ dim $G - k$. Also show that $T_e G^x = Ker\, D(\alpha_x)_e$.
(b) Show that for some neighborhood U of $e \in G$, $\alpha(U)x$ is a k-dimensional submanifold of X with $T_e(\alpha(U)x) = D(\alpha_x)_e(T_e(G))$.
(c) In particular, if $\alpha(G)x$ is a submanifold of X, then its dimension is $= k$.

6. Consider the action of $\mathbb{R}$ on $\mathbb{S}^1 \times \mathbb{S}^1$ via the left multiplication by $\alpha(t) = (e^{a_1 \imath t}, e^{a_2 \imath t})$ where $a_1, a_2 \in \mathbb{R}$. Show that if $\{a_1, a_2\}$ is linearly independent over $\mathbb{Q}$, then the orbit of any element is dense in $\mathbb{S}^1 \times \mathbb{S}^1$. Show that any one of these orbits is a (1-dimensional) submanifold iff $\{a_1, a_2\}$ is linearly dependent over $\mathbb{Q}$.

7. Generalize the above example to an action of $\mathbb{R}$ on $\mathbb{T}^n = \mathbb{S}^1 \times \cdots \times \mathbb{S}^1$ (n copies).

8. Let G be a compact Lie group acting on a manifold X. Show that every orbit is a submanifold. [Hint: Prove that $\alpha(U)x$ is an open set of $\alpha(G)x$ in X containing the point x.]

9. Let $f : G \to H$ be a homomorphism of Lie groups. If G is compact, show that $f(G)$ is a compact Lie subgroup of H.

10. **Flag Manifolds:** The above exercise is a rich source of examples of smooth manifolds. Here is an illustration: Fix a sequence of integers $1 \le k_1 < k_2 < \cdots < k_r < k_{r+1} = n$. For $\mathbb{K} = \mathbb{R}, \mathbb{C}$, or $\mathbb{H}$, let $G_{k,n} := G_{k,n}(\mathbb{K})$ denote the Grassmannian variety of k-diemnsional subspaces of $\mathbb{K}^n$. In the product of Grassmannian manifolds

$$M = G_{k_1,n} \times \cdots \times G_{k_r,n},$$

consider the point $P = (V_1, \ldots, V_r)$, where V_j is the linear span of the standard basis elements, $\{\mathbf{e}_1, \ldots, \mathbf{e}_{k_j}\}$ in $\mathbb{K}^n$. Consider the diagonal action of $\mathcal{O}(n) = \mathcal{O}(n, \mathbb{K})$ on M. From the above exercise, it follows that the orbit of P under this action is manifold. It is called the *flag manifold* and is denoted by $\mathcal{F}(k_1, k_2, \ldots, k_r)$. Show that it is diffeomorphic to the homogeneous space

$$\mathcal{O}(n)/\mathcal{O}(l_1) \times \cdots \times \mathcal{O}(l_r) \times \mathcal{O}(l_{r+1}),$$

where $l_1 = k_1$ and $l_i = k_i - k_{i-1},\ i \geq 2$.

11. Let $f : G \to H$ be a homomorphism of Lie groups. Suppose $H_1 \subset H$ is a Lie subgroup. Then show that $G_1 = f^{-1}(H_1)$ is a Lie subgroup of G with $T_e(G_1) = (Df_e)^{-1}(T_e(H_1))$. [Consider the action of G on H/H_1 via f and apply Exercise 5.(a).]

12. Show that intersection of finitely many Lie subgroups is a Lie subgroup.

13. Let H be a Lie subgroup of G. Show that any subgroup F of G such that $H_0 \subset F \subset H$ is a Lie subgroup, where H_0 denotes the connected component of H.

9.4 Lie Algebras

In this section, let $\mathbb{K}$ denote a field of characteristic $\neq 2$.

Definition 9.4.1 Let $\mathfrak{g}$ be a finite dimensional vector space over $\mathbb{K}$. A binary operation $[,] : \mathfrak{g} \times \mathfrak{g} \to \mathfrak{g}$ is called a *Lie bracket* if it satisfies the following conditions:
(LI) $[\,,\,]$ is bilinear, i.e., $[a_1\mathbf{u}_1 + a_2\mathbf{u}_2, \mathbf{v}] = a_1[\mathbf{u}_1, \mathbf{v}] + a_2[\mathbf{u}_2, \mathbf{v}]$.
(LII) $[\,,\,]$ is anti-symmetric, i.e., $[\mathbf{v}, \mathbf{u}] = -[\mathbf{u}, \mathbf{v}]$ for all $\mathbf{u}, \mathbf{v} \in \mathfrak{g}$.
(LIII) $[\,,\,]$ satisfies the Jacobi identity: $[[\mathbf{u}, \mathbf{v}], \mathbf{w}] + [[\mathbf{v}, \mathbf{w}], \mathbf{u}] + [[\mathbf{w}, \mathbf{u}], \mathbf{v}] = 0$.
A vector space $\mathfrak{g}$ together with such a Lie bracket operation is called a *Lie Algebra.*

Definition 9.4.2 By a *Lie-homomorphism* $\phi : \mathfrak{g}_1 \to \mathfrak{g}_2$ between two Lie algebras over $\mathbb{K}$ we mean a $\mathbb{K}$-linear map such that $\phi([\mathbf{u}, \mathbf{v}]) = [\phi(\mathbf{u}), \phi(\mathbf{v})]$.

Remark 9.4.1

1. If ϕ is invertible, then automatically its inverse is also a Lie homomorphism and hence ϕ is an isomorphism of Lie algebras.

2. Observe that the $[\,,\,]$ can be thought of as a multiplication rule so that $\mathfrak{g}$ becomes an algebra over $\mathbb{K}$ except that usually, the multiplication is assumed to be associative. Here, $[\,,\,]$ is perhaps highly nonassociative and instead of associativity, we have the Jacobi Identity.

3. Given an associative algebra A on $\mathbb{K}$ if we define $[\mathbf{u}, \mathbf{v}] = \mathbf{uv} - \mathbf{vu}$, it is not difficult to see that $\mathfrak{g}$ becomes a Lie algebra. On the other hand, there seems to be no way to recover the associative multiplication from the Lie bracket. Also observe that this Lie bracket measures the amount of noncommutativity of the associative multiplication. The most important example of this for us is the matrix algebra $M(n, \mathbb{K})$ from which we get the Lie bracket in the above manner, to make it into a Lie algebra. This Lie algebra will be denoted by $\mathfrak{gl}(n, \mathbb{K})$. There are several sub-algebras of this Lie algebra, which are of interest to us: viz., the space of all Hermitian matrices denoted by $\mathfrak{o}(n, \mathbb{K})$, the space of all trace zero matrices denoted by $\mathfrak{sl}(n, \mathbb{K})$, and so on.

4. Let $\mathfrak{g}$ be a 1-dimensional Lie algebra, say $\mathbf{v} \in \mathfrak{g}$ is a nonzero vector. By property (LII) it follows that $[\mathbf{v}, \mathbf{v}] = 0$. Thus, there is a unique 1-dimensional Lie Algebra. Any Lie algebra in which the Lie bracket is identically zero is called an *abelian Lie Algebra.*

5. We shall not be able to pursue the study of Lie algebras on its own here. The foremost important example of Lie algebra is something inside the the space $\mathcal{T}(X)$ of all smooth vector fields on a smooth manifold X. Recall from Section 5.6 (Exercise 2) that we have defined a "Lie bracket" operation on $\mathcal{T}(X)$ which satisfies properties (LI),(LII) (LIII) of Definition 9.4.1. Thus $(\mathcal{T}(X), [\,,\,])$ is a Lie algebra over $\mathbb{R}$. In particular, we are interested in a very special subalgebra of this Lie algebra, when $X = G$ is a Lie group.

Example 9.4.1 Let G be a Lie group. For each $g \in G$, the left-multiplication $L_g : G \to G$ is a diffeomorphism. Clearly, $D(L_g)_e : T_e(G) \to T_g(G)$ is a linear isomorphism for each g. Consider the map $TG \to G \times T_e(G)$ given by

$$(g, \mathbf{v}) \mapsto (g, (D(L_g^{-1})_g(\mathbf{v})).$$

Check that this defines an isomorphism of the vector bundles. (To see that this is a smooth map, use local coordinates and an argument similar to the one that occurred in Exercise

9.3.1(1).) In other words, the tangent bundle of a Lie group is trivial, i.e., every Lie group is parallelizable. It follows that the module $\mathcal{T}(G)$ of all smooth vector fields on G is a free $C^\infty(G)$ module of rank = dimension of G. Now let us take a closer look at this module along with its Lie algebra structure.

Given a vector field σ on Y and a diffeomorphism $f : X \to Y$ the pullback vector field or the induced vector field $f^*(\sigma)$ on X is defined by $x \mapsto D(f^{-1})_{f(x)}(\sigma(f(x)))$.

Thus, in particular, for a Lie group G, since each element $g \in G$ defines a diffeomorphism $L_g : G \to G$, we can talk about $L_g^*(\sigma)$ for each $\sigma \in \mathcal{T}(G)$. Thus, we get an action

$$G \times \mathcal{T}(G) \to \mathcal{T}(G) \quad (g, \sigma) \mapsto L_g^*(\sigma).$$

Definition 9.4.3 We say that a smooth vector field σ on G is *left-invariant*, iff $\sigma = L_g^*(\sigma)$ for all $g \in G$.

Remark 9.4.2 From (5.5), it follows that σ is left-invariant, iff σ is L_g-related to itself for all $g \in G$.

Lemma 9.4.1 A vector field σ on G is left-invariant iff

$$\sigma(g) = D(L_g)_e(\sigma(e)), g \in G. \tag{9.9}$$

$$\begin{array}{ccc} \mathcal{T}(G) & \xrightarrow{D(L_g)_e} & \mathcal{T}(G) \\ \sigma \uparrow & & \uparrow \sigma \\ G & \xrightarrow{L_g} & G \end{array}$$

Proof: Verify directly. ♠

Theorem 9.4.1 *For a Lie group G, the space $\mathfrak{g}$ of all left-invariant vector fields over G forms a Lie subalgebra of the Lie algebra of all vector fields $\mathcal{T}(G)$.*

Proof: That $\mathfrak{g}$ is a $\mathbb{R}$-linear subspace is easily verified. To see that if $\sigma_i \in \mathfrak{g}, i = 1, 2$ then $[\sigma_1, \sigma_2] \in \mathfrak{g}$, we appeal to Corollay 5.6.1 and the Remark 9.4.2 above. ♠

Definition 9.4.4 The Lie algebra $\mathfrak{g}$ of all left-invariant smooth vector fields over a Lie group G is called the *Lie algebra of G*.

Remark 9.4.3 The evaluation map $\sigma \mapsto \sigma(e)$ defines a linear map $Ev : \mathfrak{g} \to T_e(G)$. Given a vector $\mathbf{v} \in T_e(G)$ the formula

$$\gamma_{\mathbf{v}}(g) = (DL_g)_e(\mathbf{v}) \tag{9.10}$$

defines an element of $\mathfrak{g}$ such that $Ev(\gamma_{\mathbf{v}}) = \gamma_{\mathbf{v}}(e) = \mathbf{v}$. Moreover, for any $\sigma \in \mathfrak{g}$, by the invariance, we have $\sigma(g) = (DL_g)_e(\sigma(e))$. Therefore, γ is the inverse of Ev. Thus, as a vector space, $\mathfrak{g}$ is isomorphic to $T_e(G)$. It seems we have not done anything great except that the tangent space $T_e(G)$ has been now equipped with a Lie bracket. The simplest point to note at this stage is the fact that there was no way to define a Lie algebra bracket of the tangent space $T_e(G)$ directly. This Lie bracket is going to dictate a lot of algebraic and geometric behaviour of G. We shall deal with only a few of the interrelationships between G and its Lie algebra $\mathfrak{g}$.

Example 9.4.2 Consider the additive group $\mathbb{R}$ of real numbers. Its Lie algebra $\mathfrak{r}$ is a 1-dimensional vector space. The map $L : C^\infty(\mathbb{R}) \to C^\infty(\mathbb{R})$ defined by $L(f)(t_0) = \frac{df}{dt}(t_0)$ is a generator of the vector space $\mathfrak{r}$. Clearly, $[L, L] = 0$, which can be seen directly or by referring to Remark 9.4.1.(4).

Example 9.4.3 Take $G = GL(n, \mathbb{K})$ and let $A = (a_{ij}), B = (b_{ij}) \in T_{Id}(G) = M(n, \mathbb{K})$. We want to compute $[A, B]$ by interpreting A, B as left-invariant vector fields γ_A, γ_B, on G. Let x_{ij} denote the coordinate functions on $M(n, \mathbb{R}.)$ We shall compute the effect of the operator $(\gamma_B \circ \gamma_A)$ on x_{ij} at $e = Id$. For this, we have to first compute the operator $\gamma_A : C^\infty(G) \to C^\infty(G)$ at least in a neighborhood of e. Recall that γ_A at e is nothing but taking the directional derivative of a function in the direction of A at e. Similarly, $(\gamma_A)_g$ is nothing but taking the directional derivative in the direction of $D(L_g)_e(A) = gA$. In particular, we have

$$\gamma_A(g)(x_{ij}) = (gA)_{ij} = \sum_k g_{ik}a_{kj} = x_{ij}(gA).$$

We must treat the assignment $g \mapsto \sum_k g_{ik}a_{kj}$ as a function of g and compute $(\gamma_B)_e$ on this. Since $g \mapsto gA$ is a linear map (right multiplication by A), its derivative is R_A itself. Therefore the directional derivative of this map in the direction B is nothing but BA. Therefore, its $(i, j)^{\text{th}}$-coordinate is $x_{ij}(BA)$. This just means that the matrix of $\gamma_B \circ \gamma_A$ is nothing but BA. It follows that the matrix of $[\gamma_A, \gamma_B]$ is equal to $AB - BA$.

On a more canonical fashion, we may drop out the matrix notation. Thus, let V denote any finite dimensional vector space, $\mathfrak{gl}(V)$ the algebra of endomorphism of V, and $GL(V)$, the open subset of invertible elements of $\mathfrak{gl}(V)$. Clearly $GL(V)$ is a Lie group which is indeed isomorphic to $GL(n, \mathbb{K})$, where $n = \dim V$ with its Lie algebra equal to $\mathfrak{gl}(V)$. Workout the following exercises.

Exercise 9.4.1 Let V denote a finite dimensional vector space, in these exercises.

1. If X is an open subset of V, then for all $x \in X$, we have, $T_xX = V$.

2. A closed subgroup H of $GL(V)$, which is an open subspace of a subalgebra $\mathfrak{h}$ of $\mathfrak{gl}(V)$, has its Lie algebra equal to $\mathfrak{h}$.

3. Let $W \subset U \subset V$ be any subspaces. Show that

$$\{A \in GL(V) \ : \ AU \subset U\}; \ \ \{A \in GL(V) \ : \ (A - Id)U \subset W\}$$

are Lie subgroups of $GL(V)$. Describe the corresponding Lie subalgebras. [Hint: Start with a basis of W, extend it to a basis of U and then to a basis of V. You can then work inside $GL(n, \mathbb{K})$.]

4. Let $\alpha : G \to GL(V)$ be any homomorphism (i.e, linear representation), where V is a finite dimensional vector space. Let U, W be subspaces of V such that $W \subset U$. Then $G(U), G(U; W)$ respectively given by

$$\{g \in G \ : \ \alpha(g)(U) \subset U\} \ \ \& \ \ \{g \in G \ : \ (\alpha(g) - Id)(U) \subset W\}$$

are Lie subgroups of G with their respective Lie algebras

$$\{\mathbf{v} \in \mathfrak{g} \ : \ (D\alpha)_e(\mathbf{v})(U) \subset U\} \ \ \& \ \ \{\mathbf{v} \in \mathfrak{g} \ : \ (D\alpha)_e(\mathbf{v})(U) \subset W\}.$$

5. Show that the derivative of Exp : $M(n;\mathbb{K}) \to GL(n;\mathbb{K})$ at a point $A \in M(n,\mathbb{K})$ is given by the formula:

$$D_A(\text{Exp}) = R_{\text{Exp}\,A} \circ \left(\frac{\text{Exp}\,(ad_A) - Id}{ad_A}\right), \tag{9.11}$$

where $\frac{\text{Exp}\,X\ - Id}{X}$ is understood as the operator $\sum_{n=0}^{\infty} \frac{X^n}{(n+1)!}$ and R_X is the right multiplication by X.

6. (**Theorem of Nono**) Show that Exp is not a local diffeomorphism at $A \in M(n;\mathbb{K})$ iff ad_A has an eigenvalue of the form $2\pi \imath k$, where $k \neq 0$ is an integer.

Lemma 9.4.2 Let $h : G_1 \to G_2$ be a Lie homomorphism. Then given any left invariant vector field σ on G_1, there is a unique left-invariant vector field $h_*(\sigma)$ on G_2 such that σ and $h_*(\sigma)$ are h-related.

Proof: If τ is any vector field on G_2, which is h-related to σ then $\tau_e = dh_e\sigma_e$. Since τ_e is completely determined by σ_e, the uniqueness follows. It remains to show that the left-invariant vector field defined by

$$\tau(g) = d(L_g)_e(dh_e(\sigma_e)),\ g \in G_2 \tag{9.12}$$

is h-related to σ. First observe that $h : G_1 \to G_2$ is a homomorphism implies that $h \circ L_g = L_{h(g)} \circ h$ for any $g \in G_1$. Therefore, for any $g \in G_1$, putting $\mathbf{v} = \sigma(e)$, we have,

$$\begin{aligned} d(h)_g(\sigma(g)) &= d(h)_g(d(L_g)_e(\mathbf{v})) \\ &= d(h \circ L_g)_e(\mathbf{v}) \\ &= d(L_{h(g)} \circ h)_e(\mathbf{v}) \\ &= d(L_{h(g)})_e(dh_e(\mathbf{v})) \\ &= d(L_{h(g)} \circ h)_e(\mathbf{v}) = \tau(h(g)). \end{aligned}$$

We can now define $h_*(\sigma) := \tau$ as defined in (9.12). ♠

Theorem 9.4.2 *Given a Lie-homomorphism $h : G_1 \to G_2$ of two Lie groups, $d(h)_e : \mathfrak{g}_1 \to \mathfrak{g}_2$ defines a homomorphism of Lie algebras. The assignment $G \rightsquigarrow \mathfrak{g}$ is a covariant functor from the category of Lie groups and Lie homomorphisms to the category of Lie algebras over* $\mathbb{R}$.

Proof: Since $d(h)_e : T_eG_1 \to T_eG_2$ is a linear map, it follows that $h_* : \mathfrak{g}_1 \to \mathfrak{g}_2$ is a linear map. Since $\sigma_i \sim_h h_*(\sigma_i)$, it follows from Corollary 5.6.1, that $[\sigma_1, \sigma_2] \sim_h [h_*(\sigma_1), h_*(\sigma_2)]$. Therefore, from the uniqueness in the previous lemma, we have

$$h_*[\sigma_1, \sigma_2] = [h_*(\sigma_1), h_*(\sigma_2)].$$

This proves that h_* is a homomorphism of Lie algebras.

If $k : G_2 \to G_3$ is another Lie homomorphism, then clearly, by the chain rule, $d(k \circ h)_e = d(k)_e \circ d(h)_e$ and also $d(Id)_e = Id$, for the identity homomorphism of any group G. This is precisely what we mean by saying that $G \rightsquigarrow \mathfrak{g}$ defines a covariant functor. ♠

Theorem 9.4.3 *Let H be a Lie subgroup of a Lie group G. Then the inclusion map $\eta : H \to G$ induces an identification $d\eta_e : \mathfrak{h} \hookrightarrow \mathfrak{g}$ of the Lie algebra with a Lie subalgebra of $\mathfrak{g}$.*

Proof: This is clear since the only thing that we have to observe is that $d(\eta)_e$ is injective, η being an embedding. ♠

Remark 9.4.4 It becomes important to classify all finite dimensional Lie algebras up to an isomorphism. Fixing a basis $\{E_i\}$ for the vector space, the Lie-algebra structure is completely determined by the "structural constants"

$$c_{ijk} = [[E_i, E_j], E_k]$$

which satisfy a certain symmetry condition dictated by the properties (LII) and (LIII) and extended over the whole vector space via (LI). For dimensions ≤ 3, workout the following exercises. You will find that as the dimension increases the problem becomes more and more cumbersome. Cartan developed a different approach for such a classification. You may read about it from several available expositions. (See [Hum].)

Exercise 9.4.2

1. On a real vector space of dimension 2, show that any two nontrivial Lie algebras are isomorphic. (And, of course, there is a nontrivial one.)

2. Consider the set L of all strictly upper triangular 3×3 real matrices. Compute the Lie brackets of the following elements that form a basis for L.

$$E_1 = \begin{pmatrix} 0 & 1 & 0 \\ 0 & 0 & 0 \\ 0 & 0 & 0 \end{pmatrix}; \; E_2 = \begin{pmatrix} 0 & 0 & 0 \\ 0 & 0 & 1 \\ 0 & 0 & 0 \end{pmatrix}; \; E_3 = \begin{pmatrix} 0 & 0 & 1 \\ 0 & 0 & 0 \\ 0 & 0 & 0 \end{pmatrix}.$$

3. Show that the vector product $[\mathbf{u}, \mathbf{v}] = \mathbf{u} \times \mathbf{v}$ defines Lie algebra structure on $\mathbb{R}^3$.

4. Show that the Lie algebra in the above exercise is isomorphic to the Lie algebra $\mathfrak{so}(3)$ of 3×3 real skew-symmetric matrices. Indeed check that you can map $\mathbf{i}, \mathbf{j}, \mathbf{k}$ respectively to:

$$\begin{pmatrix} 0 & 1 & 0 \\ -1 & 0 & 0 \\ 0 & 0 & 0 \end{pmatrix}; \begin{pmatrix} 0 & 0 & -1 \\ 0 & 0 & 0 \\ 1 & 0 & 0 \end{pmatrix}; \begin{pmatrix} 0 & 0 & 0 \\ 0 & 0 & 1 \\ 0 & -1 & 0 \end{pmatrix}$$

5. Show that 'direct" sum of two Lie algebras is again a Lie algebra in an obvious way, where the bracket of vectors from different factors is defined to be zero. If a Lie algebra can be written as a direct sum of two or more Lie algebras of nonzero dimension then it is called *decomposable*. For example, you can write $\mathbb{R}^3 = \mathbb{R}^2 \oplus \mathbb{R}$ or $\mathbb{R}^3 = \mathbb{R} \oplus \mathbb{R} \oplus \mathbb{R}$ and take the Lie algebra structures accordingly.

6. Determine whether or not the two Lie algebras L and $\mathfrak{so}(3)$ given above on $\mathbb{R}^3$ are decomposable or not.

7. Can you find some more 3-dimensional Lie algebras?

9.5 Canonical Coordinates

We first recall some basic facts from the theory of ordinary differential equations. Let X, Y be smooth manifolds. Let $\sigma : X \times Y \to TX$ be a smooth map such that $\pi \circ \sigma = Id_X$, i.e., σ is a smooth family of vector fields on X. For a given smooth function $\alpha : X \to X$ consider the equation

$$\frac{\partial f}{\partial t}(x, t, y) = \sigma(f(x, y, t), y); \;\; f(x, 0, y) = \alpha(x) \tag{9.13}$$

where $f : X \times \mathbb{R} \times Y \to X$ is a smooth function.

By the theory of ordinary differential equations, given any point $(x, y) \in X \times Y$ there exists a unique smooth solution f of this problem in a sufficiently small neighborhood of $(x, 0, y)$. (Compare theorem 6.3.1.)

Now look at the special case when $X = GL(n, \mathbb{R})$ and $Y = M(n, \mathbb{R})$. Take $\alpha = Id_X$ and σ to be the left-invariant vector field $(g, A) \mapsto (g, gA)$ for $A \in M(n, \mathbb{R})$. Then the map f given by $(g, t, A) \mapsto ge^{tA}$ satisfies (9.13). Putting $g = e, t = 1$, we get the exponential map. Thus, we are lead to define exponential map in the case of arbitrary Lie groups via solutions of (9.13). That is what we shall do now.

Let now $X = G$ be any Lie group, $Y = \mathfrak{g}$ its Lie algebra, $\sigma : X \times Y \to TG$ be the left-invariant vector field given by $(g, A) \mapsto (g, gA)$. Let $\alpha = Id_G : G \to G$. Let f be the unique solution of (9.13) in a neighborhood $U \times (-\epsilon, \epsilon) \times V$ of $(e, 0, 0)$.

We may, and shall assume that V is a convex neighborhood of 0 in $\mathfrak{g}$.

Lemma 9.5.1 For $A \in V$, $g \in U$, $|s| < 1$ and t_1, t_2, such that $|t_1|, |t_2|$ and $|t_1 + t_2|$ all being $< \epsilon$, we have,

$$f(g, t_1 + t_2, A) = f(g, t_1, A) f(e, t_2, A) \tag{9.14}$$

and

$$f(g, st, A) = f(g, t, sA). \tag{9.15}$$

Proof: Fix t_1 and put $g_0 = \theta_A(t_1)$. Consider the two functions

$$\phi_1(g, t, A) = f(g, t_1 + t, A)g; \ \phi_2(g, t, A) = f(g, t_1, A) f(e, t, A).$$

Let us check that both are solutions of the initial value problem (9.13), with α replaced by $\beta(g) = f(g, t_1, A)g$. Then by the uniqueness of the solution, it follows that $\phi_1 = \phi_2$ from which (9.14) follows.

Clearly,

$$\phi_1(g, 0, A) = f(g, t_1, A)g = \beta(g)$$

and

$$\phi_2(g, 0, A) = f(g, t_1, A) f(e, 0, A) = f(g, t_1, A)\alpha(g) = \beta(g).$$

Next

$$\frac{d\phi_1}{dt}(g, t, A) = \frac{df}{dt}(g, t_1 + t, A) = \sigma(f(g, t_1 + t, A), A) = \sigma(\phi_1(g, t, A), A);$$

also,

$$\begin{aligned} \frac{d\phi_2}{dt}(g, t, A) &= f(g, t_1, A)\frac{df}{dt}(e, t, A) \\ &= f(g, t_1, A)\sigma(f(e, t, A), A) = \sigma(\phi_2(g, t, A), A), \end{aligned}$$

the last equality being given by the left-invariance of σ. Thus, we have verified that both ϕ_i are solutions of (9.13). The proof of (9.15) is similar. This time use the vector field $s\sigma$ and see that both the functions $f(g, st, A)$ and $f(g, t, sA)$ are solutions of (9.13). Details are left as exercise to the reader. ♠

For any $A \in V$, let us put

$$\theta_A(t) = f(e, t, A). \tag{9.16}$$

Then it follows that

$$\theta_A(t_1 + t_2) = \theta_A(t_1)\theta_A(t_2), \tag{9.17}$$

for $t_1, t_2, t_1 + t_2 \in (-\epsilon, \epsilon)$. As an immediate consequence, it follows that the map $\theta(t, A) = f(e, t, A)$ can be extended uniquely on $\mathbb{R} \times V \to G$ so that for each fixed $A \in V$, the map $\theta_A : \mathbb{R} \to G$ is a homomorphism. It also follows that θ is smooth.

Definition 9.5.1 Let G be any Lie group and $\mathfrak{g}$ be its Lie algebra. Let V be a convex neighborhood of $0 \in \mathfrak{g}$ as given above. We define the mapping $\text{Exp} : V \to G$ by the formula

$$\text{Exp}\,(A) = \theta_A(1). \tag{9.18}$$

Now given any $A \in \mathfrak{g}$, we can choose a positive integer m, such that $A/m \in V$. Using (9.15), we can define

$$\text{Exp}\,(A) = \theta_{A/m}(m) = (\theta_{A/m}(1))^m = (\text{Exp}\,(A/m))^m. \tag{9.19}$$

Verify that this way, Exp gets extended on the whole of $\mathfrak{g}$ as a smooth map.

Theorem 9.5.1 *The map Exp* $: \mathfrak{g} \to G$ *is a local diffeomorphism at* 0.

Proof: Since $D(\text{Exp}\,)_0 : T_0(\mathfrak{g}) \to T_eG$ is nothing but $Id : \mathfrak{g} \to \mathfrak{g}$, we can use the inverse function theorem to see that Exp is a diffeomorphism in a neighborhood V of 0. ♠

Remark 9.5.1 It is not true that Exp is a local diffeomorphism at other points of $\mathfrak{g}$, in general (see under Exercises 9.4.1).

Definition 9.5.2 A local inverse of Exp in a suitable neighborhood of $e \in G$ is called a *canonical coordinate system for* G.

The following lemma justifies this name.

Lemma 9.5.2 Let $f : G \to H$ be a Lie homomorphism. Then $\text{Exp} \circ df_e = f \circ \text{Exp} : \mathfrak{g} \to H$.

$$\begin{array}{ccc} G & \xrightarrow{f} & H \\ \text{Exp}\uparrow & & \uparrow\text{Exp} \\ \mathfrak{g} & \xrightarrow{df_e} & \mathfrak{h} \end{array}$$

Proof: Fix $A \in \mathfrak{g}$ and put $B = Df_e(A) \in \mathfrak{h}$. Consider the homomorphism $\theta : \mathbb{R} \to H$ given by $\theta(t) = f(\text{Exp}\,(tA))$. Then $\frac{d\theta}{dt}(0) = Df_e(\text{Exp}\,(A)) = B$. Therefore, $\theta = \theta_B$ and hence $\theta(1) = \text{Exp}\,(B)$. ♠

As an immediate corollary we have:

Theorem 9.5.2 *Let G be a connected Lie group. Then any Lie homomorphism $f : G \to H$ is completely determined by $d(f)_e : \mathfrak{g} \to \mathfrak{h}$.*

Proof: By the previous lemma and the theorem, it follows that f is completely determined in a neighborhood of $e \in G$. But then since G is connected, f is defined on a set of generators for G. ♠

Remark 9.5.2 An easy corollary of the exponential map is that any Lie group is a real analytic manifold. This follows from the fact that the canonical local charts are real analytic on the overlaps, being given as solutions of a suitable initial value problem. Further, it follows from the above theorem that any (smooth) homomorphism of Lie groups is actually real analytic.

Corollary 9.5.1 Let G be a Lie subgroup of H. Then for any neighborhood U of $0 \in \mathfrak{h}$ $\text{Exp}\,(U \cap \mathfrak{g}) = \text{Exp}\,(U) \cap G$ defines a neighborhood of $e \in G$.

One can generalize the "canonical coordinates" a little bit.

Theorem 9.5.3 *Let G be a Lie group and $\mathfrak{g}$ be its Lie algebra. Let $\mathfrak{g} = V_1 \oplus \cdots \oplus V_k$ be a direct sum decomposition of $\mathfrak{g}$ into linear subspaces. Let U be a neighborhood of $0 \in \mathfrak{g}$ and $\phi_i : U \cap V_i \to G,\ 1 \le i \le k$ be smooth maps such that $D(\phi_i)_0$ is the inclusion map $V_i \to \mathfrak{g}$. Then the map*

$$\Psi : (\mathbf{v}_1, \ldots, \mathbf{v}_k) \mapsto \phi_1(\mathbf{v}_1) \cdots \phi_k(\mathbf{v}_k) \tag{9.20}$$

defines a local coordinate system for G around e.

Proof: All that we need to see is that the derivative of Ψ at $0 \in \mathfrak{g}$ is nonsingular. Since the derivative of the product is the sum of the derivatives, it follows that $d(\Psi)_0 = Id$. By the inverse function theorem, we are through. ♠

Remark 9.5.3 In particular, if $\phi_i = \text{Exp}\,|_{V_i}$, then Ψ is called a canonical coordinate of the II-kind by some authors.

Exercise 9.5.1 Verify that $\text{Exp} : \mathfrak{g} \to G$ as given in (9.19) is well-defined and smooth.

We end this section with a computation of the power series expansion for the multiplication rule in any Lie group in terms of canonical coordinates.

Let G be any Lie group, U be some suitably small neighborhood of $e \in G$ such that $UU \subset U$ and U is diffeomorphic to a neighborhood of $0 \in \mathfrak{g}$ via the exponential map. Let $\{x_1, \ldots, x_n\}$ be the coordinate functions of the inverse of the exponential map. Then we know that x_i are analytic functions. We would like to find the power series expression for the functions $U \times U \to \mathbb{R}$ given by

$$(g, h) \mapsto x_i(gh).$$

So for the time being fix $g, h \in U$ such that $gh \in U$. Let $\mathbf{u}, \mathbf{v} \in \mathfrak{g}$ be such that $\text{Exp}\,(\mathbf{u}) = g$, $\text{Exp}\,(\mathbf{v}) = h$. We shall treat elements of $\mathfrak{g}$ as elements of $D(X)$. (See Section 5.6.) Let now f be any analytic function around $e \in G$. Put $\alpha(t) = g\text{Exp}\,(t\mathbf{v})$. Then we know that

$$\frac{d}{dt}(f \circ \alpha) = \mathbf{v}(f) \circ \alpha. \tag{9.21}$$

Applying this to the functions $\mathbf{v}^k(f)$ successively yields

$$\frac{d^k}{dt^k}(f \circ \alpha) = \mathbf{v}^k(f) \circ \alpha, \quad k \ge 1. \tag{9.22}$$

Therefore it follows that

$$f \circ \alpha(t) = \sum_{k=0}^{\infty} \mathbf{v}^k(f)(\alpha(0)) \frac{t^k}{k!}. \tag{9.23}$$

Taking $g = e$, and $t = 1$, we get

$$f(h) = f(e) + \sum_{k=1}^{\infty} \frac{\mathbf{v}^k(f)(e)}{k!}. \tag{9.24}$$

Replacing h by g and therefore $\mathbf{v}$ by $\mathbf{u}$, we have

$$f(g) = f(e) + \sum_{k=1}^{\infty} \frac{\mathbf{u}^k(f)(e)}{k!}. \tag{9.25}$$

Applying this to each $\mathbf{v}^l$ in place of f, we get

$$\mathbf{v}^l(f)(g) = \mathbf{v}^l(f)(e) + \sum_{k=1}^{\infty} \frac{\mathbf{u}^k \mathbf{v}^l(f)(e)}{k!}. \tag{9.26}$$

Putting $t = 1$ in (9.23) and substituting from (9.26), we get,

$$f(gh) = \sum_{l=0}^{\infty} \frac{\mathbf{v}^l(f)(g)}{k!} = \sum_{k,l=0}^{\infty} \frac{\mathbf{u}^k \mathbf{v}^l(f)(e)}{k!l!}. \tag{9.27}$$

We can now put $f = x_i$ in each of the above identities to obtain the power series expansion for $x_i(g), x_i(gh)$, etc. Since $x_i(e) = 0$, we get

$$x_i(g) = \sum_{k=0}^{\infty} \frac{\mathbf{u}^k(x_i)(e)}{k!}. \tag{9.28}$$

Given two analytic functions, α, β in a neighborhood of e, we shall use the notation $\alpha \sim \beta$ to denote that the terms of order ≤ 2 vanish in the power series expansion of $\alpha - \beta$. Thus, we can write,

$$x_i(gh) \sim x_i(g) + x_i(h) + \mathbf{uv}(x_i)(e). \tag{9.29}$$

It also follows that

$$x_i(gh) - x_i(hg) \sim [\mathbf{u}, \mathbf{v}](x_i)(e). \tag{9.30}$$

Lemma 9.5.3 Let $f(x, y)$ be a (smooth) analytic function in a convex neighborhood U of $(0, 0)$ such that $f(x, 0) = f(0, y) = 0$. Then there exists (smooth) analytic functions α_{ij} in U such that

$$f(x, y) = \sum_{ij} \alpha_{ij}(x, y) x_i y_j. \tag{9.31}$$

We apply this to the function $x_i(gh) - x_i(g) - x_i(h)$ and to the function $x_i(h^{-1}g^{-1}hg)$ to obtain

$$x_i(gh) - x_i(g) - x_i(h) = \sum_{jk} \alpha_{ijk}(g, h) x_j(g) x_k(h). \tag{9.32}$$

$$x_i(h^{-1}g^{-1}hg) = \sum_{jk} \beta_{ijk} x_j(g) x_k(h). \tag{9.33}$$

Therefore, replacing g by gh and h by $h^{-1}g^{-1}hg$ in (9.32) we obtain

$$\begin{aligned} & x_i(hg) - x_i(gh) - x_i(h^{-1}g^{-1}hg) \\ = & \sum_{jk} \alpha_{ijk}(gh, h^{-1}g^{-1}hg) x_j(gh) x_k(h^{-1}g^{-1}hg) \\ \sim & \ 0. \end{aligned}$$

Combining this with (9.30), we get,

$$x_i(h^{-1}g^{-1}hg) \sim [\mathbf{v}, \mathbf{u}](x_i)(e). \tag{9.34}$$

Replacing h by $g^{-1}h^{-1}gh$ in (9.32), we get

$$x_i(h^{-1}gh) - x_i(g) - x_i(g^{-1}h^{-1}gh) \sim 0.$$

Using (9.34), this gives,

$$x_i(h^{-1}gh) \sim x_i(g) + [\mathbf{u}, \mathbf{v}](x_i)(e). \tag{9.35}$$

Interchanging g and h and then g and g^{-1}, we get

$$x_i(ghg^{-1}) \sim x_i(h) + [\mathbf{u}, \mathbf{v}](x_i)(e). \tag{9.36}$$

Corollary 9.5.2 $D(C_g)_e(\mathbf{u}) = \mathbf{u} + [\mathbf{u}, \mathbf{v}]$, where $g = \text{Exp}\,\mathbf{v}$.

Proof: We use (9.36). Putting $h = \text{Exp}\,\mathbf{u}$, we have,

$$\begin{aligned} x_i(D(C_g)_e(\mathbf{u})) &= \lim_{t\to 0} \tfrac{x_i(C_g(\text{Exp}\, t\mathbf{u}))}{t} \\ &= \lim_{t\to 0} \tfrac{x_i(t\mathbf{u})}{t} \\ &= x_i(\mathbf{u}) + [\mathbf{u}, \mathbf{v}](x_i)(e) \end{aligned}$$

which proves the corollary. ♠

These derivations will become very handy for several purposes later on.

9.6 Topological Invariance

As an application of canonical coordinates of the second kind, let us prove the topological invariance of Lie groups. First we have,

Lemma 9.6.1 Let $\alpha : (-\epsilon, \epsilon) \to \mathbb{R}$ be a continuous local homomorphism. Then f is the restriction of a linear map. In particular, $\alpha(x) = x\alpha(1)$ for all $x \in \mathbb{R}$.

Proof: For any $-\epsilon < x < \epsilon$, and any integer $k \neq 0$, we have $\alpha(x) = k\alpha(x/k)$. For the same reason, if p/q is any (nonzero) rational number with $|p/q| < 1$, then $\alpha(\frac{p}{q}x) = p\alpha(\frac{x}{q}) = \frac{p}{q}\alpha(x)$. By continuity, it follows that for all $-1 \le r \le 1$, we have $\alpha(rx) = r\alpha(x)$. Now for any $0 < |x| < |y| < \epsilon$, we have $\alpha(ry) = r\alpha(y)$ implies that, (by taking $r = x/y$) $\alpha(x) = \frac{x}{y}\alpha(y)$. Therefore $\alpha(y) = \frac{y}{x}\alpha(x)$. This implies that for all $r \in \mathbb{R}$ such that $-\epsilon < rx < \epsilon$, we have $\alpha(rx) = r\alpha(x)$. Thus, this formula can be used to extend $\alpha : \mathbb{R} \to \mathbb{R}$. In particular, it follows that $\alpha(1) = \frac{1}{x}\alpha(x)$. Therefore, $\alpha(y) = \frac{y}{x}\alpha(x) = y\alpha(1)$ for all $y \in \mathbb{R}$. ♠

Theorem 9.6.1 *If $f : G_1 \to G_2$ is a continuous homomorphism of Lie groups, then it is analytic. In particular, the underlying topological group determines the structure of a Lie group completely.*

Proof: It is enough to prove that f is analytic at $e \in G_1$. First consider the case when $G_1 = \mathbb{R}$. Let $\{x_1, \ldots, x_n\}$ denote a canonical coordinate system around $e \in G_2$. It is enough to prove that $x_i(f)$ is analytic in a neighborhood of $0 \in \mathbb{R}$.

We know that for any $\mathbf{v} \in \mathfrak{g}$ and $m \in \mathbb{Z}$, $\text{Exp}\,(\mathbf{v}) = (\text{Exp}\,(\mathbf{v}/m))^m$. Thus, if g, g^m are inside a canonical neighborhood then we know that $\ln(g^m) = m\ln(g)$. Therefore, $x_i(f)(mt) = x_i((f(t))^m) = mx_i(f(t))$ for all $t \in \mathbb{R}$ and for all integers. This just means that $x_i(f)$ is a local homomorphism. From Lemma 9.6.1, we conclude that $x_i(f)$ is analytic. This completes the proof in the case $G_1 = \mathbb{R}$.

In the general case, let $\{\mathbf{v}_1, \ldots, \mathbf{v}_n\}$ be a basis for $\mathfrak{g}$ and let $\Phi : U \to G_1$ be the associated canonical coordinate system as in Theorem 9.5.3. This means that for some $\epsilon > 0$ every element belonging to $\Phi(U)$ has a unique expression

$$g = \text{Exp}\,(t_1\mathbf{v}_1) \cdots \text{Exp}\,(t_n\mathbf{v}_n); \quad |t_i| < \epsilon.$$

Since $f : G \to H$ is a homomorphism, we have,

$$f(g) = f(\text{Exp}\,(t_1\mathbf{v}_1)) \cdots f(\text{Exp}\,(t_n\mathbf{v}_n)).$$

Each $t_i \mapsto f(\text{Exp}\,(t_i\mathbf{v}_i))$ is a continuous local homomorphism and hence analytic. Being the product of finitely many analytic maps, f is also analytic. ♠

Remark 9.6.1 It is easy to see that if $\text{Exp} : B \to G$ is a diffeomorphism where B is a ball of radius r around 0 in $\mathfrak{g}$, then $U = \text{Exp}\,(\frac{1}{2}B)$ is a neighborhood e in G that does not contain any subgroup $H \neq \{e\}$ of G. This property of a Lie group is referred to as "having no small subgroups". Gleason and Yamabe proved that any Hausdorff topological group is a Lie group iff it is locally compact and has no small subgroups. In a series of papers by various authors (Montgomery, Zippin, Iwasawa, Gleason, Yamabe, etc.), it has been established that a topological group that is a topological manifold does not have small subgroups. Together with the above result of Gleason and Yamabe, this proves that a topological manifold that is a topological group is a Lie group, which answers Hilbert's 5th problem affirmatively.

9.7 Closed Subgroups

In this section, we shall see another application of generalized canonical coordinates.

Often it may happen that we do get subgroups that are not closed. On the other hand, experience tells us that closed subgroups are always better behaved. So, we begin with a basic theorem due to Lie, which justifies our definition of matrix groups.

Lemma 9.7.1 For $1 \leq i \leq k$, let $\gamma_i : (-\epsilon, \epsilon) \to G$ be smooth maps such that $\gamma_i(0) = e$ and $\frac{d\gamma_i}{dt}|_0 = \mathbf{u}_i \in T_e(G)$ are linearly independent. Let $\mathfrak{h}$ be the subspace of $T_e(G)$ spanned by $\{\mathbf{u}_1, \ldots, \mathbf{u}_k\}$. For $|t_i| < \epsilon$, consider the map

$$\Gamma\left(\sum_i t_i\mathbf{u}_i\right) = \gamma_1(t_1) \cdots \gamma_k(t_k).$$

Then Γ is smooth and $D(\Gamma)_0(\mathbf{u}) = \mathbf{u}, \quad \mathbf{u} \in \mathfrak{h}$.

Proof: Enough to show that $D(\Gamma)_0(\mathbf{u}_i) = \mathbf{u}_i$. But the left-hand side is the directional derivative of Γ in the direction of $\mathbf{u}_i$, which is equal to $\gamma_i'(0) = \mathbf{u}_i$. ♠

Theorem 9.7.1 *Let H be a closed topological subgroup of a Lie group G. Then H is a Lie subgroup.*

Proof: The only thing that we need to verify is that H is a submanifold. For this, it is enough to give a coordinate chart for H around e.

Let $\mathfrak{g}$ be the Lie algebra of G. Since we do not know whether or not H is a manifold, we cannot talk about the tangent space to H at e as such. This difficulty is overcome by using an appropriate description of the tangent space, viz., we consider the set $T(H)$ of all vectors $\gamma'(0) \in \mathfrak{g}$, where γ is a smooth curve in H such that $\gamma(0) = e$. We hope that $T(H)$ indeed plays the role of the Lie algebra of H.

We first prove that $T(H)$ is a vector subspace of $\mathfrak{g}$. Given γ such that $\gamma'(0) = \mathbf{u} \in T(H)$, and a real number r, the curve $t \mapsto \gamma(rt)$ is a curve in H with velocity vector at 0 equal to $r\mathbf{u}$. Therefore, $r\mathbf{u} \in T(H)$. If γ_1, γ_2 are two such curves in H then the curve $t \mapsto \gamma_1(t)\gamma_2(t)^{-1}$ takes values in H and has its velocity vector at 0 equal to $\gamma_1'(0) - \gamma_2'(0)$.

(We can go on to show that $T(H)$ is indeed a Lie sub-algebra. But this seems to be not necessary at all!)

Fix an ordered basis, $\{\mathbf{u}_1, \ldots, \mathbf{u}_k\}$ for $T(H)$. Let γ_i be a smooth curve in H such that $\gamma_i'(0) = \mathbf{u}_i, \gamma_i(0) = e$. Define

$$\Gamma\left(\sum_i t_i\mathbf{u}_i\right) = \gamma_1(t_1)\gamma_2(t_2)\cdots\gamma_k(t_k). \tag{9.37}$$

Then Γ makes sense for all $|t_i| < \epsilon$, for some $\epsilon > 0$ and takes values in H. Also, $\Gamma(0) = e$ and $D(\Gamma)(0)(\mathbf{u}) = \mathbf{u}$ for all $\mathbf{u} \in T(H)$ (see the lemma above). The claim is that Γ defines a parameterization of a neighborhood of e in H. For this, fix a complementary subspace W to $T(H)$, i.e., $T(H) \oplus W = \mathfrak{g}$ and let

$$\Phi(\mathbf{u} + \mathbf{v}) = \Gamma(\mathbf{u})\mathrm{Exp}\,(\mathbf{v}), \quad \mathbf{u} \in T(H); \mathbf{v} \in W. \tag{9.38}$$

Then by Theorem 9.5.3, we see that in a smaller neighborhood U around $0 \in \mathfrak{g}$, Φ is a diffeomorphism.

Clearly $\Phi(U \cap T(H)) \subset H$. The only thing that is needed now is a smaller neighborhood $U' \subset U$ of 0 in $\mathfrak{g}$, such that $H \cap \Phi(U') \subset \Phi(U' \cap T(H))$.

So, assume that this is not the case. Then there exists a sequence $\mathbf{w}_i \in \mathfrak{g} \setminus \mathfrak{h}$ such that $\mathbf{w}_i \to 0$, and $\Phi(\mathbf{w}_i) \in H$. Write $\mathbf{w}_i = \mathbf{u}_i + \mathbf{v}_i$. It follows that $\mathbf{v}_i \to 0, \mathbf{v}_i \neq 0$ and $\mathbf{v}_i \in W$. Consider the sequence $\mathbf{v}_i/\|\mathbf{v}_i\|$ of vectors of unit length. Passing to a subsequence, if necessary, we may assume that this converges to a vector $\mathbf{v}$ of unit length in W. Since $\Phi(\mathbf{w}_i) = \Phi(\mathbf{u}_i)\mathrm{Exp}\,(\mathbf{v}_i)$, it follows that $\mathrm{Exp}\,(\mathbf{v}_i) \in H$ for all i. For any $t \in \mathbb{R}$, put $n_i = \lfloor t/\|\mathbf{v}_i\| \rfloor$. Check that $n_i\mathbf{v}_i \to t\mathbf{v}$. Therefore, $\mathrm{Exp}\,(n_i\mathbf{v}_i) \to \mathrm{Exp}\,(t\mathbf{v})$. Since each $\mathrm{Exp}\,(\mathbf{v}_i) \in H$, so are $\mathrm{Exp}\,(n_i\mathbf{v}_i) = (\mathrm{Exp}\,(\mathbf{v}_i))^{n_i}$. Therefore, $\mathrm{Exp}\,(t\mathbf{v}) \in H$. Thus, we have proved that the curve $C : t \mapsto \mathrm{Exp}\,(t\mathbf{v})$ is in H. By definition, $\mathbf{v} = \frac{dC}{dt}\big|_0 \in T(H)$, which is absurd. ♠

Remark 9.7.1 We draw your attention to the fact that in the course of the proof of the above theorem, we have obtained an alternative description of the Lie subalgebra of a closed subgroup H of G, viz. $\mathfrak{h} = T(H)$. Also, Φ defines a diffeomorphism of a neighborhood of $0 \in \mathfrak{h}$ with a neighborhood of $e \in H$.

Exercise 9.7.1

1. Let H_α be a family of Lie subgroups of a Lie group G. Show that $\cap_\alpha H_\alpha$ is a Lie subgroup.

2. Give a similar description of the tangent space T_xX to a smooth submanifold X of a Euclidean space $\mathbb{R}^N$ as in Remark 9.7.1.

9.8 The Adjoint Action

Definition 9.8.1 Let G be any group and V be a vector space. By a *linear representation of G on V*, we mean a group homomorphism $\varphi : G \to Aut(V)$ where $Aut(V)$ denotes the group of all vector space isomorphisms of V to itself. The representation φ is called *faithful* if it is injective.

Definition 9.8.2 Let $\mathfrak{g}$ be a Lie algebra over $\mathbb{K}$. By a *linear representation of* $\mathfrak{g}$, we mean a Lie algebra homomorphism $\alpha : \mathfrak{g} \to \mathfrak{gl}(n, \mathbb{K})$ for some $n \geq 1$.

Example 9.8.1 Let $\mathfrak{g}$ be a (finite dimensional) Lie algebra over a field $\mathbb{K}$. Fix some basis $\{\mathbf{v}_1, \ldots, \mathbf{v}_n\}$ for $\mathfrak{g}$ over $\mathbb{K}$. For each $\mathbf{v} \in \mathfrak{g}$ consider the mapping $\mathbf{u} \mapsto [\mathbf{u}, \mathbf{v}]$. This is clearly a linear map and hence gives an element $P(\mathbf{v}) \in \mathfrak{gl}(n, \mathbb{K})$. The mapping $P : \mathfrak{g} \to \mathfrak{gl}(n, \mathbb{K})$ is itself linear. Moreover, using Jacobi-Identity and the antisymmetry property, we can easily check (exercise) that P is a Lie algebra homomorphism and hence defines a representation of $\mathfrak{g}$ in $\mathfrak{gl}(n, \mathbb{K})$. This is called the *adjoint representation* of $\mathfrak{g}$ and is denoted by $ad : \mathfrak{g} \to GL(n, \mathbb{K})$.

Example 9.8.2 Let G be a Lie group and $Aut(G)$ denote the group of all Lie automorphisms of G. Given $f \in Aut(G)$, the map $Df : \mathfrak{g} \to \mathfrak{g}$ is Lie isomorphism of $\mathfrak{g}$ to itself. By the chain rule, it follows that $f \mapsto Df$ defines a linear representation $D : Aut(G) \to GL(n, \mathbb{K})$. Now, for each $g \in G$, consider the conjugation $C_g : G \to G$, given by $h \mapsto ghg^{-1}$. Then C_g is Lie automorphism of G and hence $C_g \in Aut(G)$. Moreover, check that $g \mapsto C_g$ is a group homomorphism. Composing this with $f \mapsto Df$ considered above, yields a linear representation of G on $\mathfrak{g}$ denoted by $Ad : G \to \mathfrak{gl}(n, \mathbb{K})$ and called the *adjoint representation of* G.

Theorem 9.8.1 $D(Ad)_e = ad.$

Proof: For any two elements $\mathbf{v}, \mathbf{u} \in \mathfrak{g}$, we have to show that $D(Ad)_e(\mathbf{v})(\mathbf{u}) = ad_{\mathbf{v}}(\mathbf{u}) = [\mathbf{u}, \mathbf{v}]$. Consider the curve $t \mapsto \text{Exp}\, t\mathbf{v}$ in G passing through e and with tangent at e equal to $\mathbf{v}$. Then

$$\begin{aligned} D(Ad)(\mathbf{v})(\mathbf{u}) &= \left(\lim_{t \to 0} \frac{Ad(\text{Exp}\, t\mathbf{v}) - Id}{t} \right)(\mathbf{u}) \\ &= \left(\lim_{t \to 0} \frac{D(C_{\text{Exp}\, t\mathbf{v}}) - Id}{t} \right)(\mathbf{u}) \\ &= \lim_{t \to 0} \frac{[\mathbf{u}, t\mathbf{v}]}{t} \quad \text{(using Corollary 9.5.2)} \\ &= [\mathbf{u}, \mathbf{v}]. \end{aligned}$$

♠

Definition 9.8.3 The *center* $z(\mathfrak{g})$ of a Lie algebra $\mathfrak{g}$ is defined to be the set of all $\mathbf{u} \in \mathfrak{g}$ such that $ad_{\mathbf{u}} = 0$.

Remark 9.8.1 Clearly $z(\mathfrak{g})$ is a subalgebra of G.

Theorem 9.8.2 *Let G be a connected Lie group. Then its center is a closed Lie subgroup with its Lie algebra equal to the center $z(\mathfrak{g})$ of $\mathfrak{g}$.*

Proof: The representation $Ad : G \to GL(n, \mathbb{R})$, where $n = \dim G$ is a Lie homomorphism. The kernel of Ad is therefore a closed Lie subgroup with its Lie algebra equal to the kernel of $ad : \mathfrak{g} \to \mathfrak{gl}(n, \mathbb{R})$, which is, by definition, the center of $\mathfrak{g}$. From Theorem 9.5.2, it follows that $Ad(g) = D(C_g)_e = Id \Longleftrightarrow C_g = Id \Longleftrightarrow g \in z(G)$. ♠

Remark 9.8.2 That the center $z(G)$ is a closed Lie subgroup also follows from Theorem 9.7.1, once we observe that it is a closed topological subgroup.

Definition 9.8.4 By an *ideal in a Lie algebra* $\mathfrak{g}$ we mean a subalgebra (subspace) $\mathfrak{h}$ such that $[u, v] \in \mathfrak{h}$ for all $u \in \mathfrak{h}, v \in \mathfrak{g}$.

Remark 9.8.3 The center of a Lie algebra is an ideal.

Theorem 9.8.3 *Let H be a connected Lie subgroup of a connected Lie group G. Then H is normal in G iff its Lie algebra $\mathfrak{h}$ is an ideal in $\mathfrak{g}$.*

Proof: For any $g \in G$, consider the automorphism $C_g : G \to G$. Given any connected Lie subgroup H of G, $C_g(H) = gHg^{-1}$ is again a Lie subgroup of G with its Lie algebra equal to $D(C_g)(\mathfrak{h})$. Therefore, H is normal iff $D(C_g)(\mathfrak{h}) = \mathfrak{h}$ for all $g \in G$.

Now suppose $D(C_g)(\mathfrak{h}) = \mathfrak{h}$ for all $g \in G$. From Corollary 9.5.2, we have $D(C_g)(\mathbf{u}) = \mathbf{u} + [\mathbf{u}, \mathbf{v}]$ for $g = \text{Exp}\,\mathbf{v}$. Therefore, it follows that for all $\mathbf{u} \in \mathfrak{h}$ and all $\mathbf{v} \in \mathfrak{g}$, we have, $[\mathbf{u}, \mathbf{v}] \in \mathfrak{h}$. Conversely, if $\mathfrak{h}$ is an ideal, then for all points $g \in \text{Exp}\,(U)$, we have $D(C_g)(\mathfrak{h}) \subset \mathfrak{h}$. The set of all g for which this property holds is a subgroup that contains a neighborhood of $e \in G$. Since G is connected, this subgroup must be the whole of G. This completes the proof. ♠

9.9 Existence of Lie Subgroups

A Lie algebra of a Lie group G can have a subalgebra that does not correspond to any Lie subgroups: here is a simple example.

Example 9.9.1 Take $G = \mathbb{T}^2 = \mathbb{S}^1 \times \mathbb{S}^1$, and $\mathfrak{h}$ as the linear span of $(1, \pi)$ in $\mathfrak{g} = \mathbb{R}^2$. It follows that the corresponding algebra homomorphism is induced by the homomorphism $f : \mathbb{R} \to torus^2$ given by $t \to (e^{\imath t}, e^{\imath\pi t})$, which has dense image in $\mathbb{T}^2$. (Compare Exercise 9.3.1.(6).) Therefore, if there were a Lie subgroup $H \subset T^2$ with its Lie algebra equal to $\mathfrak{h}$ as above, then H would contain the image of f which is dense in a 2-dimensional manifold. So, H cannot be a submanifold.

However, the situation is no worse than this. Thus, we are led to make the following definition.

Definition 9.9.1 By a *virtual subgroup H of a Lie group G*, we mean a Lie group H together with a smooth injective immersive homomorphism $\phi : H \to G$. (We identify H with the underlying set $\phi(H)$ and get rid of ϕ.)

Remark 9.9.1 Thus, every Lie subgroup is a virtual subgroup. As seen above (the irrational) image of f in $\mathbb{T}^2$ is a virtual subgroup but not a Lie subgroup. The induced topology on a virtual subgroup H is often weaker than its topology. Of course, these two topologies coincide iff H is a submanifold iff H is a Lie subgroup iff H is locally closed. In particular, a virtual subgroup whose underlying set is open or closed is a Lie subgroup.

Lemma 9.9.1 Any immersive homomorphism $\alpha : H \to G$ of Lie groups defines a virtual subgroup $\alpha(H)$ of G.

Remark 9.9.2 Since $\phi : H \to G$ is immersive, it follows that we can identify the Lie algebra $\mathfrak{h}$ of H with a subalgebra of $\mathfrak{g}$. It is also clear that two connected virtual subgroups having the same Lie subalgebra are equal. So, we are now interested in the existence part.

Theorem 9.9.1 *Let $\mathfrak{h}$ be a Lie subalgebra of the Lie algebra $\mathfrak{g}$ of a Lie group G. Then there is a unique connected virtual subgroup H of G such that its Lie algebra is equal to $\mathfrak{h}$.*

We shall postpone the proof of this theorem to the end of this section. Let us see how much of the Lie theory we can build up based on this result.

As an immediate corollary, we have,

Theorem 9.9.2 *Let $\phi : \mathfrak{h} \to \mathfrak{g}$ be a Lie algebra homomorphism where $\mathfrak{h}, \mathfrak{g}$ are Lie algebras of Lie groups H, G, respectively. Then there is a local homomorphism $f : H \to G$ such that $df_e = \phi$. Further, if H is simply connected, then f extends to a homomorphism $f : H \to G$.*

Proof: Let $\Gamma_\phi \subset \mathfrak{h} \times \mathfrak{g}$ be the graph of ϕ. Check that Γ_ϕ is a sub-algebra of $\mathfrak{h} \times \mathfrak{g}$. Since $\mathfrak{h} \times \mathfrak{g}$ is the Lie algebra of $H \times G$, it follows that there is a unique connected virtual subgroup $K \subset H \times G$ whose Lie algebra is Γ_ϕ. Consider the two projections $\pi_1 : H \times G \to H$ and $\pi_2 : H \times G \to G$. The tangent space to K at (e, e) is equal to Γ_ϕ and in fact, $d(\pi_1)_{(e,e)}(\mathbf{v}, \phi(\mathbf{v})) = \mathbf{v}$. Therefore, $d(\pi_1) : T_{(e,e)}(\Gamma_\phi) \to T_e H$ is an isomorphism and hence $\pi_1 : \Gamma_\phi \to H$ is a local diffeomorphism onto a neighborhood U of $e \in H$ say. Let $f = \pi_2 \circ (\pi_1)^{-1}$ on U. It is easily verified that $df_e = \phi$. Being the local inverse of a homomorphism, π_1^{-1} is a local homomorphism. Since π_2 is a homomorphism, it follows that f is a local homomorphism. The last part follows from the Principle of Monodromy. (See Exercise 9.2.4.(5).) ♠

Corollary 9.9.1 If G, G' are any two simply connected Lie groups, then $G \approx G'$ iff their Lie algebras are isomorphic: $\mathfrak{g} \approx \mathfrak{g}'$.

Example 9.9.2 Consider a connected 1-dimensional Lie group G. By the above corollary, its universal covering group is isomorphic to the additive group of real numbers. Thus, depending on whether G were compact or not, it is either isomorphic to $\mathbb{S}^1$ or $\mathbb{R}$. In particular, we know that the multiplicative group of positive real numbers is isomorphic to $\mathbb{R}$ via the exponential map.

Corollary 9.9.2 Given any $\sigma \in \mathfrak{g}$, there is a homomorphism $h : \mathbb{R} \to G$ such that $h_*(L) = \sigma$, where $L = \frac{d}{dt}$ is the generator of the Lie algebra of $\mathbb{R}$.

Proof: For, $\mathbb{R}$ is simply connected and the association $L \mapsto \sigma$ extends uniquely to a Lie algebra homomorphism $\phi : \mathfrak{r} \to \mathfrak{g}$ and hence we can apply the latter part of Theorem 9.9.2.

Theorem 9.9.3 *Let $\mathfrak{g}$ be a Lie algebra over $\mathbb{R}$. Then there exists a connected Lie group G whose Lie algebra is isomorphic to the quotient algebra $\mathfrak{g}/z(\mathfrak{g})$.*

Proof: Consider the adjoint representation $ad : \mathfrak{g} \to \mathfrak{gl}(n, \mathbb{R})$ where $n = \dim \mathfrak{g}$. The kernel of this homomorphism is precisely equal to the center $z(\mathfrak{g})$. Therefore the image of ad is a Lie sub-algebra $\mathfrak{h}$ of $\mathfrak{gl}(n, \mathbb{R})$ isomorphic to $\mathfrak{g}/z(\mathfrak{g})$. The connected virtual subgroup H of $G \subset GL(n, \mathbb{R})$ has its Lie algebra equal to $\mathfrak{h}$. But then H is the image of a one-to-one immersion of a Lie group which is the Lie group that we are looking for. ♠

Remark 9.9.3 It is a fact (much harder to prove) that every Lie algebra is the Lie algebra of a connected, simply connected Lie group.

Lemma 9.9.2 Let $\mathfrak{g}$ be a Lie algebra. Then the linear span

$$[\mathfrak{g}, \mathfrak{g}] := L(\{[\mathbf{u}, \mathbf{v}] \ : \ \mathbf{u}, \mathbf{v} \in \mathfrak{g}\})$$

is an ideal in $\mathfrak{g}$.

Proof: Easy. ♠

Definition 9.9.2 The ideal $[\mathfrak{g}, \mathfrak{g}]$ is called the *derived sub-algebra of* $\mathfrak{g}$ and often denoted by $\mathfrak{g}'$.

Remark 9.9.4 A Lie algebra $\mathfrak{g}$ is abelian iff $\mathfrak{g}' = 0$. It is easy to see that the factor algebra $\mathfrak{g}/\mathfrak{g}'$ is always abelian. Moreover, $\mathfrak{g}'$ is contained in all ideals $\mathfrak{h}$ of $\mathfrak{g}$ such that $\mathfrak{g}/\mathfrak{h}$ is abelian.

Theorem 9.9.4 *Let G be a connected Lie group. Then the commutator subgroup $[G,G]$ is a connected Lie subgroup whose Lie algebra is equal to $[\mathfrak{g},\mathfrak{g}]$.*

Proof: First consider the case, when G is simply connected. Let $q : \mathfrak{g} \to \mathfrak{g}/[\mathfrak{g},\mathfrak{g}]$ be the quotient homomorphism of Lie algebras. Observe that $\mathfrak{g}/[\mathfrak{g},\mathfrak{g}]$ is abelian and hence is isomorphic to the Lie algebra of the additive group $\mathbb{R}^d$, where $d = \dim \mathfrak{g}/[\mathfrak{g},\mathfrak{g}]$. By Theorem 9.9.2, there is a Lie-homomorphism $f : G \to \mathbb{R}^d$ such that $df_e = q$. Clearly $Ker\, f$ is a closed Lie subgroup of G, whose Lie algebra is $[\mathfrak{g},\mathfrak{g}]$. Since $d(f)_e$ is surjective, by implicit function theorem, f defines a surjective map of some neighborhood of $e \in G$ to a neighborhood of 0 in $\mathbb{R}^d$. It follows that $f : G \to \mathbb{R}^d$ itself is surjective and hence $G/Ker\, f \approx \mathbb{R}^d$. From the homotopy exact sequence of the fibration $G \to G/Ker\, f$, viz.,

$$\pi_1(G/Ker\, f) \to \pi_0(Ker\, f) \to \pi_0(G)$$

it follows that $Ker\, f$ is connected. (See Corollary 9.3.2.)

Since $\mathbb{R}^d$ is abelian, it follows that $Ker\, f$ contains $[G,G]$. We claim that $Ker\, f = [G,G]$. It is enough to prove that a neighborhood of e in $Ker\, f$ is contained in $[G,G]$.

Choose a basis of $[\mathfrak{g},\mathfrak{g}]$ consisting of elements of the form

$$\{X_i := [\mathbf{u}_i, \mathbf{v}_i] \ : \ i = 1, \ldots, k\}$$

where $\mathbf{u}_i, \mathbf{v}_i \in \mathfrak{g}$. We may assume that $\mathbf{u}_i, \mathbf{v}_i$ are in a neighborhood V of $0 \in \mathfrak{g}$ where $\mathrm{Exp} : V \to G$ is a diffeomorphism onto a neighborhood of $e \in G$. Complete this to a basis

$$\{X_1, \ldots, X_k, X_{k+1}, \ldots, X_n\}$$

of $\mathfrak{g}$ and take canonical coordinates x_i of G around a smaller neighborhood V' of e with respect to this basis. Put $g_i(t) = \mathrm{Exp}\,(t\mathbf{u}_i); h_i = \mathrm{Exp}\,\mathbf{v}_i, i = 1, \ldots, k$ and consider $\gamma_i(t) = [g_i(t), h_i]$. Clearly, $\gamma_i(0) = e$ and $\gamma_i(t) \in [G,G]$. From equation (9.36), it follows that $\gamma_i'(0) = X_i$. Therefore, as in the proof of Theorem 9.7.1, (see the Remark 9.7.1), the map Φ defined as in (9.37), gives a diffeomorphism of V' with a neighborhood of $e \in Ker\, f$. At the same time, we know that Φ takes values in $[G,G]$. Therefore, it follows that a neighborhood of $e \in Ker\, f$ is contained $[G,G]$. Since $Ker\, f$ is connected, it follows that $Ker\, f \subset [G,G]$. This completes the proof in the case G is simply connected.

Now in the general situation, consider the simply connected covering homomorphism $\phi : \tilde{G} \to G$. Then $[\tilde{G},\tilde{G}]$ is the connected subgroup whose Lie algebra is $[\mathfrak{g},\mathfrak{g}]$. The homomorphism ϕ induces identity map $Id = D(\phi) : \mathfrak{g} \to \mathfrak{g}$ and hence $\phi([\tilde{G},\tilde{G}])$ is the subgroup of G whose Lie algebra is $[\mathfrak{g},\mathfrak{g}]$. But since ϕ is surjective, $\phi([\tilde{G},\tilde{G}]) = [G,G]$ and $\phi^{-1}([G,G]) = [\tilde{G},\tilde{G}]$. Since ϕ is a local diffeomorphism, it follows that $[G,G]$ is a submanifold of G. ♠

Remark 9.9.5 In general, $[G,G]$ may not be a closed subgroup.

We shall now take up the proof of the existence Theorem 9.9.1

Definition 9.9.3 Let G be a Lie group, $\mathfrak{g}$ its Lie algebra, and let $\mathfrak{h}$ be any subalgebra. Consider the family of all Lie subgroups F of G whose Lie algebras contain $\mathfrak{h}$. The intersection of all these Lie subgroups is again a Lie subgroup (see Exercise 9.7.1.(1)), whose Lie algebra is called the *Maltsev (Malčev) closure of* $\mathfrak{h}$ and is denoted by $\mathfrak{h}^M$. Clearly $\mathfrak{h} \subset \mathfrak{h}^M$.

Lemma 9.9.3 (Maltsev) $[\mathfrak{h}^M, \mathfrak{h}^M] = [\mathfrak{h}, \mathfrak{h}]$.

Proof: Clearly, $\mathfrak{h} \subset \mathfrak{h}^M$ and hence $[\mathfrak{h}, \mathfrak{h}] \subset [\mathfrak{h}^M, \mathfrak{h}^M]$. We have to prove the other inclusion.

We apply Exercise 9.4.1.(4) to the adjoint representation of G taking $W = [\mathfrak{h}, \mathfrak{h}]$ and U to be any subspace of $\mathfrak{g}$ containing W, to conclude that

$$H_U := \{g \in G \; : \; (Ad(g) - Id)(U) \subset [\mathfrak{h}, \mathfrak{h}]\}$$

is a Lie subgroup of G with its Lie algebra

$$\mathfrak{h}_U := \{\mathbf{v} \in \mathfrak{g} \; : \; ad(\mathbf{v})(U) \subset [\mathfrak{h}, \mathfrak{h}]\}. \tag{9.39}$$

Further, suppose that $[U, \mathfrak{h}] \subset [\mathfrak{h}, \mathfrak{h}]$. Then clearly $\mathfrak{h}$ is contained in $\mathfrak{h}_U$. Therefore, $\mathfrak{h}^M \subset \mathfrak{h}_U$. This just means that $[\mathfrak{h}^M, U] \subset [\mathfrak{h}, \mathfrak{h}]$.

In particular, taking $U = \mathfrak{h}$, we get $[\mathfrak{h}^M, \mathfrak{h}] \subset [\mathfrak{h}, \mathfrak{h}]$. Now, taking $U = \mathfrak{h}^M$, we get $[\mathfrak{h}^M, \mathfrak{h}^M] \subset [\mathfrak{h}, \mathfrak{h}]$. ♠

Proof of Existence Theorem 9.9.1

Let $\mathfrak{h}^M$ denote the Maltsev closure of $\mathfrak{h}$. Then from Lemma above we have $[\mathfrak{h}^M, \mathfrak{h}^M] = [\mathfrak{h}, \mathfrak{h}]$. Let F be the Lie subgroup of G such that its Lie algebra $\mathfrak{f} = \mathfrak{h}^M$. Let $\hat{F}$ be the universal covering group of F. Then $\hat{F}/[\hat{F}, \hat{F}]$ is a simply connected abelian group and hence is isomorphic to $\mathbb{R}^n$ for some n. (See Exercise 8 of 9.9.1.) Its Lie algebra is $\mathfrak{h}^M/[\mathfrak{h}, \mathfrak{h}]$ that contains $\mathfrak{h}/[\mathfrak{h}, \mathfrak{h}]$. Therefore, there is a (unique) Lie subgroup J of $\hat{F}/[\hat{F}, \hat{F}]$ with its Lie algebra equal to $\mathfrak{h}/[\mathfrak{h}, \mathfrak{h}]$. Let $\hat{H}$ be the inverse image of this under the quotient homomorphism $\hat{F} \to \hat{F}/[\hat{F}, \hat{F}]$. Clearly, the Lie algebra of $\hat{H}$ is equal to $\mathfrak{h}$. The required virtual subgroup H of G is the image of $\hat{H}$ in $F \subset G$. ♠

Exercise 9.9.1

1. Show that if H is a Lie subgroup of a connected Lie group G and dim H = dim G then $H = G$.

2. Show that a connected Lie group is abelian iff its Lie algebra is abelian.

3. Show that Exp : $\mathfrak{g} \to G$ is a surjective covering homomorphism for any connected abelian Lie group G.

4. By an n-dimensional torus we mean $\mathbb{T}^n := (\mathbb{S}^1)^n := \mathbb{S}^1 \times \cdots \times \mathbb{S}^1$, the Cartesian product of n copies of $\mathbb{S}^1$. Show that $\mathbb{T}^n$ is a matrix group.

5. Show that every $\mathbb{T}^n$ is isomorphic to $\mathbb{R}^n/\mathbb{Z}^n$.

6. Let H be a discrete subgroup of $\mathbb{R}^n$. Show that there exist a linearly independent set $S = \{\mathbf{v}_1, \ldots, \mathbf{v}_k\}$ in $\mathbb{R}^n$ such that the additive subgroup generated by S is equal to H.

7. Let $\{\mathbf{v}_1, \ldots, \mathbf{v}_k\}$ be a linearly independent set in $\mathbb{R}^n$ and let H be the additive subgroup generated by this set. Show that the quotient group $\mathbb{R}^n/H$ is isomorphic to $\mathbb{T}^k \times \mathbb{R}^{n-k}$.

8. Show that a connected, simply connected abelian Lie group is isomorphic to $\mathbb{R}^n$.

9. Show that any connected compact abelian Lie group is isomorphic to a torus. [Hint: Show that Exp : $\mathfrak{g} \to G$ is a surjective homomorphism and hence a covering homomorphism. Then appeal to Exercise 6.]

10. By a *torus T in G*, we mean a subgroup T of G which is isomorphic to $(\mathbb{S}^1)^k$ for some k. T is called a *maximal torus* if it is not contained in another torus in G which is larger than T. Show that every compact Lie group G contains a maximal torus.

11. Let $T \subset G$ be a torus, where G is compact. Show that T is maximal iff any element of G that commutes with every element of T is in T itself.

12. Let R_θ denote the 2×2 matrix representing the rotation through an angle θ in $\mathbb{R}^2$. For any two square matrices A, B, we shall use the notation $diag\,(A, B)$ to denote the matrix whose diagonal blocks are A and B respectively and other blocks are zero:
$$diag\,(A, B) := \begin{pmatrix} A & 0 \\ 0 & B \end{pmatrix}.$$
Let $T \subset G$ be described as below. In each case, show that T is a maximal torus of G.:
(a) $T = \{diag(e^{\imath\theta_1}, \ldots, e^{\imath\theta_n}) \;:\; \theta_i \in [0, 2\pi)\} \subset U(n) = G$.
(b) $T = \{diag(e^{\imath\theta_1}, \ldots, e^{\imath\theta_n}) \;:\; \theta_i \in [0, 2\pi)\} \subset Sp(n) = G$.
(c) $T = \{diag(e^{\imath\theta_1}, \ldots, e^{\imath\theta_{n-1}}, e^{\imath\theta_n}) \;:\; \theta_i \in [0, 2\pi), \theta_n = -\sum_{i=1}^{n-1} \theta_i\} \subset SU(n) = G$.
(d) $T = \{diag(R_{\theta_1}, \ldots, R_{\theta_n}) \;:\; \theta_i \in [0, 2\pi)\} \subset SO(2n) = G$.
(e) $T = \{diag(R_{\theta_1}, \ldots, R_{\theta_n}, 1) \;:\; \theta_i \in [0, 2\pi)\} \subset SO(2n+1) = G$.
These are called *standard maximal tori* in the respective matrix groups.

13. Use the above list to compute the center $z(G)$ in each case.

14. Use the above exercise to show that $SU(2)$ is not isomorphic to $SO(3)$ and $U(n)$ is not isomorphic to $SU(n) \times U(1)$.

15. Show that any conjugate of a maximal torus is also a maximal torus.

16. For each G as in Exercise 12, show that conjugates of the standard maximal torus T cover the entire group, i.e., for each $x \in G$ there exists $g \in G$ such that $gxg^{-1} \in T$. (This is just the diagonalization theorem for normal matrices, in case of $G = O(n), U(n)$. For other groups, you have to work a little further, especially for $Sp(n)$.

17. By a generic element (also called a generator) in a Lie group G we mean an element $g \in G$ such that the closure of the subgroup generated by g is the whole group G. Show that every torus T has a generic element. (One also says that T is monogenic.) Indeed, most of the elements in T are generic.

18. Show that if $g \in T$ is a generator of T as above, then the centralizer of g is T.

19. Let T be a maximal torus in a compact connected Lie group G. Let $N_G(T)$ be the normalizer of T in G. Show that $W := N_G(T)/T$ is finite. (W is called the *Weil group of G.*).

20. Let $T \subset G$ be a maximal torus, $\mathfrak{t}$ be the corresponding Lie subalgebra. Let $\mathbf{v} \in \mathfrak{t}$ be such that $\mathrm{Exp}\,\mathbf{v} = g$ is a generator of T. Consider the right-invariant vector field $\sigma(h) = D(R_h)_e(\mathbf{v})$. Under the quotient map $q : G \to G/T$, show that σ defines a vector field $\hat{\sigma}$ on G/T, which vanishes precisely at points of $W \subset G/T$. Show that at each of these points the index of $\hat{\sigma}$ is the same. Conclude that $\chi(G/T) \neq 0$. (Indeed, one can also prove that $\chi(G/T) = \#(W)$.)

21. Let T be a maximal torus in a connected Lie group G. Show that given any $x \in g$ there exists $g \in G$ such that $gxg^{-1} \in T$. (Hint: Show that the diffeomorphism $L_g : G/T \to G/T$ has a fixed point.]

22. Show that any two maximal tori in a connected Lie group are conjugate.

23. Describe the Lie subalgebra of the standard maximal torus T of G in each case as in Exercise 12.

24. The reflection in the hyperplane B in $\mathbb{R}^n$ perpendicular to a unit vector $\mathbf{v}$ is given by $\mathbf{u} \mapsto \mathbf{u} - 2\langle \mathbf{u}, \mathbf{v}\rangle \mathbf{v}$. Show that every reflection is an element of $O(n)$ but not $SO(n)$.

25. Every element of $O(n) \setminus SO(n)$ is a product of an element of $SO(n)$ with a reflection.

26. Conjugate of a reflection by an element of $O(n)$ is a reflection.

27. Recall from your complex analysis course that any rotation in the plane can be written as the composite of two reflections, i.e., $z \mapsto e^{\imath\theta} z$ can be written as the composite of $z \mapsto \bar{z}$ followed by reflection in the line making an angle $\theta/2$ with the x-axis. Deduce that every element of the maximal torus in $SO(n)$ is a product of reflections. Deduce that every element of $O(n)$ is a product of reflections. Also, give a direct (geometric) proof of this fact.

9.10 Foliation

In this section, we give a brief introduction to the fascinating topic of foliations and give just one application, viz., the classical proof of the existence theorem for Lie groups.

Definition 9.10.1 Let M be a smooth manifold of dimension m. By an *atlas of submersions of codimension* d *on* M, we mean a collection $\mathcal{F} = \{(U_\alpha, f_\alpha)\}$ where $\{U_\alpha\}$ is an open covering of M and $f_\alpha : U_\alpha \to \mathbb{R}^{m-d}$ are submersions satisfying the following compatibility condition: for each $x \in U_\alpha \cap U_\beta$, there exists an open neighborhood W of $f_\alpha(x)$ in $\mathbb{R}^{m-d}$ and a diffeomorphism $h : W \to f_\beta(f_\alpha^{-1}(W))$ such that $h \circ f_\alpha = f_\beta$. A maximal atlas of submersions of codimension d is called a *foliation of dimension* d or simply a d*-foliation.*

Remark 9.10.1 Two atlases are compatible with each other if their union is an atlas. Thus, it follows that given any atlas there is a unique atlas that is maximal and contains the given one. The differential structure of a smooth manifold M itself is a 0-foliation on M. A submersion $f : M \to N$ defines a d-foliation where, $d = \dim M - \dim N$. Given a d-foliation $\mathcal{F}$ on M, we get a vector sub-bundle $E(\mathcal{F})$ of the tangent bundle TM, by taking $E(\mathcal{F})_x = KerDf_\alpha$, where $x \in U_\alpha$. (Use compatibility condition to see that this is well-defined and use the implicit function theorem to see the local triviality.) Also it is clear that if $\mathcal{F}_1$ and $\mathcal{F}_2$ are compatible d-foliations then $E(\mathcal{F}_1) = E(\mathcal{F}_2)$. The following lemma will enable us to study the interrelation between subbundles of TM and foliations on M.

Lemma 9.10.1 Let U be an open subset of $\mathbb{R}^n$ and $f, g : U \to \mathbb{R}^m$ be two submersions such that $Ker\, Df = Ker\, Dg$ on U. Then for each $p \in U$, there is a neighborhood W of $f(p)$ in $\mathbb{R}^m$ and a diffeomorphism $h : W \to g(f^{-1}(W)) = W'$ such that $g = h \circ f$ on $V = f^{-1}(W)$.

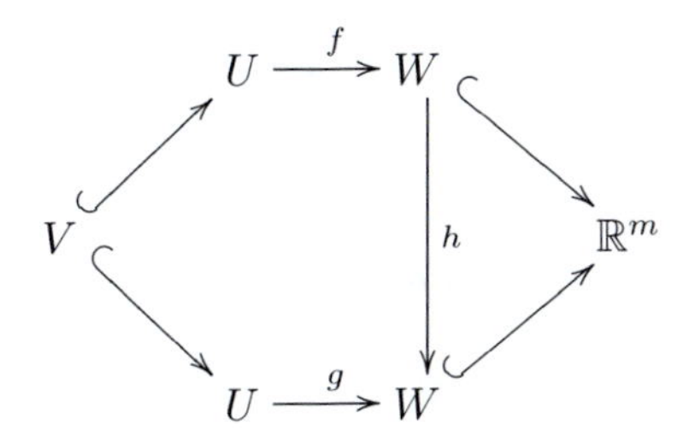

Proof: This was an exercise to you earlier (see Exercise 1.4). Since we need to use it now, here is a solution. By the implicit function theorem, we may replace f, g with $f \circ \phi$ and $g \circ \phi$ for a suitable diffeomorphism and assume that f is actually the coordinate projection

$(x_1, \ldots, x_n) \mapsto (x_1, \ldots, x_m)$. Now the condition that $Ker\, Df = Ker\, Dg$ implies that $\frac{\partial g}{\partial x_i} = 0$, for $i = m+1, \ldots, n$. Therefore, on a connected neighborhood U' of the point p, g is independent of the variables $x_{m+1}, \ldots, x_n$ and hence factors through a map $h : W' = f(U') \to g(U')$, i.e., $g = hf$. Clearly, the rank of h is equal to m on W' and hence, by the inverse function theorem, on a smaller neighborhood W of $f(p)$, it is a diffeomorphism. ♠

Corollary 9.10.1 Let $\mathcal{F}_i, i = 1, 2$ be any two foliations on M such that $E(\mathcal{F}_1) = E(\mathcal{F}_2)$. Then $\mathcal{F}_1, \mathcal{F}_2$ are compatible.

Corollary 9.10.2 Let $E \subset TM$ be a subbundle of rank d. Suppose there is an open covering $\{U_\alpha\}$ of M and submersions $f_\alpha : U_\alpha \to \mathbb{R}^{n-d}$ such that $Ker\, Df_\alpha = E_x, x \in U_\alpha$, for all α. Then the collection $\{(U_\alpha, f_\alpha)\}$ is an atlas of submersions.

Definition 9.10.2 We shall call a subbundle of E of TM *completely integrable* if there exists a foliation $\mathcal{F}$ such that $E = E(\mathcal{F})$.

Clearly the entire bundle TM and the 0-bundle are completely integrable. However, it is not at all clear, why there should be any completely integrable proper subbundles of TM for an arbitrary manifold M.

Example 9.10.1 *Every subbundle of rank* 1 *of* TM *is completely integrable.* By corollary 9.10.2, this is a local property. Thus, we may assume that U is an open neighborhood $0 \in \mathbb{R}^n$, on which a nowhere vanishing vector field σ is given. We have to find a submersion $f : W \to \mathbb{R}^{n-1}$ in a neighborhood of W of 0 such that $D(f)(\sigma) = 0$. In fact, we shall find a coordinate system $\{y_1, \ldots, y_n\}$ around 0, such that $\partial_n = \sigma$ and then take f to be the projection onto the first $(n-1)$ factors. First, choose a coordinate system such that $\sigma(0) = \partial_n$. Consider the initial value problem on U :

$$\frac{\partial g}{\partial t}\Big|_{(x,t)} = \sigma(g(x,t)), \; g(x,0) = x. \tag{9.40}$$

There exists a neighborhood W of 0, $\epsilon > 0$ and a smooth solution of this problem $g : W \times (-\epsilon, \epsilon) \to W$. Since $g(x, 0) = x$, it follows that $\frac{\partial g}{\partial x_i} = 1$ for $i = 1, 2 \ldots, n-1$. Let $V = W \cap (\mathbb{R}^{n-1} \times 0)$. It follows that $g : V \times (-\epsilon, \epsilon) \cap W \to W$ is of maximal rank at 0 and hence is a diffeomorphism in a neighborhood of 0. Now the coordinate functions of g can be taken as $y_i, i = 1, 2 \ldots, n$.

Let us denote by $\Gamma(E)$ the set of all smooth sections of a smooth vector bundle E over M. Recall that $\Gamma(E)$ forms a module over the ring $C^\infty(M)$. Also recall from Section 5.6 that on $\Gamma(TM) = \mathcal{T}(M) = \mathcal{D}(M)$, we have the Lie bracket operation $[\, ,\,]$ which makes it into a Lie algebra over $\mathbb{R}$. We now consider C^∞-submodules $\mathcal{E}$ of $\Gamma(TM)$ which are Lie subalgebras, i.e., those for which $\sigma, \tau \in \mathcal{E} \implies [\sigma, \tau] \in \mathcal{E}$. As such there is no reason to expect that given a vector subbundle $E \subset TM$, the submodule $\Gamma(E)$ will be a subalgebra of $\Gamma(TM)$.

Theorem 9.10.1 Theorem of Frobenius: *For any subbundle* $E \subset TM$, $\Gamma(E)$ *is a Lie subalgebra of* $\Gamma(TM)$ *iff* E *is completely integrable.*

Proof: Suppose $E = E(\mathcal{F})$ for some foliation $\mathcal{F}$ of M. Let $\sigma, \tau \in \Gamma(E)$. To show that $[\sigma, \tau] \in \Gamma(E)$ is purely a local problem. So, let U be a neighborhood of a point on which we have a submersion $f : U \to \mathbb{R}^{m-d}$ such that $Ker\, Df = E_x, x \in U$. This means that $\sigma(f) = 0 = \tau(f)$, on U. But then it follows that $\sigma \circ \tau(f) = \tau \circ \sigma(f) = 0$ on U. Therefore $[\sigma, \tau](f) = 0$. Therefore, $[\sigma, \tau] \in E_x, x \in U$.

To prove the converse, we induct on the rank d of E. For $d = 1$, the above example gives the result. So, let us assume that $d \geq 2$ and for every manifold Y and every subbundle $E' \subset TY$ of rank $d-1$ such that $\Gamma(E')$ is a subalgebra of $\Gamma(TY)$, E' is completely integrable. Let now E be a subbundle of rank d and that $\Gamma(E) \subset \Gamma(TM)$ is a subalgebra. Once again, by the above Corollary 9.10.2, the problem of showing that E is completely integrable is local in nature. Therefore, we may simply assume that $M = \mathbb{R}^n$ and we have a system $\{\sigma_1, \ldots, \sigma_d\}$ of vector fields that are everywhere independent in a neighborhood U of $0 \in \mathbb{R}^n$. From the example above, we may further assume that $\sigma_1 = \partial_1$ and $\sigma_i(x_1) = 0$ for $i \geq 2$ (by subtracting the appropriate x_1-components). Put $W := \{x \in U : x_1 = 0\}$ and $\tau_i = \sigma_i|_W$, $i \geq 2$. Clearly, τ_i are tangential to W and hence span a subbundle E' of rank $(d-1)$ on W. Observe that every element of $\Gamma(E')$ is a restriction of an element of $\Gamma(E)$ to W. Thus, for $i = 1, 2$, consider $\omega_i = \sum_{j=2}^d a_{ij}\tau_j$ on W. There exists smooth functions b_{ij} on $W \times (-\epsilon, \epsilon)$ such that $b_{ij}|_W = a_{ij}$. Put $\hat{\omega}_i = \sum_{j \geq 2} b_{ij}\sigma_j \in \Gamma(E)$. We have,

$$[\omega_1, \omega_2] = [\hat{\omega}_1, \hat{\omega}_2]|_W.$$

Since $\Gamma(E)$ is a subalgebra, $[\hat{\omega}_1, \hat{\omega}_2]$ is a linear combination of $\sigma_i, i \geq 1$ at every point of $W \times (-\epsilon, \epsilon)$. Since $\hat{\omega}_i(x_1) = 0$, it follows that $[\hat{\omega}_1, \hat{\omega}_2](x_1) = 0$. Therefore $[\hat{\omega}_1, \hat{\omega}_2]$ is actually a linear combination of $\sigma_i, i \geq 2$. Therefore, $[\hat{\omega}_1, \hat{\omega}_2]|_W \in \Gamma(E')$. This proves that $\Gamma(E')$ is a subalgebra of $\Gamma(TW)$. By induction hypothesis, it is completely integrable.

Let $f : W' \to \mathbb{R}^{m-d}$ be a submersion where W' is a neighborhood of 0 in W such that $Ker\, Df = E'$. We claim that the submersion $f\pi : \pi^{-1}(W') \to \mathbb{R}^{m-d}$ is the required one.

It is enough to prove that $E \subset Ker\, D(f\pi)$. Write $f\pi = (f_1, \ldots, f_{m-d})$ where each $f_j = f_j(x_2, \ldots, x_m)$. It is then enough to show that $\sigma_i(f_j) = 0$ for all $1 \leq i \leq d$, and $1 \leq j \leq n$. Note that to begin with we have $\sigma_1(f_j) = 0$ for all j. So, fix any j and for brevity write $g = f_j$. We have to show that $\alpha_i := \sigma_i(g) = 0$ for all $2 \leq i \leq d$. Again, to begin with we have for any $q \in W'$, $\alpha_i(0, q) = (\tau_i)_q(g) = 0$ since $g = f_j \in Ker\, Df$. Since $\sigma_1(g) = 0$, we have

$$\sigma_1(\alpha_i) = \sigma_1\sigma_i(g) = (\sigma_1\sigma_i - \sigma_i\sigma_1)(g) = [\sigma_1, \sigma_i](g).$$

Since $\Gamma(E)$ is a Lie subalgebra, it follows that we can write $[\sigma_1, \sigma_i] = \sum_{k=1}^d a_k\sigma_k$. Therefore, it follows that $[\sigma_1, \sigma_i](g) = \sum_{k=2}^d a_k\alpha_k$. Thus, for each fixed $q \in W'$ the $(d-1)$ functions $\alpha_i(x_1, q), 2 \leq i \leq d$ satisfy the initial value problem given by the following system of (d-1) linear differential equations:

$$\frac{\partial \alpha_i}{\partial x_1} = \sum_{k=2}^{d} a_k\alpha_k; \quad \alpha_i(0, q) = 0. \tag{9.41}$$

Since $\alpha_i(x_1, q) = 0$ is one solution of (9.41), by the uniqueness of the solution, it follows that $\alpha_i(x_1, q) = 0$. ♠

Definition 9.10.3 Let $\mathcal{F}$ be a d-foliation of a manifold X. By an *integral manifold* of $\mathcal{F}$ we mean a pair (Y, f), where Y is a smooth manifold and $f : Y \to X$ is an injective immersion such that $Df_y(T_yY) \subset E_{f(y)}(\mathcal{F})$.

On the collection of all connected integral manifolds of $\mathcal{F}$, there is an obvious partial ordering, viz., $(Y_1, f_1) \leq (Y_2, f_2)$ iff $Y_1 \subset Y_2$ and $f_1 = f_2|_{Y_1}$. A connected integral manifold of dimension d, maximal with respect to this ordering is called a *leaf* of $\mathcal{F}$.

Remark 9.10.2
(i) Often we allow ourselves to confuse image $f(Y)$ of f with the leaf (Y, f). In general, the image of a leaf need not be a submanifold of the foliated manifold X.
(ii) If $\mathcal{F}$ is the foliation of a fibration then of course the leaves are submanifolds and every

submanifold of the fiber is an integral manifold. For any foliation $\mathcal{F}$, if $(U, f) \in \mathcal{F}$ then any connected component of $f^{-1}(f(x)) \subset U$ passing through x is an integral manifold, which is also a submanifold called a *slice* of $\mathcal{F}$ at x.
(iii) It is not difficult to see that every leaf is a union of slices. Also, note that the dimension of a slice is always equal to d (where $\mathcal{F}$ is a d-foliation)

Lemma 9.10.2 Let $(U, f) \in \mathcal{F}$ and S be a slice in U.
(a) Let (Y, g) be a connected integral manifold of $\mathcal{F}$ such that $g(Y) \subset U$ and $g(Y) \cap S \neq \emptyset$. Then $g(Y) \subset S$.
(b) For any $(V, h) \in \mathcal{F}$, at most countably many slices of V will intersect S.

Proof: Since $D(f \circ g) = D(f) \circ D(g) = 0$, it follows that $f \circ g$ is locally constant. Since Y is connected, this implies $f \circ g$ is a constant. This implies (a).

Let S' be another slice of $\mathcal{F}$. Let W be a connected component of $S \cap S'$. Then $\iota : W \to S$ is an immersion and since both have same dimension, it follows that W is open in S. We conclude that $S \cap S'$ is open in both S and S'.

Therefore, $\{S \cap S' \,:\, S'\}$ forms a disjoint family Ω of open sets in S as S' varies. Now let S' vary over all slices of (V, h). Clearly, two distinct slices of (V, h) do not intersect. Therefore, by II-countability of S, Ω has only countably many members. Therefore, only countably many of S' could intersect S. ♠

Lemma 9.10.3 Every slice is contained in a d-dimensional leaf.

Proof: Let S be a slice and $\mathcal{L}$ be the family of all d-dimensional connected integral manifolds (Y', f') that contain S. If $\{(Y_k, f_k)\}$ is a chain in $\mathcal{L}$, then we can take $Y = \cup_k Y_k$ and $f : Y \to X$ to be such that $f|Y_k = f_k$. Being a countable union of an increasing family of smooth manifolds, Y is a smooth manifold, in which each Y_k is open. It follows easily that f is an injective immersion also. Therefore, every chain in $\mathcal{L}$ has an upper bound. Apply Zorn's lemma to conclude that $\mathcal{L}$ has a maximal element, (Y, f), which is easily seen to be a leaf. ♠

Theorem 9.10.2 *Let $\mathcal{F}$ be a d-foliation on X. Then X is the disjoint union of d-dimensional leaves of $\mathcal{F}$.*

Proof: The only thing that we need to see is that any two distinct d-dimensional leaves are disjoint, which follows easily from Lemma 9.10.2. ♠

Proof of Theorem 9.9.1 Let G be a Lie group and $\mathfrak{h}$ be a subalgebra of $\mathfrak{g}$. Recall that $\mathfrak{g}$ itself is the subalgebra of $\Gamma(TG)$ consisting of all left-invariant vector fields on G. If we define $E_g = D(L_g)_e(\mathfrak{h}) \subset T_g(G)$ for each $g \in G$, one can easily verify that E is a vector subbundle of TG. Moreover, $\Gamma(E)$ will be a subalgebra of $\Gamma(TG)$. (Exercise). By Frobenius Theorem 9.10.1, there exists a d-foliation $\mathcal{F}$ on G such that $E = E(\mathcal{F})$, where $d = \dim \mathfrak{h}$. Let H be the maximal leaf of $\mathcal{F}$ passing through $e \in G$. We claim that H is a virtual subgroup of G with its Lie algebra equal to $\mathfrak{h}$. Observe that $\iota : H \hookrightarrow G$ is a smooth injective immersion. Therefore, the only thing that we need to show is that H is closed under multiplication and then being the composite of $H \times H \to G \times G \to G$ the multiplication $H \times H \to H$ will be smooth. Given any $h \in H$, consider the diffeomorphism $L_{h^{-1}} : G \to G$ given by the left multiplication by h^{-1}. Since E is invariant under $D(L_h)$, it follows that $L_{h^{-1}}(H)$ is an integral submanifold of $\mathcal{F}$, which passes through e. By maximality of H this means that $L_{h^{-1}}(H) \subset H$. This just means $h^{-1}H \subset H$ for all $h \in H$, i.e., H is closed under multiplication. ♠

Remark 9.10.3 Observe that the above argument actually shows us that other leaves of $\mathcal{F}$ are nothing but left cosets of H.

Remark 9.10.4 We have seen two proofs of the existence theorem. Recall that in both the proofs, Theorem 9.9.2 on the existence of local homomorphisms was derived from Theorem 9.9.1 on virtual subgroups. An alternative idea is to first prove Theorem 9.9.2 and then derive Theorem 9.9.1. Observe that by virtue of Lemma 9.5.2, we have no choice in defining f in a canonical coordinate neighborhood, viz., $f = \mathrm{Exp}_H \circ \phi \circ (\mathrm{Exp}_G)^{-1}$.

$$\begin{array}{ccc} G & \xrightarrow{f} & H \\ {\scriptstyle \mathrm{Exp}_G}\uparrow & & \uparrow{\scriptstyle \mathrm{Exp}_H} \\ \mathfrak{g} & \xrightarrow{\phi} & \mathfrak{h} \end{array}$$

The problem is in proving that f is a local homomorphism. Indeed, this is an easy consequence of the so called Campbell-Hausdorff series that describes the canonical coordinates of the multiplication in a Lie group G completely in terms of the Lie bracket of the Lie algebra $\mathfrak{g}$. However, this is beyond the scope of this book.

Hints/Solutions to Selected Exercises

§**1.2**

10. See Lemma 8.2.1.

11. Differentiate the identity $f(t\mathbf{x}) = t^k f(\mathbf{x})$ with respect to t and put $t = 1$.

§**1.3**

1. (a) This is a linear map. Therefore, $Df_{(A,B)}(X,Y) = f(X,Y) = X + Y$.
(b) $Dg_{(A,B)}(X,Y) = AY + XB$.
(c) $Dh_A(X) = AX + XA$.

2. This is also a linear map.

3. First, do it for $m = 1$ and coordinatewise.

4. We have

$$D(g \circ f) = D(g)(f(\mathbf{x})) \circ D(f)(\mathbf{x}).$$

Note that the composition on the RHS here is indeed matrix multiplication. Therefore, Leibniz's rule applies as in Exercise 3 above and we have

$$D(D(g \circ f)) = D(D(g) \circ f) \cdot D(f) + (D(g) \circ f) \cdot D^2(f).$$

Since $D(g)$ happens to be a row vector, we convert it into a column vector by taking the transpose, before taking the derivative and then use Exercise 2. Therefore,

$$D(D(g) \circ f) = [D(D(g)^t \circ f)]^t = [D^2(g)^t \circ f \cdot D(f)]^t = D(f)^t \cdot D^2(g) \circ f.$$

The formula (1.46) follows.

§**1.4.**

1. Let $f(x,y) = (x + y^2, y)$, $g(x,y) = (x, y + x^3)$. Then the first map is $g \circ f$ and the second one is $f \circ g$.

2. See Lemma 9.10.1.

§**1.5**

1. $x^2 = y^2 = z^2 = \frac{1}{3}$.

2. Critical points are: $(\sqrt{2}, -\sqrt{2}, -1), (-\sqrt{2}, \sqrt{2}, -1), (0,0,1)$ at which the function takes values $5, 5,$ and 1, respectively. Clearly, the surface is unbounded and hence there is no maximum; the minimum is 1.

3. $(\pm\frac{2}{\sqrt{3}}, \pm\frac{4}{\sqrt{3}}, \pm\frac{2}{\sqrt{3}})$.

4. Same answer for the complex case. Use the spectral theorem.

§**1.6**

1.(a) $(-\pi/2, \pi/2)$; **(b)** $(0, \pi)$; **(c)** $(-\pi/2, \pi/2)$; **(d)** $(0, 1)$.

2. From **1.(a)**, it is clear that the map defines a diffeomorphism $\mathbb{D}^n_r \setminus \{0\} \to \mathbb{D}^n_s \setminus \{0\}$. Also the map extends to a homeomorphism of $\mathbb{D}^n_r$ to $\mathbb{D}^n_s$, sending the origin to the origin. The crux of the matter is to verify that it is continuously differentiable at 0 as well and its derivative at 0 is invertible. In this case, the derivative any point $(\mathbf{v}, t), t \neq 0$, is multiplication by $\cos t$ and hence converges to the Identity map at the origin.

3,4,5,6. Let $\hat{f}$ be any differentiable function in a neighborhood of a and which extends f on

that neighborhood. Then $D\hat{f}_a$ is completely determined by its values on any set containing n independent vectors. In each of these cases, these values in turn are completely determined by f itself, since the domain of f contains curves with n independent tangent vectors.
7. This is a special case of **8.**
8. Consider $f : \mathbb{R}^{n+1} \to \mathbb{R}^m$ given by

$$f(t_1\mathbf{e}_1 + \cdots + t_{n+1}\mathbf{e}_n) = \sum_{i=1}^{n+1} t_i(p_i - p_0).$$

Verify that f defines a diffeomorphism of the standard n simplex Δ_n spanned by $\mathbf{e}_1, \ldots, \mathbf{e}_{n+1}$ with the n simplex spanned by $\{p_0, \ldots, p_n\}$.
9. Indeed you can find a diffeomorphism of the form $z \mapsto az + b$ for some $0 \neq a \in \mathbb{C}, b \in \mathbb{C}$.
10. $\{a\}, \{b\}, \{c\}, \{d, e\}, \{f, g\}, \{h\}, \{i, j\}$.

§**1.7**

1. Take $\phi(\alpha, \beta, t) = \beta + (\alpha - \beta)\gamma_{a,b}(t)$, where γ is given by (1.78).

2. (a) Choose $b = \epsilon/2M, a = b/2$ and $\beta > 0$ such that $(1 - b)\beta < \epsilon/2$ in the above exercise and take $g(\alpha, t) = \phi(\alpha, \beta, t)$. Estimate the area.
(b) Patch two such functions.

3. Let ψ be any bump function on $[0, 1]$ with support in the interior of $[0, 1]$ and such that the area under the curve is 1. Now take h as in the exercise above and put

$$f_{\alpha,\beta}(t) = \int_0^t h(\alpha, \beta, s)ds + \left(1 - \int_0^1 h(\alpha, \beta, s)ds\right) \int_0^t \psi(s)ds.$$

4. Take $e^{-1/x} \sin \frac{\pi}{x}$.

5. Consider a smooth bump-function h as in (1.77) with $a = 0, b = 1$. Put $M = \max\{|h'(t)|\}$ and take $\eta(t) = \frac{1}{M}h$.

6. Assume first that $-\infty < \alpha < \beta < \infty$ and $-\infty < \gamma < \delta < \infty$. From Exercise 3 above it follows that there are smooth functions $\psi_i : [a_{2i}, a_{2i+1}] \to [c_{2i}, c_{2i+1}], i = 0, 1, \ldots, k$ such that

$$\psi_i(a_{2i}) = c_{2i}, \psi_i(a_{2i+1}) = c_{2i+1}, \psi_i'(a_{2i}) = \phi_i'(a_{2i}), \psi_i'(a_{2i+1}) = \phi_{i+1}'(a_{2i+1})$$

and $\psi_i'(t) > 0$ for all t. Put $\psi|_{[a_{2i}, a_{2i+1}]} = \psi_i, i = 0, 1, \ldots, k$ and
$\psi|_{[a_{2i}, a_{2i+1}]} = \phi_i, i = 1, 2 \ldots, k$.
In case any one of the $\alpha, \beta, \gamma, \delta$ is $\pm\infty$, we have only to use the open brackets at the corresponding end. The answer to the last part is NO: Take $\phi(x) = x^2$, which cannot be extended to a diffeomorphism on any interval containing zero.

7. In the construction of bump function α as in (1.79), choose $a = r/2, b = r$, where $r = min\{c - a, b - c\}$ for the given $a < c < b$. Now take $g(x) = \alpha(|t - c|)$ and $f = g/k$, where $k = \int_a^b g(t)\, dt$.

8. Choose α as in (1.79) with a, b replaced by $a + \epsilon$, and $b - \epsilon$ for a suitable $0 < \epsilon < (b - a)/2$ and scale it down by 2max $\{|\alpha'|\}$.

10. There is an open covering $\{U_j\}$ of C and smooth functions $f_j : U_j \to \mathbb{R}$ such that $f_j|_{U_j \cap C} = f$. Let $\{\theta_j\}$ be a smooth partition of unity subordinate to the cover $\{U_j\} \cup \{\mathbb{R}^n \setminus C\}$. Take $\hat{f} = \sum_j \theta_j f_j$.

11. Take $r > 0$ so that f vanishes on the closed ball $\bar{B}_r(p)$. Apply the smooth Urysohn's Lemma (Corollary 1.7.1) to the disjoint closed sets $F_0 = \bar{B}_r(p)$ and $F_1 = \text{supp}\, f$.

12. Reflexivity is obvious. For symmetry, the map $(x,t) \mapsto H(x, 1-t)$ defines a homotopy from g to f. Finally suppose $H, G : X \times \mathbb{R} \to Y$ are smooth maps such that $H_t = f, H_s = g, G_p = g, G_q = h$. For transitivity, without loss of generality we can assume that $t < s < p < q$. By composing each H, G with appropriate smooth step functions as in (1.78), we can assume that $H_r = f, r < t, H_r = g, r > s, G_r = g, r < p, G_r = h, r > q$. We can now define $J(x,r) = H(x,r), r < p$ and $J(x,r) = G(x,r), r > s$.

§**1.8**

1. Apply Rolle's theorem inductively to get a sequence $x_{n,k} \to 0$ such that $f^{(n)}(x_{n,k}) = 0$ and hence by continuity conclude that $f^{(n)}(0) = 0$. (See Exercise 1.7.4 for a nontrivial example of such a function. Also note that there does not exist any such analytic function.)
3. Each of them has a limit equal to 0 at $(0,0)$ and hence by taking the value also equal to 0 at $(0,0)$ each of them becomes continuous.
4. For (i) and (ii), use the fact addition and multiplication $\mathbb{R} \times \mathbb{R} \to \mathbb{R}$ are continuous. For (iii) and (iv), use the fact

$$\max\{\mathrm{f}, \mathrm{g}\} = \frac{\mathrm{f} + \mathrm{g} + |\mathrm{f} - \mathrm{g}|}{2}; \quad \text{and } \min\{, \mathrm{f}, \mathrm{g}\} = \frac{\mathrm{f} + \mathrm{g} - |\mathrm{f} - \mathrm{g}|}{2}.$$

If f and g are differentiable so are $f \pm g$ and fg. Maximum and minimum do not preserve differentiability as illustrated by $f(x) = x, g(x) = -x$. One can actually determine the troublesome points.
5. Every polynomial function in any number of variables is smooth.
6. Use the fact that if $\alpha(x)$ is bounded and $\beta(x) \to 0$ then $\alpha(x)\beta(x) \to 0$ to conclude that both functions are continuous everywhere.
9. Both the iterated limits exist and are equal to 0. Yet the total limit does not exist because, along the lines $y = x$ and $y = 2x$, the two limits are different.
10. Observe that taking limit as $(x,y) \to (0,0)$ is same as taking limit as $r \to 0$ in polar coordinates.
(i) and (iii) The function is independent of r and is a nonconstant. So the limit does not exist.
(ii) As $r \to 0$, the function $\frac{|x|+|y|}{x^2+y^2} \to +\infty$. The required limit is $\pi/2$.
12. $D(\tau)_I(B) = -(B + B^t)$.
13. κ is linear.
14. $D(fg)_{\mathbf{x}}(A) = D(f)_{\mathbf{x}}(A)g(\mathbf{x}) + f(\mathbf{x})D(g)_{\mathbf{x}}(A)$.
15. Each entry of $\eta(A)$ is given by a rational function of the entries of A with the denominator being the $det(A)$. Therefore, η is smooth. To compute $D(\eta)$, use Leibniz's rule and see that $D(\eta)_X(A) = -X^{-1}AX^{-1}$.
16. Let $S(X) = X^2$. Then $D(S)_X(A) = XA + AX$.
18. (i) Equating f_r and f_s to 0 yields $r - s$ and $a = 2r^2$. Therefore, if $a < 0$ then there is no extremum in the interior. For $a > 0, r = \sqrt{a/2}$ gives the maximum value of f equal to a^2.
(ii) Putting $a = 2\cos\theta$, if $\cos\theta < 0$ we see that $f(r,s) \leq 0$ for $r, s \in [0,1]$ whereas $2 + \cos\theta \geq 0$. For $\cos\theta \geq 0$, from (i) we have $f(r,s) \leq 4\cos^2\theta \leq 2 + 2\cos\theta$.
19. (a) Both $C(f)$ and $C(f^{-1})$ are continuous and inverses of each other.
(b)-(d). Let $f : \mathbb{S}^1 \to \mathbb{S}^1$ is any smooth map such that $f(1) = 1$ but $f(-1) \neq 1$. Then $C(f)$ will map the line segment $[-1,1]$ to a union of two line segments broken at 0. Hence, $C(f)$ is not differentiable at 0. Therefore, we see that it is necessary that every pair of antipodal points should be mapped onto antipodal pair of points. By continuity, it follows that $f = \pm Id$. And this condition is also enough to conclude that $C(f)$ is a diffeomorphism.

§**2.1** Choose $I_1 = (a_1, b_1)$, such that b_1 is the largest among all intervals containing a_1. Having chosen $I_k = (a_k, b_k)$ choose $I_{k+1} = (a_{k+1}, b_{k+1})$ such that b_{k+1} is the largest among all intervals containing b_k. This process clearly terminates and produces the subcovering with the required property. The latter part follows by a simple induction on the number of intervals involved.

§**2.2**
1. Let α be the bump function as in (1.79), with $0 < a < b < 1$. Enumerate the set of rationals in $\mathbb{R} : q_1, q_2, \ldots$ Take $f(x) = \sum_n \frac{q_n}{\alpha(0)}\alpha(x - 2n)$.
2. Any countable subset of $\mathbb{R}$ is of measure zero. Therefore, there is no contradiction.
4. Yes. If at all points $x \in U$, f has rank 1 means that every point in U is a critical point. Therefore, $f(U)$ is of measure zero in V and cannot be surjective since V is open.
5. Induct on n beginning with the fact that for a nonconstant analytic function $g : \mathbb{C} \to \mathbb{C}$, the zero set is discrete.

§**2.3.**
1. $\pi_1 \otimes \pi_2((1,0),(0,1)) = 1$ whereas, $\pi_2 \otimes \pi_1((1,0),(0,1)) = 0$.
3. Any independent subset of V^* can be completed to a basis. Then use Theorem 2.3.3.
4. If $\{\phi_i\}$ is dependent, then the rows of $((\phi_i(\mathbf{v}_j)))$ are dependent. Otherwise they form a basis for a vector subspace $U \subset V^*$ and we can use (2.33).
§**2.5.**
1. $df = 0 \Longrightarrow \frac{\partial f}{\partial x_i} = 0$ for each i. Given any point $\mathbf{x}$ join it 0 by line segments parallel to the axes. Then f constant along each line segment and therefore $f(\mathbf{x}) = f(0)$ for all $\mathbf{x}$.
2,3,4. In each case, join points $\mathbf{x} \in \mathbb{R}^n$ as in the above exercise to 0 and take integration from 0 to $\mathbf{x}$ of the given form along the path. The integral is well-defined because the domain is convex.
5. Statement in Exercise 1 still holds; every n-form on any open subset of $\mathbb{R}^n$ is a boundary as well a closed form, so the statement is true somewhat vacuously. Exercise 2 does not hold for $\mathbb{R}^2 \setminus \{0\}$: consider $\frac{xdy - ydx}{x^2+y^2}$. Likewise, there are closed $(n-1)$-forms on $\mathbb{R}^n \setminus \{0\}$, which are not exact.
§**2.7.**
1. Given $\tau \in \wedge^{n-1}\mathbb{R}^n$, we can view it as a linear map $\Theta : (\mathbb{R}^n)^* \to \wedge^n(\mathbb{R}^n) = \mathbb{R}$ by the formula $\Theta(\phi) = \tau \wedge \phi$. If $\tau \neq 0$ then Θ is also a nonzero linear map and hence its kernel is of dimension $n-1$. Choose a basis $\{\phi_1, \ldots, \phi_{n-1}\}$ of the $Ker\,\Theta$ and extend it to a basis $\{\phi_1, \ldots, \phi_{n-1}, \phi_n\}$ of $(\mathbb{R}^n)^*$. Now the linear map $\Theta'(\phi) = \tau' \wedge \phi$ where $\tau' = \phi_1 \wedge \cdots \wedge \phi_{n-1}$, also has the same kernel and hence $\Theta = r\Theta'$ for some $0 \neq r \in \mathbb{R}$ which gives $\tau = r\tau'$.
2. Suppose $\{\phi_i\}$ is a basis for V^* so that $\phi_1 \wedge \phi_2 + \phi_3 \wedge \phi_4 = \psi_1 \wedge \psi_2$ for some $\psi_i \in V^*, i = 1, 2$. Write $\psi_1 = \sum_i \alpha)i\phi_i, \psi_2 = \sum_i \beta_i\phi_i$ Then

$$\phi_1 \wedge \phi_2 + \phi_3 \wedge \phi_4 = \psi_1 \wedge \psi_2 = \sum_{i<j}(\alpha_i\beta_j - \alpha_j\beta_i)\phi_i \wedge \phi_j.$$

Comparing the coefficients of $\phi_1 \wedge \phi_j$ for $j = 3, 4$, we get $[\alpha_1 : \beta_1] = [\alpha_3 : \beta_3] = [\alpha_4 : \beta_4]$. But then we get $\alpha_3\beta_4 - \alpha_4\beta_3 = 0$, which is the coefficient of $\phi_3 \wedge \phi_4$. On the left-hand side, this is equal to 1 which is a contradiction.
§**3.2**
1. A local homeomorphism preserves local compactness and local path connectedness. Being locally homeomorphic to a Euclidean space, a manifold satisfies these two properties. Being a subspace of a Euclidean space, it inherits the Hausdorffness and II-countability from the Euclidean space. The last property follows since any locally path connected and connected space is path connected.
2. First find a piecewise smooth path joining any two given points such that on each piece it is one-to-one as follows: Cover a given path by finitely many coordinate charts. The problem

reduces to the case when X is replaced by $\mathbb{R}^n$, wherein you can actually join any two points via the line segment itself. Now the entire path may have self-intersections, which can be easily removed by cutting down the extra paths.
3. Apply (ii) of Remark 3.1.2, successively, backwards, beginning with $U_{n-1} \subset U_n$.
4. Consider the case when $\partial X \neq \emptyset$ and $\partial Y = \emptyset$. We have to worry about a point $(x, y) \in \partial X \times Y$. If U, V are coordinate nbds of x, y respectively, with diffeomorphisms $\phi : U \to \mathbf{H}^n$ and $\psi : V \to \mathbb{R}^m$ then clearly $\phi \times \psi : U \times V \to \mathbf{H}^n \times \mathbb{R}^m$ is a diffeomorphism which, in turn is diffeomorphic to $\mathbf{H}^{m+n}$. Thus, an atlas for $X \times Y$ can be constructed by taking the product of two atlases for X and Y, respectively. If y is also a boundary point, the product $\phi \times \psi$ gives a diffeomorphism with a "quarter" space which is not diffeomorphic to a half-space in $\mathbb{R}^{n+m}$. Indeed look at Example 3.2.1.(4).
6. By the inverse function theorem, any submanifold of the same dimension is an open subspace. If it is neat also, by definition, it is closed as well.

§**3.3.**
2. Choosing a local coordinate system $t \mapsto e^{2\pi\imath t}$ for a point $z \in \mathbb{S}^1$, the map $z \mapsto z^n$ becomes $t \mapsto nt$ whose derivative at any point is the multiplication by n. Therefore, the derivative of $z \mapsto z^n$ is also the linear map obtained by multiplication by n. However, this argument requires the knowledge that the tangent bundle of $\mathbb{S}^1$ is trivial and fixing a trivialization. One can use complex numbers and see the same thing directly as follows in a more elegant way: the complex derivative of the complex map $z \mapsto z^n$ is nz^{n-1} at any point z. The tangent space to $\mathbb{S}^1$ at z is the line spanned by $\imath z$. Restricted to this line the linear map defined by multiplication by nz^{n-1} rotates the line onto the line spanned by iz^n (which incidentally is the tangent line at z^n) and also expands it by a factor of n.
3. Let $\phi : TX \to X \times \mathbb{R}^n$ be a diffeomorphism such that $\phi(x, 0) = (x, 0)$ and such that for each x the map $\mathbf{v} \mapsto \phi(x, \mathbf{v})$ is a linear isomorphism. Set theoretically, TU coincides with all tangents to X which are drawn at points of U. Therefore, $\phi(TU) = U \times \mathbb{R}^n$ and hence it follows that ϕ itself restricts to a trivialization $TU \to U \times \mathbb{R}^n$. [Caution: It is important to note that this argument is not available if we take arbitrary submanifolds in place of an open set U. Indeed, the tangent space to a submanifold, in general, need not be trivial. However, you will have to wait to see such an example.
4. $T(V) = V \times V; D(L) = L \times L$.
5. [See Remark 1.5.1.(iii)] Since $\|x\| \to \infty$ implies $d(z, x) \to \infty$, it follows that the infimum is attained. That proves (a). Now choose smooth functions $g_1, \ldots g_{N-n}$ in a nbd of z_0 such that $X \cap U$ is given by $g_i = 0, i = 1, \ldots, N-n$. Apply Lagrange Multiplier Method (Theorem 1.5.1) to $f(x) = \|z - x\|^2$ to see that x_0 should satisfy (1.60), which is the same as saying that the vector $2(z - x_0)$ is in the linear span of vectors perpendicular to the tangent space $T_{x_0}(X)$.
6. See Lemma 8.1.2.
7. The other two vector fields are

$$(\ldots, z_j, -w_j, -x_j, y_j, \ldots); \qquad (\ldots, -w_j, -z_j, y_j, z_j, \ldots).$$

8. The other five vector fields are:
$(d, c, -b, -a, h, g, -f, -e)$; $(e, f, -g, -h, -a, -b, c, d)$;
$(f, -e, h, -g, b, -a, d, -c)$; $(g, h, e, f, -c, -d, -a, -b)$; $(h, -g, -f, e, -d, c, b, -a)$.

§**3.4**
1.
(i) Open subset of the Euclidean space

$$\{A = ((a_{ij})) \ : \ a_{ij} = 0, i < j\}.$$

(ii) $D(\det)_A(B) = tr(adj(A)B)$. Therefore, 1 is a regular value of det and hence $SL(n, \mathbb{R})$

is a codimension 1 manifold. The tangent space at Id is the kernel of the tr. Therefore, the tangent space to $SL(n,\mathbb{R})$ at Id is the space of all matrices A with $tr(A) = 0$.
(iii) This is the connected component of $O(n)$ that contains Id.
(iv) Similar to the case of $O(n)$ discussed in Example 3.4.2. $T_{Id}(O_{p,q}(\mathbb{R}))$ is the space of all A such that $AJ_{p,q} + J_{p,q}A^t = 0$.
(v) This is the connected component of $O_{p,q}(\mathbb{R})$ that contains Id.
2. Consider $\alpha : SO(n) \to V_{n-1,n}$ given by $[\mathbf{v}_1, \ldots, \mathbf{v}_n] \mapsto (\mathbf{v}_1, \ldots, \mathbf{v}_{n-1})$. Show that this defines the required diffeomorphism.
3. An affine isomorphism of $\mathbb{R}^n$ is of the form $\mathbf{x} \mapsto A\mathbf{x}+\mathbf{v}$, where $A \in GL(n,\mathbb{R})$ and $\mathbf{v} \in \mathbb{R}^n$, and therefore can be identified with the subspace of $M(n+1;\mathbb{R})$ of matrices of the form

$$\begin{bmatrix} A & \mathbf{v} \\ 0 & 1 \end{bmatrix}, \quad A \in GL(n,\mathbb{R}), \mathbf{v} \in \mathbb{R}^n.$$

This is an open subspace of the Euclidean space of all $(n+1) \times (n+1)$ matrices of the form $\begin{bmatrix} A & \mathbf{v} \\ 0 & 1 \end{bmatrix}$ and is of dimension $n^2 + n$. Among them, those that satisfy $AA^t = Id_n$ correspond to the rigid motions of $\mathbb{R}^n$. They form a submanifold of dimension $n(n+1)/2$.
4. In Example 3.4.2, replace $\mathbb{R}$ by $\mathbb{C}$, transpose by conjugate-transpose, etc., to conclude that the space $U(n)$ of unitary matrices is a complex manifold of dimension $n(n-1)/2$ and hence a real manifold of dimension $n^2 - n$.
5 Use the Gauss Elimination Method without pivoting: First, show that "sweeping" a particular column (or row) can be done homotopically. This proves that every element of $GL(n,\mathbb{K})$ can be connected to $\pm$ a permutation matrix. Next, over the complex numbers, using $e^{\imath\theta}$, connect every transposition to Id.

§**3.5**
1. If $a < 1$ the intersection is empty and hence transversal. For $a = 1$, the intersection consists of the circle $x^2 + y^2 = 1, z = 0$. At any of these points the two tangent planes coincide and hence the intersection is not transversal. For $a > 1$ the intersection consists of two circles $x^2 + y^2 = a^2, z \pm \sqrt{\frac{a^2-1}{2}}$. The tangent planes do not overlap and hence together span $\mathbb{R}^3$. Therefore, the intersection is transversal.
2. Take $f : X \to Y$ to be the inclusion map in Theorem 3.5.1. **3.** Use the fact that for $y \in W' = g^{-1}(W)$, we have $T_yW' = (dg)^{-1}(T_{g(y)}W)$.
4. Look at the circle $(x-1)^2 + y^2 = 1, (x-2)^2 + y^2 = 4$ in $\mathbb{R}^n$, $n \geq 2$. Their intersection is a singleton and hence, is of 0 dimension irrespective of what n you take. In any case, the intersection is not transversal. For $n = 2$, the intersection dimension is correct but not otherwise. Thus, one cannot say anything about the intersection dimension in case the intersection is not transversal.

§**3.7.**
1. Put $U_\pm = \mathbb{S}^1 \setminus \{(0,\pm 1)\}$ and let $\phi_\pm : U_\pm \to \mathbb{R}$ be the stereographic projections from $(0,\pm 1)$. Then the four maps $\phi_\pm \times \phi_\pm$ form an atlas for $\mathbb{S}^1 \times \mathbb{S}^1$. However, we can do better. Consider the following three open sets:
$U_1 = \mathbb{S}^1 \times \mathbb{S}^1 \setminus (1 \times \mathbb{S}^1 \cup \mathbb{S}^1 \times \omega)$; $U_2 = \mathbb{S}^1 \times \mathbb{S}^1 \setminus (\omega \times \mathbb{S}^1 \cup \mathbb{S}^1 \times \omega^2)$; and
$U_3 = \mathbb{S}^1 \times \mathbb{S}^1 \setminus (\omega^2 \times \mathbb{S}^1 \cup \mathbb{S}^1 \times 1)$.
It can also be shown that the torus cannot be covered by two charts, (each diffeomorphic to $\mathbb{R}^2$) but that needs a little more knowledge of algebraic topology. On the other hand, we have seen how to write the torus is a union two open sets each diffeomorphic to an annulus.
2 $r \neq 0$. Preimage Theorem 3.4.2.
3. By Exercise 3.3.5, it follows that if $z \in \mathbb{R}^3$ is a point at a distance b from the circle, and if p is a point on the circle at this distance, then the vector $z - p$ is perpendicular to the

tangent at p. Conversely, if we take the circle of radius b with the center at p and in the plane perpendicular to the tangent line at p then all the points on this circle are also at a distance b from a small portion of the circle around p. They would actually be at distance b if $b < a$. Therefore, in this case, the set of such points is the union of all these circles, i.e., the surface generated by rotating the circle $(x-1)^2 + z^2 = b^2$ around the z-axis. This is diffeomorphic to $\mathbb{S}^1 \times \mathbb{S}_1$, which is a torus. When $b = a$, the point $(0,0,0)$ lies on this surface and is a singular point. (The surface is not even a topological manifold.) For $b > a$, this surface is topologically a sphere but will have "corners" at $(0, 0, \pm\sqrt{b^2 - a^2})$.

4. The map $x \mapsto (x, f(x))$ is smooth with its inverse given by the restriction of the first projection $X \times Y \to X$ to Γ_f. Being diffeomorphic to a manifold, Γ_f is a manifold on its own.

5. The graph of $f : x \mapsto x^{1/3}$ is a smooth submanifold of $\mathbb{R} \times \mathbb{R}$, it is the graph of the function $y \mapsto y^3$ (interchanging the coordinates). However, we know that f is not smooth at 0.

12. Let $\phi : \mathbb{R}^n \to X$ be a parameterization at $p \in X$. Then the tangent space at p is given by $d\phi_0(\mathbb{R}^n)$. For any $\mathbf{v} \in \mathbb{R}^n$, consider the curve $t \mapsto \phi(t\mathbf{v})$ through p in X. Its velocity vector is equal to $d\phi_0(\mathbf{v})$.

13. Use the mean value theorem to see that f is one-to-one.

14. Put $g(z) = 2z^3 - 3z^2$. The derivative vanishes precisely at $z = 0, 1$. Since g is generically 3-1 mapping, it follows that $g : \mathbb{C}\setminus\{0,1\} \to \mathbb{C}$ is a surjective, local diffeomorphism. Compose this with λ to get a surjective **18.** Every submersion is an open mapping. If X is compact then $f(X)$ is compact and hence closed also.

19. Use Euler's identity, $\sum_i x_i \frac{\partial p}{\partial x_i} = kp(x_1, \ldots, x_n)$ for any homogeneous polynomial of degree k to conclude that 0 is the only value that is not regular. Consider the map

$$\mathbf{x} \mapsto \sqrt[k]{\frac{r_2}{r_1}}\mathbf{x}$$

to give the required diffeomorphism. The two surfaces $x^2 + y^2 - z^2 = \pm 1$ in $\mathbb{R}^3$ are not even homeomorphic to each other. [Use connectedness!]

20. Each ϕ_i is a proper mapping into $\mathbb{R} \setminus \{0\}$. Therefore, ϕ is a proper mapping into an appropriate open subset of $\mathbb{R}^{n+1}$. Injectivity and immersiveness follow by linear algebra and the fact that $P_1, \ldots, P_{n+1}$ are affinely independent. (Caution: ϕ is not a proper embedding into $\mathbb{R}^{n+1}$.

22. May assume that P and Q do not have common factors. Then send all zeros Q to ∞ and send ∞ to the limit $\lim_{z\to\infty} \frac{P(z)}{Q(z)}$ to obtain $\hat{f}$. To verify the smoothness of $\hat{f}$ at ∞, replace z by $1/z$; at points that are mapped to ∞, replace $\hat{f}$ by $1/\hat{f}$.

24. Consider $det : M(n;\mathbb{R}) \setminus \{0\} \to \mathbb{R}$.

26. The tangent space to the graph of a linear map f is equal to the graph Γ_f of f iteself, viz., $\Gamma_f := \{(v, f(v); v \in \mathbb{R}^n\}$. The tangent space to the diagonal is the diagonal itself. Therefore, at any point of intersection, i.e., for $f(v) = v$ the two tangent spaces span $\mathbb{R}^n \times \mathbb{R}^n$ iff their intersection is $(0,0)$. This just means that there is no vector $v \neq 0$ for which $f(v) = v$.

27. The intersection of Γ_f and Δ_X is transversal and hence is 0-dimensional and hence discrete.

29. To show that $SO(n)$ is connected, induct on n. Given any $A \in SO(n)$, show that there is a path $A(t)$ in $SO(n)$ from A to an element in $SO(n-1) \subset SO(n)$ as follows. Put $\mathbf{v} = (0, \ldots, 0, 1) \in \mathbb{R}^n$. If $A\mathbf{v} = \mathbf{v}$ then it follows that $A \in SO(n-1)$ and hence there is nothing to prove. If not, choose a plane P containing $\mathbf{v}$ and $A\mathbf{v}$. Let θ denote the angle from $A\mathbf{v}$ to $\mathbf{v}$. Let R_θ denote the rotation in the plane P that maps $A\mathbf{v}$ onto $\mathbf{v}$. Now $R_{t\theta}A, 0 \leq t \leq 1$ is a path that joins A to a point $R_\theta A \in SO(n-1)$ as required.

30. Similar to the above exercise.

31. Induct on n. For $n = 2$, we have $V_{1,2}$ is nothing but $\mathbb{S}^1$. Assume that for $n \geq 2$, $V_{k,n}$ is connected for all $1 \leq k < n$. Use the embedding $V_{k,n} \subset V_{k+1,n+1}$ via

$$(\mathbf{v}_1, \ldots, \mathbf{v}_k) \mapsto (\mathbf{v}_1, \ldots, \mathbf{v}_k, \mathbf{e}_{n+1}).$$

Now given any $F \in V_{k+1,n+1}$ use rotation to connect it to an element of the form $(\mathbf{v}_1, \ldots, \mathbf{v}_k, \mathbf{e}_{n+1})$.

§**4.1**

1. (i) Assuming that there is no such loop in X we shall show that X is orientable. Fix $z_0 \in X$ and for every $z \in X$ fix some path γ_z from z_0 to z. Fix an orientation θ_{z_0} on $T_{z_0}X$. Partition each path γ_z so that each segment is contained in one of the U_α. Obtain an orientation θ_z on T_zX as described in (b). Now given any U_α pick any point $z \in U_\alpha$ and give the constant orientation to U_α, that agrees with θ_z. That this is well-defined on each U_α and that it defines an orientation on X both follow from the assumption that there are no orientation reversing loops.
(ii) Use the classification of 1-manifolds (see Chapter 5).
(iii) Follow the procedure in Exercise 3.2.2, to approximate γ by an embedded piecewise smooth loop.
(iv) This can now be approximated by a smooth embedded loop.

3. Let $H : I \times I \to X$ be a homotopy of a constant loop to a loop γ. Divide the square into n^2 squares $[k/n, k+1/n] \times l/n, l+1/n], 0 \leq k \leq n-1, 0 \leq l \leq n-1$ so that H maps each square into a coordinate chart. Now inductively show that the loops $t \mapsto H(t, l/n)$ are orientation preserving for $l = 0, 1, \ldots, n$.

§**4.2**

2 For each $x \in U_\alpha$, we can then choose an ordered basis $\{\mathbf{v}_1(x), \ldots, \mathbf{v}_n(x)\}$ for T_xX such that $\omega_\alpha(x) = \mathbf{v}_1^* \wedge \cdots \wedge \mathbf{v}_n^*$ defines a nonvanishing smooth n-form on U_α. Take a smooth partition of unity $\theta_\alpha(x)$ subordinate to this cover and put $\omega = \sum_\alpha \theta_\alpha \omega_\alpha$.

Conversely, let ω be a nowhere vanishing smooth n-form on X. Then it follows that given any ordered basis $\{\mathbf{v}_1, \ldots, \mathbf{v}_n\}$ for T_xX either $\omega(\mathbf{v}_1, \ldots, \mathbf{v}_n)$ positive or negative. We may assume that it is positive at some point. Since X is connected, it follows that it is positive for all $x \in X$. Two such ordered bases define the same orientation class iff the values of ω on them have same sign. It follows easily that if we can choose the orientation class at each T_xX such that ω takes positive values on each of them then we get a smooth orientation on X.

§**4.3.** Let $F : X \times I \to Y$ be a smooth homotopy. By Stokes' theorem

$$0 = \int_{X \times I} F^*(d\omega) = \int_{\partial X \times I} F^*(\omega) = \int_X g^*(\omega) - \int_X f^*(\omega).$$

§**4.5.**

1. Write $\mathbb{S}^n = D_1 \cup D_2$ as the union of upper and lower hemispheres. Find $(n-1)$-forms τ_j on D_j such that $d(\tau_j) = \omega|_{D_j}$. Apply Stokes' theorem and induction.

2. The inclusion map $\iota : \partial B^n \to \mathbb{R}^n \setminus B^n$ is a homotopy equivalence. Therefore, η is exact on $\mathbb{R}^n \setminus B^n$ iff η is exact on $\partial B^n \approx \mathbb{S}^{n-1}$. Appeal to the above exercise.

3.

(a) Let η_0 be an $(n-1)$-form on $\mathbb{R}^n$ such that $d(\eta_0) = \omega$. Let $r > 0$ be such that $supp\, \omega \subset \mathbb{D}^n_r$. Then

$$0 = \int_{\mathbb{R}^n} \omega = \int_{\mathbb{D}^n_r} \omega = \int_{\partial \mathbb{D}^n_r} \eta_0.$$

By the previous exercise, η_0 is exact on $\mathbb{R}^n \setminus B^n_r$. So, let θ be an $(n-2)$-form such that

$d(\theta) = \eta_0$ on $\mathbb{R}^n \setminus B^n_r$. Let λ be a bump function $\equiv 1$ on $\mathbb{R}^n \setminus B^n_{2r}$ and such that $supp\,\lambda \subset \mathbb{R}^n \setminus B^n_r$. Put $\eta = \eta_0 - d(\lambda\theta)$. Check that $supp\,\eta \subset B^n_{2r}$ and $d(\eta) = \omega$.
(b) Let $\int_X \omega = 0$ and ω compactly supported. Using a bump function fix an n-form τ with compact support contained in $U_1 \cap U_2$ such that $\int_X \tau = 1$. (This can be done in a coordinate neighbourhood.) Let $\{\alpha, \beta\}$ be a partition of unity subordinate to the cover $\{U_1, U_2\}$. Put $t = \int_C \alpha\omega$. Then $\alpha\omega - t\tau$ is compactly supported in U_1 and $\int_X (\alpha\omega - t\tau) = 0$. Similarly, $\beta\omega + t\tau$ is compactly supported on U_2 and

$$\int_{U_2} \beta\omega + t\tau = \int_X (1-\alpha)\omega + t\int_X \tau = \int_X \omega - \int_X \alpha\omega + t = 0.$$

Therefore, there are $(n-1)$-forms ϕ_1, ϕ_2 with compact supports that are respectively contained in U_1, U_2 such that $d(\phi_1) = \alpha\omega - c\tau$ and $d(\phi_2) = \beta\omega + c\tau$. Extend by 0 both of these forms on the whole of X and take the sum.
(c) Write $X = U_1 \cup \cdots \cup U_k$, where each U_i is a coordinate chart and such that if $V_l = \cup_{i \le l} U_i$, then $V_l \cap U_{l+1}$ is nonempty. Apply (a) and (b) and induction.
4. We may assume $n \geq 2$. First consider the case when X is orientable. Check that for the orientation form ω, $\int_X \omega > 0$. Therefore, it follows that $[\tau] \mapsto \int_X \tau$ defines a surjective linear map $\Theta : H^n_{dR}(X) \to \mathbb{R}$. We need to show that $\int_X \tau = 0$ implies that τ is exact, which follows from (c) above.

Now consider the nonorientable case. One can argue as in (c) and (b) above. The important case to consider is when U_1 and U_2 are orientable but $U_1 \cup U_2$ is not. This just means that $U_1 \cap U_2 = V_1 \coprod V_2$, a disjoint union of two nonempty opens sets, such that on one of them say, V_1, the two orientations from U_1 and U_2 agree and on V_2 they don't. Choose a partition $\{\alpha, \beta\}$ subordinate to $\{U_1, U_2\}$ as before. Choose two n-forms τ_j with compact supports contained respectively in V_j and such that $\int_{V_j} \tau_j = 1$, where the orientations on V_j are taken from U_1. Put $t_1 = \frac{1}{2}\left(\int_{U_1} \alpha\omega - \int_{U_2} \beta\omega\right); t_2 = \frac{1}{2}\left(\int_{U_1} \alpha\omega + \int_{U_2} \beta\omega\right)$. Consider the n-forms $\phi_1 = \alpha\omega - t_1\tau_1 - t_2\tau_2$ on U_1 and $\phi_2 = \beta\omega + t_1\tau_1 + t_2\tau_2$ on U_2. Then $\int_{U_1} \phi_1 = 0$ and $\int_{U_2} \phi_2 = 0$ and hence there exist $(n-1)$-forms η_1, η_2 with compact supports contained in U_1 and U_2, respectively such that $d(\eta_j) = \phi_j, j = 1, 2$. It follows that $d(\eta_1 + \eta_2) = \omega$. Rest of the details are left to the reader.
§5.4. W can be expressed as $U_1 \cup U_2 \cup U_3$ where each U_1, U_3 are half-open intervals, U_2 is an open interval and $U_1 \cap U_3 = \emptyset$, and $U_1 \cap U_2, U_2 \cap U_3$ are open intervals. Now use Lemma 5.4.5 (ii) twice.
§5.6 1. If $\sigma = \sum_i \alpha_i(x)\frac{\partial}{\partial x_i}, \tau = \sum_i \beta_i(x)\frac{\partial}{\partial x_i}$ then

$$[\sigma, \tau] = \sum_j \left[\sum_i \alpha_i(x)\frac{\partial \beta_j}{\partial x_i} - \beta_i(x)\frac{\partial \alpha_j}{\partial x_i}\right]\frac{\partial}{\partial x_j}.$$

§5.8.
1. The map $(x, \mathbf{v}) \mapsto \|\mathbf{v}\|^2$ is smooth $TX \to \mathbb{R}$ and and $1 \in \mathbb{R}$ is a regular value.
7. If γ were a smooth path, then $f \circ \gamma$ will be smooth and $(f \circ \gamma)'(t) = Df_{\gamma(t)} \circ \gamma'(t) = 0$. This would mean that f is a constant on γ.
8. A biholomorphic mapping always preserves orientation. Since $\mathbb{C}$ comes with a preferred orientation so are all open subsets of $\mathbb{C}^n$.
9. Notice that there is no identification within the two open sets $\mathbb{R}^n \times \{j\} \times \{\pm 1\}$. Therefore, p defines a diffeomorphism of these two open sets onto its image. It follows that p is a double cover. We can declare the collection $\mathbb{R}^n \times \Lambda \times \{-1, 1\}$ as an atlas for $\tilde{X}$ with the quotient map restricted to each of them being a local parameterization. We orient each member of this atlas so that the image of $\mathbb{R}^n \times j \times \pm 1$ will receive an orientation so that the quotient

map will preserve the orientation or reverse it according to the sign $\pm$. Under this rule, it is clear that the atlas becomes an orientable atlas for $\tilde{X}$. Finally, if X is already oriented then you see that all the identification will occur within those members with the same signs and hence we get two disjoint copies of X itself as $\tilde{X}$. And conversely also.

10.

(i) Choose a basis for $F \in G_{k,n}$ and express them as column vectors of a matrix $A \in M(n \times k; \mathbb{R})$

(ii) Row rank = column rank implies that $U = \cup_A U_A$. Also, each U_A is open in $M(n \times k; \mathbb{R})$ itself.

(iii) This follows since each U_A is a saturated open set, i.e., if $\Theta(F_1) = \Theta(F_2)$ then $F_1 \in U_A \iff F_2 \in U_A$. To see this, consider the projection $\pi_A : \mathbb{R}^n \to \mathbb{R}^k$, which sends $x_i \to 0$ for all $i \notin A$. Then $F \in U_A$ iff $\pi_A(\Theta(F)) = \mathbb{R}^k$.

(v) Given $F \in U_A$ consider the $k \times k$ matrix $H = \tilde{F}$ formed by the rows of F corresponding to the indices in A. Then $F\tilde{F}^{-1}$ is in the image of ϕ_A.

(vi) Follows from (v) and the fact that $\Theta(F) = \Theta(F\tilde{F}^{-1})$.

12. Define $\lambda : V_{k,n} \to S$ by the formula $\lambda(F) = FF^t$. Check that λ induces a map $\tau : G_{k,n} \to S$ such that $\tau \circ \Theta = \lambda$ and τ is the inverse of η.

13. (i) In a coordinate nbd U of a point $x \in X$ first get a smooth map $\beta : U \to M(N \times n; \mathbb{R})$ such that $\beta(x)$ is a basis for $T_x(X)$. Then follow by Θ to get the Gauss map.

14. See Exercise 3.7.31.

§**6.1**

1. Given a continuous function $f : X \to \partial X$, which is identity on the boundary, by modifying it in a collar neighbourhood of ∂X, we may assume that f is smooth around each point of ∂X. Think of ∂X as embedded in some $\mathbb{R}^N$. Fix a tubular neighbourhood U of ∂X in $\mathbb{R}^N$. If ϵ is chosen sufficiently small, then a smooth approximation g to f will take its values inside U, i.e., we have a smooth map $g : X \to \mathbb{R}^N$ such that $\|f(x) - g(x)\| < \epsilon/2$ and so $g(x) \in U$. Now take $h = \pi \circ g$, where $\pi : U \to \partial X$ is the projection of the tubular neighbourhood. If g is chosen so that $g(x) = f(x) = x$ on ∂X, it follows that $h : X \to \partial X$ is a smooth function such that $h(x) = x$ on ∂X. Compactness hypothesis on X can be removed by taking proper maps, i.e., there is no proper continuous map from $X \to \partial X$, which is the identity map on the boundary. However, if we remove the properness hypothesis also, then the result does not hold: take $X = [0, 1)$.

§**6.3**

2. See corollary 8.3.1.

3. The complement of the first embedding is such that removing a single point makes it disconnected.

6. This requires the knowledge of the fundamental group of the complement of a knot: For the trefoil knot this happens to be not isomorphic to $\mathbb{Z}$. The fundamental group of the complement of the real axis is indeed isomorphic to $\mathbb{Z}$. If two proper embeddings of $\mathbb{R}$ in $\mathbb{R}^3$ are properly isotopic to each other, then their complements will have isomorphic fundamental groups, which is not the case here.

On the other hand, one can slowly pull one of the ends through the loop and "untie" the knot that gives an isotopy of the given embedding with the inclusion map $x \mapsto (x, 0, 0)$. Indeed take an isotopy of the identity map $\mathbb{R} \to \mathbb{R}$ with a diffeomorphism ϕ whose image is any tiny open interval and then compose it with the given embedding f. This example also illustrates that the isotopy extension theorem is not valid for noncompact case.

7. Use the smooth step function that we have constructed and consider a linear modification of it.

8. Use the above exercise.

9. The isotopy given in example 6.3.1 restricted to $\mathbb{D}^n$ has compact support. Using the

above exercise, "shrink" this isotopy so that the entire isotopy lies inside V. Then use the isotopy extension theorem.
10. The result is false for $n = 1$. There is no diffeomorphism of $\mathbb{R}$, which takes $-1, 0, 1$ respectively to $0, 1, -1$, because every $1-1$ continuous map $f : \mathbb{R} \to \mathbb{R}$ is monotonic.

§6.4.
1. By Corollary 6.3.1, $f : \mathbb{S}^{n-1} \to \mathbb{S}^{n-1} \subset \mathbb{R}^n$ extends to an embedding $F : \mathbb{D}^n \to \mathbb{R}^n$. Since $F(\mathbb{S}^{n-1}) = \mathbb{S}^{n-1}$ separates $\mathbb{R}^n$ into two components it follows that $F(\mathbb{D}^n)$ is contained in the closure of one of them say C. Since it is an embedding and the dimensions are the same, F is an open mapping. Since $\mathbb{D}^n$ is compact, $F(\mathbb{D}^n)$ is both closed and open in C. Therefore, $F(\mathbb{D}^n) = C$. Therefore, C is the bounded component and hence $C = \mathbb{D}^n$.
2. Combine the arguments in the proofs of the Disc Theorem 6.3.7 and the Theorem 6.3.3.
3. Cap off W at all these boundary components and then use the above exercise.
4,5,6. Indeed, ϕ_0 and ϕ_1 patch-up to define a homeomorphism ϕ, which may fail to be a diffeomorphism along $\{0, 1\} \times I$. Use smoothing lemma to modify ϕ as required.
§6.5.
1. In the Theorem 6.1.2, since X is compact, we can choose ϵ to be a positive constant instead of a continuous function. Any tubular neighbourhood looks like $\alpha(\cup_{x \in X} x \times B_\epsilon \cap N_x(X))$ where $\alpha(x, \mathbf{v}) = x + \mathbf{v}$. A reparameterization by $\mathbf{v} \mapsto \frac{\epsilon}{\epsilon'}\mathbf{v}$ will define an isotopy of the ϵ-neighbourhood with the ϵ'-neighbourhood. Since X is compact, the isotopy extends to an ambient one.
2. First find a homotopy $H : A \times I \to M$ of the inclusion map $A \hookrightarrow M$ such that for arbitrary small t, the map $H_t : A \to M$ is transversal to B. Since H_0 is an embedding by stability, it follows that there is an $\epsilon > 0$ such that H_t is an embedding for $t \le \epsilon$. This means that $H : A \times [0, \epsilon] \to M$ is an isotopy. Extend this to an ambient isotopy. Clearly $H_\epsilon(A) \cap B = \emptyset$.
3.(iii) Since f is proper, $F = f^{-1}(p)$ is a closed manifold. Therefore, we can choose $\epsilon > 0$ so that $f(N_\epsilon(F)) \subset U$ where U is a coordinate neighbourhood of the point p. Now consider the map $\Theta : N_\epsilon(F) \to F \times U$ given by $(x, \mathbf{v}) \mapsto (x, (f(x, \mathbf{v}))$. Verify that it is a submersion. Restricted to F it is injective. Therefore, by arguments similar to the ones you have seen in Theorem 3.6.1, Θ is a diffeomorphism $N_{\epsilon'}(F) \to F \times V$ for a suitable $\epsilon' > 0$ and a neighbourhood V of p, as required.
§7.2.
1. $0, 1$.
2. For any $t > 1$ the embedding $[z_1, z_2] \mapsto ([z_1, z_2], [tz_1, z_2])$ is homotopic to the diagonal. There are precisely two points of intersection $([1, 0], [1, 0])$ and $([0, 1], [0, 1])$ the intersection is transversal and the intersection number at both the points is $+1$. Therefore, the self-intersection number of the diagonal is 2.
§7.5
1. Use polar coordinates.
2. If $i \le k - 1$ and if $\alpha : \mathbb{S}^i \to M$ is a smooth map, we can homotope this so as to be transversal to N. This then implies $\alpha(\mathbb{S}^i) \subset M \setminus N$. This proves surjectivity. Likewise for $i \le k - 2$, a smooth map $\alpha : \mathbb{S}^i \to M \setminus N$ and a smooth extension $\beta : \mathbb{D}^{i+1} \to M$ of α, we can homotope β without disturbing it on the boundary sphere, to a map that is transversal to N and then again this implies $\beta(\mathbb{D}^n) \subset M \setminus N$. This proves injectivity.
3. At any point $p \in N$ we can choose an embedded $k-$disc $\alpha : \mathbb{D}^k \hookrightarrow M$ such that $D = \alpha(\mathbb{D}^k)$ intersects N transversely at a single point $\{p\}$. Now the embedding restricted to the boundary of the disc defines an element $[\alpha|_{\mathbb{S}^{k-1}}] \in \pi_{k-1}(M \setminus N)$. If this element were trivial, then we can get an extension of $\alpha|_{\mathbb{S}^{k-1}}$ to a smooth map $\beta : \mathbb{D}^k \to M \setminus N$. Thinking of these two discs as upper and lower hemispheres in $\mathbb{S}^k$, and by the smoothing lemma, we can arrange it so that α, β patch-up to define a smooth map $\gamma : \mathbb{S}^k \to M$ so that the image

γ intersects N transversely in a single point $\{p\}$. Now since $\pi_k(M) = (0)$ it follows that γ extends to a smooth map $\hat{\gamma} : \mathbb{D}^{k+1} \to M$. We can further assume that $\hat{\gamma}$ is transversal to N. Can you see the contradiction now?

§7.8.
1. You may treat σ as a smooth map $\sigma : \mathbb{D}^n \to \mathbb{R}^n$. Let $p_1, \ldots p_k \in \operatorname{int} \mathbb{D}^n$ be the zeros of σ. Choose $\epsilon > 0$ small enough so that $B_\epsilon(p_j)$ are disjoint and contained in $\mathbb{D}^n$. It follows that the winding number $W(\sigma/\mathbb{S}^{n-1}, 0)$ of $\sigma/\mathbb{S}^{n-1}$ around 0, which is equal to the degree of the map $\sigma/||\sigma|| : \mathbb{S}^{n-1} \to \mathbb{S}^{n-1}$ is zero. By Hopf-degree theorem, there is a smooth extension of $\sigma : \mathbb{S}^{n-1} \to \mathbb{R}^n \setminus \{0\}$ to a smooth map $\hat{\sigma} : \mathbb{D}^n \to \mathbb{R}^n \setminus \{0\}$.
2. Use Corollary 6.3.3 for $n \geq 2$. For $n = 1$, this is an easy consequence of the classification.
3. Follows from 1. and 2. above.
4. There is one on the cylinder that quotients down to a nowhere vanishing field on the Möbius band.

§7.9. 1. As seen in Misc. Excercises 4.5, $\omega \mapsto \int_X \omega$ defines the isomorphism $H^n(X) \to \mathbb{R}$. Now use (7.13).
2. Let $\tau : \tilde{X} \to \tilde{X}$ be the orientation revrsing diffeomorphism such that $p \circ \tau = p$. Let ω be a n-form on X. Put $\hat{\omega} = p^*(\omega)$. Then $\tau^*\hat{\omega} = \tau^* p^* \omega = p^*(\omega) = \hat{\omega}$. Since τ is orientation reversing at each point, it follows that

$$\int_{\tilde{X}} \hat{\omega} = \int_{\tilde{X}} \tau^* \hat{\omega} = -\int_{\tilde{X}} \hat{\omega} = 0.$$

Therefore, $\hat{\omega}$ is an exact form (see 4.5.4.)). Say $\hat{\omega} = d\mu$ for some $(n-1)$-form μ on $\tilde{X}$. Put $\sigma = (\mu + \tau^*\mu)/2$. Then it follows that $\sigma = p^*(\phi)$ for some $(n-1)$-form ϕ on X. It also follows that $d(\phi) = \omega$.

7.10.

1 Apply the Transversality Theorem 7.1.1 to the homotopy $F : X \times \mathbb{R}^N \to \mathbb{R}^N$ given by $(x, \mathbf{v}) \mapsto x + \mathbf{v}$.

2 In the proof of Theorem 7.1.3, we know that F_s is transversal for almost all s. By stability theorem, for all sufficiently small s, each F_s is an embedding. Therefore, if we choose $\epsilon > 0$ small enough then for $||s|| < \epsilon$ all F_s are embeddings, their image will be contained in U and some of them will be transversal to Z. We can then define the isotopy $H : X \times I \to Y$ so that $H_1(X) \pitchfork Z$. Using Isotopy Extension Theorem, we can then get an ambient isotopy as required.

3 Let $E = E(G_{k,n}) = \{(L, \mathbf{v}) \in G_{k,n} \times \mathbb{R}^n \;:\; \mathbf{v} \in L\}$. Consider the projection to the first factor $\pi_1 : E \to G_{k,n}$. (This is the projection of a vector bundle called the *tautological bundle over* $G_{k,n}$.) Show that the second projection $\pi_2 : E \to \mathbb{R}^n$ is a submersion. Now consider $W = \pi_2^{-1}(X)$ and show that $L \in G_{k,n}$ is a regular value of $\pi_1|_W$ iff $L \pitchfork X$.

4 Choose R such that $K \subset B_R(0)$. Since $n \geq 2$ it follows that $df(\mathbb{R}^n)$ is of measure zero in $M(n; \mathbb{R})$. Therefore, there exist $A \in M(n, \mathbb{R}) \setminus df(\mathbb{R}^n)$ such that $0 < \|A\| < \epsilon/R$. Take $g(x) = f(x) - Ax$.
For $n - 1$, the argument fails: take $f(x) = \sin x$, $K = [0, 2\pi]$, and $\epsilon = 1/3$.

5 (i) We know that the normal bundle of any sphere $\mathbb{S}^n \in \mathbb{R}^{n+1}$ is trivial and anyway $T(\mathbb{S}^n) \oplus N(\mathbb{S}^n) = T(\mathbb{R}^{n+1}) = \Theta^{n+1}$. Therefore $T(\mathbb{S}^n) \oplus \Theta^1 = \Theta^{n+1}$, which shows that the sphere is s-parallelizable. Inductively, assume that X is s-parallelizable. Then

$T(X\times \mathbb{S}^n)\oplus\Theta^1 = T(X)\oplus T(\mathbb{S}^n)\oplus\Theta^1 = T(X)\oplus\Theta^{n+1} = \Theta^{m+n+1}$. This shows all finite products of spheres are s-parallelizable.
(ii) Recall every odd sphere has at least one nowhere vanishing vector field [viz. $(-x_2, x_1, -x_4, x_3, \ldots, -x_{2k+2}, x_{2k+1})$]. Therefore $T(\mathbb{S}^{2k+1}) = \xi\oplus\Theta^1$. Therefore,
$T(X\times\mathbb{S}^{2k+1}) = T(X)\oplus\Theta^1\oplus\xi = \Theta^{m+1}\oplus\xi$
$= \Theta^{m-1}+\Theta^1+T(\mathbb{S}^{2k+1}) = \Theta^{m-1}\oplus\Theta^{2k+2} = \Theta^{m+2k+1}$.
(iii) By the Jordan-Brouwer separation theorem, if X is a closed codim. 1 submanifold, then X bounds a compact submanifold of dimension n in $\mathbb{R}^n$. Hence X has the unit outward normals at every point, which means that the normal bundle is trivial. Therefore $T(X)\oplus\Theta^1 = T(\mathbb{R}^n) = \Theta^n$.
(iv) Inductively having embedded a product of spheres X in $\mathbb{R}^{m+1}$, we shall show that $X\times\mathbb{S}^k$ can be embedded in $\mathbb{R}^{m+k+1}$: Consider $X\times 0\subset\mathbb{R}^{m+1}\times 0\subset\mathbb{R}^{m+1}\times\mathbb{R}^k$. The normal bundle of X in $\mathbb{R}^{m+1}\times\mathbb{R}^k$ is also trivial being the direct sum of its normal bundle in $\mathbb{R}^{m+1}$ and the trivial bundle $X\times\mathbb{R}^k$. Therefore, its unit normal bundle, which is clearly a submanifold of $\mathbb{R}^{m+1+k}$, is diffeomorphic to $X\times\mathbb{S}^k$.

6 If $\{(U_i,\phi_i)\}$ is any atlas for a manifold M, then $\{(T(U_i), D\phi_i)\}$ forms an atlas for TM. Then for any i, j, the transition function $U_i\cap U_j\times\mathbb{R}^n\to U_i\cap U_j\times\mathbb{R}^n$ is given by $(\phi_j^{-1}\circ\phi_i, d(\phi_j^{-1}\circ\phi_i))$. Now $\phi_j^{-1}\circ\phi_i$ is orientation preserving iff $d(\phi_j^{-1}\circ\phi_i)$ is and hence in either case, $(\phi_j^{-1}\circ\phi_i, d(\phi_j^{-1}\circ\phi_i))$ is orientation preserving.

7 (i) Use Exercise 7.5.1.2 and the fact that $\mathbb{R}^n$ is simply connected.
(ii) Around a point $p\in\partial M$ you can connect the 'two- sides' of $N\setminus M$.
(iii) If the normal bundle (which is of rank 1 in this case) is nontrivial, you can find an embedded loop γ in M restricted to which the bundle is non trivial, which gives an embedding of a Möbius strip S in N such that $S\cap M=\gamma$.

8 If $v\in\mathbb{S}^n$ is a regular value then $f^{-1}(v)$ consists of pairs of point $x_i, -x_i$ and hence even number of points. Therefore the mod 2 degree is 0.

9 Covering space theory tells you that there is a covering projection $p:\tilde{Y}\to Y$ such that $p_\#\pi_1(\tilde{Y}) = f_\#(\pi_1(X))$. If Y is orientable then all its coverings are also orientable and the degree of the covering map p is equal to the index of the subgroup $p_\#\pi_1(\tilde{Y}) = d$ say. Also, the covering space theory tells you that f can be lifted to a map $\tilde{f}: X\to\tilde{Y}$ such that $f = p\circ\tilde{f}$. Assuming that d is finite, it follows that $\deg f = (\deg\tilde{f})(\deg p)$. (If d is infinite then $\deg f = 0$.)

10 (i) Take a path γ from x_0 to x_1 in $\mathbb{R}^{n+1}\setminus M$ to get the homotopy $\tau_{\gamma(t)}$.
(ii) If x is in the unbounded component then τ_x makes sense on all the bounded components. Therefore, $\deg\tau_x = 0$ and hence by the Hopf degree theorem τ_x is null-homotopic. The 'if' part follows from (iii).
(iii) Choose a disc D around x and contained in the bounded component U. Then τ_x extends to $U\setminus \operatorname{int} D$ and hence $\deg(\tau_x|_M) = \deg(\tau_x|_{\partial D}) = \pm 1$.

11 (i) This is the property of the orientations on $M\times N$ and $N\times M$.
(ii) Then the map τ is null homotopic or extends to $X\times N$, where $\partial X = M$, respectively.
(iii) This goes back to the previous exercise.

12 We may assume that U is connected and $\det Df_x$ takes values in $[0,\infty)$. Let $K\subset\mathbb{R}^n$ be the set of all points at which $\det Df_x = 0$. Then K is compact. Given any point $z\in\mathbb{R}^n$ we can choose $R>0$ so that $\mathbb{D}^n_R$ contains $f(K)\cup\{z\}$ in its interior. Now consider

$M = f^{-1}(\partial \mathbb{D}^n_R)$. Since $\mathbb{S}^{n-1}_R$ separates $\mathbb{R}^n$, $M = \emptyset$ would imply that $f(U) \subset \mathbb{D}^n$. By properness of f this would imply U is compact, which is absurd. So $M \neq \emptyset$. It follows that M is a closed $(n-1)$-dimensional submanifold of U, which bounds a submanifold N of U. Put $F = f|_N$. Since at every point $x \in M$ we have Df_x is orientation preserving, it follows that the local degree of $f : M \to \mathbb{S}^{n-1}_R$ at any point is $+1$. Therefore, $\deg f > 0$. Use Theorem 7.5.1 now to conclude that for any regular value w of F in $\mathbb{D}^n_R$, the winding number $\#(F^{-1}(w)) = W(f, w) = \deg f$ is positive. This implies $F : N \to \mathbb{D}^n_R$ is surjective. In particular, z is in the image of f.

§8.2
1. (vii) $(x, y, z) \mapsto -x$ is a Morse function with two critical points, viz., $(3, 0, 0)$ of index 0 and $(1, 0, 0)$ of index 1.
(viii) The parameterization restricted to $\pi/2 \leq \psi \leq 3\pi/2$ gives the cylinder. The function $f(\theta, \psi) = (2 + \cos\theta)\cos\psi$ will then have only two critical points at $(\theta, \psi) = (\pi, \pi), (0, \pi)$ of indices 0 and 1, respectively The same domain can be used to define the Möbius band by the identification $(0, \psi) \sim (2\pi, -\psi)$. The same map factors through this identification to define $\tilde{f} : \mathbb{M} \to \mathbb{R}$, which has the same critical behavior because locally, f and $\tilde{f}$ behave the same way.
7. Take $f(x, y) = 3x^4 + 4x^3 - 12x^2 + y^2$.
§8.4. Indeed $f^{-1}(r)$ is a strong deformation retract of $f^{-1}(r) \times [r, s] = M_{r,s}$ from which the conclusion follows.
§9.1.1.
2.2e By homogeneity, it is enough to check this at $p = 1$. The tangent space to $\mathbb{S}^3$ at this point is the subspace of purely imaginary quaternions and the tangent space to $\mathbb{S}^2$ at $h(1) = \mathbf{i}$ is the space spanned by $\{\mathbf{j}, \mathbf{k}\}$. Let λ be a unit vector in $T_1\mathbb{S}^3$. Then

$$\begin{aligned} Dh(\lambda)_1 &= \lim_{t\to 0} \tfrac{h(1+t\lambda)-h(1)}{t} \\ &= \lim_{t\to 0} \tfrac{1}{t}\left(\tfrac{(1+t\lambda)\mathbf{i}(1-t\lambda)}{1+t^2} - \mathbf{i}\right) \\ &= \lambda\mathbf{i} - \mathbf{i}\lambda. \end{aligned}$$

which is easily seen to be surjective.

§9.1.4.
1. If $A \in M(n, \mathbb{H})$ is invertible, then we have $Id = \mathcal{Q}_n(AA^{-1}) = \mathcal{Q}_n(A)\mathcal{Q}_n(A^{-1})$, which gives $\det A = \det \mathcal{Q}_n(A) \neq 0$. Conversely, if B is such that $\det B \neq 0$, then B is invertible. Since $BJ_{2n} = J_{2n}\bar{B}$, it follows that $B^{-1}J_{2n} = J_{2n}\overline{B^{-1}}$ and hence there exists $A' \in M(n, \mathbb{H})$ such that $\mathcal{Q}_n(A') = B^{-1}$. It follows that $\mathcal{Q}_n(AA') = Id$ and hence $AA' = Id$.
2. $BJ_{2n} = J_{2n}\bar{B}$ implies that $\det B = \det \bar{B} = \overline{\det B}$.
5. Combine Exercises 2 and 4.
§9.1.5. $\mathcal{Q}_n$ preserves inner product. i.e., if we write $\mathbf{v}_r = z_r + w_r\mathbf{j}$, then $\langle \mathbf{v}_r, \mathbf{v}_k\rangle = \langle z_r, z_k\rangle + \langle w_r, w_k\rangle = \langle \mathcal{Q}_n(\mathbf{v}_r), \mathcal{Q}_n(\mathbf{v}_k)\rangle$. Therefore, $\{\mathbf{v}_1, \ldots, \mathbf{v}_n\}$ forms an orthonormal basis for $\mathbb{H}^n$ iff $\{\mathcal{Q}_n(\mathbf{v}_1), \mathcal{Q}_n(\mathbf{j}\mathbf{v}_1), \ldots, \mathcal{Q}_n(\mathbf{v}_n), \mathcal{Q}_n(\mathbf{j}\mathbf{v}_n)\}$ forms an orthonormal basis for $\mathbb{C}^{2n}$.
§9.3.1.
5.(a) $D(\alpha_x)_g = D(\alpha(g)) \circ D(\alpha_x)_e$ and $\alpha(g) : X \to X$ is a diffeomorphism.
10. You have to copmute the stabilizer subgroup of the element P. The elements of $\mathcal{O}_n$ which fix the subspace V_{k_r} are precisely of the subgroup $\mathcal{O}_{k_r} \times \mathcal{O}_{n-k_r}$. Now keep working backwords. **12.** Let H, K be subgroups of G. Consider the action of G on G/H and restrict to the subgroup K and use Exercise 5.(a).
13 Union of connected components of a manifold is a submanifold.
§9.4.2.
1. If $\{\mathbf{u}_1, \mathbf{u}_2\}$ is a basis, define $[\mathbf{u}_,\mathbf{u}_1] = 0 = [\mathbf{u}_2, \mathbf{u}_2]$ and $[\mathbf{u}_1, \mathbf{u}_2] = \mathbf{u}_1$. Verify that this

defines a Lie algebra. Any other 2-dimensional nonabelian Lie algebra has to be of the form in which $[\mathbf{u}_1, \mathbf{u}_2] = a\mathbf{u}_1 + b\mathbf{u}_2$ for some $a, b \in \mathbb{R}$ with $(a, b) \neq (0, 0)$. Suppose $b \neq 0$. Then take the basis $\{\mathbf{u}_1/b, \mathbf{w}\}$, where $\mathbf{w} = [\mathbf{u}_1, \mathbf{u}_2]$. It follows that $[\mathbf{u}_1, \mathbf{w}] = \mathbf{w}$.
2. $[E_1, E_2] = E_3, [E_1, E_3] = 0 = [E_2, E_3]$.
6. Both are indecomposable since both are dimension 3 and there are brackets (such as $[E_1, E_2] = E_3$, which involve three independent vectors.
7. Yes indeed. (a) There is a family of so-called nilpotent algebras (3-dimensional):

$$[E_1, E_2] = 0;\ [E_1, E_3] = aE_1 + bE_2;\ [E_2, E_3] = cE_1 + dE_2,$$

where $ad - bc \neq 0$.
(b) Also, there are two more (nonnilpotent ones):

$$[E_1, E_2] = E_3;\ [E_2, E_3] = \pm E_1;\ [E_3, E_1] = E_2.$$

§9.8.1.1. Clearly $H = \cap_\alpha H_\alpha$ is a subgroup. It will be a Lie subgroup if it is a submanifold. It will be a submanifold, if $H \cap V$ is a submanifold of V, where V is any suitable neighbourhood of $e \in G$. Since $exp : \mathfrak{g} \to G$ is a local diffeomorphism, which restricts to local diffeomorphisms of $\mathfrak{h}_\alpha \to H_\alpha$, it is enough to see that $\cap_\alpha \mathfrak{h}_\alpha$ is a subalgebra of G. But any arbitrary intersection of vector subspaces of a finite dimensional vector space is actually equal to a finite intersection. It follows that $\cap_\alpha \mathfrak{h}_\alpha$ is a Lie subalgebra.

§9.9.1.
1. If U is a nbd of e in H then U is also open in G. Therefore, U generates G.
2. G abelian iff $[G, G] = (1)$ iff $[\mathfrak{g}, \mathfrak{g}] = (0)$ iff $\mathfrak{g}$ is abelian.
3. By (9.30), it follows that if $[\mathbf{u}, \mathbf{v}] = 0$, then $\mathrm{Exp}\,(\mathbf{u} + \mathbf{v}) = \mathrm{Exp}\,(\mathbf{u})\mathrm{Exp}\,(\mathbf{v})$. Therefore, if $\mathfrak{g}$ is abelian then $\mathrm{Exp}\, : \mathfrak{a} \to G$ is a homomorphism of the additive group $\mathfrak{g}$ to the Lie group G. We know Exp is a local diffeomorphism. Therefore, $\mathrm{Exp}\,(U)$ is open for some neighbourhood of U of 0 in $\mathfrak{g}$. Since G is connected, $\mathrm{Exp}\,(U)$ generates G.
6. Choose a maximal subset of H that is linearly independent, say $\{\mathbf{v}_1, \ldots, \mathbf{v}_k\}$, where $k \leq n$. Put $H_1 = \sum_i \mathbb{Z}\mathbf{v}_i$. Clearly H_1 is a subgroup of H, which is free abelian. Let P be the parallelotope spanned by $\{\mathbf{v}_1, \ldots, \mathbf{v}_k\}$. Then P is compact and hence $P \cap H$ is finite. For each $\mathbf{x} \in H$, write $\mathbf{x} = \sum_i x_i \mathbf{v}_i$. Now for each integer m, define

$$\mathbf{x}_m = m\mathbf{x} - \sum_i \lfloor m x_i \rfloor \mathbf{v}_i.$$

Note $\mathbf{x}_m \in P \cap H$. Conclude that $P \cap H$ generates H. Since $P \cap H$ is finite, and $\mathbb{Z}$ is infinite, there exists distinct integers k, l such that $\mathbf{x}_k = \mathbf{x}_l$. This implies that $(k - l)\mathbf{x} = \sum_i (\lfloor kx_i \rfloor - \lfloor lx_i \rfloor)\mathbf{v}_i$. Therefore, H is contained in the rational linear span of H_1. Since H is finitely generated, it follows that (by taking a common denominator to all the coefficients in the expressions of all members of $P \cap H$) there is an integer d such that $dH \subset H_1$ which is a free abelian group. By the structure theorem for finitely generated abelian groups, there exists a basis $\{\mathbf{u}_1, \ldots, \mathbf{u}_k\}$ for H_1 and integers $m_1, \ldots, m_k$ such that $\{m_1\mathbf{u}_1, \ldots, m_k\mathbf{u}_k\}$ generates dH. Now $\{m_1\mathbf{u}_1/d, \ldots, m_k\mathbf{u}_k/d\}$ is a basis for H as required.
8. G is abelian and hence $\mathfrak{g}$ is abelian and therefore isomorphic to $\mathbb{R}^n$. By Exercise 3, $\mathrm{Exp}\, : \mathbb{R}^n \to G$ is a surjective homomorphism. It is also a local diffeomorphism. Therefore, its kernel is a discrete normal subgroup $N \subset \mathbb{R}^n$. But then Exp induces an isomorphism of $\mathbb{R}^n/N \to G$. On the other hand, the projection map $q : \mathbb{R}^n \to \mathbb{R}^n/N$ is a covering projection. Therefore, $(1) = \pi_1(G) = \pi_1(\mathbb{R}^n/N) = N$.
10. Every Lie algebra contains a maximal abelian subalgebra.
11. Do the same with abelian subalgebras of the Lie algebra.
13. (a) and (b) all scalar matrices; (c) n^{th} roots of unity as scalar matrices; (d) $so(2)$ for

$n = 1$ and $\pm Id$ for $n \geq 2$; (e) trivial group
14. $z(SU(2)) \approx \mathbb{Z}_2$, $z(SO(3)) = (1)$, $z(U(n)) \approx \mathbb{S}^1$ and $z(SU(n) \times U(1)) \approx \mathbb{Z}_n \times U(1)$.
17. Fix a basis $\{\mathbf{v}_1, \ldots, \mathbf{v}_r\}$ for the Lie algebra of T. Choose $t_1, \ldots, t_r \in \mathbb{R}$ such that $\{1, t_1/t_r, \ldots, t_{r-1}/t_r\}$ is linearly independent over $\mathbb{Q}$. Put $\mathbf{v} = \sum_i t_i \mathbf{v}_i$. Then show that $\{\operatorname{Exp}(n\mathbf{v}) \,:\, n \in \mathbb{Z}\}$ is dense in T.
21. If γ is a path in G connecting g and id, then L_g is homotopic to Id. Therefore, $L(L_g) = L(Id) = \chi(B/T) \neq 0$.

Bibliography

[A] J. W. Alexander, An Example of a Simply Connected Surface Bounding a Region which is not Simply Connected *Proc. National Acad. Sc.* 1924, 10(1) pp. 8–10.

[B] M. Berger, Les varietes riemanniennes (1/4)-pincees, *Ann. Scuola Norm. Sup. Pisa Cl. Sci.* 14 (1960), pp. 161–170.

[B-S] S. Brendle, and R. Schoen, Manifolds with 1/4-pinched curvature are space forms, *J. Amer. Math. Soc.* 22 (2009), no.1, pp. 287–307.

[Bd] W. Browder, *Surgery on Simply Connected Manifolds,* Ergebnisse der Mathematik und ihrer Grenzgebiete Band 65, Springer-Verlag, 1972.

[Bn] M. Brown, A proof of the generalized Schoenflies theorem, *Bull. Amer. Math. Soc.* 66 (1960). pp. 74–76.

[Ce] J. Cerf, Topologie des certains espaces de plongements, *Bull. Soc. Math. France,* 89 (1961), pp. 227–380.

[Ch] C. Chevalley, *The Theory of Lie Groups I,* Princeton University Press, 1946.

[Do] S. K. Donaldson, An application of gauge theory to four-dimensional topology, *J. Differential Geometry* 18 (1983), pp. 279–315.

[Du] J. Dugundji, *Topology,* Prentice-Hall of India Pvt Ltd. New Delhi, 1975.

[G-P] V. Guillemin and A. Pollack, *Differential Topology,* Englewood Cliff, N.J. Prentice Hall, 1974.

[H] A. Haefliger, Plongements differentiables de varietés dans varietés. *Comment. Math. Helv.* 36 (1961), pp. 47–82.

[Hat] A. Hatcher, *Algebraic Topology,* Cambridge: Cambridge University Press, 2002.

[Ham] R. Hamilton, The inverse function theorem of Nash and Moser, *Bull. Amer. Math. Soc.* 7 (1982), pp. 65–222.

[Hi] W. Hirsch, *Differential Topology,* Springer-Verlag.

[Ho] L. Hörmander, On the Nash-Moser implicit function theorem, *Annales Academie Scientiarum Fennicae, Series A. I. Mathematica* 10 (1985), pp. 255–259.

[Hum] J. E. Humphreys, *Linear Algebraic Groups,* New York: Springer-Verlag, 1975.

[H-W] W. Hurewicz and H. Wallman, *Dimension Theory,* Princeton University Press, Princeton, 1948.

[Hus] D. Husemoller, *Fibre Bundles,* New York: Springer-Verlag, 1966, 2nd Edition.

[J] K. D. Joshi, *Introduction to General Topology,* New Age International (P) Limited.

[K] M. A. Kervaire, A manifold which does not admit any differential structure, *Comment. Math Helv.* 34, 1960, pp. 257–270.

[K-M] M. A. Kervaire and J. W. Milnor, Groups of homotopy spheres I, *Ann. Math.*(2) 77 (1963), pp. 504–537.

[K-S] R. C. Kirby, and L. C. Siebenmann, *Foundational Essays on Topological Manifolds, Smoothings and Triangulations.* Annals of Mathematics Studies, No. 88, Princeton University Press, Princeton N.J., University of Tokyo Press Tokyo, 1977.

[Kl] W. Klingenberg, Ueber Riemansche Manningfaltigkeiten mit positiver Kruemmung, *Comment. Math. Helv.* 35 (1961), pp. 47–54.

[K] A. A. Kosinski, *Differential Manifolds,* Academic Press, New York.

[La] S. Lang, *Algbera,* 3rd Edition, Eddison Wesley.

[Le] J. M. Lee, *Introduction to Smooth Manifolds,* GTM 218 Springer, 2002.

[M] B. Mazur, On embeddings of spheres, *Bull. Amer. Math. Soc.* 65 1959), pp. 59–65.

[Ma] Y. Matsushima, *Differentiable Manifolds,* Marcel Dekker, New York, 1972.

[M1] J. W. Milnor, *Topology from the Differentiable Viewpoint,* University Press, Virginia.

[M2] J. W. Milnor, *Morse Theory,* Ann. Math. Studies, Princeton University Press.

[M3] J. W. Milnor, On manifolds homeomorphic to the 7-sphere, *Ann. Math.* 64 (1956), pp. 399–405.

[Moi] E. Moise, *Geometric Topology in Dimensions 2 and 3,* Springer-Verlag, New York, 1977.

[Mu] J. Munkres, Differentiable isotopies on the 2-sphere, *Mich. Math. J.* 7 (1960), pp. 193–197.

[Mu2] J. Munkres, Obstructions to smoothing of piecewise differentiable homeomorphisms, *Ann. Math.* 72 (1960), pp. 521–554.

[N] J. Nash, The embedding problem for Riemannian manifolds, *Ann. Math.* 63 (1958), pp. 20–63.

[O] A. L. Onishchik, *Lie Groups and Lie Algebras I,* Encyclopedia of Math. Sci. Vol. 20, Springer-Verlag, 1988.

[Pa] R. Palais, Local triviality of the restriction map for embeddings, *Comment. Math. Helv.* 34 (1960), pp. 305–312.

[Pe] L. Perko, *Differential Equations and Dynamical Systems,* Texts in Applied Mathematics-7, Springer-Verlag, New York, 2004.

[Po] M. Postnikov, *Lie Groups and Lie Algebras,* Lectures in Geometry Semester V. Mir Publishers Moscow, 1986.

[R] W. Rudin, *Principles of Mathematical Analysis,* 3rd Edition, McGraw-Hill, 1976.

[Sh1] A. R. Shastri, Sums are not products, *J. London Math. Soc.* (2), 15 (1977), pp. 351–368.

[Sh2] A. R. Shastri, *Basic Complex Analysis of One Variable*, Macmillan India Ltd, Delhi, 2011.

[Sm] S. Smale, Diffeomorphisms of the 2-sphere, *Proc. Amer. Math. Soc.* 10 (1959), pp. 621–626.

[Sm1] S. Smale, Generalized Poincaré conjecture in dimension greater than four, *Ann. Math.* (2) 74 (1961), pp. 391–406.

[Sm2] S. Smale, On the structure of manifolds, *Amer. J. Math.* 84 (1962), pp. 387–399.

[Spa] E. H. Spanier, *Algebraic Topology,* Springer-Verlag, New York, 1989.

[Spi] M. Spivak, *Calculus on Manifolds,* Benjamin Ink., New York, 1965.

[St] N. E. Steenrod, *Topology of Fibre Bundles,* Princeton University Press, Princeton, N.J., 1951.

[T] C. Tapp, *Matrix Groups,* Amer. Math. Soc. Publications, Student Math. Library Vol. 29, 2005.

[Th] R. Thom, La Classification des immersions, *Semin. Bourbaki* 157, 1957-58.

[Ti] V. M. Tikhomirov, *Stories about Maxima and Minima,* Mathematical World Vol. 1, AMS-MAA publication 1990.

[W] C. T. C. Wall, *Surgery on Compact Manifolds,* Academic Press London, New York 1970.

[Wd] J. H. C. Whitehead, On $\mathcal{C}^1$-complexes, *Ann of Math.* 41 (1940), pp. 809–824.

[Wy] H. Whitney, A function not constant on a connected set of critical points, *Duke Math. J.* 1 (1935), pp. 514–517.

Index